Feynman Path Integrals in Quantum Mechanics and Statistical Physics

Feynman Path Integrals in Quantum Mechanics and Statistical Physics

Professor Lukong Cornelius Fai

CRC Press
Taylor & Francis Group
Boca Raton London New York

CRC Press is an imprint of the
Taylor & Francis Group, an **informa** business

First edition published 2021
by CRC Press
6000 Broken Sound Parkway NW, Suite 300, Boca Raton, FL 33487-2742

and by CRC Press
2 Park Square, Milton Park, Abingdon, Oxon, OX14 4RN

CRC Press is an imprint of Taylor & Francis Group, LLC

Library of Congress Cataloging-in-Publication Data
Names: Fai, Lukong Cornelius, author.
Title: Feynman path integrals in quantum mechanics and statistical physics / Lukong Cornelius Fai.
Description: Boca Raton : CRC Press, 2021. | Includes bibliographical references and index.
Identifiers: LCCN 2020050051 | ISBN 9780367697853 (hardback) | ISBN 9781003145554 (ebook)
Subjects: LCSH: Feynman integrals. | Quantum theory. | Statistical physics.
Classification: LCC QC174.17.F45 F35 2021 | DDC 530.14/3--dc23
LC record available at https://lccn.loc.gov/2020050051

ISBN: 978-0-367-69785-3 (hbk)
ISBN: 978-0-367-70299-1 (pbk)
ISBN: 978-1-003-14555-4 (ebk)

Typeset in Minion
by SPi Global, India

Contents

Preface xi

1 Path Integral Formalism Intuitive Approach 1

1.1 Probability Amplitude 1

1.1.1 Double Slit Experiment 1

1.1.2 Physical State 2

1.1.3 Probability Amplitude 2

1.1.4 Revisit Double Slit Experiment 2

1.1.5 Distinguishability 3

1.1.6 Superposition Principle 3

1.1.7 Revisit the Double Slit Experiment/Superposition Principle 4

1.1.8 Orthogonality 5

1.1.9 Orthonormality 6

1.1.10 Change of Basis 7

1.1.11 Geometrical Interpretation of State Vector 8

1.1.12 Coordinate Transformation 9

1.1.13 Projection Operator 10

1.1.14 Continuous Spectrum 11

2 Matrix Representation of Linear Operators 13

2.1 Matrix Element 13

2.2 Linear Self-Adjoint (Hermitian Conjugate) Operators 13

2.3 Product of Hermitian Operators 15

2.4 Continuous Spectrum 16

2.5 Schturm-Liouville Problem: Eigenstates and Eigenvalues 17

2.6 Revisit Linear Self-Adjoint (Hermitian) Operators 20

2.7 Unitary Transformation 21

2.8 Mean (Expectation) Value and Matrix Density 22

2.9 Degeneracy 23

2.10 Density Operator 24

2.11 Commutativity of Operators 25

3 Operators in Phase Space 29

3.1 Introduction 29

3.2 Configuration Space 30

3.3 Position and Wave Function 31

3.4 Momentum Space 32

3.5 Classical Action 33

4 Transition Amplitude 35
4.1 Path Integration in Phase Space 36
4.1.1 From the Schrödinger Equation to Path Integration 36
4.1.2 Trotter Product Formula 39
4.2 Transition Amplitude 41
4.2.1 Hamiltonian Formulation of Path Integration 41
4.2.2 Path Integral Subtleties 43
4.2.2.1 Mid-point Rule 43
4.2.3 Lagrangian Formulation of Path Integration 44
4.2.3.1 Complex Gaussian Integral 44
4.2.4 Transition Amplitude 46
4.2.5 Law for Consecutive Events 51
4.2.6 Semigroup Property of the Transition Amplitude 51

5 Stationary and Quasi-Classical Approximations 53
5.1 Stationary Phase Method / Fourier Integral 53
5.2 Contribution from Non-Degenerate Stationary Points 56
5.2.1 Unique Stationary Point 58
5.3 Quasi-Classical Approximation/Fluctuating Path 60
5.3.1 Free Particle Classical Action and Transition Amplitude 60
5.3.1.1 Free Particle Classical Action 61
5.3.1.2 Free Particle Transition Amplitude 62
5.3.1.3 From Path Integrals to Quantum Mechanics 65
5.4 Free and Driven Harmonic Oscillator Classical Action and Transition Amplitude 67
5.4.1 Free Oscillator Classical Action 67
5.4.2 Driven or Forced Harmonic Oscillator Classical Action 69
5.5 Free and Driven Harmonic Oscillator Transition Amplitude 71
5.6 Fluctuation Contribution to Transition Amplitude 72
5.6.1 Maslov Correction 74

6 Generalized Feynman Path Integration 77
6.1 Coordinate Representation 77
6.2 Free Particle Transition Amplitude 79
6.3 Gaussian Functional Feynman Path Integrals 81
6.4 Charged Particle in a Magnetic Field 87

7 From Path Integration to the Schrödinger Equation 91
7.1 Wave Function 91
7.2 Schrödinger Equation 92
7.3 The Schrödinger Equation's Green's Function 94
7.4 Transition Amplitude for a Time-Independent Hamiltonian 95
7.5 Retarded Green Function 97

8 Quasi-Classical Approximation 101
8.1 Wentzel-Kramer-Brillouin (WKB) Method 101
8.1.1 Condition of Applicability of the Quasi-Classical Approximation 104
8.1.2 Bounded Quasi-Classical Motion 106
8.1.3 Quasi-Classical Quantization 109
8.1.4 Path Integral Link 111
8.2 Potential Well 112
8.3 Potential Barrier 114

8.4 Quasi-Classical Derivation of the Propagator 116
8.5 Reflection and Tunneling via a Barrier 117
8.6 Transparency of the Quasi-Classical Barrier 119
8.7 Homogenous Field 121
8.7.1 Motion in a Central Symmetric Field 125
8.7.1.1 Polar Equation 125
8.7.1.2 Radial Equation for a Spherically Symmetric Potential in Three Dimensions 129
8.7.2 Motion in a Coulombic Field 130
8.7.2.1 Hydrogen Atom 130

9 Free Particle and Harmonic Oscillator 135
9.1 Eigenfunction and Eigenvalue 135
9.1.1 Free Particle 135
9.1.2 Transition Amplitude for a Particle in a Homogenous Field 137
9.2 Harmonic Oscillator 138
9.3 Transition Amplitude Hermiticity 143

10 Matrix Element of a Physical Operator via Functional Integral 145
10.1 Matrix Representation of the Transition Amplitude of a Forced Harmonic Oscillator 147
10.1.1 Charged Particle Interaction with Phonons 150

11 Path Integral Perturbation Theory 153
11.1 Time-Dependent Perturbation 160
11.2 Transition Probability 163
11.3 Time-Energy Uncertainty Relation 164
11.4 Density of Final State 166
11.4.1 Transition Rate 166
11.5 Continuous Spectrum due to a Constant Perturbation 168
11.6 Harmonic Perturbation 169

12 Transition Matrix Element 173

13 Functional Derivative 179
13.1 Functional Derivative of the Action Functional 181
13.2 Functional Derivative and Matrix Element 183

14 Quantum Statistical Mechanics Functional Integral Approach 191
14.1 Introduction 191
14.2 Density Matrix 191
14.2.1 Partition Function 191
14.3 Expectation Value of a Physical Observable 192
14.4 Density Matrix 192
14.5 Density Matrix in the Energy Representation 194

15 Partition Function and Density Matrix Path Integral Representation 199
15.1 Density Matrix Path Integral Representation 199
15.1.1 Density Matrix Operator Average Value in Phase Space 199
15.1.1.1 Generalized Gaussian Functional Path Integral in Phase Space 201
15.1.2 Density Matrix via Transition Amplitude 202
15.2 Partition Function in the Path Integral Representation 205
15.3 Particle Interaction with a Driven or Forced Harmonic Oscillator: Partition Function 209

15.4 Free Particle Density Matrix and Partition Function 121
15.5 Quantum Harmonic Oscillator Density Matrix and Partition Function 214

16 Quasi-Classical Approximation in Quantum Statistical Mechanics 219
16.1 Centroid Effective Potential 220
16.2 Expectation Value 225

17 Feynman Variational Method 229

18 Polaron Theory 237
18.1 Introduction 237
18.2 Polaron Energy and Effective Mass 239
18.3 Functional Influence Phase 241
18.3.1 Polaron Model Lagrangian 243
18.3.2 Polaron Partition Function 243
18.4 Influence Phase via Feynman Functional Integral in The Density Matrix Representation 246
18.4.1 Expectation Value of a Physical Quantity 246
18.4.1.1 Density Matrix 246
18.5 Full System Polaron Partition Function in a 3D Structure 255
18.6 Model System Polaron Partition Function in a 3D Structure 256
18.7 Feynman Inequality and Generating Functional 257
18.8 Polaron Characteristics in a 3D Structure 259
18.8.1 Polaron Asymptotic Characteristics 264
18.9 Polaron Characteristics in a Quasi-1D Quantum Wire 265
18.9.1 Hamiltonian of the Electron in a Quasi 1D Quantum Wire 265
18.9.1.1 Lagrangian of the Electron in a Quasi-1D Quantum Wire 266
18.9.1.2 Partition Function of the Electron in a Quasi-1D Quantum Wire 267
18.10 Polaron Generating Function 269
18.11 Polaron Asymptotic Characteristics 270
18.12 Strong Coupling Regime Polaron Characteristics 273
18.13 Bipolaron Characteristics in a Quasi-1D Quantum Wire 276
18.13.1 Introduction 276
18.13.2 Bipolaron Diagrammatic Representation 278
18.13.3 Bipolaron Lagrangian 278
18.13.4 Bipolaron Equation of Motion 280
18.13.5 Transformation into Normal Coordinates 282
18.13.5.1 Diagonalization of the Lagrangian 282
18.13.6 Bipolaron Partition Function 283
18.13.7 Bipolaron Generating Function 285
18.13.8 Bipolaron Asymptotic Characteristics 286
18.14 Polaron Characteristics in a Quasi-0D Spherical Quantum Dot 289
18.14.1 Introduction 289
18.14.2 Polaron Lagrangian 290
18.14.3 Normal Modes 290
18.14.4 Lagrangian Diagonalization 291
18.14.4.1 Transformation to Normal Coordinates 291
18.14.5 Polaron Partition Function 292
18.14.6 Generating Function 293
18.15 Bipolaron Characteristics in a Quasi-0D Spherical Quantum Dot 295
18.15.1 Introduction 295
18.15.2 Model Lagrangian 296

18.15.3 Model Lagrangian ... 296
18.15.3.1 Equation of Motion and Normal Modes ... 296
18.15.4 Diagonalization of the Lagrangian ... 297
18.15.5 Partition Function ... 299
18.15.6 Full System Influence Phase ... 300
18.16 Bipolaron Energy ... 300
18.16.1 Generating Function ... 300
18.16.2 Bipolaron Characteristics ... 301
18.17 Polaron Characteristics in a Cylindrical Quantum Dot ... 304
18.17.1 System Hamiltonian ... 304
18.17.2 Transformation to Normal Coordinates ... 305
18.17.2.1 Lagrangian Diagonalization ... 305
18.17.3 Polaron Energy/Partition Function ... 306
18.17.4 Polaron Generating Function ... 307
18.17.5 Polaron Energy ... 308
18.18 Bipolaron Characteristics in a Cylindrical Quantum Dot ... 310
18.18.1 System Hamiltonian ... 310
18.18.1.1 Model System Action Functional ... 310
18.18.1.2 Equation of Motion / Normal Modes ... 311
18.18.1.3 Lagrangian Diagonalization ... 312
18.18.1.4 Bipolaron Partition Function ... 312
18.18.1.5 Bipolaron Generating Function ... 313
18.18.1.6 Bipolaron Energy ... 313
18.19 Polaron Characteristics in a Quasi-0D Cylindrical Quantum Dot with Asymmetrical Parabolic Potential ... 315
18.20 Polaron Energy ... 316
18.21 Bipolaron Characteristics in a Quasi-0D Cylindrical Quantum Dot with Asymmetrical Parabolic Potential ... 320
18.22 Polaron in a Magnetic Field ... 324

19 Multiphoton Absorption by Polarons in a Spherical Quantum Dot 337
19.1 Theory of Multiphoton Absorption by Polarons ... 337
19.2 Basic Approximations ... 338
19.3 Absorption Coefficient ... 339

20 Polaronic Kinetics in a Spherical Quantum Dot 351

21 Kinetic Theory of Gases 365
21.1 Distribution Function ... 365
21.2 Principle of Detailed Equilibrium ... 365
21.3 Transport Phenomenon and Boltzmann-Lorentz Kinetic Equation ... 369
21.4 Transport Relaxation Time ... 373
21.5 Boltzmann H-Theorem ... 375
21.6 Thermal Conductivity ... 378
21.7 Diffusion ... 380
21.8 Electron–Phonon System Equation of Motion ... 386

References ... 391

Index ... 395

Preface

This book present lectures read to graduate students at the Universities of Dschang and Bamenda, Cameroon. The foundation of the book is Chapter 13 (Functional Integration in Statistical Physics) of our book *Statistical Thermodynamics: Understanding the Properties of Macroscopic Systems* [1] as well as my doctoral thesis work at Moldova State University [2]. This book clearly develops unique techniques in great detail and is devoted to graduate students and researchers in Physics, Mathematical Physics and Applied Mathematics as well as Chemistry. Though the reader is assumed to have a good foundation in quantum mechanics, the material is treated in a manner that can be easily grasped by someone with little knowledge in quantum mechanics and no prior exposure to path integrals. A review of the fundamental ideas of quantum mechanics is done at the beginning. This presents the general construction scheme of quantum mechanics with particular emphasis placed on establishing the interconnections between the quantum mechanical path integral, Hamiltonian mechanics and statistical mechanics.

Quantum mechanics historically, began with two quite different mathematical formulations, i.e., the Schrödinger wave equation, then the Heisenberg, Born and Jordan matrix algebra. This is certainly the most fundamental and powerful formulation of quantum mechanics in a broad sense. In 1948 [3], following a suggestion by Dirac [2,4–7], R. P. Feynman proposed a new and more intuitive formulation of quantum mechanics. Broadly speaking, there are two basic approaches to the formulation of quantum mechanics. Firstly, the operator approach based on the canonical quantisation of physical observables together with the associated operator algebra, and, secondly, the Feynman path integral approach. The Feynman path integration is the most elegant formulation of quantum mechanics, presenting an alternative way of examining the subject matter than the usual approaches, with applications as vast as those of quantum mechanics itself. These include the quantum mechanics of a single particle, statistical mechanics, condensed matter physics, quantum field theory, cosmology and molecular physics as well as string theory. The Feynman path integral formulation of quantum mechanics is advantageous in that it gives a way to think about the meaning of quantum mechanics. The classical limit is not always easy to retrieve within the canonical formulation of quantum mechanics, but remains constantly visible in the path integral approach. So, the path integral makes explicit use of classical mechanics as a basic platform on which the theory of quantum fluctuations is constructed. The classical solutions of the Hamilton equation of motion remains always the central ingredient of the Feynman path formalism. The Feynman path integral approach allows for an efficient formulation of non-perturbative approaches to the solution of quantum mechanical problems. Path integral in comparison to the operator approach is more comprehensive. The operator approach gives only a local approach while path integration gives a global view. The Feynman path integral permit in an elegant manner without use of differential equations to explicitly evaluate the propagator, the energy spectrum and correct normalized wave function. From path integrals, the perturbation theory is straightforwardly incorporated.

This book describes and establishes quantum probabilities that is a central concept of the Feynman path integral, which is a simple and powerful method to depict and understand particle interaction, and then applies the method to the free particle, quantum harmonic oscillator, tunnelling of particles via potential barriers, investigate the Euclidean path integral, and discusses other applications, such as the polaron theory. It then develops a unique technique rendering the density matrix of a mixed ensemble to be decoupled to one of the products of the density matrix of pure ensembles via the functional-evolution-integral operator. The book develops an extraordinary technique of transforming real to normal coordinates, permitting an elegant diagonalization of the Lagrangian and rendering it perfectly quadratic. Consequently, the Feynman path integral is transformed to a Gaussian one that is computable exactly. A unique and effective technique is developed for computing two-fold integrals of retarded functions via the two-fold integral theorem that mimics Stokes' theorem.

The book begins with the preface and then the table of contents. Chapter 1 is devoted to an intuitive approach, where the fundamental concepts of quantum mechanics are explored and introduces path integrals where the Feynman interpretation of quantum mechanics is done via the two slit experiment. The interpretation of non-classical rules for combining probability amplitudes for particle propagation is presented.

Chapter 2 develops the matrix representations of linear operators as a useful framework for the Feynman path integral in the representation of the matrix density.

Chapter 3 introduces the concept of the Feynman path integral by developing the general construction scheme with particular emphasis on establishing the interconnections between the quantum mechanical path integral, Hamiltonian mechanics and statistical mechanics. The least action principle is well formulated, permitting an easy establishment of the equation of motion as well as the classical path.

Chapter 4 examines the transition amplitude in the Feynman path integral, which is reduced to an integral over the continuum degrees of freedom. This involves harmonic (non-interacting) systems that are reduced to the problem of multiple Gaussian integrals over potentially coupled degrees of freedom.

Chapter 5 treats the stationary and quasi-classical approximations that will be useful in the Feynman variational principle. The limit is on the most important approximations valid when quantum fluctuations are small or, equivalently, when the actions involved are large compared to Planck's constant.

Chapter 6 is devoted to Generalized Gaussian functional integrals pivotal in the exact evaluation of Feynman path integrals. It proposes the method of transformation of non-quadratic terms in a Lagrangian to quadratic forms. The chapter treats as well the quasi-classical approximation via the method of stationary phase method.

Chapter 7 shows how from the path integral we rederive Schrödinger's equation and confirms the Feynman path integration to be the formulation of quantum mechanics, presenting an alternative way of examining the subject matter rather than the usual approaches.

Chapter 8 studies another very important and powerful approximation method, known as the Wentzel-Kramer-Brillouin (WKB), method used in studying quantum mechanical systems subjected to complicated potentials.

Chapter 9 is devoted to the free particle and harmonic oscillator and, in particular, eigenfunctions, eigenvalues, transition amplitudes and Hermiticity of the transition amplitude.

Chapter 10 studies the matrix element of a physical operator via the functional integral method and shows the possibility of the basis of functional integration to obtain the formula for the evaluation of the matrix element of a physical operator while avoiding the difficulty related with the normalization factor in the evaluation of these matrix elements.

Chapter 11 is devoted to the path integral perturbation theory since the importance of the path integral formalism should be clear by now. The chapter shows the usefulness of the theory when it is used for interacting particles and for such an interaction when the non-Gaussian part of the action is weak compared to the quadratic part where the perturbation expansion is applicable. The chapter shows that this is really much easier than in the operator formalism since we deal only with c-numbers in the path integration.

Chapter 12 studies the transition matrix elements that are complementary to the perturbation theory in chapter 11 as well as the functional derivative in chapter 13, where, in path integration, one deals with classical functions and functionals rather than with operators. This leads to fairly simple classical manipulations of path integration resulting to non-trivial quantum mechanical identities.

Chapter 14 is devoted to the quantum statistical mechanics functional integral approach. This chapter establishes a rather deep relation between statistical mechanics and quantum mechanics at imaginary time. Since the most convenient object of the quantum theory to describe non-isolated systems is the density matrix it is for this reason that the chapter treats the quantum mechanical and quantum statistical density matrices that are necessary to do the investigation via path integration.

Chapter 15 studies the partition function and the density matrix path integration. It proceeds to the evaluation of the density matrix via the functional integration in a similar manner as for the transition amplitude. The chapter examines the partition function as well as the density matrix, while extrapolating them to examples of physical interest such as a system of free particles as well as a continuum of oscillators.

Chapter 16 is devoted to quasi-classical approximations in quantum statistical mechanics. We begin by showing how, by using the Feynman functional integration in the representation of the matrix density, we can make a transition to classical distribution. We show how it is interesting to approximate the quantum partition function in a classical fashion without explicitly solving the Schrödinger equation. Furthermore, the chapter shows how to find expectation values via the Feynman tips.

Chapter 17 treats the Feynman variational principle based on the Feynman functional integral method and applies it to quantum mechanical as well as statistical systems at finite temperatures.

Chapter 18 applies the Feynman path functional integration to the polaron theory. An extraordinary transformation to normal coordinates permitting an elegant diagonalization of the Lagrangian is developed which renders the Lagrangian perfectly quadratic and, consequently, the Feynman path integral to a Gaussian one is exactly computable. This technique is an elegant and effective way of computing problems of particles interacting with the environment. Applying, the Feynman path integration to the polaron problem clearly shows the polaron problem to mimic a one-particle problem where the interaction, nonlocal in time or "retarded," occurs between the electron and itself. It is shown that the formulae depend on the difference in times and indicative that the quantities (retarded functions) depend on the past. A formula of transformation of the two-fold integral of the retarded function to a single integral [1,2,5] (retarded or advanced function twofold-integration theorem) that mimics Stokes' theorem in the transformation of a surface integral to a line integral is established.

Chapters 19 and 20 once again examine the polaron problem and, in particular, respectively the multiphoton absorption by polarons and polaronic kinetics in spherical quantum dots via the Feynman path integral in the representation of the matrix density developed in chapters 14 and 15. Chapter 20 develops the kinetic theory of gases and applies to the polaron problem via the Feynman path integral in the representation of the matrix density.

In Chapter 21, in order to find the path integral link of the electron–phonon system to the kinetic theory of gases, a review of some basic notions in the kinetic theory of gases is made to enhance understanding of the subject matter.

The book ends with references and the index of key words.

The Feynman path integrals should not be confused with the Feynman integrals treated in reference [8] (Quantum Field Theory: Feynman Path Integrals and Diagrammatic Techniques in Condensed Matter) in the study of terms due to perturbation expansions in quantum field theory associated with Feynman diagrams related to finite dimensional complex integrals, and treated in relation to renormalization.

I would like to acknowledge those who have helped at various stages of the elaboration and writing, through discussions, criticism and especially, encouragement and support. I am very thankful to my wife, Prof. Dr. Mrs. Fai Patricia Bi, for all her support and encouragement and to my four children (Fai Fanyuy Nyuydze, Fai Fondzeyuv Nyuytari, Fai Ntumfon Tiysiy and Fai Jinyuy Nyuydzefon) for their understanding and moral support during the writing of this book. I am grateful for the library support received from the Abdus Salam International Centre for Theoretical Physics (ICTP), Trieste, Italy, which contributed to the writing of this book. I also acknowledge with gratitude my MSc thesis supervisors, Prof. Evgeni Petrovich Pokatilov and Prof. Stanpan Jordanovich Beril, as well as my doctoral thesis supervisors at Moldova State University, Prof. Evgeni Petrovich Pokatilov and Prof. Vladimir Mikhailovich Fomin, who equipped me with the necessary tools for my thesis work that have inspired me to write this book. I am highly indebted to Prof. Sergei Nikolayevich Klimin for the mentorship during my doctoral training and studies at Moldova State University. I acknowledge with gratitude the training by all of the professors at this institution, as well as visiting professors from Moldova Academy of Science.

1

Path Integral Formalism Intuitive Approach

1.1 Probability Amplitude

1.1.1 Double Slit Experiment

One of the important classic experiments that led to the fundamental difference between quantum and classical mechanics is the double slit experiment [9], which is interesting with respect to the path integral formalism since it leads to a conceptual motivation for introducing it. In this experiment, we examine the flux of low-intensity beam of electrons coming from an electronic source, S_e and allowed to pass through double slits a and b on a diaphragm D and detected behind the slits at a detector screen P facing the source (Figure 1.1). Each electron that hits the detector screen P is detected at a single location. So, there is no doubt the electrons should be particles.

If the slit b is closed then we measure on $P_1$the intensity J_1 of the electronic flux with amplitude $\Phi_1$and if similarly slit a is closed then we measure on P_2 the intensity J_2 of the electronic flux with amplitude Φ_2 and when slits $a$and b are both opened then we measure on P_3 the intensity

$$J \neq J_1 + J_2 \tag{1.1}$$

So, we observe that even if the beam intensity is so low that on average only a single electron at a time is in flight. Consequently, an interference pattern is built up on the detector screen P after tens of thousands of counts have been detected. This is in clear violation of classical physics since the fluxes at the sensitive screen P in the three cases mentioned above are expected to satisfy the relation

$$J = J_1 + J_2 \tag{1.2}$$

However, in reality generally, we find

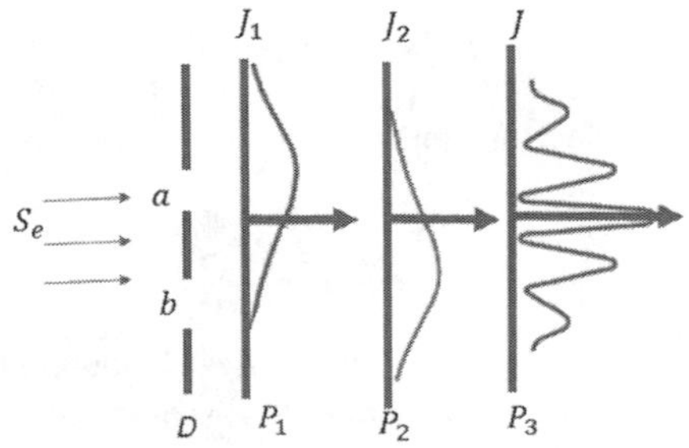

FIGURE 1.1 The double slit experiment with an electronic beam coming from an electronic source, S_e and passing via slits $\boldsymbol{a}$ and $\boldsymbol{b}$ on a diaphragm $\boldsymbol{D}$ and measured on the sensitive screen $\boldsymbol{P}$ facing the source.

$$J = J_1 + J_2 + J_{\text{int}} \tag{1.3}$$

where J_{int} is indeed an interference term due to two waves passing via slits a and b on a diaphragm D and measured on the sensitive screen P.

1.1.2 Physical State

From Quantum Mechanics, the **probability amplitude** (square of the probability function or wave function) can also be determined from the Schrödinger equation as it describes the wave properties of the particles. Note that particles described by **a probability wave** are indeed point-like discrete entities. From the double slit experiment example, the flight of an electron via a defined slit should be at the price of destroying quantum interference. This indeed presents an alternative, but fully equivalent method to compute the probability amplitude. Here, the role of the trajectory of a point-like particle, formally is revived in a way compatible with the indetermination principle and is indeed the path integral approach to quantum mechanics.

1.1.3 Probability Amplitude

The probability that a system observed initially in state $|\lambda_i\rangle$ will be observed finally in state $|\mu_l\rangle$:

$$\mathrm{P}(\lambda_i \to \mu_l) = \left|\langle \mu_l | \lambda_i \rangle\right|^2 = \langle \lambda_i | \mu_l \rangle \langle \mu_l | \lambda_i \rangle = \left\langle \lambda_i | \hat{\mathrm{P}}_{\mu_l} | \lambda_i \right\rangle \tag{1.4}$$

From the principles of quantum mechanics, there must exist an operator, say, $\hat{\mathrm{F}}$ for which one of its eigenvectors, $|\mu_l\rangle$ has μ_l as its eigenvalue otherwise measurements cannot give such a value. The projector operator P_{μ_l} corresponds to μ_l while the complex number $\langle \mu_l | \lambda_i \rangle$ is the so-called **probability amplitude** as it can be added and multiplied in the same manner as classical probabilities. We consider, for example, the following cases:

- $\lambda_i \to \mu_l$ and then $\mu_l \to \mathrm{a}$, the amplitude is $\langle \mathrm{a} | \mu_l \rangle \langle \mu_l | \lambda_i \rangle$;
- $\lambda_i \to \mu_l$ or a, the amplitude is $\langle \mu_l | \lambda_i \rangle + \langle \mathrm{a} | \lambda_i \rangle$;
- This implies that:

$$\mathrm{P}(\lambda_i \to \mu_l \to \mathrm{a}) = \left|\langle \mathrm{a} | \mu_l \rangle \langle \mu_l | \lambda_i \rangle\right|^2 = \mathrm{P}(\lambda_i \to \mu_l)\mathrm{P}(\mu_l \to \mathrm{a}) \tag{1.5}$$

1.1.4 Revisit Double Slit Experiment

From the double slit experiment, generally, the amplitudes can interfere (Figure 1.1). The probability amplitude for an electron at the initial state $|\lambda_i\rangle$ (electronic source) to have a flight via slit a, and be detected at the final state $|\mu_l\rangle$ (on screen P_1):

$$\Phi_1 = \langle \mu_l | \mathrm{a} \rangle \langle \mathrm{a} | \lambda_i \rangle \tag{1.6}$$

Similarly, the amplitude for the electron initially in state $|\lambda_i\rangle$ with flight via slit b, and detected by state $|\mu_l\rangle$:

$$\Phi_2 = \langle \mu_l | \mathrm{b} \rangle \langle \mathrm{b} | \lambda_i \rangle \tag{1.7}$$

The probability amplitude for an electron initially in state $|\lambda_i\rangle$ (source) to be detected in the final state $|\mu_l\rangle$ (screen) is obtained by adding the amplitudes due to two sources via slits, a and b:

$$\langle \mu_l | \lambda_i \rangle = \Phi_1 + \Phi_2 = \langle \mu_l | \mathrm{a} \rangle \langle \mathrm{a} | \lambda_i \rangle + \langle \mu_l | \mathrm{b} \rangle \langle \mathrm{b} | \lambda_i \rangle \tag{1.8}$$

Definitely, from the double slit experiment there is no assertion that the particle followed definite paths with certain probabilities. So, one has to obtain the probability amplitudes for various paths and add them to obtain the total amplitude and then the intensity J can be calculated:

$$J \cong \Phi^2 = \left(\Phi_1 + \Phi_2\right)^2 = \Phi_1^2 + \Phi_2^2 + 2\mathrm{Re}\left(\Phi_1 \times \Phi_2\right) \equiv J_1 + J_2 + J_{\mathrm{int}} \tag{1.9}$$

From the above we observe that each electron takes both possible paths simultaneously, and the interference between those two paths gives the observed interference pattern and other paths than the straight lines are canceled by destructive interference. From this result, should we say the wave behavior implies the electron is

- a delocalized object?
- passing through both slits?

This obviously is not the case and when detected on the sensitive screen , it is point-like. If we verify by a detector set at slits a and b where it went through, we find that it either went through slit a or b. From the above experiment, J is interpreted as a probability density and so the amplitude Φ , is interpreted as if we are dealing with waves. The intensity Φ^2 is interpreted as a probability density for a point-like particle position. The particle/wave duality is not contradictory from the indetermination principle: Attempting to detect the alternative path the electron followed, we also destroy interference. So, another formulation of the **indetermination principle**: **Any determination of the alternative achieved by a process capable of following more than one alternative destroys the interference between the alternatives.**

1.1.5 Distinguishability

For the double slit experiment, we have no knowledge of whether the flight of the particle detected at some point on the screen was via slit a or slit b – the two outcomes are **indistinguishable**. From this experiment, what adds up is the amplitude Φ and not the probability density itself and the difference between classical and quantum composition of probabilities is given by the interference between classically distinct trajectories. This can be demonstrated by the following relation:

$$\mathrm{P}\left(\lambda_i \to \mu_l\right) = \left|\left\langle \mu_l \middle| \lambda_i \right\rangle\right|^2 = \left|\Phi_1 + \Phi_2\right|^2 = \left|\left\langle \mu_l \middle| \mathrm{a}\right\rangle\left\langle \mathrm{a} \middle| \lambda_i\right\rangle + \left\langle \mu_l \middle| \mathrm{b}\right\rangle\left\langle \mathrm{b} \middle| \lambda_i\right\rangle\right|^2 \tag{1.10}$$

or

$$\mathrm{P}\left(\lambda_i \to \mu_l\right) = \left|\left\langle \mu_l \middle| \mathrm{a}\right\rangle\left\langle \mathrm{a} \middle| \lambda_i\right\rangle\right|^2 + \left|\left\langle \mu_l \middle| \mathrm{b}\right\rangle\left\langle \mathrm{b} \middle| \lambda_i\right\rangle\right|^2 + 2\mathrm{Re}\left\{\left\langle \mu_l \middle| \mathrm{a}\right\rangle\left\langle \mathrm{a} \middle| \lambda_i\right\rangle\left\langle \mu_l \middle| \mathrm{b}\right\rangle^* \left\langle \mathrm{b} \middle| \lambda_i\right\rangle^*\right\} \tag{1.11}$$

The third summand is the interference term. It is obvious from here that

$$\mathrm{P}\left(\lambda_i \to \mu_l\right) \neq \mathrm{P}\left(\lambda_i \to \mu_l \text{ via a}\right) + \mathrm{P}\left(\lambda_i \to \mu_l \text{ via b}\right) \equiv \left|\left\langle \mu_l \middle| \mathrm{a}\right\rangle\left\langle \mathrm{a} \middle| \lambda_i\right\rangle\right|^2 + \left|\left\langle \mu_l \middle| \mathrm{b}\right\rangle\left\langle \mathrm{b} \middle| \lambda_i\right\rangle\right|^2 \tag{1.12}$$

This is because the probability $\mathrm{P}(\lambda_i \to \mu_l)$ can vanish if the amplitudes for the different paths cancel.

1.1.6 Superposition Principle

We extrapolate the double slit experiment to two spinors as basis vectors:

$$|a\rangle \to \begin{bmatrix} 1 \\ 0 \end{bmatrix}, |b\rangle \to \begin{bmatrix} 0 \\ 1 \end{bmatrix} \tag{1.13}$$

The fact that the electron must follow a path via either slit a or slit b can simply be expressed via

$$\langle \lambda_i | \alpha \rangle = \sum_{\mu_n = a,b} \langle \lambda_i | \mu_n \rangle \langle \mu_n | \alpha \rangle \tag{1.14}$$

Here, α is some state and the sum is over all possible states, μ_n. The **quantum superposition principle** in equation 1.14 is one of the most important principles in quantum mechanics and it is also called **composition law of the probability amplitude**, where the amplitude $\langle \lambda_i | \alpha \rangle$ is expressed as a linear combination of the amplitudes $\langle \lambda_i | \mu_n \rangle$ with all possible values of μ_n. The equation in 1.14 also represents the transformation from the μ-representation expressing the amplitude $\langle \mu_n | \alpha \rangle$ in the λ-representation expressing $\langle \lambda_i | \alpha \rangle$ via the eigenfunction $\langle \lambda_i | \mu_n \rangle$.

1.1.7 Revisit the Double Slit Experiment/Superposition Principle

Considering 1.14 we revisit the double slit experiment to the path integral approach to quantum mechanics where we obtain two interfering alternatives for the path of an electronic beam coming from an electronic source, S_e and passing via slits a and b on a diaphragm D and measured on the particle detector sensitive screen P placed in such a manner that a number of screens, each of them containing several small holes, are placed in between the source and the detector, generalizing the result of the double slit experiment by the superposition principle [9]. The particle propagates via the holes and the amplitude of detecting the particle, Φ, is the sum over the possible ways of reaching the detector. It is the sum over rectangles from the source to the detector. For two slits then superposition principle applies

$$\Phi = \Phi_1 + \Phi_2 \tag{1.15}$$

Opening a third slit, the superposition principle:

$$\Phi = \Phi_1 + \Phi_2 + \Phi_3 \tag{1.16}$$

Consider adding an intermediate screen Q with s holes at positions (see Figure 1.2):

$$q_1, q_2, \cdots, q_s$$

This permits us to write the possible trajectories labelled by $q_{i,Q}$ through the slit at Q and by $\alpha = 1, 2, 3$, indicating the slit they went through at E. So, from the superposition principle:

$$\Phi = \sum_{i=1}^{s} \sum_{\alpha=1,2,3} \Phi\left(q_{i,Q}, \alpha\right) \tag{1.17}$$

Suppose $s \to \infty$ and the holes fill all of Q then the sum over i in 1.17 is now transformed to an integral over $q_{i,Q}$:

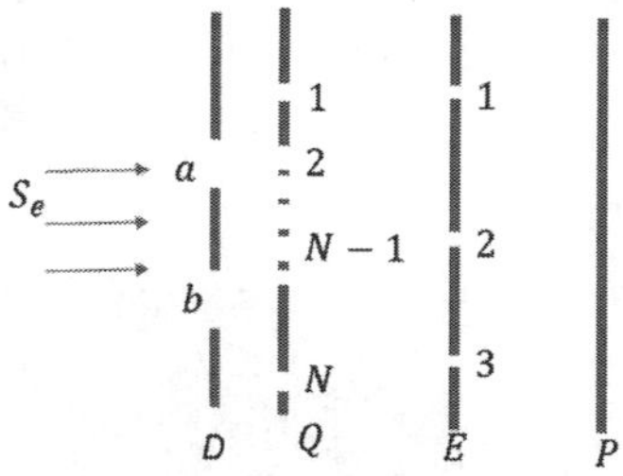

FIGURE 1.2 The multi-slit experiment with screen having **s** slits and a screen with three slits placed between the source and the detector.

$$\Phi = \sum_{\alpha=1,2,3} \int dq_Q \Phi(q_Q, \alpha) \tag{1.18}$$

We can further refine the trajectories by adding more and more screens

$$Q_1, Q_2, \cdots, Q_N$$

such that

$$\Phi = \sum_{\alpha=1,2,3} \int dq_{Q_1} dq_{Q_2} \cdots dq_{Q_N} \Phi(q_Q, \alpha) \tag{1.19}$$

In the limit when the screens Q_i and Q_{i+1} are placed closed to each other, then the particle traverses the next screen after a short time of flight. In this case, the size of the hole becomes small and close to each other. The rectangle then approaches a trajectory and we then arrive at a formal representation of the probability amplitude as a sum over all possible trajectories:

$$\Phi = \sum_{\text{all possible paths}} \Phi(\{q\}) \tag{1.20}$$

This result of the double slit interference experiment reintroduces part of classical physics into quantum mechanics and enhances our intuition towards the understanding of quantum physics. This clearly presents the **wave-particle-duality** and quantum nature for particles as emphasized by Richard Feynman [9]. This Feynman idea can be exploited in simple interference set-ups to construct novel quantum computing architectures which involves the direct use of the superposition and entanglement, to perform operations on data. A quantum computer differs from a binary digital electronic computer in that data are encoded into binary digits (bits), each of which is in one of the two definite states, 0 or 1, whereas a quantum computer is its analogue that uses quantum bits, so-called **qubits** which can be in an infinite number of superpositions of states at the same time, a property which is fundamental to quantum computing. This study of quantum computing and quantum computers will be done in detail in my next book publication.

1.1.8 Orthogonality

We consider the distribution of some physical measurement over λ where we select some arbitrary probability amplitudes $\langle\lambda_1|a\rangle$, $\langle\lambda_2|a\rangle$ and so on satisfying the following normalization condition:

$$\sum_n |\langle\lambda_n|a\rangle|^2 = 1 \tag{1.21}$$

Here, a is some state. From equation 1.21 results two important corollaries from the definition of the probability amplitude:

$$\langle\lambda_i|\lambda_k\rangle = \begin{cases} 0, i \neq k \\ 1, i = k \end{cases} \tag{1.22}$$

which may be multiplied by an arbitrary phase factor $\exp\{i\varphi\}$ that we neglect. This relation in 1.22 can be rewritten

$$\langle\lambda_i|\lambda_k\rangle = \delta_{ik} \tag{1.23}$$

where δ_{ik} is the symmetric Kronecker δ-symbol. Property 1.23 expresses the **orthogonality** of the basis vectors $|\lambda_i\rangle$and $|\lambda_k\rangle$ for $i \neq k$: **The eigenbasis states with different energies are orthogonal**. It is instructive to note that two vectors are **orthogonal** or **perpendicular** if their inner product vanishes. This implies that their projections onto each other vanish.

1.1.9 Orthonormality

Considering relations 1.14, 1.21 and 1.23 then follows the two important properties for the probability amplitude:

1. Letting $|\alpha\rangle \equiv |\lambda_i\rangle$ in 1.14 and considering 1.23 then the **normalization condition:**

$$1=\sum_{\mu_n}\langle\lambda_i|\mu_n\rangle\langle\mu_n|\lambda_i\rangle \tag{1.24}$$

From the normalization condition, 1.24 is rewritten as

$$1=\sum_{\mu_n}\left(\langle\mu_n|\lambda_i\rangle\right)^*\langle\mu_n|\lambda_i\rangle \tag{1.25}$$

The asterisk denotes the complex conjugate and from this equation

$$\sum_{\mu_n}\langle\lambda_i|\mu_n\rangle\langle\mu_n|\lambda_i\rangle=\sum_{\mu_n}\left(\langle\mu_n|\lambda_i\rangle\right)^*\langle\mu_n|\lambda_i\rangle \tag{1.26}$$

Both sums are equal for any λ and so follows the **reciprocity relation:**

$$\langle\lambda_i|\mu_n\rangle=\left(\langle\mu_n|\lambda_i\rangle\right)^* \tag{1.27}$$

This enable us to have

$$\left|\langle\mu_n|\lambda_i\rangle\right|^2=\left|\langle\lambda_i|\mu_n\rangle\right|^2 \tag{1.28}$$

This implies that the distribution over λ for given μ coincides with the distribution over μ for given λ. For the probability interpretation to be feasible it is necessary that the following completeness relation be satisfied:

$$1=\sum_{\mu_n}\left|\langle\mu_n|\lambda_i\rangle\right|^2=\sum_{\mu_n}\mathrm{P}(\lambda_i\to\mu_n) \tag{1.29}$$

In the state $|\lambda_i\rangle$, the observable physical quantity does not have a defined value. We can only talk about its probability $\mathrm{P}(\lambda_i \to \mu_n)$:

$$\sum_{\mu_n}\mathrm{P}(\lambda_i\to\mu_n)=1 \tag{1.30}$$

This is because the probability of one of the independent events is the sum of probabilities.

2. Suppose now in 1.14 we let $|\alpha\rangle = |\lambda_k\rangle$ for $i \neq k$ then from 1.23 follows

$$\sum_{\mu_n}\langle\lambda_i|\mu_n\rangle\langle\mu_n|\lambda_k\rangle=0 \tag{1.31}$$

From the orthogonality condition 1.23 of the vectors say, $|\lambda_k\rangle$ and $|\mu_n\rangle$ then

$$\sum_{\mu_n} \langle \lambda_i | \mu_n \rangle \left(\langle \lambda_k | \mu_n \rangle \right)^* = 0 \tag{1.32}$$

We swap λ and μ with their indices in this equation and we have

$$\sum_{\lambda_n} \left(\langle \lambda_n | \mu_i \rangle \right)^* \langle \lambda_n | \mu_k \rangle = 0 \tag{1.33}$$

The orthonormality condition 1.24 permits to rewrite 1.32 and 1.33 in a compact form. So, **a set of vectors is orthonormal if they are mutually orthogonal and are each normalized**:

$$\sum_{\lambda_n} \left(\langle \lambda_n | \mu_i \rangle \right)^* \langle \lambda_n | \mu_k \rangle = \delta_{ik} \tag{1.34}$$

$$\sum_{\mu_n} \left(\langle \lambda_i | \mu_n \rangle \right)^* \langle \lambda_k | \mu_n \rangle = \delta_{ik} \tag{1.35}$$

These last equations permit us to confirm the superposition principle in 1.14.

1.1.10 Change of Basis

We examine in a given representation a basis vector expressed via a matrix. Changing the representation implies representing that basis vector by a different matrix. It will be interesting to see how the two matrices relate each other. For brevity, we assume moving from one discrete orthonormal basis $\{|\lambda_n\rangle\}$ to another discrete orthonormal basis $\{|\mu_k\rangle\}$:

$$|\lambda_n\rangle = \sum_{\mu_k} |\mu_k\rangle \langle \mu_k | \lambda_n \rangle = \sum_{\mu_k} |\mu_k\rangle \mathrm{U}_{kn} \tag{1.36}$$

The vectors $|\mu_k\rangle$ in this case are used to extract physical information concerning the eigenvalue μ for a given observable quantity from arbitrary state vector $|\lambda_n\rangle$ and that is why we expand this vector in the basis $|\mu_k\rangle$ as obvious from equation 1.36. So, relation 1.36 expresses the fact that any vector $|\lambda_n\rangle$ can be expanded via the vector $|\mu_k\rangle$ which, of course, constitutes a basis. So, in 1.36 the vector $|\lambda_n\rangle$ is expanded via the basis vectors $|\mu_k\rangle$. The components $\langle \mu_k | \lambda_n \rangle$ is the probability amplitude for measuring the eigenvalue μ for the given observable quantity. The quantity U_{kn} is the transformation (or expansion) coefficient acting on the vector $|\mu_k\rangle$ and transforming it to the vector $|\lambda_n\rangle$.

It is observed from equation 1.36 that the change of basis is done by specifying $\langle \mu_k | \lambda_n \rangle$ relating the new basis to the old one. This should be the transformation matrix:

$$\langle \lambda_n | \mu_k \rangle = \left(\langle \mu_k | \lambda_n \rangle \right)^* = \mathrm{U}_{kn}^* = \mathrm{U}_{nk}^\dagger \tag{1.37}$$

permitting to write the **orthogonality relation**:

$$\langle \mu_k | \lambda_n \rangle = \delta_{kn} = \sum_{\mu_m} \langle \lambda_n | \mu_m \rangle \langle \mu_m | \mu_k \rangle = \sum_{\mu_m} \mathrm{U}_{nm}^\dagger \mathrm{U}_{mk} \tag{1.38}$$

This shows that U_{mn} is a **unitary matrix**:

$$\hat{\mathrm{U}}^\dagger \hat{\mathrm{U}} = \hat{\mathbb{I}} \tag{1.39}$$

and here, $\hat{\mathbb{I}}$ is the unit or identity operator that has only unit diagonal elements. Relation 1.38 confirms that the vector $\{|\mu_k\rangle\}$ are orthonormal. **So, we can observe that the necessary condition for an operator**

$\hat{U}$ to be unitary is that the vectors of an orthonormal basis in a given space, transformed by $\hat{U}$, constitutes another orthonormal basis.

Similarly, we find

$$\left(\langle \mu_k | \lambda_n \rangle\right)^* = \langle \lambda_n | \mu_k \rangle = \delta_{nk} = \sum_{\mu_m} \langle \mu_m | \lambda_n \rangle \langle \mu_k | \mu_m \rangle = \sum_{\mu_m} \mathrm{U}_{mn} \mathrm{U}_{km}^{\dagger} \tag{1.40}$$

So, from here,

$$\hat{\mathrm{U}}\ \hat{\mathrm{U}} = \hat{\mathrm{U}}\hat{\mathrm{U}}\ = \hat{\mathbb{I}} \tag{1.41}$$

Considering the reciprocity relation 1.27 then the operator $\hat{\mathrm{U}}$ is unitary if its inverse $\hat{\mathrm{U}}^{-1}$ is equal to its **adjoint** $\hat{\mathrm{U}}$ then

$$\hat{\mathrm{U}}^{-1} = \hat{\mathrm{U}} \tag{1.42}$$

The matrix $\hat{\mathrm{U}}$, satisfying the relation in 1.42 or 1.41, is called a unitary matrix. The operator corresponding to the quantity U* is denoted by $\hat{\mathrm{U}}$ (**read: operator** U **dagger**). It is the conjugate operator of $\hat{\mathrm{U}}$ and it is, however, different from the complex conjugate of $\hat{\mathrm{U}}^*$. Swapping rows into columns and columns into rows of the matrix $\hat{\mathrm{U}}$ and taking the complex conjugate of the resultant matrix then

$$\mathrm{U}_{kn}^{\dagger} = \left(\mathrm{U}_{nk}\right)^* = \left(\langle \lambda_n | \mu_k \rangle\right)^* \tag{1.43}$$

It is obviously shown that $\langle \lambda_k | \mu_n \rangle$ forms a unitary matrix. These unitary operators are not in general observables since they do not have real eigenvalues.

We can simply write 1.36 in operator form as

$$|\lambda_n\rangle = \hat{\mathrm{U}} |\mu_k\rangle \tag{1.44}$$

From this relation, it is obvious that $|\lambda_n\rangle$ has two different components in two different bases. Relation 1.44 is a unitary transformation and seen as a linear transformation transforming a vector to another vector in the same space. In addition, unitary operators:

- transform an orthonormal basis to another orthonormal basis;
- time evolve the basis, spatially rotate them and so on;
- are used to transform to another basis which is unitary as they preserve the scalar product as well as the trace, determinant, and algebraic equations involving matrices and vectors;
- are like complex numbers with a unit modulus, exp $\{i\phi\}$ and conjugating such a number results in its multiplicative inverse. This is same as taking the adjoint of a unitary operator gives its operator product inverse.

1.1.11 Geometrical Interpretation of State Vector

The set of vectors $\mathrm{e} = \{e_i\}_{i=1}^{n}$ is an **orthonormal** basis in the n-dimensional space, $\mathbb{R}^n$ when the **orthonormality** can indeed be expressed as the inner or dot product

$$\hat{\mathrm{P}}_{ik} = \mathrm{e}_i^* \cdot \mathrm{e}_k = \mathrm{e}_i \cdot \mathrm{e}_k = \delta_{ik} \tag{1.45}$$

In this case we say, such a set consists of n linearly independent vectors and defines an **orthonormal basis** in the n -dimensional space, $\mathbb{R}^n$.

So, if $e = \{e_i\}_{i=1}^n$ is an **orthonormal** basis in the n-dimensional space $\mathbb{R}^n$, then an arbitrary vector a with an inner product can be expressed uniquely in the given orthonormal basis as:

$$a = \sum_i e_i a_i \tag{1.46}$$

Here, a_i the components of the transformed vector in the basis e_i can be obtained as:

$$a \cdot e_i = \sum_j a_j e_j \cdot e_i = \sum_j a_j \hat{P}_{ij} = \sum_j a_j \delta_{ij} = a_i \tag{1.47}$$

This implies the components of the given vector in a given orthonormal basis can be obtained by taking the inner product of the vector with the appropriate basis vectors. From here, it is obvious that the vector a can be expanded as:

$$a = \sum_i a_i e_i = \sum_{i,j} a_j e_i \hat{P}_{ij} = \sum_{i,j} a_j e_i \delta_{ij} = \sum_{i,j} (a \cdot e_j) e_i \delta_{ij} \tag{1.48}$$

This shows that the bases can be changed to make the problem simpler. We observe that relations 1.6 and 1.3 are the generalization of relations 1.45 and 1.46 respectively. The quantity $\hat{P}_{ik}$ should also have the sense of the projection operator. Operators have to be transformed also, under similar transformation as in relation 1.48.

So, to conclude, any such set of say *n* linearly independent vectors is called a basis for the vector space and said to span the vector space. The coefficients $\{a_i\}$ for a particular vector a are called the components of a. Equations 1.46 or 1.48 is said to be the (linear) expansion of a in terms of the basis $\{e_i\}$ and the vector space is said to be the space spanned by the basis. We observe from 1.47 that an inner or dot product space is defined such that component values can only be real numbers.

From the aforementioned, the probability amplitude $\langle \lambda_k | \mu_n \rangle$, can be examined as the scalar product of the vector $|\mu_n\rangle$ (state vector) on the basis vector $|\lambda_k\rangle$. Generally, in quantum mechanics we write in compact form general relations via state vector description without making reference to a particular representation. So, relations are described by symbols that mimic vector quantities in vector algebra as well as vector analysis and the concrete description of these relations can only be unveiled by using a defined representation.

1.1.12 Coordinate Transformation

Moving to another system of coordinates, equation 1.46 can be transformed:

$$a = \sum_i e'_i a'_i \tag{1.49}$$

Comparing 1.46 with 1.49 then we have the transforming of the projection a'_i onto a_i under the action of the unitary operator $\hat{U}$:

$$a'_k = \sum_i \hat{U}_{ki} a_i \tag{1.50}$$

where,

$$\hat{U}_{ki} = e'_k e_i \tag{1.51}$$

We find the inverse transform of 1.50:

$$a_k = \sum_i \left(\hat{U}^{-1}\right)_{ki} a'_i \tag{1.52}$$

Comparing expression U_{ki} with $\left(\hat{U}^{-1}\right)_{ki}$ we find that

$$\hat{U}^{-1} = \hat{U}^{\dagger} \tag{1.53}$$

So, an operator is unitary if its inverse equal to its adjoints and equation 1.53 confirms the unitarity of the matrix $\hat{U}$. From equation 1.52, the unitary transformation is seen as a linear transformation transforming a vector to another vector in the same space. The projections a_i and a'_i may be examined as the geometrical interpretation of the amplitudes $\langle\lambda_k|a\rangle$ and $\langle\mu_n|a\rangle$. We examine 1.50 as the geometrical interpretation of 1.14. It is instructive to note that a unitary transformation does not change the physics of a system but merely transforms one description of the system to another physically equivalent description.

1.1.13 Projection Operator

We examine the measurement process as projecting ket vector $|a\rangle$ onto the ket $|\mu_n\rangle$ where the projection operator on the subspace spanned by the single orthonormal basis vector $|\mu_n\rangle$:

$$\hat{P}_{\mu_n} = |\mu_n\rangle\langle\mu_n| \tag{1.54}$$

Geometrically, $\hat{P}_{\mu_n}$ can be considered as an orthogonal projection operator onto to the ket vector $|\mu_n\rangle$ confirmed by the following properties:

$$\hat{P}^2_{\mu_n} = |\mu_n\rangle\langle\mu_n|\mu_n\rangle\langle\mu_n| \tag{1.55}$$

Since, the ket vector $|\mu_n\rangle$ is normalized to unity:

$$\langle\mu_n|\mu_n\rangle = \delta_{nn} = 1 \tag{1.56}$$

then

$$\hat{P}^2_{\mu_n} = |\mu_n\rangle\langle\mu_n| = \hat{P}_{\mu_n} \tag{1.57}$$

So, projecting twice in succession onto a given vector is equivalent to projecting once where operators with such properties are called **idempotent operators**. Relation 1.57 does not imply $\hat{P}_{\mu_n}$ is its own inverse. It is instructive to note that projection operators are in general noninvertible.

In equation 1.54, $\{|\mu_n\rangle\}$ are members of an orthonormal basis. The projection operator $\hat{P}_{\mu_n}$ satisfies the **Dirac formula**:

$$\sum_{\mu_n} \hat{P}_{\mu_n} = \sum_{\mu_n} |\mu_n\rangle\langle\mu_n| = 1 \tag{1.58}$$

This is also known as a **closure relation**. This formula may be generalized:

$$\sum_{\mu_n\eta_m} |\mu_n\eta_m\cdots\rangle\langle\mu_n\eta_m\cdots| = 1 \tag{1.59}$$

Relation 1.58 is for the discrete spectrum. For a mixed spectrum:

$$\sum_{\mu_n} |\mu_n\rangle\langle\mu_n| + \int d\mu |\mu\rangle\langle\mu| = 1 \tag{1.60}$$

For any measurement, the state changes:

$$|a\rangle \mapsto \hat{P}_{\mu_n} |a\rangle = |\mu_n\rangle\langle\mu_n|a\rangle \tag{1.61}$$

with a probability

$$\left\|\hat{P}_{\mu_n} |a\rangle\right\|^2 = \langle a|\mu_n\rangle\langle\mu_n|\mu_n\rangle\langle\mu_n|a\rangle = \left|\langle\mu_n|a\rangle\right|^2 \tag{1.62}$$

The projection operator $\hat{P}_{\mu_n}$ acting on an arbitrary ket vector $|a\rangle$ yields a ket proportional to $|\mu_n\rangle$ with the coefficient of proportionality being $\langle\mu_n|a\rangle$ that is the scalar product of $|a\rangle$ by $|\mu_n\rangle$.

Similarly,

$$\langle a|\hat{P}_{\mu_n} = \langle a|\mu_n\rangle\langle\mu_n| = \left(\langle\mu_n|a\rangle\right)^* \langle\mu_n| \tag{1.63}$$

So, **the geometrical interpretation of $\hat{P}_{\mu_n}$ is the orthogonal projection operator onto the vector $|\mu_n\rangle$. It is instructive to note that $\hat{P}_{\mu_n}$ is specified when the basis $|\mu_n\rangle$ is specified**. So, the meaning of $\hat{P}_{\mu_n}$ will always depend on context.

1.1.14 Continuous Spectrum

So far, we have dealt with the discrete basis and now we examine a continuum basis. We consider the probability amplitude $\langle\xi|\mu_n\rangle$ that is a function of the continuous variable ξ and μ_n corresponding to a discrete spectrum. The quantity $|\langle\xi|\mu_n\rangle|^2$ for given μ_n is the probability amplitude and

$$dP = \left|\langle\xi|\mu_n\rangle\right|^2 d\xi \tag{1.64}$$

is the probability in the state μ_n of the finding the quantity ξ within the interval ξ and $\xi + d\xi$. The **ortho-normalization condition** is written:

$$\int d\xi \left(\langle\xi|\mu_n\rangle\right)^* \langle\xi|\mu_k\rangle = \delta_{nk} \tag{1.65}$$

while the superposition principle in 1.14 is now written:

$$\langle\xi|a\rangle = \int d\mu \langle\xi|\mu\rangle\langle\mu|a\rangle \tag{1.66}$$

and viewed as the expansion of a component of $|\xi\rangle$ in one basis, $|a\rangle$, into those of another basis, $|\mu\rangle$. From here we have

$$\langle\xi|\eta_k\rangle^* \langle\xi|\eta_n\rangle = \int d\mu d\mu' \langle\xi|\mu\rangle\left(\langle\xi|\mu'\rangle\right)^* \left(\langle\mu'|\eta_k\rangle\right)^* \langle\mu|\eta_n\rangle \tag{1.67}$$

Considering the normalization condition 1.64 and integrating over ξ then

$$\delta_{kn} = \int d\mu d\mu' \left(\langle\mu'|\eta_k\rangle\right)^* \langle\mu|\eta_n\rangle \int d\xi \langle\xi|\mu\rangle\langle\xi|\mu'\rangle^* \tag{1.68}$$

Considering that

$$\int d\mu \left(\langle \mu | \eta_k \rangle\right)^* \langle \mu | \eta_n \rangle = \delta_{kn} \tag{1.69}$$

Then for 1.67 to be equal to the LHS (Left-Hand_Side) when in the continuum basis we have

$$\int d\xi \langle \xi | \mu \rangle \left(\langle \xi | \mu' \rangle\right)^* = \delta(\mu - \mu') \tag{1.70}$$

Here, $\delta(\mu - \mu')$ is called the Dirac δ-function. A similar result can be obtained for a parameter λ_n of a discrete spectrum resulting in the **closure relation**:

$$\sum_{\lambda_n} \left(\langle \lambda_n | \mu \rangle\right)^* \langle \lambda_n | \mu' \rangle = \delta(\mu - \mu') \tag{1.71}$$

2

Matrix Representation of Linear Operators

2.1 Matrix Element

It is instructive to note that the principal advantage of linear operators is that their action on any vector is defined purely by their action on a set of basis vectors. Let us consider the linear operator $\hat{F}$ transforming the vector a via its components a_k:

$$\left(\hat{F}a\right)_i = \sum_k F_{ik} a_k \tag{2.1}$$

Similarly, in quantum mechanics the action of the linear operator $\hat{F}$ on the vector of state $|a\rangle$ is symbolically denoted as $\hat{F}|a\rangle$ and defined by the relation:

$$\left\langle \lambda_i|\hat{F}|a\right\rangle = \sum_{\lambda_k}\left\langle \lambda_i|\hat{F}|\lambda_k\right\rangle\langle\lambda_k|a\rangle \equiv \sum_{\lambda_k} F_{ik}\langle\lambda_k|a\rangle \tag{2.2}$$

This represents $\left\langle \lambda_i|\hat{F}|a\right\rangle$ via the probability amplitude $\langle\lambda_k|a\rangle$ where $\left\langle \lambda_i|\hat{F}|\lambda_k\right\rangle$ is the **matrix element** of the operator $\hat{F}$ in the λ-representation and can be defined as the projection of $\hat{F}|\lambda_k\rangle$ onto $|\lambda_i\rangle$. So, if all the vectors $\hat{F}|\lambda_k\rangle$ are known, consequently all the matrix elements F_{ik} of the operator $\hat{F}$ are known in the particular basis. When F_{ik} is known, then the transformation property of any arbitrary vector can be easily established.

Relation 2.1 can be examined as the matrix multiplication of an $n \times n$ matrix against a single-column n-row matrix, where n is the dimensionality of the vector space. From 2.73, it is observed that once an orthonormal basis is specified, any operator can be written as a bilinear expression in the basis kets and bras, with the coefficients simply being all the matrix elements in that basis.

2.2 Linear Self-Adjoint (Hermitian Conjugate) Operators

In quantum mechanics it is required that all operators be self-adjoint and linear. This requirement is such that the superposition principle holds. In order to have meaningful and measurable observables with operators, it is necessary that the expectation (mean) values of these operators be real. We find if there is some sense in the complex conjugate of relation 2.2:

$$\left\langle \lambda_i|\hat{F}|a\right\rangle = \sum_{\lambda_k}\left\langle \lambda_i|\hat{F}|\lambda_k\right\rangle\langle\lambda_k|a\rangle = \left(\sum_{\lambda_k}\langle a|\lambda_k\rangle\left\langle \lambda_k|\hat{F}|\lambda_i\right\rangle\right)^* = \left(\left\langle a|\hat{F}|\lambda_i\right\rangle\right)^* \tag{2.3}$$

For an arbitrary operator $\hat{F}$ we may select its **adjoint** $\tilde{\hat{F}}$ defined:

$$\left\langle \lambda_i|\hat{F}|a\right\rangle = \left\langle a|\tilde{\hat{F}}|\lambda_i\right\rangle \tag{2.4}$$

Comparing 2.3 with 2.4 while substituting in place of $|a\rangle$the state vector $\langle\lambda_i|$ then we have

$$\left\langle \lambda_i|\tilde{\hat{F}}|\lambda_i\right\rangle = \left\langle \lambda_i|\hat{F}^*|\lambda_i\right\rangle \tag{2.5}$$

from where

$$\tilde{\hat{F}} = \hat{F}^* \tag{2.6}$$

The operators satisfying condition 2.6 are **Hermitian.** So, **the operators for which the physical quantities are real are Hermitian.**

We examine the case for which the physical quantities are complex. Let that quantity be F, then we may take its complex conjugate F*. The operator corresponding to the quantity F* is denoted by $\hat{F}^\dagger$. It is the conjugate operator of $\hat{F}$ and it is, however, different from the complex conjugate of $\hat{F}^*$. **So, finally one simply transposes and complex conjugates a matrix representation of the operator $\hat{F}$ to get the corresponding matrix representation of $\hat{F}^\dagger$:**

$$F^\dagger_{mn} = \left(F_{nm}\right)^* \tag{2.7}$$

If we want to get the *m*, *n* element of $\hat{F}^\dagger$ we go to the *n*, *m* element of $\hat{F}$ (the indices are reversed) and we take its complex conjugate. So, a Hermitian operator is thus represented by a Hermitian matrix. This implies one in which any two elements that are symmetric with respect to the principal diagonal are complex conjugates of each other. If we consider our condition, then

$$\overline{F^*} = \overline{(F)}^*, \hat{F}^\dagger = \tilde{\hat{F}}^* \tag{2.8}$$

From here it is obvious that $\hat{F}^\dagger$ however coincides with $\hat{F}^*$. For real physical quantities, i.e., the operator coincides with its conjugate (the Hermitian operator is also called **self-adjoint**). From here, considering 2.3 then

$$\left\langle a|\hat{F}\,|\lambda_i\right\rangle = \sum_{\lambda_k} \left\langle a|\lambda_k\right\rangle \left\langle \lambda_k|\hat{F}\,|\lambda_i\right\rangle \tag{2.9}$$

Condition 2.8 is the reason for which the operator of a physical quantity should be **self-adjoint** and it also shows that the mean value of a physical quantity should be real. **So, the necessary condition for the self-adjoint of operators of physical quantities stems from the condition that the mean values of these quantities should be real and so the diagonal elements of a Hermitian matrix are thus always real numbers.** Hence **the operator that is linear and self-adjoint is Hermitian.** It is instructive to note that in quantum mechanics observables are associated with Hermitian operators.

If the operator has to be transformed as

$$\hat{F}' = \hat{U}\hat{F}\hat{U}^{-1} = \hat{U}\hat{F}\hat{U}^\dagger \tag{2.10}$$

where $\hat{U}$ is a unitary operator then if $\hat{F}$ is Hermitian so $\hat{F}'$ is also Hermitian with $\hat{F}$ and $\hat{F}'$ having the same eigenvalues. It is instructive to note that complex numbers remain unchanged under this unitary transformation. Typically, we use a unitary transformation to diagonalize an operator: The new operator $\hat{F}'$ is diagonal in the original basis and has the same eigenvalues, in the same order, as $\hat{F}$.

2.3 Product of Hermitian Operators

We find the operator that is Hermitian conjugate of the product of the operators $\hat{F}$ and $\hat{L}$. So, we consider

$$\left\langle \lambda_i|\hat{F}\hat{L}|a\right\rangle = \sum_{\lambda_k}\left\langle \lambda_i|\hat{F}\hat{L}|\lambda_k\right\rangle\langle\lambda_k|a\rangle = \left(\sum_{\lambda_k}\langle a|\lambda_k\rangle\left\langle \lambda_k|\hat{L}\hat{F}|\lambda_i\right\rangle\right)^* = \left(\left\langle a|\hat{L}\hat{F}|\lambda_i\right\rangle\right)^* \tag{2.11}$$

Letting,

$$\hat{L}|\lambda_k\rangle = |\eta_k\rangle \tag{2.12}$$

then

$$\left\langle \lambda_i|\hat{F}\hat{L}|\lambda_k\right\rangle = \left\langle \lambda_i|\hat{F}|\eta_k\right\rangle = \left(\left\langle \eta_k|\hat{F}\,|\lambda_i\right\rangle\right)^* \tag{2.13}$$

Letting,

$$\hat{F}^{\dagger}|\lambda_i\rangle = |\mu_i\rangle \tag{2.14}$$

then

$$\left(\left\langle \eta_k|\hat{F}^{\dagger}\,|\lambda_i\right\rangle\right)^* = \left\langle \mu_i|\hat{L}|\lambda_k\right\rangle = \left(\left\langle \lambda_k|\hat{L}^{\dagger}\,|\mu_i\right\rangle\right)^* = \left(\left\langle \lambda_k|\hat{L}^{\dagger}\hat{F}^{\dagger}\,|\lambda_i\right\rangle\right)^* \tag{2.15}$$

So,

$$\left(\left\langle \lambda_i|\hat{F}\hat{L}|\lambda_n\right\rangle\right)^* = \left(\left\langle \lambda_n|\left(\hat{F}\hat{L}\right)^{\dagger}|\lambda_i\right\rangle\right) = \sum_{\lambda_m}\left(\left\langle \lambda_i|\hat{F}|\lambda_m\right\rangle\right)^*\left(\left\langle \lambda_m|\hat{L}|\lambda_n\right\rangle\right)^* \tag{2.16}$$

or

$$\left(\left\langle \lambda_n|\left(\hat{F}\hat{L}\right)^{\dagger}|\lambda_i\right\rangle\right) = \sum_{\lambda_m}\left\langle \lambda_n|\hat{L}^{\dagger}\,|\lambda_m\right\rangle\left\langle \lambda_m|\hat{F}^{\dagger}\,|\lambda_i\right\rangle = \left\langle \lambda_n|\hat{L}^{\dagger}\hat{F}^{\dagger}\,|\lambda_i\right\rangle \tag{2.17}$$

Hence, taking the adjoint of the product of matrix operators corresponds to reverse the order of the matrices, complex conjugate and transpose each matrix, then matrix multiply them. So, the resulting expression will be to complex conjugate each of the two matrices, matrix multiply them, and then transpose:

$$\left(\hat{F}\hat{L}\right)^{\dagger} = \hat{L}^{\dagger}\hat{F}^{\dagger} \tag{2.18}$$

Summarily, taking the adjoint of the product of operators reverses the order of all the factors in a product and takes the adjoint of each factor independently. The adjoint of a sum is simply the sum of the adjoints.

2.4 Continuous Spectrum

For the case of a continuous spectrum then 2.1 can be rewritten:

$$\left\langle \xi|\hat{F}|a \right\rangle = \int d\xi' \left\langle \xi|\hat{F}|\xi' \right\rangle \langle \xi'|a\rangle \tag{2.19}$$

In this case, $\hat{F}$ is an **integral operator** and

$$\hat{F}(\xi,\xi') = \left\langle \xi|\hat{F}|\xi' \right\rangle \tag{2.20}$$

the **kernel of the operator** that should be understood as the generalized matrix operator for the case of a continuous spectrum. Letting in 2.19,

$$\hat{F} = 1 \tag{2.21}$$

then

$$\langle \xi|a\rangle = \int d\xi' \delta(\xi - \xi')\langle \xi'|a\rangle \tag{2.22}$$

We find the action of the operator $\hat{F}$ on the amplitude $\langle \xi | a\rangle$:

$$\hat{F}\langle \xi|a\rangle = \hat{F}\int d\xi' \delta(\xi - \xi')\langle \xi'|a\rangle \tag{2.23}$$

Suppose we swap the integral with the operator $\hat{F}$ then

$$\hat{F}\langle \xi|a\rangle = \int d\xi' \hat{F}\delta(\xi - \xi')\langle \xi'|a\rangle \tag{2.24}$$

We can then examine this as an integral with the kernel:

$$\left\langle \xi|\hat{F}|\xi' \right\rangle = \hat{F}\delta(\xi - \xi') \tag{2.25}$$

So, further we understand the product of the operators 2.25 in the sense of 2.24.

2.5 Schturm-Liouville Problem: Eigenstates and Eigenvalues

Letting F_n, be an observable that in basis states $|\mu_n\rangle$ then we construct an operator $\hat{F}$, corresponding to the given observable:

$$\hat{F} = \sum_{\mu_n} F_n \hat{P}_{\mu_n} = \sum_{\mu_n} F_n |\mu_n\rangle\langle\mu_n| \tag{2.26}$$

This equation is called the **spectral representation** of the operator $\hat{F}$.

We consider the operator $\hat{F}$ defined in 2.26 and acting on the basis states$|\mu_n\rangle$:

$$\hat{F}|\mu_n\rangle = \sum_{\mu_m} F_m \hat{P}_{\mu_m} |\mu_n\rangle = \sum_{\mu_m} F_m |\mu_m\rangle\langle\mu_m|\mu_n\rangle \tag{2.27}$$

From 1.23 of chapter 1, then 2.27 becomes the **Schturm-Liouville problem:**

$$\hat{F}|\mu_n\rangle = \sum_{\mu_m} F_m |\mu_m\rangle \delta_{mn} = F_n|\mu_n\rangle \tag{2.28}$$

or

$$\hat{F}|\mu_n\rangle = F_n|\mu_n\rangle \tag{2.29}$$

If an operator of a physical quantity acts on an eigenstate of that quantity (i.e., on the eigenstate of that state for which the physical quantity has a defined value say F_m) then the resultant is the product of the eigenvalue and the eigenstate of that quantity.

The Schturm-Liouville problem in equation 2.29 may simply be rewritten:

$$\hat{F}|\mu\rangle = F|\mu\rangle \equiv \mu|\mu\rangle \tag{2.30}$$

The parameters μ are called **eigenvalues or the proper values of the operator** $\hat{F}$. They may form a countable set of **discrete spectra of eigenvalues.** We may also have a **continuous spectrum of eigenvalues.** Considering 2.29 or 2.30, the eigenvector direction in the inner product space is left invariant by the action of the operator $\hat{F}$. It is instructive to note that an operator can have multiple eigenvectors, and the eigenvectors need not all be different.

CONCLUSION: The observable value of a physical quantity is the eigenvalue **of the operator of that quantity. The corresponding eigenfunction of the operator of the physical quantity is the wave function of that state for which that quantity has a defined value.**

We write relation 2.30 in different representations:

1. Choosing the representation, say $\{|\lambda_n\rangle\}$ and then project the vector equation 2.30: onto the various orthonormal basis vectors $|\mu\rangle$ for a discrete spectrum:

$$\left\langle \lambda_n|\hat{F}|\mu \right\rangle = \mu\langle\lambda_n|\mu\rangle \tag{2.31}$$

We insert the closure relation between $\hat{F}$ and $|\lambda_n\rangle$:

$$\sum_{\lambda_k}\left\langle \lambda_n|\hat{F}\hat{P}_{\lambda_k}|\mu\right\rangle \equiv \sum_{\lambda_k}\left\langle \lambda_n|\hat{F}|\lambda_k\right\rangle\langle\lambda_k|\mu\rangle = \mu\langle\lambda_n|\mu\rangle \quad (2.32)$$

or

$$\sum_{\lambda_k}\left\langle \lambda_n|\hat{F}|\lambda_k\right\rangle\langle\lambda_k|\mu\rangle = \mu\langle\lambda_n|\mu\rangle \quad (2.33)$$

We observe from here that $\{|\mu\rangle\}$ is the set of eigenstates of the Hermitian operator $\hat{F}$ with eigenvalues μ. When the particle is in the arbitrary state $|\lambda_n\rangle$ then measurement of the variable corresponding to the operator $\hat{F}$ will yield only the eigenvalues $\{\mu\}$ of $\hat{F}$. This measurement will yield the particular value μ for that variable with relative probability $|\langle\lambda_n|\mu\rangle|^2$. The system will then change from state $|\lambda_n\rangle$ to state $|\mu\rangle$ as a result of the measurement taken. That is to say, the eigenvalues of $\hat{F}$ are the only measurable quantities with the measurement outcome being fundamentally probabilistic. The relative probability of the particular allowed outcome μ is obtained by finding the projection of $|\lambda_n\rangle$ onto the corresponding eigenstate $|\mu\rangle$. This implies that, if $|\lambda_n\rangle$ is an eigenstate of $\hat{F}$, the measurement will always yield the corresponding eigenvalue. The measurement process itself changes the state of the particle to the eigenstate $|\mu\rangle$ corresponding to the measurement outcome μ.

Suppose

$$\left\langle \lambda_n|\hat{F}|\lambda_k\right\rangle = F_{nk}, \langle\lambda_n|\mu\rangle = C_n \quad (2.34)$$

So, 2.104 becomes

$$\sum_{\lambda_k} F_{nk}C_k = \mu C_n \quad (2.35)$$

or

$$\sum_{\lambda_k}\left[F_{nk} - \mu\delta_{nk}\right]C_k = 0 \quad (2.36)$$

This is a set of linear homogeneous equations (**characteristic equation**) with the unknown coefficients C_k being the components of the eigenvector in the chosen representation. The quantity F_{nk} is the matrix element of the matrix $\hat{F}$. Relation 2.36 is a system of N equations,

$$n = 1, 2, \cdots, N \quad (2.37)$$

with k unknown coefficients C_k,

$$k = 1, 2, \cdots, N \quad (2.38)$$

As 2.36 is a linear and homogenous equation, then it has non-trivial solutions when the determinant of the coefficient matrix vanishes yielding the following **characteristic or secular equation**:

$$\det\left[\hat{F} - \mu\hat{\mathbb{I}}\right] = 0 \quad (2.39)$$

Here, $\hat{F}$ is the $N \times N$ matrix constituting the matrix elements F_{nk} and $\hat{\mathbb{I}}$, the unit matrix. As we are working in the k-dimensional space, then 2.39 is an k-th order polynomial equation in μ and so would possess k solutions for μ corresponding to all the eigenvalues of the operator $\hat{F}$. The equation 2.39 has k roots that may be real or imaginary, distinct or identical. **Since the characteristic equation is independent of the given representation then the eigenvalues of the operator are obviously roots of its characteristic equation.** From the eigenvalues it is then easy to find the eigenstates from equation 2.36 that for Hermitian operators $\hat{F}$ will be a linear combination of k independent eigenstates.

2. The representation that relates a continuous spectrum:

$$\int d\xi' \left\langle \xi|\hat{F}|\xi' \right\rangle \langle \xi'|\mu\rangle = \mu\langle \xi|\mu\rangle \tag{2.40}$$

As the kernel $\left\langle \xi|\hat{F}|\xi' \right\rangle$ has the form in 2.25 then

$$\hat{F}\langle \xi|\mu\rangle = \mu\langle \xi|\mu\rangle \tag{2.41}$$

We consider 2.33 for the case of a discrete spectrum where the matrix element is diagonalized:

$$\left\langle \mu_n|\hat{F}|\mu_k \right\rangle = \mu_n\delta_{nk} \tag{2.42}$$

Similarly, for the case of a continuous spectrum then

$$\left\langle \mu|\hat{F}|\mu' \right\rangle = \mu\delta(\mu - \mu') \tag{2.43}$$

Suppose in 2.97 and 2.29 we consider the eigenvalue F_n to have the value λ_n then we have the following operator for a discrete spectrum:

$$\hat{F} = \sum_{\mu_n} |\mu_n\rangle \mu_n \langle \mu_n| \tag{2.44}$$

and for the continuous spectrum

$$\hat{F} = \int d\mu |\mu\rangle \mu \langle \mu| \tag{2.45}$$

and for a mixed spectrum

$$\hat{F} = \sum_{\mu_n} |\mu_n\rangle \mu_n \langle \mu_n| + \int d\mu |\mu\rangle \mu \langle \mu| \tag{2.46}$$

Example:

We consider an example of an **eigenvalue** problem where the operator:

$$\hat{F} = \sin\frac{d}{d\phi} \tag{2.47}$$

We solve the Schturm-Liouville equation:

$$\hat{F}|\phi\rangle = F|\phi\rangle \tag{2.48}$$

where

$$\hat{F} = \sin\frac{d}{d\phi} \tag{2.49}$$

We represent the operator $\sin\frac{d}{d\phi}$ in the form of a series:

$$\sin\frac{d}{d\phi}|\phi\rangle = \left[\frac{d}{d\phi} - \frac{1}{3!}\frac{d^3}{d\phi^3} + \frac{1}{5!}\frac{d^5}{d\phi^5} - \cdots\right]|\phi\rangle = \sum_{k=0}^{\infty}\frac{(-1)^k}{(2k+1)!}\frac{d^{2k+1}}{d\phi^{2k+1}}|\phi\rangle \tag{2.50}$$

If we consider this and the Schturm-Liouville equation, then we may find the solution in the form

$$|\phi\rangle = \exp\{-\lambda\phi\} \tag{2.51}$$

From here considering that the wave function should be univalent then we let

$$\lambda = im, \quad m = 0, \pm 1, \cdots \tag{2.52}$$

Thus,

$$\sin\frac{d}{d\phi}|\phi\rangle = \sum_{k=0}^{\infty}\frac{(-1)^k (im)^k}{(2k+1)!}|\phi\rangle = \sin(im)|\phi\rangle \tag{2.53}$$

and

$$F = \sin(im), \quad m = 0, \pm 1, \cdots \tag{2.54}$$

2.6 Revisit Linear Self-Adjoint (Hermitian) Operators

The self-adjoint of physical quantities of operators stems from the mean values of these quantities that should be real. For this case **two** eigenstates **of a Hermitian operator** $\hat{F}$ **corresponding to two different eigenvalues should be orthogonal**. This can as well be shown via equation 2.44 where we take its complex conjugate:

$$\int d\xi' \left(\left\langle \xi|\hat{F}|\xi'\right\rangle\right)^* \left(\langle\xi'|\mu'\rangle\right)^* = \mu'\left(\langle\xi|\mu'\rangle\right)^* \tag{2.55}$$

Multiplying 2.44 by $(\langle\xi|\mu'\rangle)^*$ and 2.55 by $\langle\xi|\mu\rangle$ then we take the difference while integrating over ξ so that

$$\int d\xi' \left|\left(\left\langle \xi|\hat{F}|\xi'\right\rangle\right)^* - \left\langle \xi'|\hat{F}|\xi\right\rangle\right| \langle\xi|\mu\rangle\left(\langle\xi|\mu'\rangle\right)^* = (\mu'-\mu)\int d\xi \langle\xi|\mu\rangle\left(\langle\xi|\mu'\rangle\right)^* \tag{2.56}$$

For the self-adjoint operator, the LHS of 2.56 is equal to zero where for $\mu' \neq \mu$ we have:

$$\int d\xi \langle \xi | \mu \rangle \left(\langle \xi | \mu' \rangle \right)^* = 0 \tag{2.57}$$

This confirms the orthogonality of the eigenstates.
For the discrete spectrum then

$$\int d\xi \left| \langle \xi | \mu_n \rangle \right|^2 = 1 \tag{2.58}$$

This is the condition of orthonormalization of the **eigenstate**.

2.7 Unitary Transformation

We examine the transformation of the matrix element of an operator $\hat{F}$ from the μ-representation to the λ-representation. We will be inspired by equation 2.1 rewritten:

$$\left\langle \lambda_i | \hat{F} | \lambda_m \right\rangle = \sum_{\mu_k} \left\langle \lambda_i | \hat{F} | \mu_k \right\rangle \langle \mu_k | \lambda_m \rangle \tag{2.59}$$

From the closure relation in 2.57 we rewrite 2.59:

$$\left\langle \lambda_i | \hat{F} | \lambda_m \right\rangle = \sum_{\mu_k, \mu_i} \left\langle \lambda_i | \hat{P}_{\mu_k} \hat{F} \hat{P}_{\mu_i} | \lambda_m \right\rangle = \sum_{\mu_k, \mu_i} \langle \lambda_i | \mu_k \rangle \left\langle \mu_k | \hat{F} | \mu_i \right\rangle \langle \mu_i | \lambda_m \rangle \tag{2.60}$$

We observe that for this transformation, the LHS of 2.60 is invariant relative to the change of the representation. Letting $i = m$ and summing over λ_m we observe that

$$\sum_{\lambda_m} \left\langle \lambda_m | \hat{F} | \lambda_m \right\rangle = \sum_{\mu_k} \left\langle \mu_k | \hat{F} | \mu_k \right\rangle \tag{2.61}$$

Revisiting equation 2.37 for the representation of the unitary operator $\hat{U}$ then 2.60 can be rewritten:

$$\left\langle \lambda_i | \hat{F} | \lambda_m \right\rangle = \sum_{\mu_k, \mu_i} U^{\dagger}_{ik} \left\langle \mu_k | \hat{F} | \mu_i \right\rangle U_{im} \tag{2.62}$$

or

$$\hat{F}' = \hat{U}^{\dagger} \hat{F} \hat{U} \tag{2.63}$$

This is a unitary transformation and leaves all operator relations invariant. So, $\hat{F}$ can be derived in a suitable chosen representation. It is the matrix operator in the μ-representation that has the same matrix elements as the operator $\hat{F}'$ in the λ-representation. Equation 2.63 is the transform $\hat{F}'$ of the operator $\hat{F}$ by the unitary operator $\hat{U}$. In the same manner considering as well

$$\sum_{\lambda_n} \hat{P}_{\lambda_n} = \sum_{\lambda_n} | \lambda_n \rangle \langle \lambda_n | = 1 \tag{2.64}$$

we find the inverse transformation of 2.62:

$$\left\langle \mu_i|\hat{F}|\mu_n\right\rangle = \sum_{\lambda_k,\lambda_i}\left\langle \mu_i|\hat{P}_{\lambda_k}\hat{F}\hat{P}_{\lambda_i}|\mu_n\right\rangle = \sum_{\lambda_k,\lambda_i}\langle\mu_i|\lambda_k\rangle\left\langle\lambda_k|\hat{F}|\lambda_i\right\rangle\langle\lambda_i|\mu_n\rangle \tag{2.65}$$

or

$$\left\langle \mu_i|\hat{F}|\mu_n\right\rangle = \sum_{\lambda_k,\lambda_i}U_{ik}\left\langle\lambda_k|\hat{F}|\lambda_i\right\rangle U^{\dagger}_{in} \tag{2.66}$$

or

$$\hat{F} = \hat{U}\hat{F}'\hat{U}^{\dagger} \tag{2.67}$$

Considering the continuous spectrum then 2.63 becomes:

$$\left\langle\lambda_i|\hat{F}|\lambda_m\right\rangle = \int\langle\lambda_i|\mu\rangle\left\langle\mu|\hat{F}|\mu'\right\rangle\langle\mu'|\lambda_m\rangle d\mu d\mu' \tag{2.68}$$

For

$$\left\langle\mu|\hat{F}|\mu'\right\rangle = \hat{F}\delta(\mu-\mu') \tag{2.69}$$

then

$$\left\langle\lambda_i|\hat{F}|\lambda_m\right\rangle = \int\langle\lambda_i|\mu\rangle\hat{F}\langle\mu|\lambda_m\rangle d\mu = \int\left(\langle\mu|\lambda_i\rangle\right)^*\hat{F}\langle\mu|\lambda_m\rangle d\mu \tag{2.70}$$

2.8 Mean (Expectation) Value and Matrix Density

Before introducing the matrix density, which will be a very useful mathematical tool that will facilitate the simultaneous application of the postulates of quantum mechanics and statistical physics, we examine first the physical quantity F and introduce the notion of its **mean (expectation) value.** Let

$$F_1, F_2, \cdots, F_k \tag{2.71}$$

be a series of measurement of the physical quantity F in the states $|\mu_k\rangle$ for which there are $N \to \infty$ experiments (measurements):

$$N_1 + N_2 + \cdots + N_k = N \to \infty \tag{2.72}$$

Thus, the mean (expectation) value $\bar{F}$ of the quantity F in the state $|\mu_k\rangle$:

$$\bar{F} = \sum_{\mu_k}\frac{N_k}{N}F_k \tag{2.73}$$

and

$$\left.\frac{N_k}{N}\right|_{N\to\infty} = W_k \tag{2.74}$$

is the probability that we get the quantity $F_k$which is also equal to $|\langle\mu_k| a\rangle|^2$. So, we can say we are dealing with a statistical mixture of states with given probabilities. This implies that the results of measurement must be weighted by the probability W_k and then summed over the various values μ_k, i.e., over all the states of the statistical mixture.

Hence the mean (expectation) value $\bar{F}$:

$$\bar{F} = \sum_{\mu_k} \left|\langle\mu_k|a\rangle\right|^2 F_k = \sum_{\mu_k} \langle a|\mu_k\rangle\langle\mu_k|a\rangle F_k = \left\langle a|\hat{F}|a\right\rangle \tag{2.75}$$

Applying relation 2.57 then equation 2.75 can be rewritten:

$$\left\langle a|\hat{F}|a\right\rangle = \sum_{\mu_k,\mu_i} \langle a|\mu_k\rangle\left\langle \mu_k|\hat{F}|\mu_i\right\rangle\langle\mu_i|a\rangle \tag{2.76}$$

Substituting the expression for $\left\langle \mu_k|\hat{F}|\mu_i\right\rangle$ via 2.42 then 2.76 becomes

$$\left\langle a|\hat{F}|a\right\rangle = \sum_{\mu_k,\mu_i} \langle a|\mu_k\rangle \mu_k \delta_{ki} \langle\mu_i|a\rangle = \sum_{\mu_k} \langle a|\mu_k\rangle \mu_k \langle\mu_k|a\rangle \tag{2.77}$$

or

$$\left\langle a|\hat{F}|a\right\rangle = \sum_{\mu_k} \left(\langle\mu_k|a\rangle\right)^* \mu_k \langle\mu_k|a\rangle \tag{2.78}$$

The mean (expectation) value of $\hat{F}$ in the λ-representation may be rewritten:

$$\left\langle a|\hat{F}|a\right\rangle = \sum_{\lambda_k,\lambda_i} \left(\langle\lambda_k|a\rangle\right)^* \left\langle \lambda_k|\hat{F}|\lambda_i\right\rangle\langle\lambda_i|a\rangle \tag{2.79}$$

The matrix element $\left\langle \lambda_k|\hat{F}|\lambda_i\right\rangle$ in this case is not in diagonalized form.

2.9 Degeneracy

The operator $\hat{F}$ may have degeneracies:

$$\hat{F}\left|\mu_n,\mu_m\right\rangle = F_n\left|\mu_n,\mu_m\right\rangle \tag{2.80}$$

where F_n is a certain set and so measurement of F_n then projects onto a degenerate subspace:

$$|a\rangle \mapsto \hat{P}_n|a\rangle = \sum_m \left|\mu_n,\mu_m\right\rangle\left\langle\mu_n,\mu_m|a\right\rangle \tag{2.81}$$

Projecting onto a definite eigenstate involves measuring further observables that commute with $\hat{F}$ and leads to the notion of a maximally commuting set of observables. The basis states may then be organized into irreducible representations of discrete or continuous symmetries of the system with the observables corresponding to generators of these symmetries.

2.10 Density Operator

Suppose we are concerned with an ensemble of identical quantum mechanical systems. It is obvious that each system in the ensemble should be in a different eigenstate of the Hamiltonian and so there will be a statistical distribution of the system at various eigenstates of the given Hamiltonian. We now introduce the density operator $\hat{\rho}$ of the system and find its matrix elements in the state $|a\rangle$ i.e., the amplitude $\langle \lambda_k | a\rangle$:

$$\left\langle \lambda_k \middle| \hat{\rho} \middle| \lambda_i \right\rangle = \langle \lambda_k | a\rangle \left(\langle \lambda_i | a\rangle\right)^* = \langle \lambda_k | a\rangle \langle a | \lambda_i\rangle \tag{2.82}$$

It is obvious from here that this is in principle the matrix element in the λ-representation of the projecting operator:

$$\hat{P}_a = |a\rangle\langle a| \tag{2.83}$$

The expectation value of the operator $\hat{F}$ in the state $|a\rangle$ can then be expressed via the density operator $\hat{\rho}$:

$$\left\langle a \middle| \hat{F} \middle| a \right\rangle = \mathrm{Tr}\left(\hat{\rho}\hat{F}\right) \tag{2.84}$$

Here $\mathrm{Tr}\,\hat{A}$ means the trace of the operator $\hat{A}$ and is evaluated via any complete set of states say $|a\rangle$. So, given any operator $\hat{A}$, its ensemble average $\bar{A}$ can be defined as in 2.84. If in 2.84 we set $\hat{F} = 1$ then we examine the following properties of the density operator:

$$\hat{\rho} = \hat{\rho}\ , \mathrm{Tr}\left(\hat{\rho}\right) = 1, \mathrm{Tr}\left(\hat{\rho}^2\right) = 1 \tag{2.85}$$

The first property shows that the density operator is **Hermitian** while the last equation that stems from the fact that $\hat{\rho}$ is a projecting operator holds only for the pure state.

Consider the case of the interaction of an object with some system, then the state $|a\rangle$ may probably be undefined. In such a case, the parameter of the given state may be subject to some fluctuation about a certain mean value. So, we find the matrix density defined by the following expression:

$$\left\langle i \middle| \hat{\rho} \middle| k \right\rangle = \overline{\langle i | a\rangle \left(\langle k | a\rangle\right)^*} \tag{2.86}$$

and with properties

$$\hat{\rho} = \hat{\rho}^\dagger, \mathrm{Tr}\left(\hat{\rho}\right) = 1, \hat{\rho}^2 \neq \hat{\rho} \tag{2.87}$$

The last equation in 158 shows that $\hat{\rho}$ is no longer a projecting operator while the first equation considering relation 2.86 shows that $\hat{\rho}$ is a Hermitian operator, and, therefore, it is obviously diagonalized. This condition in 2.87 is for the so-called **mixed state** that differs from the pure state mentioned above when $\hat{\rho}^2 = \hat{\rho}$. It is obvious that from 2.86, the averaging operation contributes to some loss of information of the full system. So, the mixed density matrix in 2.86 does not have all information of the full system. When the averaging operation is not feasible then it is impossible to evaluate the density matrix and so its parameters can only be obtained from experimentation. Therefore, to have full knowledge of the parameters of the full system, the expression 2.84 of the pure system should be the tailoring factor.

Suppose we want to evaluate the mean value of some operator acting on the parameter λ_k then the role of the density matrix in 2.84 will be played by the expression:

$$\left\langle i\middle|\hat{\rho}\middle|k\right\rangle = \sum_{\mu_n} \left\langle \lambda_i, \mu_n \middle| a\right\rangle \left\langle \lambda_k, \mu_n \middle| a\right\rangle \tag{2.88}$$

2.11 Commutativity of Operators

Consider two linear operators $\hat{L}$ and $\hat{F}$ acting on the state $|\mu_n, \lambda_n\rangle$ then from 2.29 we have

$$\hat{F}\left|\mu_n\right\rangle = F_n\left|\mu_n\right\rangle \tag{2.89}$$

$$\hat{L}\left|\mu_n\right\rangle = L_n\left|\lambda_n\right\rangle \tag{2.90}$$

and so,

$$\hat{L}\hat{F}\left|\mu_n,\lambda_n\right\rangle = \hat{L}F_n\left|\mu_n,\lambda_n\right\rangle = F_n\hat{L}\left|\mu_n,\lambda_n\right\rangle = F_nL_n\left|\mu_n,\lambda_n\right\rangle = \hat{F}\hat{L}\left|\mu_n,\lambda_n\right\rangle \tag{2.91}$$

So, two linear operators $\hat{L}$ and $\hat{F}$ commute when

$$\left[\hat{L},\hat{F}\right] = 0 \tag{2.92}$$

The commutator $\left[\hat{L},\hat{F}\right]$ of $\hat{L}$ and $\hat{F}$ is defined:

$$\left[\hat{L},\hat{F}\right] = \hat{L}\hat{F} - \hat{F}\hat{L} \tag{2.93}$$

So, from 2.91 when two operators commute then they have a common eigenstate say for this case, $|\mu_n, \lambda_n\rangle$.

If we are given the function $f(L)$ of the physical quantity, say L then we may also have $f\left(\hat{L}\right)$.

Example

1. We consider the example where the operators $\hat{L}$ and $\hat{F}$ satisfy the relation

$$\hat{L}\hat{F}-\hat{F}\hat{L}=1 \tag{2.94}$$

and we have to find

$$\hat{L}\hat{F}^2-\hat{F}^2\hat{L} \tag{2.95}$$

Adding to the expression $\hat{L}\hat{F}^2-\hat{F}^2\hat{L}$ the quantity $\pm\hat{F}\hat{L}\hat{F}$ then

$$\hat{L}\hat{F}^2-\hat{F}^2\hat{L}+\hat{F}\hat{L}\hat{F}-\hat{F}\hat{L}\hat{F}=\left(\hat{L}\hat{F}-\hat{F}\hat{L}\right)\hat{F}+\hat{F}\left(\hat{L}\hat{F}-\hat{F}\hat{L}\right) \tag{2.96}$$

As,

$$\hat{L}\hat{F}-\hat{F}\hat{L}=1 \tag{2.97}$$

then from 2.96 we have

$$\hat{L}\hat{F}^2-\hat{F}^2\hat{L}=2\hat{F} \tag{2.98}$$

2. We consider the case of the operators $\hat{L}$ and $\hat{F}$ satisfying the relation

$$\hat{L}\hat{F}-\hat{F}\hat{L}=1 \tag{2.99}$$

and we have to find

$$f\left(\hat{L}\right)\hat{F}-\hat{F}f\left(\hat{L}\right) \tag{2.100}$$

We show that if it is true for $n = 1$and $n = 2$ then it should be true for $n + 1$. From example 2.1 we have

$$\hat{L}\hat{F}-\hat{F}\hat{L}=1\cdot\hat{F}^{1-1}=1,\quad n=1 \tag{2.101}$$

$$\hat{L}\hat{F}^2-\hat{F}^2\hat{L}=2\cdot\hat{F}^{2-1}=2\hat{F},\quad n=2 \tag{2.102}$$

Then we show for $n + 1$ by letting

$$\hat{F}\hat{L}^n-\hat{L}^n\hat{F}=-n\hat{L}^{n-1} \tag{2.103}$$

and we use it to find

$$\hat{F}\hat{L}^{n+1} - \hat{L}^{n+1}\hat{F} = \hat{F}\hat{L}^{n+1} - \hat{L}\left(\hat{F}\hat{L}^{n} + n\hat{L}^{n-1}\right) = \left(\hat{F}\hat{L} - \hat{L}\hat{F}\right)\hat{L}^{n} - n\hat{L}^{n} = -(n+1)\hat{L}^{n} \tag{2.104}$$

So, the relation

$$\hat{F}\hat{L}^{n} - \hat{L}^{n}\hat{F} = -n\hat{L}^{n-1} \tag{2.105}$$

is proven for any n. From definition

$$f\left(\hat{L}\right) = \sum_{n=0}^{\infty} \frac{f^{(n)}(0)}{n!}\hat{L}^{n} \tag{2.106}$$

So,

$$f\left(\hat{L}\right)\hat{F} - \hat{F}f\left(\hat{L}\right) = \sum_{n=0}^{\infty} \frac{f^{(n)}(0)}{n!}\left(\hat{L}^{n}\hat{F} - \hat{F}\hat{L}^{n}\right) = \sum_{n=1}^{\infty} \frac{f^{(n)}(0)}{(n-1)!}\hat{L}^{n-1} \tag{2.107}$$

But if we let $n - 1 = m$ then

$$\sum_{m=0}^{\infty} \frac{f^{(m+1)}(0)}{m!}\hat{L}^{m} = \sum_{m=0}^{\infty} \frac{\left[f'(0)\right]^{m}}{m!}\hat{L}^{m} = f'\left(\hat{L}\right) \tag{2.108}$$

Thus

$$f\left(\hat{L}\right)\hat{F} - \hat{F}f\left(\hat{L}\right) = f'\left(\hat{L}\right) \tag{2.109}$$

3

Operators in Phase Space

3.1 Introduction

Ordinarily, the formulation of quantum mechanics is a blend of Heisenberg matrix mechanics and Schrödinger wave mechanics. This, of course, is the most fundamental, having its own strength that is, broadly speaking, non-classical. The Feynman path integral formulation of quantum mechanics reveals new aspects of physical and mathematical construction that might suggest new hints for development of the quantum theory in addition to its added practical applications.

This heading introduces the concept of the Feynman path integral by developing the general construction scheme with particular emphasis on establishing the interconnections between quantum mechanical path integral, Hamiltonian mechanics and statistical mechanics. Basically, there are two approaches to the formulation of quantum mechanics, i.e., the operator (based on the canonical quantization of physical observables together with the associated operator algebra) and the Feynman1 path integral approach. The path integral formulation has the following merits:

- It makes explicit use of classical mechanics as a basic tool on which to construct the theory of quantum fluctuations.
- It is an efficient formulation of non-perturbative approaches to the solution of quantum mechanical problems.
- It is a prototype of higher dimensional functional field integrals.
- It is of relevance to a wide variety of applications in many–body physics.

The path-integral formalism was originally formulated by Richard P. Feynman in his 1942 PhD Thesis [1,9] in the mid-1940s, following a hint from an earlier paper by Dirac [6,7] and became the Feynman new formulation of quantum mechanics [3,9,10] as a new space-time approach to non-relativistic quantum mechanics. This serves as an answer to the question of how to extend the classical principle of stationary action, the so-called principle of least action to quantum mechanics. It is for this reason that the principle of stationary action is first introduced [1,2] **The principle of least action in quantum physics**, introduced path integrals tailored by the **classical action functional**, the integral of the Lagrangian over time.

The essence of Feynman functional integration is to transform the Schrödinger differential equation into an integral equation. So, this Feynman functional integration should be the main object of quantum mechanics describing the real world via the classical Hamiltonian operator $\hat{H}$ or Lagrangian L. This may be done without the use of operators or states in the Hilbert space. The main object in quantum mechanics is referred to as the transition amplitude between those states, matrix elements of physical operators and, in particular, the **S**-matrix in the problem of quantum scattering. On the basis of quantum mechanics, the operator method is closed, well formulated mathematically, and so is an appropriate universal language describing quantum field theory, the quantum physics of condensed states and quantum relativistic statistical physics. In more complex problems, it is appropriate to introduce the method of functional integration in the language of quantum mechanics principally based on the variational method. The transition amplitude is derived based on minimization of the action of a classical system and so the initial discussion is the classical notion of action [1,2].

3.2 Configuration Space

In classical mechanics, the state of a system with s degrees of freedom is represented by $2s$ time-dependent generalized coordinates $\{q(t)\}$ and generalized velocities $\{\dot{q}(t)\}$:

$$q_1(t), q_2(t), \cdots, q_s(t) = \{q(t)\} \tag{3.1}$$

Here, the dot on $\dot{q}(t)$, denotes the time derivative $\frac{d}{dt}$. The space formed from the $q(t)$ values is called the **configuration space**. During the time evolution from time moment t_a to time moment t_b, the system will follow a certain path $q(t)$ in configuration space. The starting point of classical mechanics is the concept of the state of motion, that is, the set of information specifying the history of a point particle as the function of the time. The differential equation defining the path of the system in the configuration space is called the **equation of motion** of the system. Newton's equation of motion is the second order in the time derivative. So, we need generalized coordinates $\{q(t)\}$ and generalized velocities $\{\dot{q}(t)\}$ to identify the time evolution, described by the trajectory, $q(t)$. It should be noted that this classical description is connected with the notion of the path or trajectory. The Schrödinger equation is first order in time derivative. So, it is sufficient to specify the wave function at an initial time moment and the quantum mechanical state can be specified via the coordinate alone. From the Heisenberg uncertainty principle,

$$\Delta q \Delta p \geq \frac{\hbar}{2}, \hbar = \frac{h}{2\pi} = 1.0545919(80) \times 10^{-27} \text{erg sec} \tag{3.2}$$

it is therefore very possible to use the coordinate or the momentum or both to define the state of motion with the main object of the restriction being the trajectory, $q(t)$. In relation 3.2, $\hbar$ is the universal Planck constant with dimension equal that of an action and the number in parentheses indicates the experimental uncertainty of the last two digits before it. Knowledge of the trajectory of a particle by the continuous monitoring of its location will permit knowledge of the coordinate and momentum simultaneously.

The aim of classical as well as quantum mechanics is to predict the time evolution of the system, starting from knowledge of the state of the system at a given point in time. We consider the simplest quantum mechanical system with one degree of freedom of a non-relativistic particle described by the Hamiltonian operator $\hat{H}\left(\hat{p}, \hat{q}\right)$:

$$\hat{H}\left(\hat{q}, \hat{p}\right) = \frac{\hat{p}^2}{2m} + U\left(\hat{q}\right) \tag{3.3}$$

Here $\hat{q} = \left\{\hat{q}_i\right\}_{i=1}^{s}$ and $\hat{p} = \left\{\hat{p}_i\right\}_{i=1}^{s}$ are respectively the operators of the generalized coordinate and momentum with s being the number of degrees of freedom and m, the mass of the particle; $U\left(\hat{q}\right)$, the operator of the potential energy with $q = \{q_i\}_{i=1}^{s}$ being the generalized coordinate.

3.3 Position and Wave Function

Consider again the two slit experiment shown in Figure 3.1, generalized to an N-slit experiment:

$$P(\lambda_i \to \mu_l) = \left| \sum_{\mu_n=1}^{N} \langle \mu_l | \mu_n \rangle \langle \mu_n | \lambda_i \rangle \right|^2 \equiv \left| \langle \mu_l | \lambda_i \rangle \right|^2 \tag{3.4}$$

Letting $N \to \infty$, then it is necessary to introduce a continuous label q:

$$|\mu_n\rangle \to |q\rangle, \sum_n \{\cdots\} \to \int dq \{\cdots\}, \delta_{nm} \to \delta(q-q') \tag{3.5}$$

So, the component of say, $|\Psi\rangle$ in the basis $|q\rangle$ is now a function of the continuous variable :

$$|\Psi\rangle = \int_a^b dq \Psi(q) |q\rangle \tag{3.6}$$

Considering the orthonormalization relation

$$\langle q' | q \rangle = \delta(q-q') \tag{3.7}$$

then

$$\langle q' | \Psi \rangle = \int_a^b dq \Psi(q) \langle q' | q \rangle = \Psi(q') \tag{3.8}$$

or

$$\Psi(q) = \langle q | \Psi \rangle, \Psi^*(q) = \langle \Psi | q \rangle \tag{3.9}$$

Relation 3.9 is the probability amplitude for measuring the eigenvalues q of the position operator $\hat{q}$. This implies that $|\Psi(q)|^2$ should be the probability density in q-space. From here

$$|\Psi\rangle = \int_a^b dq |q\rangle \langle q | \Psi \rangle \tag{3.10}$$

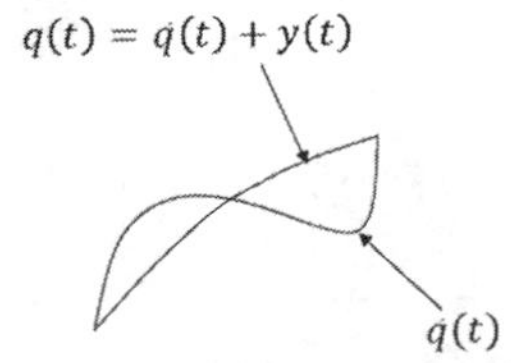

FIGURE 3.1 The classical or true path $\bar{q}(t)$, the false path $q(t)$ and the quantum fluctuation $y(t)$ around the classical path.

So, the closure relation of the eigenvector $|q\rangle$ (showing the particle must be somewhere) can be written in the following form:

$$\int_a^b dq|q\rangle\langle q| = \hat{\mathbb{I}} \tag{3.11}$$

Here, $\hat{\mathbb{I}}$ is a unit matrix. For the probability interpretation to be feasible as seen earlier it is necessary the completeness relation:

$$1 = \langle\Psi|\Psi\rangle - \int_a^b dq\langle\Psi|q\rangle\langle q|\Psi\rangle = \int_a^b dq\Psi^*(q)\Psi(q) = \int_a^b dq\mathrm{P}(\Psi(q)) \tag{3.12}$$

Here, the probability that the particle will be detected in the interval between q and $q + dq$:

$$\mathrm{P}(\Psi(q))dq \equiv |\Psi(q)|^2 dq \tag{3.13}$$

So, $\Psi(q)$ is the spatial wave function of wave mechanics and $|\Psi(q)|^2$ is the relative probability of the particle position located in the interval between q and $q + dq$.

We rewrite the operator in 2.80 via the continuum position operator form:

$$\hat{q} = \int_a^b dqq|q\rangle\langle q| \tag{3.14}$$

Then, considering 7, we have

$$\hat{q}|q\rangle = \int_a^b dq'q'|q'\rangle\langle q'|q\rangle = \int_a^b dq'q'|q'\rangle\delta(q-q') = q|q\rangle \tag{3.15}$$

From equation 3.14, we evaluate the expectation value of q in the state $|\Psi\rangle$:

$$\left\langle\Psi|\hat{q}|\Psi\right\rangle = \int_a^b dq\langle\Psi|q\rangle q\langle q|\Psi\rangle = \int_a^b dq\Psi^*(q)q\Psi(q) \tag{3.16}$$

Rather than limiting ourselves to the finite interval (a, b) we set $a \to -\infty$ and $b \to \infty$ so that $q \in (-\infty, \infty)$. So, the transition amplitude for the continuously-infinite slit experiment can be rewritten:

$$\langle\mu_l|\lambda_i\rangle = \int_{-\infty}^{\infty} dq\langle\mu_l|q\rangle\langle q|\lambda_i\rangle \tag{3.17}$$

3.4 Momentum Space

From the above developments, we move to momentum space by using the Fourier transform basis:

$$|p\rangle = \int_{-\infty}^{\infty} dq|q\rangle\langle q|p\rangle \tag{3.18}$$

From here, the following amplitudes are referred to as **momentum space wave functions**:

$$\langle q|p\rangle = \exp\left\{\frac{i}{\hbar}pq\right\},\ \langle p|q\rangle = \exp\left\{-\frac{i}{\hbar}pq\right\} \tag{3.19}$$

These are just unnormalized plane waves describing free particles of momentum p with the momentum states $|p\rangle$ satisfying the orthogonality relation:

$$\langle p'|p\rangle = \int_{-\infty}^{\infty} dq\langle p'|q\rangle\langle q|p\rangle = \int_{-\infty}^{\infty} dq \exp\left\{\frac{i}{\hbar}(p-p')q\right\} = 2\pi\hbar\delta(p-p') \tag{3.20}$$

Here, $\delta(p - p')$ is the Dirac δ-function. The completeness relation:

$$\int \frac{dp}{2\pi\hbar}|p\rangle\langle p| = \hat{\mathbb{I}} \tag{3.21}$$

We construct a Hermitian operator $\hat{p}$:

$$\hat{p} = \int_{-\infty}^{\infty} \frac{dp}{2\pi\hbar} p|p\rangle\langle p| \tag{3.22}$$

Then

$$\hat{p}|p = \int_{-\infty}^{\infty} \frac{dp'}{2\pi\hbar} p'|p'\rangle\langle p'|p\rangle = \int_{-\infty}^{\infty} \frac{dp'}{2\pi\hbar} p'|p'\rangle 2\pi\hbar\delta(p-p') \equiv p|p\rangle \tag{3.23}$$

These states will be generalized to eigenstates of Heisenberg representation-operators by selecting

$$\hat{q}(t)|q,t\rangle = q|q,t\rangle,\ \hat{p}(t)|p,t\rangle = p|p,t\rangle \tag{3.24}$$

Note that these are eigenstates of the Heisenberg representation $q(t)$, not the result of allowing $|q, t\rangle$ to evolve with time, where t_a is an initial time moment.

3.5 Classical Action

To connect classical physics to quantum physics, we start with a review of the concept of **action** and its variation [1,2]. The **Lagrangian** L of the system is defined as the difference between the kinetic energy and the potential energy U of the system integrated over time resulting in a functional S, called the **action** of the system, for that particular path [1,2]:

$$S = S[q] \equiv \int_{t_a}^{t_b} L(q,\dot{q},t)dt \tag{3.25}$$

$$L(q,\dot{q},t) = \sum_{i=1}^{s} \frac{m_i\dot{q}_i^2(t)}{2} - U(q_1(t),q_2(t),\cdots,q_s(t)) \tag{3.26}$$

The action 3.25 is a functional that takes as argument a function $q(t)$ and returns a number, the action $S[q]$. Here, $q(t_a) \equiv q_a$ and$q(t_b) \equiv q_b$ are the starting and ending coordinates of the path; m_i is the mass associated with the ith generalized coordinate. The path $\bar{q}(t)$ actually chosen by the system is called the **classical path or true path** and nearby path not chosen by the system denoted by $q(t)$ is called the **false path**. The classical path has the property of extremizing the action S in comparison with all neighbouring paths,

$$q(t) = \bar{q}(t) + y(t) \tag{3.27}$$

having the same end points $q(t_a) \equiv q_a$ and$q(t_b) \equiv q_b$. Here, $\delta q(t) \equiv y(t)$ may be interpreted as quantum fluctuations around the classical path (see Figure 3.1). The representation in 3.27 will be shown to separate the action into a classical and a fluctuating part. For force-driven systems, each of the terms contains a force-free and the other a force term.

We see that the kinetic and potential energies are functions of time t and so for each possible path we obtain a different action. Hence, there are many possible paths in configuration space. However, in classical mechanics only one path is actually realized: the **classical path** or **true path** that is determined from the principle of stationary action, so-called **calculus of variations** [1,2]:

$$\delta S[q] = S[\bar{q} + y] - S[\bar{q}] = 0 \tag{3.28}$$

This variation of the action is done over all possible paths of the system in configuration space during the time evolution when the fluctuating part $y(t)$ vanishes at the boundaries, that in mathematics are referred to as **Dirichlet boundary conditions:**

$$y(t_a) = y(t_b) = 0 \tag{3.29}$$

This yields the Euler-Lagrange equations of motion for $\bar{q}(t)$:

$$\frac{d}{dt}\frac{\partial L}{\partial \dot{\bar{q}}(t)} - \frac{\partial L}{\partial \bar{q}(t)} = 0 \qquad 3.30)$$

It is a set of s equations, one for each $(q,\dot{q})$-pair with solution being the classical path $\bar{q}(t)$. So, it is instructive to note that the **particle's motion is governed by the principle that the action is stationary, the so-called the Least Action Principle from which the Euler-Lagrange equations of classical mechanics are derived.**

4

Transition Amplitude

We examine under this heading the transition amplitude in the Feynman path integral, which we will discuss shortly and will be reduced to an integral over the continuum degrees of freedom, such as a set of $\left\{p_i\right\}_{i=1}^{s}$ and $\left\{q_i\right\}_{i=1}^{s}$. This will involve harmonic (non-interacting) systems that are reduced to the problem of multiple Gaussian integrals over potentially coupled degrees of freedom, $\left\{q_i\right\}_{i=1}^{s}$ that we develop appropriate Feynman path integral tools for their solution. Considering that the harmonic oscillator is one of the central objects of theoretical physics then Gaussian integrals are likely to be the key tools of theoretical physics. This will be demonstrated utilizing Feynman path-integral formulation of quantum mechanics where Gaussian integrals will be pivotal for computation in quantum mechanics, quantum statistical mechanics and generally in quantum field theory. It will be interesting to introduce the Maslov correction [11,12] to the wave function which is a jump of $\frac{\pi}{2}$ in the phase when a system passes through a caustic point , a phenomenon related to the second variation as well as to the geometry of paths. This phenomenon is applicable to any system using the quasi-classical approximation. The harmonic oscillator will be one of the objects amongst others to illustrate this.

Path integrals are used in fields such as quantum mechanics, quantum field theories, quantum gravity, field theories, string theory stochastic dynamics, polymer physics, protein folding. In classical mechanics, only one path in configuration space is realized when the action achieves its extremal value:

$$\delta S\left[q\right]=0 \tag{4.1}$$

Nonetheless, in path-integral formulation of quantum mechanics all paths in configuration space contribute and interfere with each other, resulting in a probability transition distribution amplitude to find a particle at a given time *t* and position *q*(*t*). We define the transition amplitude by first expressing the Heisenberg operator $\hat{q}\left(t\right)$, via the Schrödinger operator $\hat{q}$:

$$\hat{q}\left(t\right)=\hat{\mathrm{U}}^{\dagger}\left(t\right)\hat{q}\hat{\mathrm{U}}\left(t\right) \tag{4.2}$$

Here, $\hat{\mathrm{U}}^{\dagger}$ and $\hat{\mathrm{U}}$ are unitary operators:

$$\hat{\mathrm{U}}^{\dagger}\left(t\right)=\exp\left\{\frac{i}{\hbar}\hat{\mathrm{H}}t\right\},\ \hat{\mathrm{U}}\left(t\right)=\exp\left\{-\frac{i}{\hbar}\hat{\mathrm{H}}t\right\} \tag{4.3}$$

We denote by $|q',t'\rangle$ the eigenvector of the Heisenberg operator $\hat{q}\left(t'\right)$ at the time moment t':

$$\hat{q}\left(t'\right)\left|q',t'\right\rangle=q'\left|q',t'\right\rangle \tag{4.4}$$

The transition from the Schrödinger to the Heisenberg representation of the state vectors is via the relation:

$$|q',t'\rangle = \hat{U}(t)|q'\rangle \tag{4.5}$$

Considering equation 4.5, we examine a particle moving from the localized state$|q',t'\rangle$ with coordinate q' at time moment t' to another localized $|q'',t''\rangle$ with coordinate q'' at time moment t'', then the **celebrated Feynman propagator** or **transition amplitude** between coordinate eigenstates is defined:

$$K(q'',t'';q',t') \equiv \langle q'',t''|q',t'\rangle \equiv q''|\hat{U}(t''-t')|q', \quad t''>t' \tag{4.6}$$

It is precisely this **propagator that is the central object of Feynman's formulation of quantum mechanics.** Before studying the path integral representation of the propagator, it is therefore useful to examine some of its properties.

4.1 Path Integration in Phase Space

4.1.1 From the Schrödinger Equation to Path Integration

The physical sense of the transition amplitude in the Schrödinger representation may be seen by examining the Schrödinger state vector $|q,t''\rangle$ of a particle at time moment t''. Then the evolution of the state of the particle is described by the non-relativistic time-dependent Schrödinger equation:

$$i\hbar\frac{\partial}{\partial t''}|q'',t'' = \hat{H}|q'',t''\rangle \tag{4.7}$$

The formal solution of equation 4.7, starting from some initial state at time t', has the form of the **time evolution of states in the Schrödinger picture:**

$$|q'',t''\rangle = \hat{U}(t''-t')|q',t'\rangle \equiv \hat{U}(t'',t')|q',t'\rangle, \quad t''>t' \tag{4.8}$$

Here $\hat{U}(t'',t')$ is the **time evolution operator** with the initial condition:

$$\hat{U}(t',t') = \hat{\mathbb{I}} \tag{4.9}$$

The evolution operator $\hat{U}$ is unitary, since $\hat{H}$ is Hermitian. We assume that the Hamiltonian $\hat{H}$ does not depend explicitly on time and we assume this throughout the derivation of the transition amplitude so only time differences $(t'' - t')$occur.

Consider the Hamiltonian $\hat{H}\left(\hat{p},\hat{q}\right)$ to be time independent and of the standard form, i.e. it is a sum of the kinetic and the potential energies. Substitute 8 into the Schrödinger equation 4.7 then we have

$$i\hbar\frac{\partial}{\partial t''}\hat{U}(t'',t') = \hat{H}\hat{U}(t'',t') \tag{4.10}$$

that completely defines $\hat{U}(t'',t')$, taking into consideration the initial condition 4.9. The formal solution of 4.10 considering 4.9 is written as a chronological-ordered-product:

$$\hat{U}(t'',t') = \hat{T}\exp\left\{-\frac{i}{\hbar}\int_{t'}^{t''}\hat{H}(t)dt\right\} \tag{4.11}$$

Here, $\hat{T}$ is the operator of the chronological-ordered-product that orders the times chronologically with the latest time to the left:

$$t'' \equiv t_N > t_{N-1} > t_{N-2} > \cdots > t_0 \equiv t' \tag{4.12}$$

So, in the Hamiltonian $\hat{H}(t)$, the time, t, plays the role of an argument as well as an ordering parameter.

We construct the finite time evolution operator $\hat{U}$ from short time slice the interval $(t'' - t')$ into N infinitesimal parts of equal widths, $\frac{\Delta t}{N} \equiv \varepsilon$ then via the Riemann sum, using points

$$t_i = t' + \frac{\Delta t}{N} i, \Delta t \equiv t'' - t', i = 0,\cdots,N \tag{4.13}$$

at the centers of the time slices we have:

$$\lim_{\substack{N\to\infty \\ \max|\Delta\xi_i|\to 0}} \sum_{i=0}^{N} \hat{H}(t_i)\Delta t_i \equiv \int_{t'}^{t''}\hat{H}(t)dt \tag{4.14}$$

So,

$$\hat{U}_N(t'',t') = \hat{T}\exp\left\{-\frac{i}{\hbar}\sum_{i=0}^{N}\hat{H}(t_i)\Delta t_i\right\} \tag{4.15}$$

or

$$\hat{U}_N(t'',t') = \hat{F}_N(t'',t_{N-1})\cdots\hat{F}_i(t_i,t_{i-1})\cdots\hat{F}_1(t_1,t') \tag{4.16}$$

Here,

$$\hat{F}_i(t_i,t_{i-1}) = \exp\left\{-\frac{i}{\hbar}\hat{H}(t_i)\Delta t_i\right\}, i = 1,\cdots,N \tag{4.17}$$

From here,

$$\hat{U}(t'',t') = \lim_{\substack{N\to\infty \\ \max|\Delta t_i|\to 0}} \hat{U}_N(t'',t') \tag{4.18}$$

Consider the system to be prepared at time moment t' with coordinate q' at the localized state:

$$|q',t'\rangle = |q'\rangle \tag{4.19}$$

Subsequently, the system evolves to a state $|q'',t''\rangle$at time moment t''. This permits us to rewrite the solution in 8 in the following form

$$|q'',t''\rangle = \hat{U}(t'',t')|q', \quad t''>t' \tag{4.20}$$

So, considering a particle moving from the localized state$|q',t'\rangle$ with coordinate q' at time moment t' to another localized $|q'',t''\rangle$ with coordinate q'' at time moment t'', the **transition amplitude** is rewritten exactly as in 6 being the overlap or scalar product:

$$K(q'',t'';q',t') \equiv \langle q'',t''|q',t'\rangle \equiv q''|\hat{U}(t'',t')|q'\rangle, \quad t''>t' \tag{4.21}$$

It is obvious from here that equations 4.7 and 4.8 are connected. The equation 4.21 has the full dynamics of the system and to use it as the definition of the propagator we solve the Heisenberg equation of motion for K.

Remarks

We therefore make the following remarks:

1. Relation 4.21 is the transition probability amplitude of a particle moving from the localized state $|q',t'\rangle$ at the point q' at time moment t' to localized $|q'',t''\rangle$ with coordinate q'' at time moment ″. This is independent of the representation in which we work. For the quantum mechanics of non-relativistic particles, we restrict our attention to evolution forwards in time for relevance as we consider the **causal propagator** or **retarded propagator** and so equation 4.21 may as well be represented as follows:

$$K(q'',t'';q',t') = q''|\hat{U}(t'',t')|q'\theta(t''-t') \tag{4.22}$$

   This permits to write the exact definition of $K(q'',t'';q',t')$ in exactly the form 4.10 via the **Heaviside step function**:

$$\theta(t''-t') = \begin{cases} 1 & , \quad t''>t' \\ 0 & , \quad t''<t' \end{cases} \tag{4.23}$$

   The introduction of $\theta(t''-t')$ serve both a physical and mathematical interest where the physical interest entails compelling the system at a starting point to be evolving towards the future and so also reflecting causality. The mathematical interest is that $K(q'',t'';q',t')$, because of the factor $\theta(t''-t')$, obeys a partial differential equation with the right-hand side (source term) being a delta function. That differential equation defines a **Green function** that in this case is $K(q'',t'';q',t')$. With the step function $\theta(t''-t')$, the given Green function will be the retarded solution to that differential equation.
2. The state vector $|q'',t''\rangle$ at time moment $t=t''$ can be expressed via a complete set of position eigenstates by inserting a resolution of the identity:

$$\hat{\mathbb{I}} = \int dq''|q''\rangle\langle q''| \tag{4.24}$$

So,

$$\left|q'',t''\right\rangle = \int dq''\left|q''\right\rangle\left\langle q''\right|\hat{U}\left(t'',t'\right)\left|q'\right\rangle \tag{4.25}$$

From 4.10, then the eigenstate evolves as

$$\left|q'',t''\right\rangle = \int dq''\left|q''\right\rangle K\left(q'',t'';q',t'\right) \tag{4.26}$$

This is nothing but the **Huygens(-Fresnel) principle**: at time moment t' every point is the source of an elementary wave, whose propagation in the time interval $(t'' - t')$ yields the new wave amplitude $K(q'',t'';q',t')$ called the **probability transition amplitude** for the propagation from $|q'\rangle$ at time moment t' to $|q''\rangle$ at time moment t''. It is also called the **time evolution kernel**. So, the probability transition amplitude (propagator) $K(q'',t'';q',t')$ is the amplitude for the system to be in position state $|q''\rangle$ at time moment t'', given that it started in position state $|q'\rangle$ at time moment t'. So, K governs how a wave function $|q,t\rangle$ evolves with time t. The probability transition amplitude being the kernel K of the Schrödinger equation, therefore, completely gives the dynamics of the given system.

We observe that the transition amplitude $K(q'',t'';q',t')$ is completely a quantum object on which basis we may construct the Feynman functional integral. Equation 4.26 is the integral equation of quantum mechanics, equivalent to the time-dependent Schrödinger differential equation 4.7. This equation is readily generalized to many-body systems. So, all information about any autonomous quantum mechanical system is contained in the matrix elements of its time evolution operator, $K(q'',t'';q',t')$.

4.1.2 Trotter Product Formula

From equations 4.21 and 4.22, the transition amplitude $K(q'',t'';q',t')$ is observed to be the matrix element of the evolution operator $\hat{U}\left(t'',t'\right)$ by two Schrödinger state vectors. We move to functional integration by re-deriving equation 4.21 via the classical Hamiltonian function $H(q,p)$ that maps some quantum operator $\hat{H}\left(\hat{q},\hat{p}\right)$. We simplify the problem of finding the transition amplitude $K(q'',t'';q',t')$, by computing it first for short and equidistance time slices when it takes a simpler form. This procedure imitates adding more gratings to the slit experiment mentioned earlier. The i grating is passed at position q_i at time t_i, and the state is $|q_i,t_i\rangle$. This permit us to construct the finite time transition amplitude from short time slice via Feynman's procedure of time-slicing the operator $\hat{U}\left(t''-t'\right)$. Since, our system is classified as autonomous as its Hamiltonian does not explicitly depend on time then the so-called **Trotter product formula:**

$$\exp\left\{-\frac{i}{\hbar}\hat{H}\Delta t\right\} = \lim_{\substack{N\to\infty\\ \Delta t\to 0}}\left[\exp\left\{-\frac{i}{\hbar}\hat{H}\frac{\Delta t}{N}\right\}\right]^N \equiv \lim_{\substack{N\to\infty\\ \varepsilon\to 0}}\left[\exp\left\{-i\frac{\varepsilon}{\hbar}\hat{H}\right\}\right]^N \tag{4.27}$$

The representation 4.27 has the advantage in that the factors $\exp\left\{-i\frac{\varepsilon}{\hbar}\hat{H}\right\}$ as well as their expectation values are small. In particular, if ε is much smaller than the reciprocal of the eigenvalues of the Hamiltonian in the regime of physical interest, then the factors are exponentially small compared with unity and can then be treated perturbatively. A first simplification from this fact is that the exponentials can be factorised into two pieces each of which can be readily diagonalized. To apply these facts to 4.27 we write first the transition amplitude:

$$\mathrm{K}(q'',t'';q',t') \equiv \langle q'',t''|q',t'\rangle \equiv \lim_{\substack{N\to\infty \\ \varepsilon\to 0}} \langle q''|\hat{\mathrm{U}}(N\varepsilon)|q'\rangle \equiv \lim_{\substack{N\to\infty \\ \varepsilon\to 0}} \langle q''|\left(\hat{\mathrm{U}}(\varepsilon)\right)^N|q'\rangle \tag{4.28}$$

By inserting the resolution of the identity 4.24 between each operator in the above relation for the transition amplitude then we evaluate the matrix elements by taking a multiple integral over all values of q_i for i between 1 and $N - 1$ (all integrals extend over the entire real line):

$$\mathrm{K}(q'',t'';q',t') \equiv \lim_{\substack{N\to\infty \\ \varepsilon\to 0}} \prod_{k=1}^{N-1} \int_{-\infty}^{+\infty} dq_k \langle q''|\hat{\mathrm{U}}(\varepsilon)|q_{N-1}\rangle\langle q_{N-1}|\hat{\mathrm{U}}(\varepsilon)|q_{N-2}\rangle\cdots\times\langle q_2|\hat{\mathrm{U}}(\varepsilon)|q_1\rangle\langle q_1|\hat{\mathrm{U}}(\varepsilon)|q' \tag{4.29}$$

We do not integrate over the fix end points, q'or q''. This relation 4.29 holds for any path integral as $N \to \infty$. This entails from the double slit experiment adding more and more gratings so that the time intervals get smaller and smaller and we fix each path between q' and q'' more and more precisely. In addition, we obtain more and more integrals in order to integrate over all the paths. In effect, any trajectory between the given initial $q'$and final point q'' can be approximated by a piecewise constant function when the length of the time interval $\Delta t = \varepsilon \to 0$ when the function is constant tends to zero.

So, the right-hand side of 4.29 can be considered as a summation over paths, made up of piecewise smooth linear functions that becomes an integral over paths in the continuum limits, $N \to \infty$ and $\varepsilon \to 0$, from where we analyse one of the following matrix elements in 4.29 re-expressed as:

$$\langle q_{i+1}|\exp\left\{-i\frac{\varepsilon}{\hbar}\hat{\mathrm{H}}\right\}|q_i\rangle \cong \langle q_{i+1}|\left(\hat{\mathbb{I}} - i\frac{\varepsilon}{\hbar}\hat{\mathrm{H}} + o(\varepsilon^2)\right)|q_i\rangle \tag{4.30}$$

and

$$\langle q_{i+1}|\left(\hat{\mathbb{I}} - i\frac{\varepsilon}{\hbar}\hat{\mathrm{H}} + o(\varepsilon^2)\right)|q_i\rangle = \langle q_{i+1}|q_i\rangle\left(1 - i\frac{\varepsilon}{\hbar}\frac{q_{i+1}|\hat{\mathrm{H}}|q_i\rangle}{\langle q_{i+1}|q_i\rangle} + o(\varepsilon^2)\right) \tag{4.31}$$

or

$$\langle q_{i+1}|q_i\rangle\left(1 - i\frac{\varepsilon}{\hbar}\frac{q_{i+1}|\hat{\mathrm{H}}|q_i\rangle}{\langle q_{i+1}|q_i\rangle} + o(\varepsilon^2)\right) \cong \exp\left\{-i\frac{\varepsilon}{\hbar}\frac{q_{i+1}|\hat{\mathrm{H}}|q_i\rangle}{\langle q_{i+1}|q_i\rangle}\right\} \tag{4.32}$$

However, here, the problem is the orthogonality of the basis vectors

$$\langle q_{i+1}|q_i\rangle = \delta(q_{i+1} - q_i) \tag{4.33}$$

The small parameter in the expansion should be $\frac{\varepsilon}{\langle q_{i+1}|q_i\rangle}$ and is diverging for $q_{i+1} \neq q_i$. This problem can be resolved if we use two overlapping bases in an alternating manner. For the case of continuous space, the choice of the other, overlapping basis is a momentum basis, $|p_i\rangle$ with

$$\hat{p}_i|p_i\rangle = p_i|p_i\rangle \tag{4.34}$$

The corresponding resolution of the identity as defined in 4.24 should be

$$\int_{-\infty}^{+\infty} \frac{dp_i}{2\pi\hbar} |p_i\rangle\langle p_i| = \hat{\mathbb{I}} \tag{4.35}$$

inserted in equation 4.30 yields

$$q_i\left|\left(\hat{\mathbb{I}} - i\frac{\varepsilon}{\hbar}\hat{\text{H}} + o(\varepsilon^2)\right)\right|q_i\rangle = \int_{-\infty}^{+\infty} \frac{dp_i}{2\pi\hbar} \langle q_{i+1}|p_i\rangle\langle p_i\left|\left(\hat{\mathbb{I}} - i\frac{\varepsilon}{\hbar}\hat{\text{H}} + o(\varepsilon^2)\right)\right|q_i\rangle \tag{4.36}$$

From 4.19 we have

$$\langle q_{i+1}|p_i\rangle = \exp\left\{\frac{i}{\hbar} q_{i+1} p_i\right\} \tag{4.37}$$

$$p_i\left|\left(\hat{\mathbb{I}} - i\frac{\varepsilon}{\hbar}\hat{\text{H}} + o(\varepsilon^2)\right)\right|q_i\rangle = \langle p_i|q_i\rangle - i\frac{\varepsilon}{\hbar}\langle p_i|\hat{\text{H}}\left(\hat{p},\hat{q}\right)|q_i\rangle + o(\varepsilon^2) \tag{4.38}$$

From the infinitesimally small function (ε^2) , the computation is limited to the linear term in ε. We consider in 4.38 that the Hamiltonian $\hat{\text{H}}\left(\hat{p},\hat{q}\right)$ is composed out of position $\hat{q}$ and momentum $\hat{p}$ operators in such a way that it turns into the Hamilton function when acting on either $\langle p_i|$ or $|q_i\rangle$:

$$\langle p_i|\hat{\text{H}}\left(\hat{p},\hat{q}\right)|q_i\rangle = \langle p_i\left|\left(\frac{\hat{p}^2}{2m} + U\left(\hat{q}\right)\right)\right|q_i\rangle = \left(\frac{p_i^2}{2m} + U(q_i)\right)\langle p_i|q_i\rangle \tag{4.39}$$

then

$$\langle p_i\left|\left(\hat{\mathbb{I}} - i\frac{\varepsilon}{\hbar}\hat{\text{H}} + o(\varepsilon^2)\right)\right|q_i = \langle p_i|q_i\rangle\left(1 - i\frac{\varepsilon}{\hbar}\text{H}(p_i,q_i) + o(\varepsilon^2)\right) \tag{4.40}$$

From

$$1 - i\frac{\varepsilon}{\hbar}\text{H}(p_i,q_i) + o(\varepsilon^2) \cong \exp\left\{-i\frac{\varepsilon}{\hbar}\text{H}(p_i,q_i)\right\} \tag{4.41}$$

then to the approximation of $o(\varepsilon^2)$ we have

$$\langle q_{i+1}\left|\exp\left\{-i\frac{\varepsilon}{\hbar}\text{H}(p_i,q_i)\right\}\right|q_i\rangle \cong \int_{-\infty}^{+\infty} \frac{dp_i}{2\pi\hbar} \exp\left\{\frac{i\varepsilon}{\hbar}\left[p_i\left(\frac{q_{i+1}-q_i}{\varepsilon}\right) - \text{H}(p_i,q_i)\right]\right\} \tag{4.42}$$

4.2 Transition Amplitude

4.2.1 Hamiltonian Formulation of Path Integration

We find the transition amplitude by substituting the expression 4.42 into 4.29 considering that q' and q'' should be identified with q_0 and q_N that are necessary for the exponential terms to be well-defined at the edges. So, the transition amplitude to the approximation of $o(\varepsilon^2)$ as a path integral in phase space has the form:

$$K(q_N,t_N;q_0,t_0) \cong \int_{-\infty}^{+\infty} \frac{dp_0}{2\pi\hbar} \prod_{i=1}^{N-1} \frac{dp_i dq_i}{2\pi\hbar} \exp\left\{ \frac{i\varepsilon}{\hbar} \sum_{i=0}^{N-1} \left[p_i \left(\frac{q_{i+1}-q_i}{\varepsilon} \right) - H(p_i,q_i) \right] \right\} \tag{4.43}$$

It is interesting that we have now got c-numbers instead of operators. The exact expression of the transition amplitude is computed as the Hamiltonian path integral:

$$K(q_N,t_N;q_0,t_0) = \lim_{\substack{\varepsilon\to 0 \\ N\to\infty}} \int_{-\infty}^{+\infty} \frac{dp_0}{2\pi\hbar} \prod_{i=1}^{N-1} \frac{dp_i dq_i}{2\pi\hbar} \exp\left\{ \frac{i\varepsilon}{\hbar} \sum_{i=0}^{N-1} \left[p_i \left(\frac{q_{i+1}-q_i}{\varepsilon} \right) - H(p_i,q_i) \right] \right\} \tag{4.44}$$

This relation is the formal expression for the path integral where the exponent in the integrand imitates the action written in both the coordinate and momentum, the so-called phase space representation. In this phase space representation of the transition amplitude 4.44, there is always one more momentum than position integral and so the integration measure is therefore not invariant under a canonical transformation. This is a consequence that each short-time-slice transition amplitude contains one momentum integral while the position integrals are inserted between the short-time-slice transition amplitudes and the two end points q_0 and q_N are not integrated over. Nevertheless, often equation 4.44 can be used as an operational definition of a Hamiltonian path integral.

The argument of the exponent in 4.44 has the Riemann sum that may be transformed into an integral if the functions $q(t)$and $p(t)$ are piecewise continuous and the Hamilton function $H(p(t), q(t))$ is piecewise continuous. So,

$$\lim_{\substack{N\to\infty \\ \max|\Delta t_i|\to 0}} \sum_{i=0}^{N-1} \left[p_i \left(\frac{q_{i+1}-q_i}{\varepsilon} \right) - H(p_i,q_i) \right] \varepsilon = \int_{t_0}^{t_N} \left[p\dot{q} - H(p,q) \right] dt \tag{4.45}$$

where

$$\lim_{\varepsilon\to 0} \frac{q_{i+1}-q_i}{\varepsilon} = \dot{q}(t) \tag{4.46}$$

The Riemann limit of the transition amplitude 4.44 may be rewritten in symbolic form as follows:

$$K(q_N,t_N;q_0,t_0) \equiv \int Dp(t) Dq(t) \exp\left\{ \frac{i}{\hbar} S[p,q] \right\} \tag{4.47}$$

where the **Hamiltonian action**

$$S[p,q] = \int_{t_0}^{t_N} \left[p\dot{q} - H(p,q) \right] dt \tag{4.48}$$

depends on the full phase space path from t_0 to t_N and is therefore a **functional.** The integration in 4.47 for the transition amplitude extends over all possible paths through the classical phase space of the system that begins and ends at the same **configuration** points $q(t_0) \equiv q_0$ and $q(t_N) \equiv q_N$ while the path $p(t)$ is completely unconstrained and is not related to $q(t)$ (or $\dot{q}(t)$) in any way; the symbols $Dp(t)Dq(t)$ is a functional measure viewed as the measure of paths denoting a finite multiple integral:

$$Dp(t)Dq(t) \equiv \lim_{\substack{\varepsilon \to 0 \\ N\to\infty}} \frac{dp_0}{2\pi\hbar} \prod_{i=1}^{N-1} \frac{dp_i dq_i}{2\pi\hbar} \tag{4.49}$$

This functional measure is descriptive and easily recalled as partitioning phase space into cells of the size $2\pi\hbar$ (volume of one quantum state in 1D) as in our book [1]. The contribution of each path is weighted by its **Hamiltonian action** $S[p,q]$. It is important to note that $p(t)$ and $q(t)$ are defined by integrating over the values of $p(t)$ and $q(t)$ independently at each value of t, so the paths that contribute to the functional integral are, in general, highly discontinuous. The paths contributing most to the functional integral are those satisfying the variational principle defined by equation 4.28. These are the phase-space trajectories $p(t)$ and $q(t)$ satisfying the initial conditions $q(t_0) \equiv q_0$ and $q(t_N) \equiv q_N$ with no restrictions on the initial and final values of momentum or energy. The variational principle defined by equation 4.28, singles out the classical trajectories obeying the Hamilton equations of motion with the specified initial and final conditions.

Relation 4.47 is called the transition amplitude in the Hamilton formulation written in terms of the momenta and positions that are conceptually or intrinsically impossible in quantum mechanics to have simultaneous transition probability functions for the momenta and positions by virtue of Heisenberg uncertainty principle.

4.2.2 Path Integral Subtleties

4.2.2.1 Mid-point Rule

The derivation of the path integral for a Hamiltonian with a velocity-dependent potential, such as the magnetic interaction $\frac{\left(\vec{\mathrm{p}}-\vec{\mathrm{A}}\right)^2}{2m}$, may be more subtle where $\vec{\mathrm{A}}$ is the vector potential. The discretized version requires a **mid-point rule** for the Hamiltonian. Evaluation along some space-time path $q(t)$ starting from the point (q_i, t_i) to the point $(q_{i+1}, t_i + \varepsilon)$ then

$$q(t_i + \varepsilon) - q(t_i) \equiv q_{i+1} - q_i = \Delta q \tag{4.50}$$

where, the path satisfies the requirement

$$(\Delta q)^2 \approx \Delta t \equiv \varepsilon \tag{4.51}$$

The paths are continuous and the velocity $\dot{q}(t)$, associated to them is infinite as

$$\dot{q}(t) = \lim_{\varepsilon\to 0} \frac{q_{i+1} - q_i}{\varepsilon} \approx \lim_{\varepsilon\to 0} \frac{\sqrt{\varepsilon}}{\varepsilon} = \lim_{\varepsilon\to 0} \frac{1}{\varepsilon} \to \infty \tag{4.52}$$

This subtle nature of the paths may be observed for a particle in a magnetic field described by the Lagrangian:

$$L = \frac{m\dot{q}^2}{2} + \frac{e}{c}\left(\vec{v}, \vec{\mathrm{A}}\right) \tag{4.53}$$

Here, e, c, $\vec{A}$ and $\vec{v} = \dot{q}$ are, respectively, the electronic charge, speed of light, vector potential and velocity of the particle. For the correct quantization of the Lagrangian 4.53, the finite N discretized form of the vector potential contribution can be written:

$$\int\left(\vec{v},\vec{A}\right)dt \cong \sum_i A\frac{1}{2}\left(q_{i+1}+q_i\right)\left(q_{i+1}-q_i\right) \tag{4.54}$$

where $\vec{A}$ is evaluated at the mid-point of the interval. The principal objective is the selection of an appropriate point where the variables are evaluated.

4.2.3 Lagrangian Formulation of Path Integration

We represent path integration 4.47 in an alternative form convenient in various applications and physically instructive and motivated by the observation that 4.47 imitates the Hamiltonian formulation of classical mechanics. Since, classically, the Hamiltonian and **Lagrangian** mechanics can be equally employed to describe dynamical evolution, then it is instructive to find a Lagrangian analogue of 4.47. Focusing on Hamiltonians where the dynamics is free then the kinetic energy dependence is quadratic in p and so, the Lagrangian formulation of path integration can be inferred from 4.47 by Gaussian integration.

The development of the transition amplitude via path integration as a new space-time approach to non-relativistic quantum mechanics was developed by Richard Feynman in his 1942 PhD Thesis [5] in the mid-1940s, following a hint from an earlier paper by Dirac [7,13]. Apparently, Dirac's motivation was to formulate quantum mechanics from the Lagrangian rather than the Hamiltonian formulation of classical mechanics. So, making an attempt to understand the transition amplitude in a purely quantum mechanical sense then for the replacement

$$\hat{H}\left(\hat{p},\hat{q}\right) \to H\left(p,q\right) \tag{4.55}$$

the classical Hamiltonian function in 4.47 has the following form in the generic case:

$$H\left(p,q\right) = \frac{p^2}{2m} + U\left(q\right) \tag{4.56}$$

Here, $U(q)$ is the potential energy associated with the particle.

Our goal now is to integrate over the momenta in 4.47, withholding only integrations over coordinates with the aid of expression 4.43. Transforming the finite multiple integral over the momenta with the ith momentum integral takes the form where exponent of the integral is quadratic in the momentum variable p (integral is Gaussian in p):

$$\int_{-\infty}^{+\infty}\frac{dp_i}{2\pi\hbar}\exp\left\{\frac{i}{\hbar}\varepsilon\left[p_i\left(\frac{q_{i+1}-q_i}{\varepsilon}\right)-\frac{p_i^2}{2m}-U\left(q_i\right)\right]\right\} = \exp\left\{-\frac{i}{\hbar}\varepsilon U\left(q_i\right)\right\}\int_{-\infty}^{+\infty}\frac{dp_i}{2\pi\hbar}\exp\left\{\frac{i\varepsilon}{\hbar}\left[p_i\left(\frac{q_{i+1}-q_i}{\varepsilon}\right)-\frac{p_i^2}{2m}\right]\right\} \tag{4.57}$$

4.2.3.1 Complex Gaussian Integral

The integral on the right-hand side of 4.57 may be easily evaluated via the standard Gaussian integral:

$$\int_{-\infty}^{+\infty}dq\exp\left\{-aq^2+bq\right\} = \left(\frac{\pi}{a}\right)^{\frac{1}{2}}\exp\left\{\frac{b^2}{4a}\right\} \quad , \; a \neq 0, \mathrm{Re}\, a \geq 0 \tag{4.58}$$

This integral in equation 4.57 is an (extended) Fresnel integral:

$$\mathrm{I}(a)=\int_{-\infty}^{+\infty}dq\exp\{iaq^2\}=\lim_{\delta\to0+}\int_{-\infty}^{+\infty}dq\exp\{-(\delta-ia)q^2\}=\lim_{\delta\to0+}\left(\frac{\pi}{\delta-ia}\right)^{\frac{1}{2}},\mathrm{Re}\,a\geq0 \tag{4.59}$$

or

$$\mathrm{I}(a)=\left(\frac{\pi}{|a|}\right)^{\frac{1}{2}}\exp\left\{\frac{i}{2}\lim_{\delta\to0+}\tan^{-1}\frac{a}{\delta}\right\}=\left(\frac{\pi}{|a|}\right)^{\frac{1}{2}}\exp\left\{i\frac{\pi}{4}\mathrm{sgn}\,a\right\}=\left(\frac{\pi}{2|a|}\right)^{\frac{1}{2}}[1+i\,\mathrm{sgn}\,a] \tag{4.60}$$

It should be noted that the real parameters a can take any sign.

The root of a complex variable z is understood as the principal value. This implies the branch of the two-valued square root function that has a positive real part:

$$\sqrt{z}=\sqrt{|z|}\exp\left\{\frac{1}{2}i\arg z\right\},-\pi<\arg z\leq\pi \tag{4.61}$$

So, the integrand in 4.61, however, is imaginary and hence the necessity to introduce a supplementary procedure for its evaluation which should be the analytic continuation of the integral 4.61 by a rotation of the integration variable through the angle $-\frac{\pi}{2}$ in the complex plane. This rotation transforms the integral into Gaussian form in 4.59 that may be easily evaluated. The rotation of the integration variable through $\frac{\pi}{2}$ in the complex plane yields the same result. From this analytic continuation, we arrive at the result from integrating out the momenta only:

$$\int_{-\infty}^{+\infty}\frac{dp_i}{2\pi\hbar}\exp\left\{\frac{i\varepsilon}{\hbar}\left[p_i\left(\frac{q_{i+1}-q_i}{\varepsilon}\right)-\frac{p_i^2}{2m}-U(q_i)\right]\right\}=\left(\frac{m}{2\pi\hbar i\varepsilon}\right)^{\frac{1}{2}}\exp\left\{\frac{i\varepsilon}{\hbar}\left[\frac{m}{2}\left(\frac{q_{i+1}-q_i}{\varepsilon}\right)^2-U(q_i)\right]\right\} \tag{4.62}$$

This result has complex amplitude $\left(\frac{m}{2\pi\hbar i\varepsilon}\right)^{\frac{1}{2}}$ modified by a phase factor with the dominant controlling factor being the phase argument of the exponential function in 4.62.

Another way of calculating the Fresnel integral with infinite limits is to examine it as the limit of finite Fresnel integrals:

$$\mathrm{I}(a)=\int_{-\infty}^{+\infty}dq\exp\{iaq^2\}=\lim_{\in\to\infty}\int_{-\in}^{+\in}dq\left(\cos(aq^2)+i\sin(aq^2)\right) \tag{4.63}$$

From the change of variable

$$q=\left(\frac{\pi}{2|a|}\right)^{\frac{1}{2}}t,\ \zeta=\sqrt{\frac{\pi}{2}} \tag{4.64}$$

then

$$\mathrm{I}(a)=\sqrt{\frac{2\pi}{|a|}}\lim_{\in\to\infty}\left[\frac{2\zeta}{\sqrt{2\pi}}\int_0^{+\in}dt\cos(\zeta^2t^2)+i\,\mathrm{sgn}\,a\frac{2\zeta}{\sqrt{2\pi}}\int_0^{+\in}dt\sin(\zeta^2t^2)\right] \tag{4.65}$$

Since for the Fresnel integrals

$$\lim_{\epsilon\to\infty}\frac{2\zeta}{\sqrt{2\pi}}\int_0^{+\epsilon}dt\cos\left(\zeta^2t^2\right)=\lim_{\epsilon\to\infty}\frac{2\zeta}{\sqrt{2\pi}}\int_0^{+\epsilon}dt\sin\left(\zeta^2t^2\right)=\frac{1}{2} \tag{4.66}$$

then

$$\mathrm{I}(a)=\sqrt{\frac{\pi}{2|a|}}\left(1+i\,\mathrm{sgn}\,a\right) \tag{4.67}$$

and coincides exactly with 4.58.

4.2.4 Transition Amplitude

Considering 6, the form of 4.67 is not exactly like the transition amplitude:

$$\mathrm{K}\left(q_{i+1},t_i+\varepsilon;q_i,t_i\right)=\langle q_{i+1}|\exp\left\{-\frac{i\varepsilon}{\hbar}\hat{\mathrm{H}}\right\}|q_i\rangle \tag{4.68}$$

If we consider the RHS of 4.68 then for

$$|q_{i+1}-q_i|>>\left(\frac{\hbar\varepsilon}{m}\right)^{\frac{1}{2}} \tag{4.69}$$

the phase of the transition amplitude oscillates extremely rapidly and sum over neighboring trajectories may tend to cancel out by interference or destructive interference occurs. So, the type of the coordinates q_{i+1} and q_i contributing most to the transition amplitude in 4.36 are those that satisfy the relation:

$$|q_{i+1}-q_i|\leq\left(\frac{\hbar\varepsilon}{m}\right)^{\frac{1}{2}} \tag{4.70}$$

This condition tailors the phase of the transition amplitude to have an extremal (or stationary) value where sum over neighboring trajectories tend to interfere constructively as their phases are nearly equal. This can be viewed differently when the phases of the transition amplitude are much larger than the quantum of the action $\hbar$ (**quasi-classical situation**).

Consider the limit for which the time step becomes infinitesimal small

$$\varepsilon=\Delta t\to dt,\varepsilon\to 0,\quad N\to\infty \tag{4.71}$$

Then

$$\lim_{\varepsilon\to 0}\frac{q_{i+1}-q_i}{\varepsilon}=\dot{q}(t) \tag{4.72}$$

$$\lim_{\varepsilon\to 0}\frac{i\varepsilon}{\hbar}\left[\frac{m}{2}\left(\frac{q_{i+1}-q_i}{\varepsilon}\right)^2-U\left(q_i\right)\right]=\frac{i}{\hbar}L\left(q,\dot{q},t\right)dt \tag{4.73}$$

where

$$L(q,\dot{q},t) = \frac{m}{2}\dot{q}^2 - U(q) \tag{4.74}$$

is the classical Lagrange function of the particle. The integral in 4.47 involves the Lagrangian integrated over time (is the action). Let us examine the classical action corresponding to one-time step:

$$S[q_i,q_{i+1}] = \int_{t_i}^{t_i+\varepsilon} dt' L(q(t'),\dot{q}(t'),t') \tag{4.75}$$

$$q(t_i) \equiv q_i,\quad q(t_i+\varepsilon) \equiv q_{i+1} \tag{4.76}$$

The evaluation is done along some space-time path $q(t)$ starting from the point (q_i, t_i) to the point $(q_{i+1}, t_i+\varepsilon)$. For 4.70 as $\varepsilon \to 0$, the path $q(t)$ giving a considerable contribution to the transition amplitude should lie within the neighbourhood of $\left(\frac{\hbar\varepsilon}{m}\right)^{\frac{1}{2}}$ of the coordinate q_i. For this reason, when evaluating the action 4.75 from the starting and end points, the condition 4.70 should be satisfied so that the path $q(t)$ should almost coincide with the line:

$$q(t) \cong \left(1 - \frac{t-t_i}{\varepsilon}\right)q_i + \frac{t-t_i}{\varepsilon}q_{i+1} \tag{4.77}$$

If we substitute 4.77 into 4.75 while using 4.74, then we have this step's contribution to the action:

$$S[q_i,q_{i+1}] \cong \frac{m}{2}\left(\frac{q_{i+1}-q_i}{\varepsilon}\right)^2 \varepsilon - \int_{t_i}^{t_i+\varepsilon} dt' U(q(t')) \cong \left[\frac{m}{2}\left(\frac{q_{i+1}-q_i}{\varepsilon}\right)^2 - U(q_i)\right]\varepsilon \tag{4.78}$$

In this way, the transition amplitude between the states $|q_i(t_i)\rangle$and $|q_{i+1}(t_i+\varepsilon)\rangle$ as $\varepsilon \to 0$ may be written in the form:

$$K(q_{i+1},t_i+\varepsilon;q_i,t_i) = \left(\frac{m}{2\pi i\hbar\varepsilon}\right)^{\frac{1}{2}} \exp\left\{\frac{i}{\hbar}S[q_i,q_{i+1}]\right\} \tag{4.79}$$

This is an important fundamental result and the original path integral by Feynman transition amplitude. It gives the transition amplitude that expresses all possible path/histories reaching the final position of the particle from an initial position after an infinitesimal time step. The momentum degree of freedom has been integrated out. There is a normalization factor and a phase. We may see from 4.65 that the contribution to the phase from a given path is the action functional S for that path in units of the quantum of action, $\hbar$. In summary, the probability P_{ab} to go from a point (q_a, t_a) to the point (q_b, t_b) is the absolute square of the transition amplitude,

$$P_{ab} = |K(q_b,t_b;q_a,t_a)|^2 \tag{4.80}$$

The argument is the amplitude $K(q_b, t_b; q_a, t_a)$ to go from point (q_a, t_a) to point (q_b, t_b). This amplitude is the sum of the contributions of the functions $K(q_{i+1}, t_i + \varepsilon; q_i, t_i)$ from each path connecting the desired end points. The contribution of each path has a phase proportional to its action 4.65. In this way, quantum mechanics has been reduced to a simple procedure, in principle. Each amplitude for a chosen path can be thought to represent an event (the particular propagation along that path). The contributions of different paths are summed to generate the total amplitude, $K(q_b, t_b; q_a, t_a)$. The contributions of different paths interfere when converted into a probability by squaring the total amplitude. From the conceptual viewpoint, it is equivalent to saying that a quantum particle propagates along all possible (classical) paths — although the resulting interference effects tends to lead to a certain dominance of some paths over others. The quantum of action, $\hbar$, determines the practical range of paths that actually contribute. As expected, that range corresponds to the limits allowed within the Heisenberg uncertainty principle.

Thus, for a finite interval $(t_b - t_a)$ if we consider 4.67 and 4.65 then the transition amplitude for a sequence of many time steps, with the momenta integrated out, is [1,2]:

$$K\left(q_b,t_b;q_a,t_a\right)=\lim_{\substack{\varepsilon\to 0\\ N\to\infty}}\left(\frac{m}{2\pi i\hbar\varepsilon}\right)^{\frac{N}{2}}\prod_{i=1}^{N-1}\int_{-\infty}^{+\infty}dq_i\exp\left\{\frac{i}{\hbar}\sum_{i=0}^{N-1}S\left[q_i,q_{i+1}\right]\right\} \tag{4.81}$$

This is the expression of a Feynman path integral for the particle. It is expressed only in terms of the particle's coordinate, although it contains the effects of its momentum. The path integral 4.81 is a natural generalization of the two slit experiment. Even if we have knowledge where the particle originates from and where it hit on the screen, we have no knowledge which slit the particle came through. This path integral should be an infinite slit experiment. As we do not specify where the particle goes through, we can sum them up. So, the action in 4.81, considering 4.49, is evaluated in the direction of the piecewise line shown in Figure 4.1. We consider here a particle traveling from point (q_a, t_a) to (q_b, t_b) through a series of intermediate points with $q_1, q_2, \cdots, q_{N-1}$ that define a **path**. The transition amplitude $K(q_b, t_b; q_a, t_a)$ for the particle to start at (q_a, t_a) and end up at (q_b, t_b) is given by the sum over all possible paths. This implies that the particle seeks all possible values of the intermediate points(q_i, t_i) (see Figure 4.1).

Taking the limits in expression 4.81 as the time increment ε approaches zero, the number of integrations over the intermediate points become infinite and the transition amplitude (propagator or type of Green's function) $K(q_b, t_b; q_a, t_a)$ in 4.81 is a **configuration space path integral** written symbolically as a the kernel describing a wave function to propagate from (q_a, t_a) to (q_b, t_b) :

$$K\left(q_b,t_b;q_a,t_a\right)\equiv\int Dq\left(t\right)\exp\left\{\frac{i}{\hbar}S\left[q\right]\right\} \tag{4.82}$$

where,

$$S\left[q\right]=\int_{t_a}^{t_b}L\left(q,\dot{q},t\right)dt \tag{4.83}$$

associated with the Lagrangian $L\left(q,\dot{q},t\right)$; t is time, the action 4.83 is a functional that takes as argument a function $q(t)$ (describing a trajectory or a path in space-time) and returns a number, the action $S[q]$ then

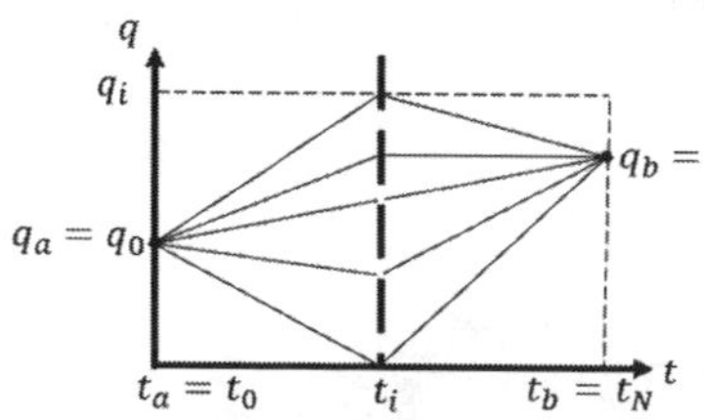

FIGURE 4.1 The propagators between space-time points are depicted by straight lines connecting corresponding two points and at the intermediate time t_i integration is done over all positions q_i.

$$Dq(t) = \frac{1}{A} \lim_{\substack{\varepsilon \to 0 \\ N \to \infty}} \prod_{i=1}^{N-1} \frac{dq_i}{A} \tag{4.84}$$

is the functional measure over the space of trajectories (or paths) and A is an appropriate normalizing factor as taken from 4.81:

$$\frac{1}{A} = \left(\frac{m}{2\pi i\hbar\varepsilon} \right)^{\frac{1}{2}} \tag{4.85}$$

The quantity $\int Dq(t)$ denotes summation over all possible paths (**so-called path or functional integral**) satisfying the boundary conditions $q(t_a) = q_a$ and $q(t_b) = q_b$.

The transition amplitude 4.82 is called the **Feynman transition amplitude**. The RHS (Right-Hand-Side) is expressed as a Feynman path (or functional) integral which has the Lagrangian version of the path integral. In essentially all cases, this Lagrangian formulation will be used as our starting point.

Considering 4.70 or the classical limit $\hbar \to 0$, the quantum mechanical amplitude 4.82 from a sufficiently small class of piecewise functions would become increasingly dominated by the contribution to the sum from the classical path $q_c(t)$. These functions may be varied on a piecewise interval that is close to the classical path $q_c(t)$, on which the action achieves its extremal (or stationary) value. This is because non–extremal configurations would be weighted by a rapidly oscillating amplitude associated with the large phase $\frac{S}{\hbar}$ and would, then, average to zero. In addition, quantum mechanical tunneling would be a natural element of the theory. Also, non-classical paths do contribute to the net amplitude, but at the cost of a damping factor specified by the imaginary action. Intuitively from the double slit experiment, the paths that move close to the classical path will all have nearly the same phase. Hence, they interfere constructively in the transition amplitude and make the major contribution. Paths farther from the classical path (in the sense of different values of the action), interfere destructively and their contributions tend to cancel out.

In the functional limit 4.82, a piecewise path is transformed to a smooth path. Thus, in evaluating the transition amplitude the dominant contribution in the action is from a sufficiently small class of functions that are all smooth—those near the classical path. From this reasoning, the most approximate method of the evaluation of 4.82 is called the **stationary phase method**.

Thus, we have obtained two types of functional integrals for the evaluation of the transition amplitude:

1. The evaluation through the classical Hamiltonian $\mathrm{H}(p,q)$ (the Hamilton transition amplitude in 4.47). This involves integrations over coordinate and momenta.
2. The evaluation through the classical Lagrangian $L(q,\dot{q},t)$ (Feynman transition amplitude 4.82). This involves integrations only over coordinates.

It should be noted that 4.82 is obtained from 4.47 considering the fact that

$$\hat{\mathrm{H}}\left(\hat{p},\hat{q}\right) \to \mathrm{H}\left(p,q\right) \tag{4.86}$$

From this argument the first evaluation of the transition amplitude is the Hamiltonian form. However, it is easily evaluated via the action using the notion of paths. This method is that of the stationary phase—the Feynman transition amplitude 4.82. Thus, in order to evaluate the transition amplitude, it is first written in the Hamiltonian form and then transformed into the Feynman form. We then observe that in the path-integral formalism, only two postulates are needed:

1. If an ideal measurement is performed to determine if a particle has a path lying in a region of space-time, then the probability that the result of measurement is affirmative will be the absolute square of a sum of complex contributions, one from each path in the region.
2. The paths contribute equally in magnitude, but the phase of their contribution is the classical action S in units of $\hbar$. This implies the time integral of the Lagrangian taken along the path.

The above derivations apply to all spatial dimensions where p and q may be treated as vectors. So, from the Feynman approach, follow the following postulates:

1. For any path $q(t)$, we can find a complex valued functional $\Phi[q(t)]$ called **the transition amplitude** from the initial point q_a to the final point q_b respectively for the time momentst$_a$ and t_b.
2. The transition amplitude from the initial point q_a to the final point q_b for the time interval $t_b - t_a$, from any of the possible paths is equal to the sum of the transition amplitude for any of the possible paths:

$$\mathrm{K}\left(q_b,t_b;q_a,t_a\right)=\sum_{\text{all possible paths}}\Phi_{ab}\left[q(t)\right] \tag{4.87}$$

3. The complex valued functional $\Phi[q(t)]$ is the contribution from the given path $q(t)$:

$$\Phi\left[q(t)\right]=\exp\left\{\frac{i}{\hbar}S\left[q(t)\right]\right\} \tag{4.88}$$

Here, the action functional $S[q(t)]$ for each path:

$$S\left[q(t)\right]=\int_{t_a}^{t_b}L\left(q,\dot{q},t\right)dt \tag{4.89}$$

So, the quantum amplitude is obtained as:

$$\mathrm{K}\left(q_b,t_b;q_a,t_a\right)=\sum_{\text{all possible paths}}\exp\left\{\frac{i}{\hbar}S\left[q(t)\right]\right\} \tag{4.90}$$

This equation is the heart of the path integral formulation. From here we observe that the paths contribute equally in magnitude where the phase of their contribution is the action functional $S[q(t)]$ in units of $\hbar$ and all possible paths enter and interfere with one another in the convolution of probability amplitudes.

Consider again 4.81 then

$$t_{i+1}-t_i=\varepsilon\equiv\frac{t_b-t_a}{N} \tag{4.91}$$

$$L=L\left(\frac{q_{i+1}-q_i}{\varepsilon},\frac{q_{i+1}+q_i}{2},\frac{t_{i+1}-t_i}{2}\right) \tag{4.92}$$

and

$$K\left(q_b,t_b;q_a,t_a\right)=\sum_{\text{all possible paths}}\exp\left\{\frac{i\varepsilon}{\hbar}\sum_i L\left(\frac{q_{i+1}-q_i}{\varepsilon},\frac{q_{i+1}+q_i}{2},\frac{t_{i+1}-t_i}{2}\right)\right\} \tag{4.93}$$

or

$$K\left(q_b,t_b;q_a,t_a\right)=\lim_{\substack{\varepsilon\to 0\\ N\to\infty}}\left(\frac{m}{2\pi i\hbar\varepsilon}\right)^{\frac{N}{2}}\prod_{i=1}^{N-1}\int_{-\infty}^{+\infty}dq_i\exp\left\{\frac{i\varepsilon}{\hbar}\sum_{i=0}^{N-1}L\left(\frac{q_{i+1}-q_i}{\varepsilon},\frac{q_{i+1}+q_i}{2},\frac{t_{i+1}-t_i}{2}\right)\right\} \tag{4.94}$$

This yields the following transition amplitude

$$K\left(q_b,t_b;q_a,t_a\right)=\int Dq(t)\exp\left\{\frac{i}{\hbar}\int_{t_a}^{t_b}L(q,\dot{q},t)dt\right\}\equiv\int Dq(t)\exp\left\{\frac{i}{\hbar}S[q]\right\} \tag{4.95}$$

4.2.5 Law for Consecutive Events

We look back to the path integral representation again by examining the property of the transition amplitude K following the rather trivial factorization of time evolution operators in the Lagrangian. In this manner, the action for a path obtained joining two points (q_a, t_a) and (q_b, t_b) by the intermediate point (q_c, t_c) satisfies the relation (**law for consecutive events**):

$$S=\int_{t_a}^{t_b}L(q,\dot{q},t)dt=\int_{t_a}^{t_c}L(q,\dot{q},t)dt+\int_{t_c}^{t_b}L(q,\dot{q},t)dt \tag{4.96}$$

So,

$$\begin{aligned}K\left(q_b,t_b;q_a,t_a\right)&=\lim_{\substack{\varepsilon\to 0\\ M\to\infty}}\left(\frac{m}{2\pi i\hbar\varepsilon}\right)^{\frac{M}{2}}\prod_{i=1}^{M-1}\int_{-\infty}^{+\infty}dq_i\exp\left\{\frac{i\varepsilon}{\hbar}\sum_{j=0}^{M-1}L\left(\frac{q_{j+1}-q_j}{\varepsilon},\frac{q_{j+1}+q_j}{2},\frac{t_{j+1}-t_j}{2}\right)\right\}\\&\times\lim_{\substack{\varepsilon\to 0\\ N\to\infty}}\left(\frac{m}{2\pi i\hbar\varepsilon}\right)^{\frac{N}{2}}\prod_{s=M+1}^{M+N-1}\int_{-\infty}^{+\infty}dq_s\exp\left\{\frac{i\varepsilon}{\hbar}\sum_{k=M}^{M+N-1}L\left(\frac{q_{k+1}-q_k}{\varepsilon},\frac{q_{k+1}+q_k}{2},\frac{t_{k+1}-t_k}{2}\right)\right\}\end{aligned} \tag{4.97}$$

This translates into the key property of the transition amplitude that is the **convolution property**:

$$K\left(q_b,t_b;q_a,t_a\right)=\int K\left(q_b,t_b;q_c,t_c\right)K\left(q_c,t_c;q_a,t_a\right)dq_c \tag{4.98}$$

Here, we have set $q_c\equiv q_M$. From here follows the:

4.2.6 Semigroup Property of the Transition Amplitude

From the semigroup property 4.98 the transition amplitude, $K(q_b, t_b; q_a, t_a)$ to go from point q_a to q_b may be decomposed into transition amplitudes of arrival at some time t_c at an intermediate point q_c and propagation continuing from there to the final point q_b at time moment t_b.

Relation 4.98 is also called the **composition property**. It is also called the **Chapman- Kolmogoroff equation** in probability theory, and the **Smoluchovsky equation** in diffusion theory. This is represented in Figure 4.2.

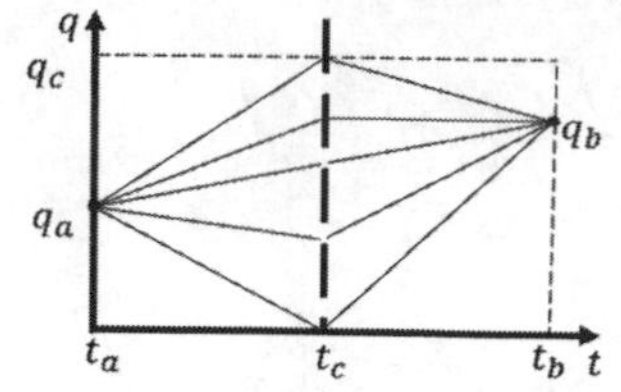

FIGURE 4.2 The semigroup property where the propagator $K(q_b, t_b; q_a, t_a)$ may be decomposed into propagators arriving at some time $t_c$at an intermediate point q_c and propagators continuing from there to the final point q_b at time moment t_b.

Formula 4.98 is a quantum mechanical rule for combining amplitudes, i.e., if a process can occur several ways, then the amplitudes for each of these ways should add. A particle propagating from point q_a to point q_b must be **somewhere** at the intermediate position q_c for the intermediate time moment t_c. So, we compute the amplitude for propagation via the point q_c corresponding to the product of two propagators and then integrate over all possible intermediate positions. This result is reminiscent of that of the double slit experiment, where the amplitudes for passing through each of the two slits combine and interfere.

The path integral formula 4.98 can be used N times for small $t \to \Delta t$, such that K can be approximated at short times:

$$K(q_b,t_b;q_a,t_a) = \int\cdots\int K(q_b,t_b;q_{N-1},t_{N-1})K(q_{N-1},t_{N-1};q_{N-2},t_{N-2})\cdots K(q_1,t_1;q_a,t_a)dq_{N-1}dq_{N-2}\cdots dq_1 \quad (4.99)$$

This corresponds to the coherent superposition of the transition amplitudes that are associated with all possible paths starting from the point q_a at time t_a and ending at a point q_b at time t_b and successively passing via all the intermediate points q_i at time moments t_i. So, the sum over paths is a sum over intermediate states. This will be very useful when investigating the path integral representation of the propagator. It is instructive to note that the initial q_a and final positions q_b are **not** integrated over. To verify our understanding of the above notions we consider the free particle example and we compute its transition amplitude.

5

Stationary and Quasi-Classical Approximations

It is interesting to know when to expect the wave nature of a particle propagation to be well approximated by the properties of the trajectory of a classical particle. For the propagation of light, say, the wavelength $\lambdabar$ is much smaller than the characteristic physical size L of the system. In this case, the diffraction phenomena are unimportant, and we can employ geometric optics. Similar results are expected in quantum mechanics when the wavelength $\lambdabar$ is short enough compared to the typical length L over which the potential varies. We can then expect a wave packet with size within $\lambdabar$ and L, with the motion approximated by that of a classical particle.

Usually, whenever there is an integral expression for a given quantity, it is often easier to employ approximate methods compared to a differential equation. Good examples include the perturbative expansion and the steepest descent method. We could equally employ change of variables to simplify the problem. Most of the quantum mechanical systems with the potential, at most quadratic in the coordinate are exactly solvable. However, in most cases of interest, the potential is more complicated and apart from a few exceptions an exact calculation of the path integral may be impossible. Such systems may be solvable only via approximate schemes. We will limit ourselves to the most important approximations valid when the quantum fluctuations are small or, equivalently, when the actions involved are large compared to Planck's constant, $\hbar$:

$$\frac{|S|}{\hbar} >> 1 \tag{5.1}$$

It is simply a matter of convenience representing a path by a classical path and fluctuations around it. It is really immaterial how to express a path satisfying the boundary conditions for the exactly solvable problem of a driven harmonic oscillator, say. For the quasi-classical approximation, however, it is critical to expand around the path leading to the dominant contribution, i.e. the classical path. From a mathematical approach, we have to evaluate a path integral over $\exp\left\{\frac{i}{\hbar}S\right\}$ for small $\hbar$ and this can be done systematically by the stationary phase method where the exponent has to be expanded about the extrema of the action S.

5.1 Stationary Phase Method/Fourier Integral

Further, in chapter 8 we will be studying an important and powerful approximation method known as the **Wentzel-Kramer-Brillouin (WKB) method**, where interest will be oscillatory integrals of the form:

$$\mathcal{F}\left[\hbar\right] = \int_a^b dq f\left(q\right) \exp\left\{\frac{i}{\hbar} S\left(q\right)\right\} \tag{5.2}$$

Here, $f(q)$ and $S(q)$ are assumed smooth functions and independent of $\hbar$. The integral of the form in 5.2 mimics the Fourier integral and $S(q)$ the phase or phase function. We examine the asymptotic behaviour of the integral $\mathcal{F}\left[\hbar\right]$ for small $\hbar$, ($\hbar \to 0$) , i.e., for the quasi-classical approximation. We will not consider the trivial cases when

$$f(q) = 0 \tag{5.3}$$

or

$$S(q) = \text{const} \tag{5.4}$$

If,

$$S(q) = q \tag{5.5}$$

then the integral 5.2 becomes the Fourier transform:

$$\mathcal{F}\left[\hbar\right] = \int_a^b dq f(q) \exp\left\{\frac{i}{\hbar} q\right\} \tag{5.6}$$

The function

$$\text{Re}\left(f(q) \exp\left\{\frac{i}{\hbar} q\right\}\right) \tag{5.7}$$

for small $\hbar$, ($\hbar \to 0$) rapidly oscillates and yields two neighboring half-waves of approximately equal areas but opposite in sign. So, the sum of the areas is an infinitesimal quantity and consequently the following integral is also an infinitesimal quantity according to the Riemann-Lebesgue lemma:

$$\int_a^b dq \,\text{Re}\left(f(q) \exp\left\{\frac{i}{\hbar} q\right\}\right) \tag{5.8}$$

For the finite integral

$$\int_a^b dq \left| f(q) \right| \tag{5.9}$$

then

$$\lim_{\hbar \to 0} \mathcal{F}\left[\hbar\right] = \lim_{\hbar \to 0} \int_a^b dq f(q) \exp\left\{\frac{i}{\hbar} q\right\} \to 0 \tag{5.10}$$

The rate at which the given integral tends to zero is dependent solely on the properties of the derivatives of the function $f(q)$.

Suppose the functions $f(q)$ and $S(q)$ are infinitely differentiable and $S(q) \neq 0$ for any q belonging to the interval $[a, b]$, then as $\hbar \to 0$, we investigate on the principal term in the asymptotic expansion $\mathcal{F}\left[\hbar\right]$ from equation 5.2. Letting $\lambda = \frac{1}{\hbar}$, we do integration by parts in 5.2:

$$\mathcal{F}\left[\hbar\right] \equiv \mathcal{F}\left[\lambda\right] = \frac{1}{i\lambda} \exp\left\{i\lambda S(q)\right\} \frac{f(q)}{S'(q)}\Bigg|_{q_a}^{q_b} + \frac{1}{i\lambda} \mathcal{F}_1\left[\lambda\right] \tag{5.11}$$

Here,

$$\mathcal{F}_1[\lambda] = -\int_a^b dq \exp\{i\lambda S(q)\} \frac{d}{dq}\left[\frac{f(q)}{S'(q)}\right] \tag{5.12}$$

then from

$$\lim_{\hbar\to 0} \mathcal{F}_1[\hbar] \equiv \lim_{\lambda\to\infty} \mathcal{F}_1[\lambda] = o(1) \tag{5.13}$$

it follows from Riemann-Lebesgue lemma that

$$\mathcal{F}_1[\lambda] = o(\lambda^{-1}) \tag{5.14}$$

Letting,

$$\mathrm{f}_1(q) = \frac{d}{dq}\left[-\frac{f(q)}{S'(q)}\right] \tag{5.15}$$

then

$$\mathcal{F}_1[\lambda] = \int_a^b dq \mathrm{f}_1(q) \exp\{i\lambda S(q)\} \tag{5.16}$$

and carrying out integration by parts then

$$\mathcal{F}_1[\lambda] = \frac{1}{i\lambda} \exp\{i\lambda S(q)\} \frac{\mathrm{f}_1(q)}{S'(q)}\bigg|_{q_a}^{q_b} + \frac{1}{i\lambda}\mathcal{F}_2[\lambda] \tag{5.17}$$

Here,

$$\mathcal{F}_2[\lambda] = -\int_a^b dq \exp\{i\lambda S(q)\} \frac{d}{dq}\left[\frac{\mathrm{f}_1(q)}{S'(q)}\right] \tag{5.18}$$

then from

$$\lim_{\lambda\to\infty} \mathcal{F}_2[\lambda] = o(1) \tag{5.19}$$

it follows from Riemann-Lebesgue lemma that

$$\mathcal{F}_2[\lambda] = o(\lambda^{-1}) \tag{5.20}$$

So,

$$\mathcal{F}[\lambda] = \frac{1}{i\lambda} \exp\{i\lambda S(q)\} \frac{f(q)}{S'(q)}\bigg|_{q_a}^{q_b} + o(\lambda^{-2}) \tag{5.21}$$

The last term should be the correction to the formula.

5.2 Contribution from Non-Degenerate Stationary Points

The above study does not give room for the phase $S(q)$ to have a stationary point within an interval. This implies that

$$S(q) \neq 0 \quad (5.22)$$

for q defined within the interval $[a, b]$. If such a stationary point exists then the asymptotic behavior of the integral $\mathcal{F}[\lambda]$ will be of a different nature than that shown above.

Consider the phase

$$S(q) = q^2 \quad (5.23)$$

So,

$$\mathcal{F}[\lambda] = \int_0^a dqf(q)\exp\left\{i\frac{\alpha\lambda q^2}{2}\right\} \quad (5.24)$$

This has a stationary point $q = 0$ and is depicted in Figure 5.1. At the neighborhood of this stationary point, the function $\cos\lambda q^2$ is not oscillatory. So, the sum of the areas of the waves in the cosine is sufficiently less than that at the stationary point. Hence, the value of the integral $\mathcal{F}[\lambda]$ is of order $\lambda^{-\frac{1}{2}}$.

If the function $f(q)$ should be infinitely differentiable in the interval $[0, a]$ then

$$\mathcal{F}[\lambda] = \int_0^a dqf(q)\exp\left\{i\frac{\alpha\lambda q^2}{2}\right\} = \frac{1}{2}\sqrt{\frac{2\pi}{|\alpha|\lambda}}\exp\left\{i\frac{\pi}{4}\text{sgn}\,\alpha\right\} f(0) + o(\lambda^{-1}), \lambda \to \infty,\ \alpha \neq 0 \quad (5.25)$$

Here, $\text{sgn}\,\alpha$ denotes the sign of α.

We show the truthfulness of this equation by letting

$$f(q) = 1, \alpha > 0 \quad (5.26)$$

then,

$$\mathcal{F}[\lambda] = \int_0^a dq\exp\left\{i\frac{\alpha\lambda q^2}{2}\right\} = \frac{1}{\sqrt{\alpha\lambda}}\int_0^{a\sqrt{\alpha\lambda}} d\tau\exp\left\{\frac{i\tau^2}{2}\right\} \quad (5.27)$$

or

$$\mathcal{F}[\lambda] = \frac{1}{\sqrt{\alpha\lambda}}\left[\int_0^{\infty} d\tau\exp\left\{\frac{i\tau^2}{2}\right\} - \int_{a\sqrt{\alpha\lambda}}^{\infty} d\tau\exp\left\{\frac{i\tau^2}{2}\right\}\right] \quad (5.28)$$

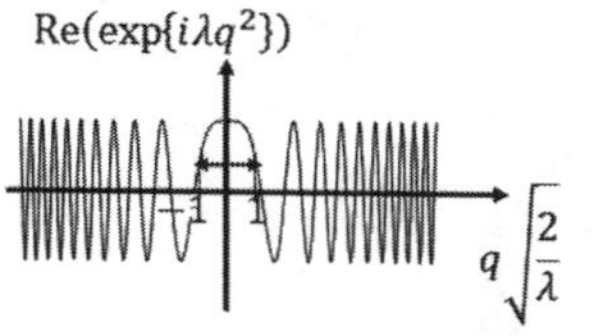

FIGURE 5.1 showing only a small region at the neighborhood of the extremum $q = 0$ contributes to the integral of equation 5.24 in stationary phase approximation.

The first integral yields

$$\int_0^\infty d\tau \exp\left\{\frac{i\tau^2}{2}\right\} = \sqrt{i\pi}\exp\left\{i\frac{\pi}{4}\right\} \tag{5.29}$$

then the second integral yields

$$\int_{a\sqrt{\alpha\lambda}}^\infty d\tau \exp\left\{\frac{i\tau^2}{2}\right\} = o\left(\lambda^{-\frac{1}{2}}\right), \lambda \to \infty \tag{5.30}$$

So,

$$\int_0^a dq \exp\left\{i\frac{\alpha\lambda q^2}{2}\right\} = \frac{1}{2}\sqrt{\frac{2\pi}{\alpha\lambda}}\exp\left\{i\frac{\pi}{4}\right\} + o\left(\lambda^{-1}\right), \lambda \to \infty \tag{5.31}$$

Let,

$$f(q) = f(0) + \left(f(q) - f(0)\right) = f(0) + qf_1(q) \tag{5.32}$$

where, we introduce an infinitely differentiable function$f_1(q)$ for q belonging to the interval $[0, a]$:

$$f_1(q) = \frac{f(q) - f(0)}{q} \tag{5.33}$$

So,

$$\mathcal{F}[\lambda] = \frac{1}{2} f(0)\sqrt{\frac{2\pi}{|\alpha|\lambda}}\exp\left\{i\frac{\pi}{4}\operatorname{sgn}\alpha\right\} + o\left(\lambda^{-1}\right) + \int_0^a dq q f_1(q)\exp\left\{i\frac{\alpha\lambda q^2}{2}\right\} \tag{5.34}$$

We study the last term:

$$\mathcal{F}_r[\lambda] = \left|\int_0^a dq q f_1(q)\exp\left\{i\frac{\alpha\lambda q^2}{2}\right\}\right| = \left|\frac{1}{i\alpha\lambda}\int_0^a f_1(q) d\left[\exp\left\{i\frac{\alpha\lambda q^2}{2}\right\}\right]\right| \tag{5.35}$$

or

$$\mathcal{F}_r[\lambda] = \left|\frac{1}{i\alpha\lambda}\left(\left[f_1(q)\exp\left\{i\frac{\alpha\lambda q^2}{2}\right\}\right]_0^a - \int_0^a dq f'_1(q)\exp\left\{i\frac{\alpha\lambda q^2}{2}\right\}\right)\right| \tag{5.36}$$

or

$$\mathcal{F}_r[\lambda] \leq \frac{1}{\alpha\lambda}\left|\left[f_1(q)\right]_0^a\right| + \frac{1}{\alpha\lambda}\int_0^a dq\left|f'_1(q)\right| = o\left(\lambda^{-1}\right), \lambda \to \infty \tag{5.37}$$

So, from 5.34 we have

$$\mathcal{F}[\lambda] = \frac{1}{2}\sqrt{\frac{2\pi}{|\alpha|\lambda}}\exp\left\{i\frac{\pi}{4}\operatorname{sgn}\alpha\right\} f(0) + o\left(\lambda^{-1}\right), \lambda \to \infty, \alpha \neq 0 \tag{5.38}$$

From here and established by analyses of Figure 5.1, the dominant contribution to the integral comes from a region of size $\frac{1}{\sqrt{\lambda}}$, about the **extremal** (**critical** or **stationary** or **saddle**) **point** of the function $S(q)$. Outside of this region, the integrand is rapidly oscillating and so gives to leading order a negligible contribution.

5.2.1 Unique Stationary Point

Suppose the functions $f(q)$ and $S(q)$ are infinitely differentiable on the interval $[a, b]$ and for the function $S(q)$ there exist only a unique **stationary** (**critical** or **saddle**) **point** $\bar{q}$ belonging to $[a, b]$ and

$$S''(\bar{q}) \neq 0 \tag{5.39}$$

then from 5.2 we have the relation:

$$\mathcal{F}[\lambda] = \exp\{i\lambda S(\bar{q})\}\mathrm{F}(\lambda)\exp\left\{i\frac{\pi}{4}\operatorname{sgn} S''(\bar{q})\right\}f(0) + o(\lambda^{-1}), \lambda \to \infty \tag{5.40}$$

or

$$\mathcal{F}[\hbar] = \exp\left\{\frac{i}{\hbar}S(\bar{q})\right\}\mathrm{F}(\lambda)\exp\left\{i\frac{\pi}{4}\operatorname{sgn} S''(\bar{q})\right\}f(0) + o(\hbar), \hbar \to 0 \tag{5.41}$$

Here,

$$\mathrm{F}(\lambda) = \sqrt{\frac{2\pi}{|S''(\bar{q})|\lambda}} \tag{5.42}$$

We show this and for definiteness we let

$$S''(\bar{q}) > 0 \tag{5.43}$$

So,

$$S'(\bar{q}) > 0 \tag{5.44}$$

for q belonging to the interval $[\bar{q}, b]$. Hence,

$$S(q) > S(\bar{q}) \tag{5.45}$$

for q belonging to the interval $[\bar{q}, b]$. Suppose $\mathcal{F}[\lambda]$ is examined on the intervals $[a, \bar{q}]$ and $[\bar{q}, b]$ giving respectively the integrals $\mathcal{F}_1[\lambda]$ and $\mathcal{F}_2[\lambda]$. For the integral $\mathcal{F}_2[\lambda]$, we set

$$q = \phi(y) \tag{5.46}$$

such that the fluctuation

$$S(q) - S(\bar{q}) = y^2 \tag{5.47}$$

and

$$\mathcal{F}_2[\lambda] = \exp\{i\lambda S(\bar{q})\}\int_0^{b_1} dy f(\phi(y))\phi'(y)\exp\{i\lambda y^2\} \tag{5.48}$$

Here,

$$b_1 = \sqrt{S(b) - S(\bar{q})} \tag{5.49}$$

So, from above,

$$\mathcal{F}_2[\lambda] = \frac{1}{2}\exp\{i\lambda S(\bar{q})\}F(\lambda)\exp\left\{i\frac{\pi}{4}\right\}f(\bar{q}) + o(\lambda^{-1}), \lambda \to \infty \tag{5.50}$$

Similarly, we do for $\mathcal{F}_1[\lambda]$ and taking the sum of $\mathcal{F}_1[\lambda]$ and $\mathcal{F}_2[\lambda]$, relation 5.50 is confirmed. So, the **stationary phase approximation** (or **saddle point approximation**) to the integral implies neglecting higher order terms encoded by $o(\lambda^{-1})$. We have shown that this approximation is true when the contribution due to $o(\lambda^{-1})$ for $\lambda \to \infty$ is indeed negligible. The exponential factor having the function $S(\bar{q})$ is the dominant contribution to the function $\mathcal{F}[\lambda]$ from the classical path. The factor $F(\lambda)$ describes the contribution from the quantum corrections of the order $o(\lambda^{-1})$. Relation 5.50 in the stationary phase method, imitates the standard quantum mechanics and for large $\lambda \to \infty$, the quantity $\exp\{i\lambda S(\bar{q})\}$ oscillates rapidly when the function $S(q)$ changes. In this case, $F(\lambda)$ smoothly depends on $\hbar$. So, regions without critical points $\bar{q}$ contribute $o(\lambda^{-1})$terms to the integral $\mathcal{F}[\lambda]$. Relation 5.50 shows irrespective of the neglected terms, there are some contributions due to small neighborhoods of the critical points $\bar{q}$ to the total integral of order $o(\lambda^{-1})$. So, in the limit $\lambda \to \infty$, it is sufficient to evaluate at least the first quantum correction and hence, via the stationary phase approximation with reasonable accuracy the integral $\mathcal{F}[\lambda]$ can be evaluated.

We could also consider the case

$$S''(\bar{q}) < 0 \tag{5.51}$$

where we have to evaluate the integral

$$\overline{\mathcal{F}[\lambda]} = -\int_a^b dq \overline{f(q)}\exp\{i\lambda\tilde{S}(q)\} \tag{5.52}$$

Here,

$$\tilde{S}(q) = -S(q) \tag{5.53}$$

and

$$\tilde{S}''(\bar{q}) > 0 \tag{5.54}$$

The reader is expected to study this case.

We thus observe that for large λ, the region determining the integral is very small and so we may expand the function $S(q)$ locally around the extremum $\bar{q}$ and replace $f(q)$ by $f(\bar{q})$. If $S(q)$ has more than one extremum then we sum over the contributions of all extrema except one extremum is observed to be dominant.

5.3 Quasi-Classical Approximation/Fluctuating Path

The knowledge of the stationary phase approximation is very appropriate to apply to path integration. The transition from its quantum mechanical formulation to the classical limit can be achieved for $\hbar \to 0$. For this limit we will observe that path integral will be dominated by paths that are extremal points $\bar{q}$ of the action. This implies the classical solutions to the equations of motion. This is in contrast with the tedious eikonal quasi-classical approximation to the Schrödinger equation.

In the equations above we replace the function $S(q)$ by the action functional $S[q]$. The quasi-classical approximation entails expanding $S[q]$ about the extremal point $\bar{q}$ which is the classical solution of the Euler-Lagrange equation in 3.30 of chapter 3. This procedure will be very appropriate for non-Gaussian path integrals and we see further that this yields an expansion of the path integral in powers of $\hbar$.

We will show further in this book the concordance of the stationary phase approximation to the path integral for appropriate cases. Around the classical path, a quantum particle **explores** its vicinity. The path can deviate from the classical one when the difference in the action is roughly within $\hbar$. If a classical particle is confined with a potential well, then the quantum particle can perform an excursion and tunnel through the potential barrier. We apply now the stationary phase approximation to path integrals where $\lambda = \frac{1}{\hbar}$ plays the role of the large parameter. As the action is stationary at classical paths, then considering the excursion of the particle, for the new path we select a fluctuating path

$$\delta q(t) \equiv y(t) \tag{5.55}$$

about the classical one:

$$q(t) = \bar{q}(t) + y(t) \tag{5.56}$$

Then we Taylor series expand the action function about the classical path $\bar{q}(t)$ and we find that the action separates into a classical and a fluctuating part:

$$S[q(t)] = S[\bar{q}(t)] + S[y(t)] \tag{5.57}$$

Thus, for the stationary phase or quasi-classical approximation to the path integral the action is separable in the sense that the propagator factorizes into the product:

$$\int_{q_a}^{q_b} Dq(t) \exp\left\{\frac{i}{\hbar} S[q]\right\} \cong F(T) \exp\left\{\frac{i}{\hbar} S[\bar{q}(t)]\right\} \tag{5.58}$$

Here, the factor (**reduced propagator**) $F(T)$ that is a path integral taken over all variations and due to the vanishing of $y(t)$ at the end points, it is independent of q_a and q_b but only dependent on the initial and final times t_a and t_b. The time translational invariance will be observed to reduce this dependence further to the time difference $T = t_b - t_a$.

5.3.1 Free Particle Classical Action and Transition Amplitude

This heading examines some specific systems of particular interest in our investigation. These include the free particle and the oscillator in addition to others that will provide an avenue for the evaluation of the path integrals. For brevity, we consider the case of particles moving in one dimension labelled by the position coordinate q.

5.3.1.1 Free Particle Classical Action

We begin first with the free particle where the particle is associated with the Lagrange function:

$$L = \frac{m\dot{q}^2}{2} \tag{5.59}$$

Here, the mass of the particle is m. Assuming only those paths quite near to the classical path $\bar{q}(t)$ are important then for the new path we select a fluctuating path $y(t)$ about the classical one (Figure 5.2):

$$q(t) = \bar{q}(t) + y(t) \tag{5.60}$$

Here, $\bar{q}(t)$ is the classical path that connects the space-time points (q_a, t_a) and (q_b, t_b):

$$\bar{q}(t) = q_a + \frac{q_b - q_a}{t_b - t_a}(t - t_a) \tag{5.61}$$

It is important to note that for

$$q(t_a) = \bar{q}(t_a) = q_a \text{ , } q(t_b) = \bar{q}(t_b) = q_b \tag{5.62}$$

So, follows the **Dirichlet boundary conditions**:

$$y(t_a) = y(t_a) = 0 \tag{5.63}$$

From the equation of motion for the free particle considering 5.59 then

$$\frac{d}{dt}\frac{\partial L}{\partial \dot{q}} - \frac{\partial L}{\partial q} = 0 \tag{5.64}$$

and

$$\frac{d}{dt}\frac{\partial L}{\partial \dot{q}} = \frac{d}{dt}(m\dot{q}) = m\ddot{q}, \quad \frac{\partial L}{\partial q} = 0 \tag{5.65}$$

This shows that the acceleration is zero,

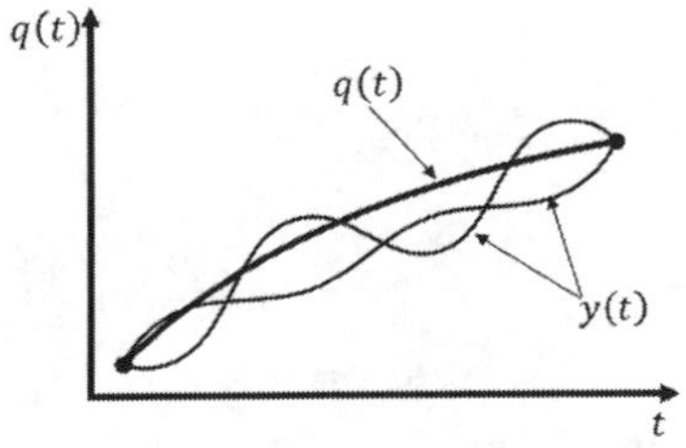

FIGURE 5.2 The thick line represents the classical path $\bar{q}$ and the thinner lines associated with quantum fluctuations y.

$$\ddot{q} = 0 \tag{5.66}$$

So, from here

$$\dot{q} = v = \text{const}, \quad dq = vdt \tag{5.67}$$

and

$$(q_b - q_a) = v(t_b - t_a) \tag{5.68}$$

So, the velocity of the classical path is constant:

$$v = \frac{q_b - q_a}{t_b - t_a} = \dot{q} \tag{5.69}$$

We can then find the action integral $S[q(t)]$ for any path $q(t)$ via an action integral over the path $y(t)$ relative to the classical path $\bar{q}(t)$. One obtains

$$S[q(t)] = \frac{m}{2}\int_{t_a}^{t_b} \left(\dot{\bar{q}}^2 + 2\dot{\bar{q}}\dot{y} + \dot{y}^2\right)dt \tag{5.70}$$

From here, the first integrand is the contribution of the classical path to the free particle action $S_{cl}[q_b, t_b; q_a, t_a]$:

$$S_{cl}\left[q_b, t_b; q_a, t_a\right] = \frac{m}{2}\int_{t_a}^{t_b} \dot{q}^2 dt = \frac{m\dot{q}^2}{2}\int_{t_a}^{t_b} dt = \frac{m}{2}\left(\frac{q_b - q_a}{t_b - t_a}\right)^2 (t_b - t_a) \tag{5.71}$$

or

$$S_{cl}\left[q_b, t_b; q_a, t_a\right] = \frac{m}{2T}(q_b - q_a)^2, \quad T = t_b - t_a \tag{5.72}$$

The second integrand gives

$$2\dot{\bar{q}}\int_{t_a}^{t_b} \dot{y}dt = y(t_b) - y(t_a) = 0 \tag{5.73}$$

The third integrand is the contribution of the fluctuation to the action:

$$S[y(t)] = \frac{m}{2}\int_{t_a}^{t_b} \dot{y}^2 dt \tag{5.74}$$

So, we observe the contribution of only a quadratic fluctuation to the action.

5.3.1.2 Free Particle Transition Amplitude

We show, as already noted by Dirac [7], that the quantum mechanical propagator and the classical properties of a free particle are closely related. So, we now evaluate the path integral for the transition amplitude of a free particle following the algorithm introduced above considering the action has contributions from the classical path $\bar{q}$ as well as the fluctuation $y(t)$ to the classical path:

$$S\left[q(t)\right]=S_{cl}\left[q_b,t_b;q_a,t_a\right]+S\left[y(t)\right] \tag{5.75}$$

Here, there is no linear term in $y(t)$ and no higher than quadratic terms because the free particle action itself is quadratic. So, in the path integral, instead of integrating over all paths $q(t)$ with boundary conditions 5.62 we now integrate over all paths $y(t)$ with the boundary conditions 5.63. So, the path integral expression for the transition amplitude for the free particle can be evaluated via 5.58:

$$K\left(q_b,t_b;q_a,t_a\right)=\exp\left\{\frac{i}{\hbar}S_{cl}\left[q_b,t_b;q_a,t_a\right]\right\}\int_0^0 Dy(t)\exp\left\{\frac{im}{2\hbar}\int_{t_a}^{t_b}\dot{y}^2dt\right\} \tag{5.76}$$

We observe from 5.75 that the action is a sum of two terms and so the amplitude factorizes into the product of a classical amplitude and a fluctuation factor. We rewrite the Gaussian integral around the classical path (**reduced propagator or fluctuation factor**) for the free particle as

$$F(T)=\int_0^0 Dy(t)\exp\left\{\frac{i}{\hbar}\int_0^T\frac{m}{2}\dot{y}^2dt\right\} \tag{5.77}$$

Due to the vanishing of $y(t)$ at the end points, this integral is independent of q_a and q_b but only on the initial and final times t_a and t_b with the time translational invariance reducing this dependence further to the time difference $T=t_b-t_a$.

The path integral in 5.77 is that of an oscillator with frequency $\omega=0$. So, we can therefore more explicitly expand $y(t)$ in normalized eigenmodes $y_n(t)$ of some differential operator for the free particle:

$$y(t)=\sum_n a_n y_n(t) \tag{5.78}$$

with, respectively, the conditions of orthonormality and those at end points of the path:

$$\int_{t_a}^{t_b} y_n(t)y_m(t)dt=\delta_{nm},\quad y_n(t_a)=y_n(t_b)=0 \tag{5.79}$$

where, the first equality shows that the expansion functions are orthogonal while the second gives the **Dirichlet boundary conditions (conditions at end points of the paths)**.

From the conditions 5.79, as $y_n(t)$ vanishes at the initial time $t=t_a$, the expansion can be constrained to the sine functions while the vanishing at the final time $t=t_b$ is imposed by a constraint of the frequencies ω_n with discrete values:

$$y_n(t)=\sqrt{\frac{2}{T}}\sin\omega_n\left(t-t_a\right),\quad \omega_n=\frac{\pi n}{T} \tag{5.80}$$

So, the expansion of $y(t)$ in terms of eigenmodes is synonymous to diagonalizing the operator and path integral over all $y(t)$ equal integral over the real Fourier coefficients a_n:

$$\int_0^0 Dy(t)=J\prod_n\int_{-\infty}^{\infty}da_n \tag{5.81}$$

We evaluate exponential function with the fluctuation $y(t)$ in 5.77:

$$\int_0^T \dot{y}^2(t)dt = \int_0^T \frac{2}{T}\sum_{nn'} a_n a_{n'}\omega_n\omega_{n'}\cos\omega_n t\cos\omega_{n'} t dt \tag{5.82}$$

So, from the first condition in 5.79 we have

$$\int_0^T \dot{y}^2(t)dt = \sum_n \frac{2}{T}a_n^2\omega_n^2\int_0^T \cos^2\omega_n t dt \tag{5.83}$$

As,

$$\int_0^T \cos^2\omega_n t dt = \int_0^T \sin^2\omega_n t dt \tag{5.84}$$

then

$$\int_0^T \sin^2\omega_n t dt = \frac{1}{2}\int_0^T\left(\sin^2\omega_n t + \cos^2\omega_n t\right)dt = \frac{1}{2}\int_0^T dt = \frac{T}{2} \tag{5.85}$$

and

$$\int_0^T \dot{y}^2(t)dt = \sum_n a_n^2\omega_n^2 \tag{5.86}$$

So,

$$S = \frac{m}{2T}\left(q_b - q_a\right)^2 + \frac{m}{2}\sum_n \omega_n^2 a_n^2 \tag{5.87}$$

The first term is nothing but the classical action and then the path integral reduces to an infinite collection of Fresnel integrals:

$$J\prod_n\int_{-\infty}^{\infty} da_n \exp\left\{\frac{im}{2\hbar}\left[\frac{1}{T}\left(q_b - q_a\right)^2 + \sum_n \omega_n^2 a_n^2\right]\right\} \tag{5.88}$$

So,

$$\mathrm{K}\left(q_b,t_b;q_a,t_a\right) = J\prod_n\left(-\frac{im}{2\pi\hbar}\omega_n^2\right)^{-\frac{1}{2}}\exp\left\{\frac{im}{2\hbar T}\left(q_b - q_a\right)^2\right\} \tag{5.89}$$

This may be simply represented by

$$\mathrm{K}\left(q_b,t_b;q_a,t_a\right) = J'\left(T\right)\exp\left\{\frac{im}{2\hbar T}\left(q_b - q_a\right)^2\right\} \tag{5.90}$$

The normalization constant J can depend only on the time interval T determined from the requirement that

$$\int \mathrm{K}\left(q_b,t_b;q,t\right)\mathrm{K}\left(q,t;q_a,t_a\right)dq = \mathrm{K}\left(q_b,t_b;q_a,t_a\right) \tag{5.91}$$

and so,

$$J'\left(t_b - t\right)J'\left(t_a - t\right)\left(\frac{2\pi\hbar i\left(t_b - t\right)\left(t_a - t\right)}{mT}\right)^{\frac{1}{2}} = J'\left(T\right) \tag{5.92}$$

Hence, the **reduced propagator**:

$$J'\left(T\right) = \left(\frac{m}{2\pi\hbar iT}\right)^{\frac{1}{2}} \equiv \mathrm{F}\left(T\right) \tag{5.93}$$

and multiplying the **reduced propagator** (fluctuation factor) with the classical amplitude, then the transition amplitude for the free particle:

$$\mathrm{K}\left(q_b,t_b;q_a,t_a\right) = \left(\frac{m}{2\pi i\hbar T}\right)^{\frac{1}{2}} \exp\left\{\frac{im}{2\hbar T}\left(q_b - q_a\right)^2\right\} \tag{5.94}$$

From here we observe that the quantum mechanical propagator and the classical properties of a free particle are closely related [7]. So, $S_{\mathrm{cl}}[q_b, t_b; q_a, t_a]$, is the classical action along the classical path, with the condition that $q(0) = q_a$ and $q(T) = q_b$. The absolute square of K :

$$\frac{d\mathrm{i}}{dq} = \left|\mathrm{K}\right|^2 = \frac{m}{2\pi\hbar T} \tag{5.95}$$

This gives the probability distribution per unit length that the particle will be found around position q_b, given that it was known to be at q_a at the start of the time interval T. This is simply a diffusive motion or spreading of a localized wave packet that began, as a Dirac delta-function centered at q_a which corresponds in momentum space to an exactly flat distribution of momenta. We note that the wave packet width (in probability) increases linearly with the time and one might suspect that it is due to a peculiarity of the free particle. However, this result in 5.94 may indeed be employed to construct a representation of the propagator where the classical action appears in the exponent. This is because, the propagation in a general potential (in the absence of delta-function contributions) may be decomposed into a series of short-time propagations of a free particle. The type of action $S_{\mathrm{cl}}[0, T]$, appears only within the quasi-classical approximation or for potentials where the given approximation turns out to be exact.

5.3.1.3 From Path Integrals to Quantum Mechanics

From equation 4.99 chapter 4, we can find the link between the coordinates and the wave parameters. The type of the coordinates q_{i+1} and q_i contributing most to the transition amplitude are those satisfying the relation in 5.70 and tailor the phase of the transition amplitude to have an extremal (or stationary) value where sum over neighbouring trajectories tend to interfere constructively as their phases are nearly equal. From here, considering

$$q_b - q_a = q \tag{5.96}$$

and supposing that

$$\Delta q \ll q \tag{5.97}$$

then

$$(q + \Delta q)^2 \cong q^2 + 2q\Delta q \tag{5.98}$$

We can introduce the notion of the wave length λ. The coordinate q experiences a shift of λ so that the phase changes to 2π:

$$\frac{m}{2\hbar T}(q+\lambda)^2 - \frac{m}{2\hbar T}q^2 = 2\pi \tag{5.99}$$

Considering

$$\lambda^2 \to 0 \tag{5.100}$$

then

$$\frac{mq\lambda}{\hbar T} = 2\pi \tag{5.101}$$

But,

$$\frac{q}{T} = v \ , \ p = \frac{2\pi\hbar}{\lambda} \tag{5.102}$$

These formulae relate the corpuscular and wave nature of the particle.

We examine the factor $\exp\left\{\frac{imq^2}{2\hbar t}\right\}$ then

$$\frac{mq^2}{2\hbar t} - \frac{mq^2}{2\hbar(t+T)} = 2\pi \tag{5.103}$$

From

$$\frac{1}{t+T} \cong \frac{1}{t}\left(1 - \frac{1}{T}\right), \ \frac{mq^2}{2\hbar t}T = 2\pi, \ \frac{q^2}{t^2} = v^2 \tag{5.104}$$

Then the energy of the system

$$E = \frac{2\pi\hbar}{T} = \hbar\omega \tag{5.105}$$

So, we have the corpuscular and wave dualism that is inherent in the Feynman approach.

5.4 Free and Driven Harmonic Oscillator Classical Action and Transition Amplitude

5.4.1 Free Oscillator Classical Action

We devote this example to a study of an important physical system called the **one-dimensional harmonic oscillator which is some system having one degree of freedom and executing in its direction a harmonic oscillation**. We examine a one-dimensional motion of a point mass that is attracted to a fixed center by a restoring force F_q: This restoring force is proportional to the displacement q from that center:

$$F_q = -m\omega^2 q \tag{5.106}$$

where m and ω are the mass and the classical frequency respectively. This provides one of the fundamental problems of classical dynamics. This study is very important in physics as a large number of systems may always be analyzed in terms of normal modes of motion. In the limit of small oscillations, the normal modes are formally equivalent to harmonic oscillators. The harmonic oscillator has practical applications in a variety of domains of modern physics, such as molecular spectroscopy, solid state physics, nuclear structure, quantum field theory, quantum statistical mechanics and so on. The harmonic oscillator also provides the path in the study of the electromagnetic field. The vibrations of the electromagnetic field in a cavity provide an infinite number of stationary waves, i.e., normal modes of the cavity. We may expand the electromagnetic field in terms of these modes. If we consider Maxwell's equations, it can be seen that each coefficient of expansion conforms to a differential equation. This equation is identical to that of a harmonic oscillator having angular frequency ω associated with a normal mode. This follows that the vibrations of the electromagnetic field in a cavity may be analyzed into harmonic normal modes each of which energy levels is of the oscillator type. Quantizing the given oscillators associated with various normal modes of the cavity may quantize the field. So, a detail investigation of the problem of the harmonic oscillator in quantum mechanics is very important.

We examine the harmonic oscillator with the procedure in reference [1] considering the stationary phase method. The Lagrangian for this harmonic oscillator has the form:

$$L(q,\dot{q}) = \frac{m}{2}\dot{q}^2 - V(q) \tag{5.107}$$

The aim here is to compute the leading contribution to the path integral in the formal limit $\hbar \to 0$. As seen above, this will correspond to deriving a general expression for the **Gaussian, or quasi-classical approximation**. So, the main contribution to the path integral for the transition amplitude will come from paths at the vicinity of the classical path as earlier indicated. This implies those paths with an action within about $\hbar$ of the action of the classical path where the least action principle is satisfied for

$$\delta S = 0 \tag{5.108}$$

From here, we arrive at the Euler-Lagrange equation for the classical path:

$$\frac{d}{dt}\frac{\partial L}{\partial \dot{q}} - \frac{\partial L}{\partial q} = m\ddot{\bar{q}} + \left.\frac{dV(q)}{dq}\right|_{q=\bar{q}(t)} = 0, \quad \bar{q}(0) = q_a, \quad \bar{q}(T) = q_b \tag{5.109}$$

Consider a particle of mass m executing a harmonic oscillation in the direction of the q-axis with a cyclical frequency ω. From the restoring force 5.106 we find the potential energy $V(q)$:

$$V(q)=\frac{m}{2}\omega^2\bar{q}^2 \tag{5.110}$$

then from equation 5.109 follows

$$m\ddot{\bar{q}}+m\omega^2\bar{q}=0 \tag{5.111}$$

with solution in the form:

$$\bar{q}(t)=A\cos(\omega t+\alpha) \tag{5.112}$$

Here, α is the initial phase. Letting

$$A\cos\alpha=X \quad , \quad A\sin\alpha=Y \tag{5.113}$$

Then

$$\bar{q}(t)=X\cos\omega t-Y\sin\omega t \tag{5.114}$$

and

$$q_a=X\cos\omega t_a-Y\sin\omega t_a \quad q_b=X\cos\omega t_b-Y\sin\omega t_b \tag{5.115}$$

The classical action for the harmonic oscillator can be found:

$$S_{cl}[0,T]=\int_{t_a}^{t_b}\left[\frac{m\dot{q}^2}{2}-\frac{m\omega^2q^2}{2}\right]dt=\int_{t_a}^{t_b}\left[\frac{m}{2}\left(\frac{d}{dt}(q\dot{q})-\ddot{q}q\right)-\frac{m\omega^2q^2}{2}\right]dt \tag{5.116}$$

or

$$S_{cl}[0,T]=\frac{m}{2}q\dot{q}\Big|_{t_a}^{t_b}-\frac{m}{2}\int_{t_a}^{t_b}q\left[\ddot{q}+\omega^2q\right]dt=\frac{m}{2}\left(q_b\dot{q}_b-q_a\dot{q}_a\right) \tag{5.117}$$

The second term in 5.117 vanishes due to the equation of motion in 5.109. From

$$q(t)=\begin{cases}q_a, & t=t_a\\ q_b, & t=t_b\end{cases} \tag{5.118}$$

then this permit to find for X and Y in 5.115:

$$X=q_a\frac{\sin\omega t_b}{\sin\omega T}-q_b\frac{\sin\omega t_a}{\sin\omega T} \quad , \quad Y=q_a\frac{\cos\omega t_b}{\sin\omega T}-q_b\frac{\cos\omega t_a}{\sin\omega T} \tag{5.119}$$

Letting,

$$\Gamma_a(T-t)\equiv\frac{\sin\omega(T-t)}{\sin\omega T},\ \Gamma(t)\equiv\frac{\sin\omega t}{\sin\omega T}\ ,\ T=t_b-t_a \tag{5.120}$$

So, the following solution only makes sense

$$q(t) \equiv q_{\mathrm{H}}(t) = q_a \Gamma_a (T-t) + q_b \Gamma(t) \tag{5.121}$$

provided

$$\sin \omega T \neq 0 \tag{5.122}$$

This implies, if

$$\omega T \neq n\pi \tag{5.123}$$

So, if T is not an integral multiple of half the period τ,

$$T \neq n \times \frac{\tau}{2}, \; \tau = \frac{2\pi}{\omega} \tag{5.124}$$

then there exists a unique classical path that starts from the initial point q_a and ends at q_b within the time interval T. From here and 5.117, then follows the classical action of a harmonic oscillator that is, of course, force free:

$$S_{\mathrm{Hcl}}\left[0,T\right] = \frac{m\omega}{2\sin \omega T}\left[\left(q_a^2 + q_b^2\right)\cos \omega T - 2q_a q_b\right] \tag{5.125}$$

5.4.2 Driven or Forced Harmonic Oscillator Classical Action

This involves the problem of interaction of a particle with a harmonic oscillator where now the potential energy:

$$V(q) = \frac{m\omega^2 q^2}{2} + \gamma(t) q \tag{5.126}$$

Here, the oscillator of mass m and frequency ω is subjected to a time-dependent force $\gamma(t)$ that changes the state of the self-consistent system. This force $\gamma(t)$ may be due to an external field, e.g. an electric field coupling via dipole interaction to a charged particle. Considering dissipative quantum mechanics, the harmonic oscillator could represent a degree of freedom of the environment under the influence of a force exerted by the system. This type of problem may imitate the interaction of an electron with phonons. Here, we have the electronic as well as the phonon subsystems.

Substitute 5.126 into 5.109 then the classical path has to satisfy the equation of motion:

$$m\ddot{q} + m\omega^2 q - \gamma(t) = 0 \tag{5.127}$$

The homogenous solution $q_H(t)$ for $\gamma(t) = 0$ is already written in 5.121 and the non-homogeneous $q_{\mathrm{NH}}(t)$ solution for $\gamma(t) \neq 0$ can be obtained by writing the full solution in the form:

$$q(t) = q_{\mathrm{H}}(t) + \frac{1}{m \sin \omega T}\int_{t_a}^{T} \mathcal{G}(t,s)\gamma(s) ds \tag{5.128}$$

Substituting this into 5.127 yields the equation for the function $\mathrm{G}(t, s)$:

$$\frac{1}{m\sin\omega T}\int_{t_a}^{T}\left(\ddot{\mathcal{G}}+\omega^2\mathcal{G}\right)\gamma(s)ds=\frac{\gamma(t)}{m} \tag{5.129}$$

From here

$$\frac{1}{\sin\omega T}\left(\ddot{\mathcal{G}}+\omega^2\mathcal{G}\right)=\delta(s-t) \tag{5.130}$$

The solution for $\mathcal{G}(t,s)$ permits to find the non-homogenous solution:

$$q_{\mathrm{NH}}(t)=-\int_0^T\varepsilon(s,t)\gamma(s)ds \tag{5.131}$$

where

$$\varepsilon(s,t)\equiv\mathcal{F}(t,s)\theta(t-s)+\mathcal{F}(s,t)\theta(s-t) \tag{5.132}$$

The solution for 5.127 is then

$$q(t)=q_{\mathrm{H}}(t)+q_{\mathrm{NH}}(t) \tag{5.133}$$

and the classical action of the forced harmonic oscillator:

$$S_{\mathrm{FHO}}=\frac{m\omega}{2\sin\omega T}\begin{bmatrix}\left(q_a^2+q_b^2\right)\cos\omega T-2q_aq_b+2\tilde{\mathcal{F}}(T)-\\-2\tilde{\mathcal{F}}_{\mathrm{NH}}(T)\end{bmatrix} \tag{5.134}$$

where

$$\tilde{\mathcal{F}}(T)=\int_0^T q_{\mathrm{H}}(s)\gamma(s)ds \tag{5.135}$$

and

$$\tilde{\mathcal{F}}_{\mathrm{NH}}(T)=\frac{1}{m\omega}\int_0^T\int_0^s\gamma(s)\mathcal{F}(s,t)\gamma(t)dtds \tag{5.136}$$

are force-driven terms contributing to the action and

$$\mathcal{F}(s,t)=\frac{\sin\omega(T-s)\sin\omega(t-t_a)}{m\omega\sin\omega T} \tag{5.137}$$

For the free oscillator $\gamma=0$ then we recover the result in 5.125 and for the free particle, $=\omega=0$, we recover the result in 5.72:

$$S_{\mathrm{cl}}[0,T]=\frac{m\omega}{2\omega T}\left[q_a^2+q_b^2-2q_aq_b\right]=\frac{m\left(q_b-q_a\right)^2}{2T} \tag{5.138}$$

5.5 Free and Driven Harmonic Oscillator Transition Amplitude

Following the same procedure as for the free particle, for the functional integral in 5.89, only those paths quite near to the classical path $\bar{q}(t)$ are important. So, we decompose the path into the classical path $\bar{q}(t)$ and the fluctuation *y*(*t*) as in 5.89, determine the classical action. The potential *V*(*q*) being a smooth function of the coordinate can be Taylor series expanded about the classical path $\bar{q}(t)$:

$$V(q) = U(\bar{q} + y) \cong U(\bar{q}) + U'(\bar{q})y + \frac{1}{2!}U''(\bar{q})y^2 + \frac{1}{3!}U'''(\bar{q})y^3 + \quad (5.139)$$

All differentials are evaluated along the classical path $\bar{q}(t)$, and the prime on *U* shows differentiation with respect to its coordinate argument. After some transformations we arrive at the following for the Lagrangian:

$$L(q,\dot{q}) = \frac{m}{2}(\dot{\bar{q}} + y)^2 - U(\bar{q} + y) \cong \left(\frac{m}{2}\dot{\bar{q}}^2 - U(\bar{q})\right) + \left(m\dot{\bar{q}}\dot{y} - U'(\bar{q})y\right) + \left(\frac{m}{2}\dot{y}^2 - \frac{1}{2!}U''(\bar{q})y^2 - \frac{1}{3!}U'''(\bar{q})y^3 - \cdots\right) \quad (5.140)$$

The contribution of the second term in the action is seen to vanish:

$$\int_0^T \left(m\dot{\bar{q}}\dot{y} - U'(\bar{q})y\right)dt = m\int_0^T \frac{d}{dt}(\dot{\bar{q}}y)dt - \int_0^T \left(m\ddot{\bar{q}} + U'(\bar{q})\right)ydt = m\dot{\bar{q}}y\Big|_0^T = 0 \quad (5.141)$$

The contribution of the first term to the action is equal to the action of a classical path that has already been evaluated for two cases mentioned above:

$$S_{cl}[0,T] = \int_0^T \left[\frac{m}{2}\dot{\bar{q}}^2 - U(\bar{q})\right]dt \quad (5.142)$$

Substituting these results into the expression of the transition amplitude 5.76 and considering 5.140, we obtain the following transition amplitude 5.58 with the contribution of the fluctuations determined from the **reduced propagator or fluctuation factor**:

$$F(T) = \int_0^0 Dy(t)\exp\left\{\frac{i}{\hbar}\int_0^T \left(\frac{m}{2}\dot{y}^2 - \frac{1}{2!}U''(\bar{q})y^2\right)dt\right\}\exp\left\{-\frac{i}{\hbar}\int_0^T \left(\frac{1}{3!}U'''(\bar{q})y^3 + \cdots\right)dt\right\} \quad (5.143)$$

The relation 5.58 is an approximate formula for the transition amplitude in the stationary phase method. The exponential factor having the classical action is the dominant contribution to the transition amplitude from the classical path. The factor F(*T*), which is the functional integral, describes the contribution from the quantum corrections of the order $o(\hbar^2)$. Generally, the relation 5.143 in the stationary phase method, imitates the standard quantum mechanics. In the small limit $\hbar \to 0$, the "**classical**" part of the transition amplitude, $\exp\left\{\frac{i}{\hbar}S_{cl}[0,T]\right\}$ oscillates rapidly when the classical action changes, while the prefactor F(*T*) depends smoothly on $\hbar$ in this limit. So, in the limit $\hbar \to 0$, it is sufficient to evaluate at least the first quantum correction that describes the complete quadratic action in *y*:

$$\int_0^T \left(\frac{m}{2}\dot{y}^2 - \frac{1}{2!}U''(\bar{q})y^2\right)dt \quad (5.144)$$

This action provides a Gaussian integral and can be evaluated to the approximation of a normalization constant. The next term in the action should be the nontrivial quantum corrections:

$$\int_0^T \left(\frac{1}{3!} U'''(\bar{q}) y^3 + \cdots \right) dt \cong o\left(\hbar^2\right) \tag{5.145}$$

This quantity in F(T) can be evaluated via the perturbation theory. So, the integrand in 5.144 is quadratic in y and, thus, the integral in 5.143 is a Gaussian and may be evaluated conveniently. In this way the stationary phase method evaluates properly the first quantum correction $\approx \hbar$ in the transition amplitude. Finally, dropping orders 5.145 higher than the second in the Taylor expansion of the action of the function F(T) contributing to the fluctuations 5.143, we obtain the **Gaussian (or quasi-classical approximation)** to the transition amplitude for the quantum oscillator 5.58 where now, the **reduced propagator or fluctuation factor** becomes

$$\mathrm{F}(T) = \int_0^0 Dy(t) \exp\left\{ \frac{i}{\hbar} \int_0^T \left(\frac{m}{2} \dot{y}^2 - \frac{1}{2!} U''(\bar{q}) y^2 \right) dt \right\} \tag{5.146}$$

5.6 Fluctuation Contribution to Transition Amplitude

The stationary phase method applied to a simple case of an oscillator when

$$U(q) = \frac{m}{2} \omega^2 q^2, \quad U'' = m\omega^2 \tag{5.147}$$

$$U^{(n)}(q) = 0, \quad n \geq 3 \tag{5.148}$$

From the conditions 5.147 and 5.148 we consider now the Gaussian path integral (**reduced propagator or fluctuation factor**) easily evaluated:

$$\mathrm{F}(T) = \int_0^0 Dy(t) \exp\left\{ \frac{i}{\hbar} \int_0^T \frac{m}{2} \left(\dot{y}^2 - \omega^2 y^2 \right) dt \right\} \tag{5.149}$$

This describes an oscillatory motion between the spaces (q_a, t_a) and (q_b, t_b) with period T and frequency ω. This permits us to Fourier series expand $y(t)$ via the harmonic frequency equal ω_n as in the case of free particle seen above:

$$y(t) = \sum_n a_n y_n(t) = \sqrt{\frac{2}{T}} \sum_n a_n \sin \omega_n t \ , \quad \omega_n = \frac{\pi n}{T} \tag{5.150}$$

with, respectively, the conditions of orthonormality and those at ends points of the path:

$$\int_{t_a}^{t_b} y_n(t) y_m(t) dt = \delta_{nm}, \ y_n(t_a) = y_n(t_b) = 0 \tag{5.151}$$

Letting $t_a = 0$ for convenience then we select the Fourier coefficients a_n as a discrete set of integration variables:

$$a_n = \int y(t) \sin \omega_n t dt \tag{5.152}$$

with each path having its own coefficients a_n and results in a set of coefficients $\{a_n\}$.

We evaluate the term proportional to the kinetic energy:

$$\int_0^T \dot{y}^2(t)dt = \int_0^T \frac{2}{T}\sum_{nn'} a_n a_{n'}\omega_n\omega_{n'} \cos\omega_n t \cos\omega_{n'} t dt \tag{5.153}$$

If $n \neq n'$, then the integral is equal to zero, while, for $n = n'$ we have

$$\int_0^T \dot{y}^2(t)dt = \sum_n \frac{2}{T} a_n^2\omega_n^2 \int_0^T \cos^2 \omega_n t dt \tag{5.154}$$

Since,

$$\int_0^T \cos^2 \omega_n t dt = \int_0^T \sin^2 \omega_n t dt \tag{5.155}$$

then

$$\int_0^T \sin^2 \omega_n t dt = \frac{1}{2}\int_0^T \left(\sin^2 \omega_n t + \cos^2 \omega_n t\right)dt = \frac{1}{2}\int_0^T dt = \frac{T}{2} \tag{5.156}$$

So

$$\int_0^T \dot{y}^2(t)dt = \sum_n a_n^2\omega_n^2 \tag{5.157}$$

Similarly, the term proportional to the potential energy:

$$\int_0^T y^2(t)dt = \int_0^T \frac{2}{T}\sum_{nn'} a_n a_{n'} \sin\omega_n t \sin\omega_{n'} t dt \tag{5.158}$$

or

$$\int_0^T y^2(t)dt = \int_0^T \frac{2}{T}\sum_n a_n^2 \sin^2 \omega_n t dt = \sum_n a_n^2 \tag{5.159}$$

and

$$\int_0^T \frac{m}{2}\left(\dot{y}^2 - \omega^2 y^2\right)dt = \frac{m}{2}\sum_n \left[\omega_n^2 - \omega^2\right]a_n^2 \tag{5.160}$$

This implies a sum over all paths, which is a sum over the coefficients$\{a_n\}$. So, the expansion of $y(t)$ in terms of eigenmodes in 5.150 is synonymous to diagonalizing the operator and path integral over all $y(t)$ equal integral over a_n. As $y \to a_n$ then the path integration reduces to an infinite product of finite-dimensional Gaussian integrals that is the path integral which is an infinite number of Fresnel integrals over a_n:

$$\int_0^0 Dy(t)\exp\left\{\frac{i}{\hbar}\int_0^T \frac{m}{2}\left(\dot{y}^2 - \omega^2 y^2\right)dt\right\} = J\prod_n \int_{-\infty}^{\infty} da_n \exp\left\{-\frac{m}{2\hbar i}\left[\omega_n^2 - \omega^2\right]a_n^2\right\} \tag{5.161}$$

Here, J is the Jacobian transformation. Thus, the **reduced propagator** (fluctuation factor):

$$\mathrm{F}(T)=J\prod_n\left(\frac{2\pi i\hbar}{m}\right)^{\frac{1}{2}}\left[\omega_n^2-\omega^2\right]^{-\frac{1}{2}} \tag{5.162}$$

To interpret this result in 5.162, we first make sense out of the infinite product that seems divergent for times equal to $\frac{\pi}{\omega}$. Besides, the value of the Jacobian J is yet to be properly determined. These difficulties can be resolved by exploiting the transition amplitude of a free particle system corresponding to that of the harmonic oscillator when $\omega = 0$. This implies, had we computed the transition amplitude for the free particle system via the path integral, we would have obtained the same Jacobian J and, particularly, an infinite product like the one in 5.162, but with $\omega = 0$. So, for motion over an integer number of periods then $q_b = q_a$ and the Jacobian J is:

$$J=\left(\frac{m}{2\pi i\hbar}\right)^{\frac{1}{2}}\frac{1}{\prod_n\left(\frac{2\pi i\hbar}{m}\right)^{\frac{1}{2}}\omega_n^{-1}} \tag{5.163}$$

So,

$$\mathrm{F}(T)=J\prod_n\left(\frac{2\pi i\hbar}{m}\right)^{\frac{1}{2}}\omega_n^{-1}\left[1-\frac{\omega^2}{\omega_n^2}\right]^{-\frac{1}{2}} \tag{5.164}$$

Using the following infinite-product formula [1,14]:

$$\prod_n\left[1-\frac{\omega^2}{\omega_n^2}\right]^{-\frac{1}{2}}=\left(\frac{\omega T}{\sin\omega T}\right)^{\frac{1}{2}} \tag{5.165}$$

then,

$$\mathrm{F}(T)=\left(\frac{m\omega}{2\pi i\hbar\sin\omega T}\right)^{\frac{1}{2}} \tag{5.166}$$

This is indeed the transition amplitude from $y(t_a) = 0$ to $y(t_b) = 0$. Due to time translation invariance the reduced propagator $\mathrm{F}(T)$ can only depend on T. For $\omega = 0$, then the result for the free particle:

$$\mathrm{F}(T)=\left(\frac{m}{2\pi i\hbar T}\right)^{\frac{1}{2}} \tag{5.167}$$

5.6.1 Maslov Correction

We consider an interesting phenomenon [11,12] that is due to a phase-jump of light by $-\frac{\pi}{2}$ after passing via a **focal point**. When a light beam is split into two parts, one part passes via a focal point and the other not. If the two partial waves are recombined, a destructive interference is observed, a phenomenon known as the **Maslov correction [11,12]** . The expressions 5.162 or 5.166 can only be valid when T is within the range

$$0 < \omega T < \pi \tag{5.168}$$

For larger time moments:

$$T = \frac{n\pi}{\omega} + \tau \quad , 0 < \tau < \frac{\pi}{\omega} \quad , n \in \mathbb{N} \tag{5.169}$$

then F(T) acquires an additional factor $\exp\left\{-i\frac{n\pi}{2}\right\}$ **[11,12]** where the so-called **Maslov-Morse index** n counts how many times the trajectory has passed through a **focal point** where the prefactor diverges. At the focal points the quantum mechanical probability amplitude achieves singularity that in real systems is prevented by anharmonic terms, so that the intensity achieves its maximal value. Spatial accumulation of such points is called a **caustic phenomenon.** This is often achieved in optics, for example when a water glass is illuminated by sunlight.

So, this can be applied to the transition amplitude for the free and driven oscillator with

$$\mathrm{K}\left(q_b,t_b;q_a,t_a\right) = \left(\frac{m\omega}{2\pi\hbar\left|\sin\omega T\right|}\right)^{\frac{1}{2}} \exp\left\{-i\frac{\pi}{4}\right\} \exp\left\{\frac{i}{\hbar} S_{\mathrm{cl}}\left[q_b,t_b;q_a,t_a\right]\right\} \exp\left\{-i\frac{n\pi}{2}\right\} \tag{5.170}$$

The absolute square of the transition amplitude gives the probability per unit length of finding the mass back to where it started at the beginning of an oscillation with period T:

$$\left|\mathrm{K}\left(q_b,t_b;q_a,t_a\right)\right|^2 = \frac{m\omega}{2\pi\hbar\sin\omega T} \tag{5.171}$$

6

Generalized Feynman Path Integration

6.1 Coordinate Representation

The transition amplitudes 4.47 and 4.82 of chapter 4 may be generalized for a problem with s degrees of freedom that may be described by a set of operators $\hat{q} = \left\{\hat{q}_\alpha\right\}_{\alpha=1}^{s}, \hat{p} = \left\{\hat{p}_\alpha\right\}_{\alpha=1}^{s}$:

$$K\left(q'',t'';q',t'\right) = \int \prod_{\alpha=1}^{s} Dp_\alpha Dq_\alpha \exp\left\{\frac{i}{\hbar}\int_{t'}^{t''}\left[\sum_{\alpha=1}^{s} p_\alpha \dot{q}_\alpha - H\left(p,q\right)\right]dt\right\} \tag{6.1}$$

Here, the Hamiltonian:

$$H\left(p,q\right) = \sum_{\alpha=1}^{s} \frac{p_\alpha^2}{2m} + U\left(q\right) \tag{6.2}$$

We evaluate the integral 4.1 over the momenta and this gives the Feynman transition amplitude:

$$K\left(q'',t'';q',t'\right) = \int \prod_{\alpha} Dq_\alpha \exp\left\{\frac{i}{\hbar}\int_{t'}^{t''} L\left(q_\alpha,\dot{q}_\alpha,t\right)dt\right\} \tag{6.3}$$

The functional measures in equations 6.1 and 6.3 are defined respectively as in equations 4.49 and 4.70 of chapter 4.

We examine the transformation from the Hamiltonian to the Feynman transition amplitude by generalizing the Hamiltonian in 6.2. Consider an example of a system with s degrees of freedom described by the following Lagrangian quadratic in $\dot{q}_\alpha$:

$$L\left(q_\alpha,\dot{q}_\alpha\right) = \frac{1}{2}\sum_{\alpha,\beta=1}^{s} \dot{q}_\alpha \mathbf{m}_{\alpha\beta}\left(q\right)\dot{q}_\beta + \sum_{\alpha=1}^{s} b_\alpha\left(q\right)\dot{q}_\alpha - U\left(q\right) \tag{6.4}$$

Here, the real nondegenerate matrix $\mathbf{m}$, is symmetrical relative to transposition:

$$\mathbf{m}_{\alpha\beta} = \mathbf{m}_{\beta\alpha}\ ,\ \mathbf{m}^{\mathrm{T}} = \mathbf{m} \tag{6.5}$$

The matrix, $\mathbf{m}$, in the given Lagrangian 6.4, does not only describe the general case of the quadratic expression for the given variable but as well its derivative. For the relativistic field system, where the variable q is understood as the field degree of freedom, the non-degenerate matrix $\mathbf{m}_{\alpha\beta}$ describes the gauge of the field.

From equation 6.4 we construct the Hamiltonian of the system via the canonically conjugate coordinates q_α relating canonical momenta:

$$p_\alpha = \frac{\partial L}{\partial \dot{q}_\alpha} = \sum_\beta \mathbf{m}_{\alpha\beta}\dot{q}_\alpha + b_\alpha \tag{6.6}$$

Introduce the following matrices:

$$\mathrm{q} \equiv \begin{bmatrix} q_1 \\ q_2 \\ \vdots \\ q_s \end{bmatrix}, \quad \mathrm{p} \equiv \begin{bmatrix} p_1 \\ p_2 \\ \vdots \\ p_s \end{bmatrix}, \quad \mathbf{b} \equiv \begin{bmatrix} b_1 \\ b_2 \\ \vdots \\ b_s \end{bmatrix} \tag{6.7}$$

then

$$\mathrm{p} = \mathbf{m}\dot{\mathrm{q}} + \mathbf{b} \tag{6.8}$$

If the matrix **m** has an inverse then

$$\mathbf{m}\mathbf{m}^{-1} = \hat{\mathbf{I}} \tag{6.9}$$

and the speed in the matrix form is represented:

$$\dot{\mathrm{q}} = \mathrm{m}^{-1}(\mathrm{p} - \mathrm{b}) \tag{6.10}$$

We now conveniently arrive at the following Hamiltonian:

$$\mathrm{H}(\mathrm{p},\mathrm{q}) = \mathrm{p}\dot{\mathrm{q}} - L = \frac{1}{2}\tilde{\mathrm{p}}^{\mathrm{T}}\mathbf{m}^{-1}\tilde{\mathrm{p}} + U(\mathrm{q}) \tag{6.11}$$

where $\tilde{\mathrm{p}}^{\mathrm{T}}$ denoting the transposed of the matrix $\tilde{\mathrm{p}}$ and

$$\tilde{\mathrm{p}} \equiv \mathrm{p} - \mathbf{b}(\mathrm{q}) \tag{6.12}$$

We rewrite the transition amplitude of the quantum system in the Hamilton form:

$$\mathrm{K}(\mathrm{q}'',t'';\mathrm{q}',t') = \int D\mathrm{p}D\mathrm{q}\exp\left\{\frac{i}{\hbar}\int_{t'}^{t''}\left[\mathrm{p}^{\mathrm{T}}\dot{\mathrm{q}} - \mathrm{H}(\mathrm{p},\mathrm{q})\right]dt\right\} \tag{6.13}$$

where,

$$D\mathrm{p}D\mathrm{q} \equiv \prod_{\alpha=1}^{s} Dp_\alpha Dq_\alpha \tag{6.14}$$

The evaluation of the functional integral may be done based on the functional analogue of the change of variables in the integrand:

$$\tilde{\mathrm{p}}(t) = \mathrm{p}(t) - \mathbf{b}(\mathrm{q}), \quad D\mathrm{p}(t) = D\tilde{\mathrm{p}}(t) \tag{6.15}$$

This yields the following transition amplitude:

$$K(q_b,t_a;q_a,t_a) = \int Dq \exp\left\{\frac{i}{\hbar}\int_{t_a}^{t_b}\left[\mathbf{b}^T\dot{q} - U(q)\right]dt\right\}\int D\tilde{p} \exp\left\{\frac{i}{\hbar}\int_{t_a}^{t_b}\left[-\frac{1}{2}\tilde{p}^T\mathbf{m}^{-1}\tilde{p} + \tilde{p}^T\dot{q}\right]dt\right\} \tag{6.16}$$

We evaluate the integral over the momenta with the aid of the following:

$$\lim_{\substack{\varepsilon\to 0\\ N\to\infty}}\int\prod_{k=1}^{N-1}\frac{d\tilde{p}_k}{(2\pi\hbar)^s}\exp\left\{\frac{i\varepsilon}{\hbar}\sum_{k=1}^{N-1}\left[\tilde{p}_k^T\left(\frac{q_{k+1}-q_k}{\varepsilon}\right) - \frac{1}{2}\tilde{p}_k^T\mathbf{m}^{-1}\tilde{p}_k\right]\right\} \tag{6.17}$$

The evaluation of each of the above multiple integrals yields the following:

$$\int\frac{dp_k}{(2\pi\hbar)^s}\exp\left\{\frac{i\varepsilon}{\hbar}\left[p_k^T\left(\frac{q_{k+1}-q_k}{\varepsilon}\right) - \frac{1}{2}p_k^T\mathbf{m}^{-1}p_k\right]\right\} = \left(\frac{1}{2\pi i\hbar\varepsilon}\right)^{\frac{s}{2}}\left[\det\mathbf{m}^{-1}\right]^{-\frac{1}{2}}$$
$$\exp\left\{\frac{i\varepsilon}{\hbar}\left[\frac{1}{2}\left(\frac{q_{k+1}-q_k}{\varepsilon}\right)^T\mathbf{m}\left(\frac{q_{k+1}-q_k}{\varepsilon}\right)\right]\right\} \tag{6.18}$$

Thus, we arrive at the Feynman transition amplitude:

$$K(q_b,t_b;q_a,t_a) = \int Dq(t)\exp\left\{\frac{i}{\hbar}\int_{t'}^{t''}L(q,\dot{q})dt\right\} \tag{6.19}$$

Here, now

$$Dq(t) = \lim_{\substack{\varepsilon\to 0\\ N\to\infty}}\frac{1}{A}\prod_{i=1}^{N-1}\frac{dq_i}{A_i} \tag{6.20}$$

and

$$\frac{1}{A_i} = \frac{1}{(2\pi i\hbar\varepsilon)^{\frac{s}{2}}}\prod_{q_\alpha(t)}\left[\det\mathbf{m}^{-1}(q_\alpha(t))\right]^{-\frac{1}{2}} \tag{6.21}$$

So, for the matrix representation 6.7, the Hamiltonian 6.11 permits the transformation of the Feynman transition amplitude equation 6.3 to the modified form 6.13. The quantity $\left[\det\mathbf{m}^{-1}\right]^{-\frac{1}{2}}$ is common for the Feynman transition amplitude transformation seen in reference [8].

6.2 Free Particle Transition Amplitude

We apply the generalized Feynman representation to the case of a free particle with the Lagrangian:

$$L(\dot{q}) = \frac{1}{2}\dot{q}\mathbf{m}\dot{q} \tag{6.22}$$

The transition amplitude for a free particle can be written via Feynman path integration with the help of relations 4.6, 4.21 and 4.22 of chapter 4:

$$K(q,t;q_0,t_0) \equiv \langle q|\hat{U}(t,t_0)|q_0\rangle = \int Dq \exp\left\{\frac{i}{\hbar}S(q,t,t_0)\right\} \tag{6.23}$$

Here, the integration measure:

$$Dq = \lim_{\substack{\max|\Delta t_k|\to 0 \\ N\to\infty}} \frac{1}{A}\prod_{k-1}^{N-1}\frac{dq_k}{A_i} \tag{6.24}$$

and the action functional $S(q_N, t, t_0)$:

$$S(q_N,t,t_0) = \frac{1}{2}\sum_{k=0}^{N-1}\frac{1}{\Delta t_k}(q_{k+1}-q_k)^T \mathbf{m}(q_{k+1}-q_k) \tag{6.25}$$

The transition amplitude 6.23 considering 6.25 represents a set of Gaussian integrals. The integral of a Gaussian is a Gaussian. So, the integrations in 6.23 can be carried out on one variable after the other and when completed then the limit $N\to\infty$ is applied. We therefore start first when $k=0$ and we perform integration over q_1 then computing the integral

$$\int \exp\left\{\frac{i}{2\hbar}\sum_{k=0}^{1}\frac{1}{\Delta t_k}(q_{k+1}-q_k)^T \mathbf{m}(q_{k+1}-q_k)\right\}\frac{1}{A}\frac{dq_1}{A_1} \tag{6.26}$$

yields

$$\left[\det m\frac{1}{(2\pi i\hbar 2\Delta t)^s}\right]^{\frac{1}{2}} \exp\left\{\frac{i}{2\hbar 2\Delta t}(q_2-q_0)\mathbf{m}(q_2-q_0)\right\} \tag{6.27}$$

This procedure is continued similarly for $q_2, \cdots, q_{N-1}$ and we see a recursion process is established, so that after $(N-1)$ steps we have:

$$K(q,t;q_0,t_0) = \left[\det\mathbf{m}\frac{1}{(2\pi i\hbar T)^s}\right]^{\frac{1}{2}} \exp\left\{\frac{i}{2\hbar T}(q-q_0)\mathbf{m}(q-q_0)\right\} \tag{6.28}$$

where

$$t - t_0 = T \tag{6.29}$$

Relation 6.28 coincides with the quantum mechanical transition amplitude and the phase of the exponential factor is the classical action along the classical path, with the condition that $q(0) = q_0$ and $q(T) = q$.

6.3 Gaussian Functional Feynman Path Integrals

We consider the following generalized Feynman path integral:

$$\langle q|\hat{U}(t,t_0)|q_0\rangle = \int Dq \exp\left\{\frac{i}{\hbar}\int_{t_0}^{t} L(q,\dot{q})dt'\right\} = \int Dq \exp\left\{\frac{i}{\hbar}S(q,t,t_0)\right\} \tag{6.30}$$

where, the **Lagrangian:**

$$L(q,\dot{q}) = \frac{1}{2}\dot{q}\mathbf{m}\dot{q} - q\mathbf{b}\dot{q} - \frac{1}{2}q\mathbf{k}q + \mathbf{f}(t')q \tag{6.31}$$

is **quadratic both in the set of generalized coordinates** q **as well as generalized velocities** $\dot{q}$ and $\mathbf{f}(t')$ **is the driving force. Such an integral in 6.30 is therefore Gaussian** and so may be exactly solvable analytically with

$$q(t_0) \equiv q_0 \; , \; q(t) \equiv q \tag{6.32}$$

being, respectively, the starting and ending coordinates of the path. All the matrices in 6.31 are real-valued with **k** and **b** being respectively symmetric and anti-symmetric matrices.

We write the Euler-Lagrange equation described by the relation in 3.30 of chapter 3 with the help of 6.31 in the matrix form:

$$\mathbf{m}\ddot{q} + 2\mathbf{b}\dot{q} + \mathbf{k}q = \mathbf{f}(t') \tag{6.33}$$

The solution of this equation gives the classical path q_{cl} with the property of extremizing the action S.

We observe two non-quadratic terms in the Lagrangian described in 6.31. So, it is necessary to transform the Lagrangian to quadratic form for the integral in 6.30 to be easily evaluated. We consider the quasi-classical approximation via the method of stationary phase method as mentioned earlier. For this, we assume that only those paths quite near to the classical path $q_{cl}(t)$ are important. So, from the quasi-classical approximation, we select a fluctuating path $q'(t')$ and interpreted as the quantum fluctuations around the classical path $q_{cl}(t')$:

$$q(t') = q_{cl}(t') + q'(t') \tag{6.34}$$

So, the classical action:

$$S(q_{cl},t,t_0) = \frac{1}{2}\int_{t_0}^{t} q_{cl}(t')\mathbf{f}(t')dt' + \frac{1}{2}\left[q_{cl}(t)\mathbf{m}\dot{q}_{cl}(t) - q_{cl}(t_0)\mathbf{m}\dot{q}_{cl}(t_0)\right] \tag{6.35}$$

Then considering 6.34 we have

$$\langle q|\hat{U}(t,t_0)|q_0\rangle = F(T)\exp\left\{\frac{i}{\hbar}S(q_{cl}(t'),t,t_0)\right\} \tag{6.36}$$

where, the Gaussian integral around the classical path is the **reduced propagator**:

$$F(T) = \int_0^0 Dq'(t')\exp\left\{\frac{i}{\hbar}S(q'(t'),t,t_0)\right\} \tag{6.37}$$

The interpretation of the expression in 6.36 is that the quantum transition amplitude of a particle can be separated into two factors:

- The first factor $\exp\left\{\frac{i}{\hbar}S\left(q_{cl}\left(t'\right),t,t_0\right)\right\}$ is written in terms of the action of the classical path q_{cl}. Indeed, the first factor contains often most of the useful information contained in the propagator.
- The second factor F(T) is the path integral over fluctuations $q'(t')$ from the classical path q_{cl}. This second factor is independent of the initial $q(t_0)$ and final points $q(t)$.

So, the quasi-classical limit should then correspond to the possibility to reduce the path integral to a Gaussian integral. Hence our problem is reduced to determining the classical action, once relation 6.36 is shown to be true for the system. This follows that the normalization F(T) of the propagator becomes the only challenging task.

Setting $\mathbf{f}\left(t'\right)\equiv 0$ in equation 6.36 and considering $q_0 = q$, we then integrate both sides of the resultant equation by q and this yields the **reduced propagator**:

$$F\left(T\right)=\frac{F_1\left(T\right)}{F_2\left(T\right)} \tag{6.38}$$

where,

$$F_1\left(T\right)=\int dq\int_{q}^{q}Dq\,\exp\left\{\frac{i}{\hbar}S\left(q,t,t_0\right)\right\} \tag{6.39}$$

and

$$F_2\left(T\right)=\int dq\,\exp\left\{\frac{i}{\hbar}S\left(q_{cl}\left(t'\right),t,t_0\right)\right\} \tag{6.40}$$

The **reduced propagator** 6.37 or 6.38 describes a cyclical path from q and back to q with period T and frequency ω_n and permits us to Fourier series expand the generalized coordinates q via the fundamental harmonic frequency equal to $\frac{2\pi}{T}$:

$$q\left(t'\right)=2\,\mathrm{Re}\sum_{n}\mathbf{a}_n\exp\left\{i\omega_n\left(t'-t_0\right)\right\}+\mathbf{y} \tag{6.41}$$

with the frequency

$$\omega_n=\frac{2\pi n}{T},n=1,2,\cdots, \tag{6.42}$$

The fluctuation vector satisfies cyclical boundary conditions and is expressed:

$$\mathbf{y}=q-2\,\mathrm{Re}\sum_{n}\mathbf{a}_n \tag{6.43}$$

The Fourier coefficients $\mathbf{a}_n$ will be observed to serve as a discrete set of integration variables with each path having its own coefficients $\mathbf{a}_n$. The expansion of $\mathbf{y}$ in terms of the Fourier coefficients $\mathbf{a}_n$ will result in the diagonalizing of the integrand of the action which is Lagrangian and path integral over all $\mathbf{y}$ equal integral over $\mathbf{a}_n$.

From

$$\int_{t_0}^{t} \exp\{i(\omega_n - \omega_m)(t' - t_0)\} dt' = T\delta_{nm} \tag{6.44}$$

then substituting 6.41 into 6.38 and considering the Hermitian matrix

$$\mathbf{M}^{\dagger} = \mathbf{M} \tag{6.45}$$

denoted by

$$\mathbf{M} = \omega_n^2 \mathbf{m} - 2i\omega_n \mathbf{b} - \mathbf{k} \tag{6.46}$$

then we have the following quadratic action functional as a result of the diagonalizing procedure:

$$S(\mathrm{q},t,t_0) = T\left[\sum_n \mathbf{a}_n^{\mathrm{T}} \mathbf{M} \mathbf{a}_n - \frac{1}{2}\mathbf{y}^{\mathrm{T}} \mathbf{k} \mathbf{y}\right] \tag{6.47}$$

We consider extremizing the action function and we find the Fourier coefficient corresponding to the classical path:

$$\mathbf{a}_{n\mathrm{cl}}^{\mathrm{T}} = -\mathbf{y}_{\mathrm{cl}} \mathbf{k} \mathbf{M}^{-1}, \ \mathbf{a}_{n\mathrm{cl}} = -\mathbf{M}^{-1} \mathbf{k} \mathbf{y}_{\mathrm{cl}} \tag{6.48}$$

Substituting into 6.41 and letting

$$\mathbf{x} = \hat{\mathbf{1}} - \mathbf{o}\mathbf{k}, \quad \mathbf{o} = \sum_n \left[\mathbf{M}^{-1} + \left(\mathbf{M}^{-1}\right)^{\mathrm{T}}\right] \tag{6.49}$$

then

$$\mathbf{y}_{\mathrm{cl}} = \mathbf{x}^{-1} \mathrm{q} \tag{6.50}$$

From equations 6.48 to 6.50 then the action functional 6.47 for the classical path:

$$S(\mathrm{q}_{\mathrm{cl}},t,t_0) = -\frac{1}{2} T \mathrm{q} \mathbf{x}^{-1} \mathbf{k} \mathrm{q} \tag{6.51}$$

This permits us to elegantly compute $\mathrm{F}_2(T)$:

$$\mathrm{F}_2(T) = \left[\left(\frac{2\pi\hbar}{iT}\right)^s \det \mathbf{x} \times \det \mathbf{k}^{-1}\right]^{\frac{1}{2}} \tag{6.52}$$

We find now $\mathrm{F}_1(T)$ by time-slicing the time axis:

$$\mathrm{q}(t') = \lim_{N \to \infty} \mathrm{q}^{(N)}(t') \tag{6.53}$$

with

$$\mathrm{q}^{(N)}(t') = 2\,\mathrm{Re} \sum_n \mathbf{a}_n \exp\{i\omega_n (t' - t_0)\} + \mathbf{y}^{(N)} \tag{6.54}$$

and

$$\mathbf{y}^{(N)} = \mathrm{q} - 2\mathrm{Re}\sum_n \mathbf{a}_n \tag{6.55}$$

From 6.53 to 6.55 we compute the path integral in 6.39:

$$\Phi(T) \equiv \int_{\mathrm{q}_0}^{\mathrm{q}} D\mathrm{q}\exp\left\{\frac{i}{\hbar}S(\mathrm{q},t,\mathrm{t}_0)\right\} = \lim_{N\to\infty}\Phi^{(N)}(T) \tag{6.56}$$

where, the integral over all Fourier coefficients $\mathbf{a}_n$:

$$\Phi^{(N)}(T) \equiv \int \exp\left\{\frac{i}{\hbar}S\left(\mathrm{q}^{(N)}(t'),t,\mathrm{t}_0\right)\right\}\mathrm{J}\prod_n \lambda_n d\,\mathrm{Re}\,\mathbf{a}_n d\mathrm{Im}\,\mathbf{a}_n \tag{6.57}$$

and the factor $\mathrm{J}\prod_n \lambda_n$ is the Jacobian transformation for the manifold integral.

For a free particle then

$$\mathbf{b} = \mathbf{k} = 0 \tag{6.58}$$

and

$$\Phi(T)\Big|_{\mathbf{b}=\mathbf{k}=0} \equiv \left[\frac{1}{(2\pi i\hbar T)^s}\det\mathbf{m}\right]^{\frac{1}{2}} \tag{6.59}$$

So, the following quadratic action functional:

$$S\left(\mathrm{q}^{(N)}(t'),t,\mathrm{t}_0\right) = T\left[\sum_n \mathbf{a}_n^{\mathrm{T}}\mathbf{M}\mathbf{a}_n - \frac{1}{2}\left(\mathbf{y}^{\mathrm{T}}\right)^{(N)}\mathbf{k}\mathbf{y}^{(N)}\right] \tag{6.60}$$

and for fixed $\mathbf{y}^{(N)}$ then 6.56 becomes:

$$\Phi^{(N)}(T) = \exp\left\{-\frac{iT}{2\hbar}\left(\mathbf{y}^{\mathrm{T}}\right)^{(N)}\mathbf{k}\mathbf{y}^{(N)}\right\}\mathrm{J}\prod_n \lambda_n\left(\frac{i\pi\hbar}{T}\right)^s \det\mathbf{M}^{-1} \tag{6.61}$$

From 6.46 and 6.58 then

$$\Phi^{(N)}(T)\Big|_{\mathbf{b}=\mathbf{k}=0} \equiv \mathrm{J}\prod_n \lambda_n\left(\frac{i\pi\hbar}{T\omega_n^2}\right)^s \det\mathbf{m}^{-1} \tag{6.62}$$

So, equation 6.56 is satisfied when the Jacobian J takes the value

$$\mathrm{J} = \left[\frac{1}{(2\pi i\hbar T)^s}\det\mathbf{m}\right]^{\frac{1}{2}},\quad \lambda_n = \left(\frac{T\omega_n^2}{i\pi\hbar}\right)^s \det\mathbf{m} \tag{6.63}$$

So,

$$\Phi^{(N)}(T) = \left[\frac{1}{(2\pi i\hbar T)^s}\det\mathbf{m}\right]^{\frac{1}{2}} \exp\left\{-\frac{iT}{2\hbar}\left(\mathbf{y}^{\mathrm{T}}\right)^{(N)}\mathbf{k}\mathbf{y}^{(N)}\right\}\prod_n\left[\omega_n^{2s}\det\mathbf{m}\times\det\mathbf{M}^{-1}\right] \tag{6.64}$$

and

$$F_1(T) = \left[\frac{1}{(iT)^{2s}}\det\mathbf{m}\times\det\mathbf{k}^{-1}\right]^{\frac{1}{2}}\prod_n\left[\omega_n^{2s}\det\mathbf{m}\times\det\mathbf{M}^{-1}\right] \tag{6.65}$$

with

$$F(T) = \left[\frac{1}{(2\pi i\hbar T)^s}\det\mathbf{m}\times\det\mathbf{x}^{-1}\right]^{\frac{1}{2}}\prod_n\left[\omega_n^{2s}\det\mathbf{m}\times\det\mathbf{M}^{-1}\right] \tag{6.66}$$

We compute the product in 6.66 supposing the matrix **m** to be positive definite. So, **v** is a symmetric and positive definite matrix:

$$\mathbf{v} = \mathbf{m}^{\frac{1}{2}} \tag{6.67}$$

From here, considering 6.46 then the following Hermitian matrix:

$$\mathbf{M}' = \omega_n^2\hat{\mathbf{1}} - 2i\omega_n\mathbf{b}' - \mathbf{k}' \tag{6.68}$$

and consequently,

$$\mathbf{x}' = \hat{\mathbf{1}} - \mathbf{o}'\mathbf{k}' \tag{6.69}$$

where,

$$\{\cdots\}' = \mathbf{v}\{\cdots\}\mathbf{v}^{-1} \tag{6.70}$$

From 6.67 to 6.70 it is obvious that

$$\det\mathbf{M}' = \det\left[\mathbf{M}'\right]^* = \det\mathbf{M}\times\det\mathbf{m}^{-1} \tag{6.71}$$

$$\det\mathbf{x}' = \det\mathbf{x} \tag{6.72}$$

Then,

$$F(T) = \left[\frac{1}{(2\pi i\hbar T)^s}\det\mathbf{m}\times\det\left[\mathbf{x}'\mathbf{u}\right]^{-1}\right]^{\frac{1}{2}} \tag{6.73}$$

Here,

$$\mathbf{u} = \prod_n \omega_n^{-4} \mathbf{M}' \left[\mathbf{M}' \right]^* \tag{6.74}$$

and differentiating this by $\mathbf{k}'$ then

$$\frac{\partial}{\partial \mathbf{k}'} \mathbf{u} = -\mathbf{o}' \tag{6.75}$$

Substituting this into 6.69 then

$$\mathbf{x}' = \hat{\mathbf{I}} + \frac{\partial}{\partial \mathbf{k}'} \ln \mathbf{u}\mathbf{k}' \tag{6.76}$$

Considering 6.68 and 6.74 then

$$\mathbf{u} = \prod_n \prod_{k=\pm} \left[\hat{\mathbf{I}} - \frac{\left(\mathbf{k}' - \mathbf{b}'^2 \right)}{\left(\omega_n \hat{\mathbf{I}} + ki\mathbf{b}' \right)^2} \right] \left[\hat{\mathbf{I}} + \frac{1}{\omega_n^2} \mathbf{b}' \right]^2 \tag{6.77}$$

This product can be simplified via the formula:

$$\prod_n \prod_{k=\pm} \left[\left(1 - \left(\frac{x}{n\pi + ky} \right)^2 \right) \left(1 - \left(\frac{y}{n\pi} \right)^2 \right) \right]^2 = \frac{\cos 2x - \cos 2y}{2\left(y^2 - x^2 \right)} \tag{6.78}$$

and consequently,

$$\mathbf{u} = \frac{2}{T} \mathbf{k}'^{-1} \left[\cosh \mathbf{b}'T - \cos \mathbf{w}T \right], \ \mathbf{w} = \left(\mathbf{k}' - \mathbf{b}'^2 \right)^{\frac{1}{2}} \tag{6.79}$$

Considering 6.76, then

$$\mathbf{x}' = \frac{1}{T} \mathbf{w} \sin \mathbf{w}T \tag{6.80}$$

and consequently,

$$\mathrm{F}(T) = \left[\frac{1}{\left(2\pi i\hbar T \right)^s} \det \mathbf{m} \times \det \mathbf{x}'^{-1} \right]^{\frac{1}{2}} \tag{6.81}$$

This is indeed the transition amplitude describing a cyclical path from q and back to q with period T and scaled frequency $\mathbf{w}$.

This problem may be applied to the case of a harmonic oscillator of frequency ω and driving force $\gamma(t)$ under chapter 5 where we set

$$\mathbf{k} = \mathrm{m}\omega^2, \mathbf{k}' = \omega^2, \ \mathbf{b} = \mathbf{b}' = 0 \tag{6.82}$$

6.4 Charged Particle in a Magnetic Field

The problem of a charged particle in a magnetic field was first solved by L.D. Landau in 1930 in the Schrodinger theory [15, 16]. We consider the case of the magnetic field with the magnetic field induction $\vec{B}$ pointing along the z -axis direction:

$$\vec{B}=\left(0,0,B_z\right) \tag{6.83}$$

The Lagrangian for such a charged particle e in the uniform magnetic field and subjected to the force $\vec{\gamma}(t)$:

$$L\left(\vec{r},\dot{\vec{r}},t\right)=\frac{1}{2}m\dot{\vec{r}}^2+\frac{eB}{2c}\left(\dot{y}x-\dot{x}y\right)+\vec{\gamma}(t)\vec{r} \tag{6.84}$$

Compare equation 6.84 with 6.31 and apply these equations to 6.70 then

$$\mathbf{k}=\mathbf{k}'=0,\mathbf{b}=\frac{eB}{2c}\begin{bmatrix}0&-1&0\\1&0&0\\0&0&0\end{bmatrix},\mathbf{b}'=\omega_B\begin{bmatrix}0&-1&0\\1&0&0\\0&0&0\end{bmatrix} \tag{6.85}$$

The expression , can be obtained:

$$\mathbf{w}=\frac{\omega_c}{2}\begin{bmatrix}1&0&0\\0&1&0\\0&0&0\end{bmatrix} \tag{6.86}$$

Here, we introduce the field-dependent magnetic frequency:

$$\omega_B=\frac{\omega_c}{2} \tag{6.87}$$

with ω_c, being the **Landau frequency** or **cyclotron frequency** defined as

$$\omega_c=\frac{eB}{mc} \tag{6.88}$$

Here, c is the speed of light. To calculate the reduced propagator F(T), we substitute 6.86 into 6.80 and consequently 6.81:

$$F(T)=\left[\left(\frac{m}{2\pi i\hbar T}\right)^3\left(\frac{\omega_c T}{2\sin\frac{\omega_c T}{2}}\right)^2\right]^{\frac{1}{2}} \tag{6.89}$$

From the Lagrangian in 6.84 we can construct a Feynman path integral that will be Gaussian and exactly solvable analytically with

$$\vec{r}(t_0)\equiv\vec{r}_0\ ,\ \vec{r}(t)\equiv\vec{r} \tag{6.90}$$

being respectively, the starting and ending position vectors. We write the Euler-Lagrange equation described by the relation in 3.30 of chapter 3 via 6.84:

$$\ddot{\vec{r}}\mathbf{m} + 2\dot{\vec{r}}\mathbf{b} = \vec{\gamma}(t) \tag{6.91}$$

The solution of this equation gives the classical path $\vec{r}_{\text{cl}}$ with the property of extremizing the action S:

$$\vec{r}_{\text{cl}}(t) = \vec{r}_0 \mathbf{t}_0^{\text{T}}(T-t) + \vec{r}\mathbf{t}^{\text{T}}(t) - \int_{t_0}^{T} \vec{\gamma}(s)\left[\theta(s-t)\mathbf{f}^{\text{T}}(s,t) + \theta(t-s)\mathbf{f}(t,s)\right]ds \tag{6.92}$$

Here, $\theta(s-t)$ is the Heaviside step function and the antisymmetric matrices:

$$\mathbf{t}^{\text{T}}(t) = \begin{bmatrix} \text{F}^{11}(t) & \text{F}^{12}(t) & 0 \\ \text{F}^{21}(t) & \text{F}^{22}(t) & 0 \\ 0 & 0 & \text{F}^{33}(t) \end{bmatrix}, \quad \mathbf{t}_0^{\text{T}}(T-t) = \begin{bmatrix} \text{F}_0^{11}(t) & \text{F}_0^{12}(t) & 0 \\ \text{F}_0^{21}(t) & \text{F}_0^{22}(t) & 0 \\ 0 & 0 & \text{F}_0^{33}(t) \end{bmatrix} \tag{6.93}$$

and

$$\mathbf{f}^{\text{T}}(s,t) = \begin{bmatrix} \mathcal{F}^{11}(s,t) & \mathcal{F}^{12}(s,t) & 0 \\ \mathcal{F}^{21}(s,t) & \mathcal{F}^{22}(s,t) & 0 \\ 0 & 0 & \mathcal{F}^{33}(s,t) \end{bmatrix} \tag{6.94}$$

with

$$\text{F}^{11}(t) = \text{F}^{22}(t) = \frac{\sin\frac{\omega_c}{2}(t-t_0)\cos\frac{\omega_c}{2}(T-t)}{\sin\frac{\omega_c T}{2}},$$

$$\text{F}^{12}(t) = -\text{F}^{21}(t) = -\text{F}_0^{12}(t) = \text{F}_0^{21}(t) = \frac{\sin\frac{\omega_c}{2}(t-t_0)\sin\frac{\omega_c}{2}(T-t)}{\sin\frac{\omega_c T}{2}} \tag{6.95}$$

$$\text{F}_0^{11}(t) = \text{F}_0^{22}(t) = \frac{\sin\frac{\omega_c}{2}(T-t)\cos\frac{\omega_c}{2}(t-t_0)}{\sin\frac{\omega_c T}{2}}, \quad \text{F}^{33}(t) = \frac{t-t_0}{T}, \text{F}_0^{33}(t) = \frac{T-t}{T} \tag{6.96}$$

$$\text{F}^{11}(s,t) = \text{F}^{22}(s,t) = \frac{2\sin\frac{\omega_c}{2}(T-t)\sin\frac{\omega_c}{2}(s-t_0)\cos\frac{\omega_c}{2}(t-s)}{m\omega_c \sin\frac{\omega_c T}{2}},$$

$$\text{F}^{12}(s,t) = -\text{F}^{21}(s,t) = \frac{2\sin\frac{\omega_c}{2}(T-t)\sin\frac{\omega_c}{2}(s-t_0)\sin\frac{\omega_c}{2}(t-s)}{m\omega_c \sin\frac{\omega_c T}{2}},$$

$$\text{F}^{33}(s,t) = \frac{(T-t)(s-t_0)}{mT} \tag{6.97}$$

From here, then substituting the classical path $\vec{r}_{\text{cl}}(t)$ in equation 6.92 into 6.35 then we have the action functional:

$$S(\vec{r}_{\text{cl}},t,t_0)=S_{\perp}(\vec{r}_{\text{cl}},t,t_0)+\frac{m}{2T}(z-z_0)^2+\int_{t_0}^{T}\vec{r}_{\text{H}}(t)\vec{\gamma}(t)dt-m\omega_c\tilde{\mathbf{f}}_{\text{NH}}(T) \tag{6.98}$$

where the orthogonal part of the action:

$$S_{\perp}(\vec{r}_{\text{cl}},t,t_0)=\frac{m\omega_c}{4}\left[2(yx_0-xy_0)+\left[(x-x_0)^2+(y-y_0)^2\right]\cot\left(\frac{\omega_c T}{2}\right)\right] \tag{6.99}$$

and

$$\vec{r}_{\text{H}}(t)=\vec{r}\mathbf{t}^{\text{T}}(t)+\vec{r}_0\mathbf{t}_0^{\text{T}}(T-t) \tag{6.100}$$

$$\tilde{\mathbf{f}}_{\text{NH}}(T)=\frac{1}{m\omega_c}\int_{t_0}^{T}\int_{t_0}^{t}\vec{\gamma}(t)\mathbf{f}(t,s)\vec{\gamma}(s)dsdt \tag{6.101}$$

The magnetic interaction of a particle of charge e for the 3D case has the transition amplitude:

$$\text{K}(\vec{r},t;\vec{r}_0,t_0)=\left[\left(\frac{m}{2\pi i\hbar T}\right)^3\left(\frac{\omega_c T}{2\sin\frac{\omega_c T}{2}}\right)^2\right]^{\frac{1}{2}}\exp\left\{\frac{i}{\hbar}S(\vec{r}_{\text{cl}},t,t_0)\right\} \tag{6.102}$$

7

From Path Integration to the Schrödinger Equation

We show how from the path integral we rederive the Schrödinger equation. This will confirm the Feynman path integration to be the formulation of quantum mechanics presenting an alternative way of examining the subject matter than the usual approaches with applications as vast as those of quantum mechanics itself.

7.1 Wave Function

We show how the wave function $\Psi(q, t) = \langle q|\Psi(t)\rangle$ is the total amplitude for the particle to be found at (q, t), arriving from the past in some situation relates the transition amplitude $K(q_b, t_b; q_a, t_a)$ describing a particle coming from (q_a, t_a) and arriving at (q_b, t_b). The transition amplitude 4.98 relates $\langle q_b|\Psi(t_b)\rangle$ when t_c the intermediate time falls between the times t_a and t_b:

$$\langle q_b|\Psi(t_b)\rangle \equiv \Psi(q_b, t_b) = \int dq_c K(q_b, t_b; q_c, t_c) K(q_c, t_c; q_a, t_a) \tag{7.1}$$

Since the states have delta-normalization, then

$$\Psi(q_b, t_b) = \int dq \delta(q_b - q) K(q, t_b; q_a, t_a) = K(q_b, t_b; q_a, t_a) \tag{7.2}$$

so, the transition amplitude $K(q_b, t_b; q_a, t_a)$ plays the role of the wave function when the initial condition is rigorously defined and this wave function must obey the following integral equation:

$$\Psi(q_b, t_b) = \int dq_c K(q_b, t_b; q_c, t_c) \Psi(q_c, t_c) \tag{7.3}$$

In fact, equation 7.3 is an integral equation of quantum mechanics, which is equivalent to the following time-dependent Schrödinger differential equation that we show later:

$$\hat{H}\Psi(q, t) = i\hbar \frac{\partial}{\partial t} \Psi(q, t) \tag{7.4}$$

So, the transition amplitude K is the evolution operator in the representation of functional integrals and is also the wave function that has all the history of incident of the particle on the point q_b. Usually, the wave function in ordinary quantum mechanics gives the probability that we find the particle at a given point and does not give the answer on where initially was found the particle. However, K gives the information on the point from where the particle left before being incident on the point q_b. This implies K is the wave function where correspondingly is rigorously defined the initial state.

If the action functional in 4.82 is selected to be equal to the classical action 4.83 and equation 7.3 becomes the integral representation of the Schrödinger equation. The path-integral formulation of quantum mechanics fully is equivalent to the Schrödinger and Heisenberg pictures. When $|\delta S|$ is large compared to $\hbar$, i.e., when small changes in the path induce large changes in the action, then the phase factor $\exp\left\{\frac{i}{\hbar}S\left[q\right]\right\}$ changes rapidly and nearby paths will cancel each other's contributions to the path integral due to destructive interference. So, for paths with the action more than $\hbar$ different from the classical path, interference becomes more and more destructive and there is lesser net contribution. Paths where the action is stationary ($\delta S = 0$) will experience the least destructive interference, so the largest contribution to the path integral is given by these classical paths. This implies paths close to the classical path have nearly the same action and thus the same phase and so will interfere constructively and make the dominating contributions to the transition amplitude. So, we expect paths whose action is within about $\hbar$ of the classical path will contribute strongly. This links the path-integral formalism with the principle of stationary action in classical mechanics. This presents an elegant manner of thinking about quantum dynamics, and, at the same time, illustrating how the Heisenberg uncertainty limit comes about via wave (or path) interference.

7.2 Schrödinger Equation

We examine again, the formal solution of equation 4.7 of chapter 4 that expresses the wave function at an infinitesimal later time ($t + \varepsilon$) via the wave function at the time t by applying the time-evolution operator:

$$\left|\Psi\left(t+\varepsilon\right)\right\rangle = \hat{U}\left(t+\varepsilon,t\right)\left|\Psi\left(t\right)\right\rangle \tag{7.5}$$

By inserting identity operators this can be rewritten as:

$$\left|\Psi\left(t+\varepsilon\right)\right\rangle = \int dq\left|q\right\rangle\left\langle q\right|\hat{U}\left(t+\varepsilon,t\right)\int dq'\left|q'\right\rangle\left\langle q'\middle|\Psi\left(t\right)\right\rangle \tag{7.6}$$

From relation 4.21, then 7.6 can be rewritten

$$\left|\Psi\left(q,t+\varepsilon\right)\right\rangle = \int dq'\,\mathrm{K}\left(q,t+\varepsilon;q',t\right)\left|\Psi\left(q',t\right)\right\rangle \tag{7.7}$$

This is a consequence of the superposition law via the integral equation connecting the wave function at time t to the wave function at time $t + \varepsilon$. Physically, the amplitude of the particle at position q at time moment ($t + \varepsilon$) depends on amplitudes at all locations q' at earlier times t. For brevity, to relate path integration for one time moment to its value at a later short time moment, we consider two states $|\Psi(q',t)\rangle \to |\Psi(q,t+\varepsilon)\rangle$ separated by an infinitesimal time interval: $(t + \varepsilon) - t = \varepsilon \to 0$. The transition amplitude for such particle of mass m with a one-dimensional motion subjected to a potential U:

$$\mathrm{K}\left(q,t+\varepsilon,q',t\right) = \frac{1}{A}\exp\left\{\frac{i\varepsilon}{\hbar}\left(\frac{m\left(q-q'\right)^2}{2\varepsilon^2} - U\left(\frac{q+q'}{2},\frac{\left(t+\varepsilon\right)+t}{2}\right)\right)\right\} \tag{7.8}$$

The wave function for such a particle is

$$\left|\Psi\left(q,t+\varepsilon\right)\right\rangle = \int \mathrm{K}\left(q,t+\varepsilon,q',t\right)\left|\Psi\left(q',t\right)\right\rangle dq' \tag{7.9}$$

Suppose the coordinate:

$$q = q' + \xi, \tag{7.10}$$

then 7.9 can be rewritten

$$|\Psi(q,t+\varepsilon)\rangle \cong \frac{1}{A}\exp\left\{-\frac{i\varepsilon}{\hbar}U(q,t)\right\}\int_{-\infty}^{\infty}\exp\left\{\frac{im\xi^2}{2\hbar\varepsilon}\right\}|\Psi(q-\xi,t)\rangle d\xi \tag{7.11}$$

The term $\varepsilon U\left(\frac{q+q}{2},\frac{2t+\varepsilon}{2}\right)$ is safely replaced by $\varepsilon U(q,t)$ as the error is of a higher order than ε. As ξ is a small parameter, then the exponential function $\exp\left\{\frac{im\xi^2}{2\hbar\varepsilon}\right\}$ in 7.11 can oscillate very slowly. If it oscillates extremely rapidly around a unit circle in a complex plane, then it causes almost complete destructive interference. So, the only significant contribution to the integral 7.11 comes from values of ξ at the neighbourhood of 0. We observe the phase of the function $\exp\left\{\frac{im\xi^2}{2\hbar\varepsilon}\right\}$ to change by the order of a radian while ξ is of the order of $\left(\frac{\varepsilon\hbar}{m}\right)^{\frac{1}{2}}$. Significant contributions to the integral come from values of ξ of that order. From $q' = q - \xi$, then the Taylor series expansion of the functions $\Psi(q,t+\varepsilon)$ and $\Psi(q-\xi,t)$ to first order in ε and second order in ξ:

$$|\Psi(q,t)\rangle + \varepsilon\frac{\partial}{\partial t}|\Psi(q,t)\rangle = \frac{1}{A}\exp\left\{-\frac{i\varepsilon}{\hbar}U(q,t)\right\}\int_{-\infty}^{\infty}\exp\left\{\frac{im\xi^2}{2\hbar\varepsilon}\right\}\left[|\Psi(q,t)\rangle - \xi\frac{\partial}{\partial q}|\Psi(q,t)\rangle + \frac{1}{2}\xi^2\frac{\partial^2}{\partial q^2}|\Psi(q,t)\rangle + \cdots\right]d\xi \tag{7.12}$$

For both sides of 7.12 to agree as ε approaches 0, the normalization factor A must take the value:

$$A = \left(\frac{2\pi\hbar\varepsilon i}{m}\right)^{\frac{1}{2}} \tag{7.13}$$

To simplify equation 7.12, we evaluate the following three Gaussian integrals where the integral over odd powers and symmetrical limits vanish while the integrals over even powers of ξ yield:

$$\int_{-\infty}^{\infty}\exp\left\{\frac{im\xi^2}{2\hbar\varepsilon}\right\}d\xi = \left(\frac{2\pi\hbar\varepsilon i}{m}\right)^{\frac{1}{2}}, \quad \int_{-\infty}^{\infty}\exp\left\{\frac{im\xi^2}{2\hbar\varepsilon}\right\}\xi d\xi = 0 \tag{7.14}$$

$$\int_{-\infty}^{\infty}\exp\left\{\frac{im\xi^2}{2\hbar\varepsilon}\right\}\xi^2 d\xi = \frac{\partial}{\partial\nu}\int_{-\infty}^{\infty}\exp\{\nu\xi^2\}d\xi = \left(\frac{\hbar\varepsilon i}{m}\right)\left(\frac{2\pi\hbar\varepsilon i}{m}\right)^{\frac{1}{2}}, \nu \equiv \frac{im}{2\hbar\varepsilon} \tag{7.15}$$

The last integral is easily obtained by differentiation of the first Gaussian integral with respect to the parameter ν. So,

$$|\Psi(q,t)\rangle + \varepsilon\frac{\partial}{\partial t}|\Psi(q,t)\rangle = \left[|\Psi(q,t)\rangle + \frac{i\hbar\varepsilon}{2m}\frac{\partial^2}{\partial q^2}|\Psi(q,t)\rangle\right]\exp\left\{-\frac{i\varepsilon}{\hbar}U(q,t)\right\} \tag{7.16}$$

As $\varepsilon \to 0$ we Taylor expand the exponential function:

$$\left|\Psi(q,t)\right\rangle + \varepsilon\frac{\partial}{\partial t}\left|\Psi(q,t)\right\rangle = \left[\left|\Psi(q,t)\right\rangle + \frac{i\hbar\varepsilon}{2m}\frac{\partial^2}{\partial q^2}\left|\Psi(q,t)\right\rangle\right]\left[1 - \frac{i\varepsilon}{\hbar}U(q,t)\right] \tag{7.17}$$

From here

$$i\hbar\frac{\partial}{\partial t}\left|\Psi(q,t)\right\rangle = -\frac{\hbar^2}{2m}\frac{\partial^2}{\partial q^2}\left|\Psi(q,t)\right\rangle + U(q,t)\left|\Psi(q,t)\right\rangle \equiv \hat{\mathrm{H}}\left|\Psi(q,t)\right) \tag{7.18}$$

This is the time-dependent Schrödinger equation for a particle in the potential $U(q,t)$ and, by design, evolution of the wave function in time via the transition amplitude is equivalent to evolution via the Schrödinger equation. **Caution should be taken when dealing with velocity-dependent interactions. For example, say, an interaction of a charged particle in a magnetic field that contains a term $\vec{\mathrm{A}}(q)\cdot\vec{v}$. Here, $\vec{\mathrm{A}}(q)$ is the corresponding vector potential and $\vec{v}$ is the velocity of the particle. So, it is important to know which discrete form of the path integral should be employed for the given magnetic term and is not considered in this book as the mid-point rule has already been mentioned.**

7.3 The Schrödinger Equation's Green's Function

The transition amplitude

$$\mathrm{K}(q_b,t_b;q_a,t_a) = \begin{cases} \mathrm{K}(q_b,t_b;q_a,t_a), & t_b > t_a \\ 0, & t_b < t_a \end{cases} \equiv \theta(t_b - t_a)\mathrm{K}(q_b,t_b;q_a,t_a) \tag{7.19}$$

satisfies the wave function:

$$\left|\Psi(q_b,t_b)\right) = \int \mathrm{K}(q_b,t_b;q_a,t_a)\left|\Psi(q_a,t_a)\right\rangle dq_a \tag{7.20}$$

and the time-dependent Schrödinger equation:

$$i\hbar\frac{\partial}{\partial t_b}\left|\Psi(q_b,t_b)\right) = \hat{\mathrm{H}}\left|\Psi(q_b,t_b)\right\rangle \tag{7.21}$$

from where

$$i\hbar\frac{\partial}{\partial t_b}\left[\theta(t_b - t_a)\mathrm{K}\right] = i\hbar\theta(t_b - t_a)\frac{\partial}{\partial t_b}\mathrm{K} + i\hbar\mathrm{K}\frac{\partial}{\partial t_b}\theta(t_b - t_a) \tag{7.22}$$

In the above relation, the factor in the last term:

$$\frac{\partial}{\partial t_b}\theta(t_b - t_a) = \delta(t_b - t_a) \tag{7.23}$$

For $t_b > t_a$,

$$i\hbar\frac{\partial}{\partial t_b}\mathrm{K} = \hat{\mathrm{H}}\mathrm{K} \tag{7.24}$$

Also, in the limit $t_b = t_a$ from equation 7.20 then

$$K\left(t_b \to t_a\right) \cong \left(\frac{m}{2\pi i\varepsilon\hbar}\right)^{\frac{1}{2}} \exp\left\{\frac{im}{2\hbar\varepsilon}\left(q_b - q_a\right)^2\right\} \tag{7.25}$$

So, if

$$q_b \to q_a \neq 0, \varepsilon \to 0, K \to 0 \tag{7.26}$$

$$q_b \neq q_a, \frac{1}{\varepsilon^{\frac{1}{2}}} \to \infty, K \to \infty \tag{7.27}$$

$$\int K\left(q_b, t_a + \varepsilon; q_a, t_a\right) dq_b = 1 \tag{7.28}$$

then

$$\lim_{\varepsilon \to 0} K\left(q_b, t_a + \varepsilon; q_a, t_a\right) = \left\{\begin{matrix} \infty & , & q_b = q_a \neq 0 \\ 0 & , & q_b \neq q_a \end{matrix}\right\} \to \delta\left(q_b - q_a\right) \tag{7.29}$$

Hence, the **retarded propagator** satisfies the equation whose right-hand side is proportional to a four-dimensional delta function:

$$i\hbar\frac{\partial}{\partial t_b}\left[\theta K\right] - \hat{H}\left[\theta K\right] = i\hbar\delta\left(q_b - q_a\right)\delta\left(t_b - t_a\right) \tag{7.30}$$

From here, $\theta(t_b - t_a)K(q_b, t_b; q_a, t_a)$ the **retarded propagator** is a **Green's function** for the Schrödinger equation. Physically, in equation 7.20 , the wave function $\Psi(q_a, t_a)$ acts as a source term, and the amplitude $K(q_b, t_b; q_a, t_a)$ propagates the effect of that source to the space-time point (q_b, t_b), where it generates its contribution to $\Psi(q_b, t_b)$. Integration over all such sources gives the total wave function at (q_b, t_b).

7.4 Transition Amplitude for a Time-Independent Hamiltonian

The propagator contains the complete information about the eigenenergies E_n and the corresponding eigenstates or eigenvectors $|\Phi_n\rangle$. So, we will write the propagator in the basis formed from the eigenfunctions of the time-independent Schrödinger. We assume a system with a time-independent Hamiltonian $\hat{H}$ corresponding to any situation where the action S does not depend explicitly on time. Here, the transition amplitude $K(q_b, t_b; q_a, t_a)$ is a function only of the time interval $T = t_b - t_a$ and maps the state vector $|\Psi(t)\rangle$ described by the time-dependent Schrödinger equation:

$$i\hbar\frac{\partial}{\partial t}\left|\Psi(t)\right\rangle = \hat{H}\left|\Psi(t)\right\rangle \tag{7.31}$$

This has the solution

$$\left|\Psi(t)\right\rangle = \exp\left\{-i\frac{Et}{\hbar}\right\}\left|\Phi\right\rangle \tag{7.32}$$

with the eigenstate $|\Phi\rangle$ satisfying the stationary eigenvalue problem:

$$\hat{H}\left|\Phi_n\right\rangle = E_n\left|\Phi_n\right\rangle \tag{7.33}$$

This permits the obtaining of the energy spectrum E_n and the eigenvectors $|\Phi_n\rangle$, or, equivalently, the eigenfunctions represent a complete orthonormal set of eigenfunctions of the Hamiltonian $\hat{H}$:

$$\Phi_n(q) = \langle q | \Phi_n \rangle \tag{7.34}$$

in the coordinate representation. The general solution of 7.31 can be expressed using a linear combination of the complete basis eigenvectors $|\Phi_n\rangle$:

$$|\Psi(t)\rangle = \sum_n C_n \exp\left\{-i\frac{E_n t}{\hbar}\right\}|\Phi_n\rangle \equiv \sum_n C_n |\Phi_n(t)\rangle \tag{7.35}$$

where

$$|\Phi_n(t)\rangle \equiv \exp\left\{-i\frac{E_n t}{\hbar}\right\}|\Phi_n\rangle \tag{7.36}$$

We find the transition amplitude $K(q_b, t_b; q_a, t_a)$ in terms of the states in 7.36 and suppose the system starts at time t_a in the position eigenstate $|q_a\rangle$. We can find C_n via a scalar product with $\langle \Phi_m|$ from the left:

$$\langle \Phi_m | \Psi(t_a)\rangle = \sum_n C_n \exp\left\{-i\frac{E_n t_a}{\hbar}\right\}\langle \Phi_m | \Phi_n\rangle = C_m \exp\left\{-i\frac{E_m t_a}{\hbar}\right\} \tag{7.37}$$

and so

$$C_n = \exp\left\{i\frac{E_n t_a}{\hbar}\right\}\langle \Phi_n | \Psi(t_a)\rangle = \exp\left\{i\frac{E_n t_a}{\hbar}\right\}\langle \Phi_n | q_a, t_a\rangle \tag{7.38}$$

From here, the state vector at a later time t_b is found from 4.8:

$$|\Psi(t_b)\rangle = \sum_n \exp\left\{i\frac{E_n t_a}{\hbar}\right\}\langle \Phi_n | q_a\rangle \exp\left\{-i\frac{E_n t_b}{\hbar}\right\}|\Phi_n\rangle \tag{7.39}$$

So, we have the following identity connecting the Schrödinger eigenfunctions and eigenenergies to Feynman path integral in the form of the transition amplitude:

$$K(q_b, t_b; q_a, t_a) = \langle q_b | \Psi(t_b)\rangle = \sum_n \exp\left\{i\frac{E_n t_a}{\hbar}\right\}\langle \Phi_n | q_a\rangle \exp\left\{-i\frac{E_n t_b}{\hbar}\right\}\langle q_b | \Phi_n\rangle \tag{7.40}$$

or

$$K(q_b, t_b; q_a, t_a) = \sum_n \langle \Phi_n(q_b)| \exp\left\{-i\frac{E_n(t_b - t_a)}{\hbar}\right\}|\Phi_n(q_a)\rangle \tag{7.41}$$

or

$$K(q_b, t_b; q_a, t_a) = \sum_n \exp\left\{-i\frac{E_n(t_b - t_a)}{\hbar}\right\}\langle \Phi_n(q_b) | \Phi_n(q_a)\rangle \tag{7.42}$$

or summarily

$$K(q_b,t_b;q_a,t_a)=\sum_n \exp\left\{-\frac{iE_n(t_b-t_a)}{\hbar}\right\}\langle\Phi_n(q_b)|\Phi_n(q_a)\rangle\theta(t_b-t_a) \tag{7.43}$$

Here, the **Heaviside step function is:**

$$\theta(t_b-t_a)=\begin{cases}1, & t_b>t_a\\ 0, & t_b<t_a\end{cases} \tag{7.44}$$

The expression might be viewed as if the particle moves around in all possible locations in the energy space (acquiring different phase factors) as it propagates from the selected initial position to the final position. The representation in 7.43 can be used as the definition of the transition amplitude K, provided the spectrum is known and the n -sum converges.

Suppose in 7.43 we set t_b to an arbitrary time t then the spectral representation of the transition amplitude:

$$K(q,t;q_a,t_a)=\sum_n \exp\left\{-\frac{iE_n(t-t_a)}{\hbar}\right\}|\Phi_n(q)\rangle\langle\Phi_n(q_a)| \tag{7.45}$$

where E_n are the eigenvalues and $|\Phi_n(q)\rangle$ the eigenvectors of the stationary states. From here, we again confirm that K totally governs the time evolution of the initial state and so, the time evolution is perfectly causal: the time evolution is unique if the system is left alone. However, as soon as a measurement is made the system changes in an uncontrollable manner into one of the eigenstates of the operator consistent with the observable being measured.

We apply the transition amplitude 7.45 to the case of a free particle in 5.94 then this permits us to read off the normalized eigenfunctions

$$\frac{1}{(2\pi\hbar)^{\frac{1}{2}}}\exp\left\{\frac{i}{\hbar}pq\right\} \tag{7.46}$$

and eigenenergies

$$E_p=\frac{p^2}{2m} \tag{7.47}$$

7.5 Retarded Green Function

It is instructive to note that, we can extract from the propagator information on the eigenenergies and eigenstates. To this end, we can introduce the retarded Green function or retarded propagator $G_R(q,q',E)$ that is a Fourier transform of $\theta(t-t')K(q,t;q',t')$. It is instructive to note that all information about any autonomous quantum mechanical system is inherent in the matrix elements of its time evolution operator $U(t''-t')$. Since a system is classified as autonomous if its Hamiltonian does not explicitly depend on time then the time evolution operator $U(t''-t')$ can be obtained from the time-dependent Schrödinger equation

$$i\hbar\frac{\partial}{\partial t''}|q'',t''\rangle=\hat{H}|q'',t''\rangle \tag{7.48}$$

that has the formal solution:

$$|q'',t''\rangle = \hat{U}(t''-t')|q',t'\rangle \equiv \hat{U}(t'',t')|q',t'\rangle, \quad t''>t' \tag{7.49}$$

$$\hat{q}(t')|q',t'\rangle = q'|q',t'\rangle \tag{7.50}$$

The transition from the Schrödinger to the Heisenberg representation of the state vectors is via the relation:

$$|q',t'\rangle = \hat{U}(t)|q'\rangle \tag{7.51}$$

The Feynman propagator between coordinate eigenstates is defined:

$$K(q'',t'';q',t') = \langle q''|\hat{U}(t''-t')|q'\rangle, \quad t''>t' \tag{7.52}$$

The operator $\hat{U}(t''-t')$ describes dynamical evolution under the influence of the Hamiltonian from a time t' to time t'' and **causality** will imply that $t'' > t'$ as indicated by the step or Heaviside function , $\theta(t'' - t')$. Since we can restrict our attention to evolution forwards in time, we consider the **causal propagator** or **retarded propagator** and so equation 7.52 becomes:

$$K(q'',t'';q',t') \equiv G_R(q,q',E) = \langle q''|\hat{U}(t'',t')|q'\rangle\theta(t''-t') \tag{7.53}$$

Substituting the following eigenfunction into 7.48 then

$$|q'',t''\rangle = \int dq''|q''\rangle K(q'',t'';q',t') \tag{7.54}$$

Considering 7.53 then

$$i\hbar\frac{\partial}{\partial t''}\langle q''|\hat{U}(t'',t')|q'\rangle = \hat{H}\langle q''|\hat{U}(t'',t')|q'\rangle \tag{7.55}$$

For $t'' = t'$, then

$$\langle q''|\hat{U}(t'',t')|q'\rangle = \delta(q''-q') \tag{7.56}$$

Considering,

$$\frac{d}{dt}\theta(t) = \delta(t) \tag{7.57}$$

So,

$$i\hbar\frac{\partial}{\partial t''}G - \hat{H}G = i\hbar\delta(t''-t')\delta(q''-q') \tag{7.58}$$

and, thus, the evolution operator relates the position space representation of the inverse of the Schrödinger operator, the so-called **resolvent operator**:

$$\hat{K} = \frac{i\hbar}{E - \hat{H} + i\delta} \tag{7.59}$$

that stems from the Fourier transform of the **causal propagator** or **retarded propagator**:

$$K(q,q',E) = \int_0^\infty K(q,t;q',0)\exp\left\{\frac{iEt}{\hbar}\right\}dt = \int_0^\infty \left\langle q\middle|\exp\left\{\frac{i}{\hbar}\left(E - \hat{H}\right)t\right\}\middle|q'\right\rangle \tag{7.60}$$

or

$$K(q,q',E) = \lim_{\delta\to 0^+}\int_0^\infty \left\langle q\middle|\exp\left\{\frac{i}{\hbar}\left(E - \hat{H} + i\delta\right)t\right\}\middle|q'\right\rangle dt = \lim_{\delta\to 0^+}\left\langle q\middle|\hat{K}\middle|q'\right\rangle \tag{7.61}$$

The factor $\exp\left\{-\frac{\delta t}{\hbar}\right\}$, with $\delta \to 0^+$ is an infinitesimally small positive number and introduced to ensure that for positive real energies the integral converges at the upper limit $t \to \infty$. As integration is done over $t > 0$, then the function $\exp\left\{\frac{iEt}{\hbar}\right\}$ guarantees strong convergence of the integral in 7.61. This is for any E in the upper complex plane, i.e., for $\mathrm{Im}E > 0$ where the retarded propagator $K(q, q', E)$ is an analytic function of the complex variable E. It is an important property of retarded Green's functions:

- **Analyticity in the upper energy plane is the Fourier space manifestation of causality which is of retardation while for the advanced propagator there is analyticity in the lower complex E plane.**

The operator $\hat{K}$ satisfies the relation:

$$\lim_{\delta\to 0^+}\left(E - \hat{H} + i\delta\right)\hat{K} = i\hbar\hat{\mathbb{I}} \tag{7.62}$$

from where the coordinate representation

$$\left\langle q\middle|\lim_{\delta\to 0^+}\left(E - \hat{H} + i\delta\right)\middle|q'\right\rangle = i\hbar q''\left|\hat{\mathbb{I}}\right|q'\rangle = i\hbar\delta\left(q - q'\right) \tag{7.63}$$

or

$$\left(-\frac{\hbar^2}{2m}\frac{\partial^2}{\partial q^2} + V(q) - E\right)K(q,q',E) = -i\hbar\delta\left(q - q'\right) \tag{7.64}$$

From here, we write the time-independent Schrödinger equation of a particle:

$$\frac{d^2}{dq^2}\psi + \frac{2m}{\hbar^2}\left(E - V(q)\right)\psi = 0 \tag{7.65}$$

Assuming that $E > V$ and letting

$$\kappa^2 = \frac{2m}{\hbar^2}\left(E - V(q)\right) \equiv \frac{p^2}{\hbar^2} \tag{7.66}$$

while considering the potential V to be constant then the solutions of 7.65:

$$\psi_{1,2}(q) = A\exp\left\{\pm\frac{i}{\hbar}pq\right\}, p = \hbar\kappa \tag{7.67}$$

The de Broglie wavelength associated with this motion is

$$\lambda = \frac{\hbar}{p} \tag{7.68}$$

while the phase of the wave is $\frac{p}{\hbar}$.

If now the potential V is a smooth and slow-varying function then within a small region compared to that over which the potential varies, the solution of 7.65 are still plane waves with wavelength:

$$\lambda = \frac{\hbar}{\sqrt{2m(E - V(q))}} \tag{7.69}$$

So, between $q = q_0$ to q the phase shift per unit length $\frac{p(q)}{\hbar}$ will no longer be a constant but an accumulated phase shift:

$$\int_{q_0}^{q} \exp\left\{\frac{p(q')}{\hbar}\right\} dq' \tag{7.70}$$

The solution of equation 7.65 in this case can now be

$$\psi_{1,2}(q) = A\exp\left\{\pm\frac{i}{\hbar}\int_{q_0}^{q} \exp\{p(q')dq'\}\right\} \tag{7.71}$$

The above solutions are only true when the change of wavelength over a cycle is small compared to the wavelength itself:

$$\left|\frac{d\lambda}{dq}\right| \ll 1 \tag{7.72}$$

From here we find the following function satisfying equation 7.64:

$$K(q,q',E) = \frac{2mi}{\hbar W}\left[\theta(q-q')\psi_1(q)\psi_2(q') + \theta(q'-q)\psi_2(q)\psi_1(q')\right] \tag{7.73}$$

Here, the Wronskian is constant over q:

$$W = \psi'_1(q)\psi_2(q) - \psi'_2(q)\psi_1(q) \tag{7.74}$$

and $\psi_1(q)$ with $\psi_2(q)$ are the solutions of the homogenous equation derived from 7.64. The prime in ψ' is the derivative over the coordinate q. The Wronskian W is constant over q considering the homogenous equation derived from 7.64. Equation 7.73 coincides with the retarded propagator, when $\psi_1(q)$ and $\psi_2(q)$ satisfy appropriate boundary conditions taking simple physical considerations into account.

We examine a more rigorous solution of equation 7.65 quasi-classically for an arbitrary potential changing slowly.

8

Quasi-Classical Approximation

8.1 Wentzel-Kramer-Brillouin (WKB) Method

In this chapter we study another very important and powerful approximation method, the so-called **eikonal approximation** or **Wentzel-Kramer-Brillouin (WKB) method,** which is used in studying quantum mechanical systems that are subjected to complicated potentials. This method can be used to find approximate solutions to differential equations where the highest derivative is multiplied by a small parameter. The WKB method gives approximate solution to the Schrödinger equation irrespective of its complicated potential and is mostly used to study one-dimensional systems and also the radial equations for higher-dimensional systems with rotational symmetry.

We revisit the homogenous equation derived from 8.64. To find the total energy, we review the **WKB method** from quantum mechanics via the Schrödinger equation:

$$\left(-\frac{\hbar^2}{2m}\frac{\partial^2}{\partial q^2}+V(q)-E\right)\psi(q)=0 \tag{8.1}$$

Here the wave function in the quasi-classical approximation is given as:

$$\psi=\exp\left\{\frac{i}{\hbar}S\right\} \tag{8.2}$$

The exponential action function, S, is the so-called the **eikonal**. If $\hbar \to 0$ then $S \to S_0$, where S_0 is the classical action that satisfies the Jacobi-Hamilton equation. The criterion of applicability of the classical system is that:

$$|S_0| \gg \hbar \tag{8.3}$$

We recover classical mechanics when we take the limit $\hbar \to 0$. In this case, quantum effects can be neglected. In real life, systems are such that

$$|S_0| \oplus \hbar \tag{8.4}$$

This follows the necessity for the use of quasi-classical approximations – **the Wentzel-Kramer-Brillouin (WKB) method.** In this case, $\hbar$ is assumed to be an infinitesimal parameter.

Substituting 8.2 into 8.1 then

$$\frac{d}{dq}\psi=\exp\left\{\frac{i}{\hbar}S\right\}\frac{i}{\hbar}\frac{d}{dq}S \tag{8.5}$$

$$\frac{d^2}{dq^2}\psi = \exp\left\{\frac{i}{\hbar}S\right\}\frac{i}{\hbar}\frac{d^2}{dq^2}S + \left(\frac{i}{\hbar}\frac{d}{dq}S\right)^2 \exp\left\{\frac{i}{\hbar}S\right\} \tag{8.6}$$

and

$$\frac{1}{2m}\left(\frac{d}{dq}S\right)^2 - \frac{i\hbar}{2m}\frac{d^2}{dq^2}S = E - V(q) \tag{8.7}$$

We Taylor series expand the action S in powers of $\hbar$, the so-called **quasi-classical expansion of the eikonal**:

$$S = S_0 + \frac{\hbar}{i}S_1 + \left(\frac{\hbar}{i}\right)^2 S_2 + \cdots \tag{8.8}$$

We consider the zero order in the expansion:

$$S = S_0 \tag{8.9}$$

So, equation 8.7 now becomes an equation independent of $\hbar$, meaning a classical equation that mimics the **Hamilton-Jacobi differential equation** of classical mechanics:

$$\frac{1}{2m}\left(\frac{d}{dq}S_0\right)^2 = E - V(q) \tag{8.10}$$

or

$$\frac{d}{dq}S_0 \equiv S'_0 = \pm\sqrt{2m(E - V(q))} \tag{8.11}$$

From where the solution

$$S_0 = \pm\int\sqrt{2m(E - V(q))}dq \equiv \pm\int p(q)dq \tag{8.12}$$

and

$$p(q) = \sqrt{2m(E - V(q))} \tag{8.13}$$

is the **local classical momentum** of the particle.

We consider the first order expansion of S:

$$S = S_0 + \frac{\hbar}{i}S_1 \tag{8.14}$$

Substituting this into equation 8.7 and considering 8.13 then

$$\frac{1}{2m}\left(\frac{d}{dq}\left(S_0 + \frac{\hbar}{i}S_1\right)\right)^2 - \frac{i\hbar}{2m}\frac{d^2}{dq^2}\left(S_0 + \frac{\hbar}{i}S_1\right) = E - V \tag{8.15}$$

or

$$\frac{1}{2m}\left[\left(\frac{dS_0}{dq}\right)^2+\frac{2\hbar}{i}\frac{dS_0}{dq}\frac{dS_1}{dq}-\hbar^2\left(\frac{dS_1}{dq}\right)^2-\frac{i\hbar}{2m}\frac{d^2S_0}{dq^2}-\frac{i\hbar}{2m}\frac{\hbar}{i}\frac{d^2S_1}{dq^2}\right]=E-V \quad (8.16)$$

Separating the imaginary part from the real part of this equation then we have

$$-2S'_0S'_1-S''_0=0 \quad (8.17)$$

or

$$\frac{dS_1}{dq}=-\frac{S''_0}{2S'_0}=-\frac{p'}{2p} \quad (8.18)$$

From here,

$$S_1=-\frac{1}{2}\ln p \quad (8.19)$$

or

$$S_1=-\ln\sqrt{p} \quad (8.20)$$

then

$$\psi=\exp\left\{\frac{i}{\hbar}S\right\}\cong\exp\left\{\frac{i}{\hbar}\left(S_0+\frac{\hbar}{i}S_1\right)\right\} \quad (8.21)$$

or

$$\psi\cong\exp\{S_1\}\exp\left\{\frac{i}{\hbar}S_0\right\} \quad (8.22)$$

So, substituting in the above expression the formulae 8.12 and 8.19 then

$$\psi(q)=\frac{1}{\sqrt{p}}\exp\left\{\pm\frac{i}{\hbar}\int p\,dq\right\} \quad (8.23)$$

This is the **WKB solution of the Schrödinger equation** when the potential changes only slowly enough where in the classically accessible region $V(q) < E$, it is an oscillating wave function while in the inaccessible region $V(q) > E$, it decreases or increases exponentially. The transition from one region to the other is nontrivial as for $V(q) = E$, the WKB approximation breaks down. The solution in the classical accessible region is a linear superposition of two functions which is the quasi-classical wave function:

$$\psi(q)=\frac{C_1}{\sqrt{p}}\exp\left\{\frac{i}{\hbar}\int p\,dq\right\}+\frac{C_2}{\sqrt{p}}\exp\left\{-\frac{i}{\hbar}\int p\,dq\right\} \quad (8.24)$$

What should be the physical significance of the factor $\frac{1}{\sqrt{p}}$? If $p = m\nu(q)$, where m and $\nu(q)$ are, respectively, the mass and velocity then the probability is mass- as well as position-dependent. The position dependence is inherent in the factor $\nu(q)$. This merely recalls the classical fact that the dwell time in a small interval about point q is inversely proportional to the velocity $\nu(q)$ at that point. This could also be explained in terms of the probability of finding a given particle at point q which should be inversely proportional to its (classical) momentum so its velocity at the given point. In this case, the dwell time is very short in places where the particle is moving very rapidly. So, the probability of the particle of being caught there is small.

8.1.1 Condition of Applicability of the Quasi-Classical Approximation

The result for the wave function in equation 8.23 or 8.24 is called the quasi-classical approximation (or the Wentzel-Kramer-Brillouin or WKB approximation after its originators) due to the following assumption:

$$\hbar \frac{d^2S}{dq^2} \lll \left(\frac{dS}{dq}\right)^2 \tag{8.25}$$

or

$$\left|\frac{d}{dq}\left(\frac{\hbar}{\dot{S}}\right)\right| = \left|\frac{d}{dq}\left(\frac{\hbar}{p}\right)\right| \lll 1 \tag{8.26}$$

or

$$\left|\frac{d}{dq}\left(\frac{\hbar}{p}\right)\right| \cong \left|\frac{d}{dq}\lambda\right| \lll 1, \lambda = \frac{2\pi\hbar}{p} \tag{8.27}$$

where, the classical momentum is

$$\dot{S} = \frac{dS}{dq} = p \tag{8.28}$$

and the de Broglie wavelength λ varies by a relatively small amount over a distance of order λ. The fractional change in the momentum must be small in a wavelength compared to 1. The potential must change so slowly enough that the momentum of the particle is a constant on a given wavelength. So, keeping the classical parameters for the given problem fixed, this requirement is satisfied by formally setting $\hbar \to 0$. This limit, with the wavelength tending to zero, is a quantum mechanical analogue of geometrical optics limit for the propagation of electromagnetic waves. For this limit, the propagation of light can be described by the propagation of particles along rays. As the wavelength is small, then diffraction phenomena are negligible.

We consider another variant that can be written as a constraint on the force

$$\mathrm{F} = -\frac{dV}{dq} \tag{8.29}$$

acting on the particle and

$$\frac{dp}{dq} = \frac{d}{dq}\sqrt{2m\left(E - V(q)\right)} = -\frac{1}{2\sqrt{2m\left(E - V(q)\right)}} 2m\frac{dV}{dq} = -\frac{m}{p}\frac{dV}{dq} \tag{8.30}$$

or

$$\frac{dp}{dq} = \frac{mF}{p} \tag{8.31}$$

and

$$\left|\frac{\hbar p'}{p^2}\right| = \left|\frac{m\hbar F}{p^3}\right| \ll 1 \tag{8.32}$$

This implies we need a big value of the momentum p for a given potential energy. **So, at the points where the classical momentum is very small the WKB approximation breaks down** particularly, in equations 8.23 or 8.24 at the turning points of the classical motion, $E = V$ and $p = 0$. Implying, **for the classical turning point, the quasi-classical approximation does not work for** $p = 0$. Physically, this implies the wave function normalizing factor $\frac{1}{\sqrt{p}}$ in 8.22 or 8.23 blows up, the dwell time of the particle becomes infinite and the de Broglie wave wavelength becomes infinite:

$$\lim_{p\to 0}\frac{\hbar}{p} = \lambda \to \infty \tag{8.33}$$

The assumption that the potential changes only slowly enough over a wavelength is no longer valid.

It is worth noting that in all interesting applications the wave function at the turning points may not be directly needed as at those points the wave function is infinite. To compute the propagator or the wave function at points where the WKB approximation applies we still have to deal with matching conditions close to the turning points where the approximation breaks down. The problem arises for the path integral formulation as we have to consider the quasi-classical computation of the propagator trajectories going through the turning points. To resolve the breakdown of the WKB approximation at the turning points we deform the trajectory into the complex plane for the coordinate q as well as time t in such a manner as to avoid the point with $V(q) = 0$.

We consider the second order in the expansion of S:

$$S = S_0 + \frac{\hbar}{i}S_1 + \left(\frac{\hbar}{i}\right)^2 S_2 \tag{8.34}$$

So, from 8.7 we have

$$\frac{1}{2m}\left[\frac{d}{dq}\left(S_0 + \frac{\hbar}{i}S_1 + \left(\frac{\hbar}{i}\right)^2 S_2\right)\right]^2 - \frac{i\hbar}{2m}\frac{d^2}{dq^2}\left(S_0 + \frac{\hbar}{i}S_1 + \left(\frac{\hbar}{i}\right)^2 S_2\right) = E - V(q) \tag{8.35}$$

Similarly, as done above we have

$$\frac{1}{2m}\left[\left(\frac{dS_1}{dq}\right)^2 + 2\frac{dS_0}{dq}\frac{dS_2}{dq} + \frac{d^2S_1}{dq^2}\right] = 0 \tag{8.36}$$

or

$$S'_1 S'_2 + \frac{1}{2}\left(S'_1\right)^2 + \frac{1}{2}S''_1 = 0 \tag{8.37}$$

or

$$S'_2 = -\frac{1}{2}\frac{\left(S'_1\right)^2 + S''_1}{S'_0} \tag{8.38}$$

But,

$$S'_0 = p, S'_1 = -\frac{p'}{2p}, S''_1 = -\frac{p''}{2p} + \frac{\left(p'\right)^2}{2p^2} \tag{8.39}$$

then

$$S'_2 = \frac{p''}{4p^2} - \frac{3\left(p'\right)^2}{8p^3} \tag{8.40}$$

So,

$$S_2 = \int\left(\frac{p''}{4p^2} - \frac{3\left(p'\right)^2}{8p^3}\right)dq \tag{8.41}$$

Consider,

$$\mathrm{F} = \frac{p}{m}\frac{dp}{dq} = \frac{pp'}{m} \tag{8.42}$$

from where

$$p' = \frac{m\mathrm{F}}{p} \tag{8.43}$$

Substituting into 8.41 then

$$S_2 = \frac{m\mathrm{F}}{4p^3} + \frac{m^2}{8}\int\frac{\mathrm{F}}{p^5}dq \tag{8.44}$$

8.1.2 Bounded Quasi-Classical Motion

For a particle moving in a potential, we can partition the entire space into various regions considering the energy of the particle. We examine in detail the case with only one turning point q_0 that separates the classical accessible region from the inaccessible region. Consider a particle with energy $V = E$ depicted in Figure 8.1. One turning point, q_0, separates the classical accessible region from the inaccessible region. The classical accessible region $q < q_0$ (I) lies left of the turning point while the classical inaccessible region $q > q_0$ (II) is to the right of the turning point.

We write the wave functions for regions (I) and (II):

1. For $q < q_0$, we have the wave function which is the superposition of two waves:

$$\psi_{\mathrm{I}}(q) = \frac{C_1}{\sqrt{p}}\exp\left\{\frac{i}{\hbar}\int pdq\right\} + \frac{C_2}{\sqrt{p}}\exp\left\{-\frac{i}{\hbar}\int pdq\right\} \tag{8.45}$$

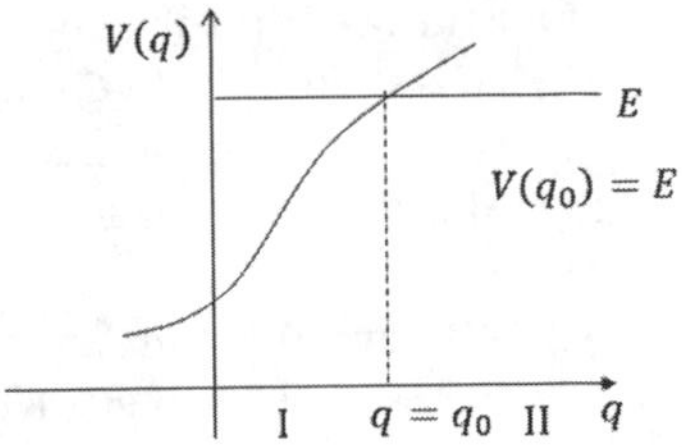

FIGURE 8.1 One turning point q_0 separating the classical accessible region from the inaccessible region. The classical accessible region $q < q_0$ lies left of the turning point while the classical inaccessible region $q > q_0$ is to the right of the turning point.

2. For $q > q_0$, which is the classical inaccessible region $E < V$, the validity of the WKB solution remains true and the momentum p is purely imaginary:

$$\pm\sqrt{2m\left(E-V(q)\right)} = \pm i\sqrt{2m\left(V(q)-E\right)} \tag{8.46}$$

and

$$\psi_{\text{II}}(q) = \frac{C_1}{\sqrt{|p|}}\exp\left\{-\frac{1}{\hbar}\int|p|dq\right\} + \frac{C_2}{\sqrt{|p|}}\exp\left\{\frac{1}{\hbar}\int|p|dq\right\} \tag{8.47}$$

From the condition of boundedness of the wave function, $C_2 = 0$. In this case we have a damped wave function that vanishes at $q \to \pm\infty$. So, the condition of boundedness of the wave function permit the achievement of an evanescent wave that permits the particle to have a non-zero probability of the presence in classical inaccessible region.

It is instructive to note that from the solution of the Schrödinger equation, the wave function is supposed to be continuous across a boundary. This continuity, in addition to giving the physical conservation of particles, leads to quantization of energy levels for the case of bound states. **For the case of WKB, matching $\psi_{\text{I}}(q)$ and $\psi_{\text{II}}(q)$ at the turning point $q = q_0$ separating the classical accessible and inaccessible regions is impossible since at that point the quasi-classical approximation does not work. At the turning point the wave function becomes infinite.** The solution of the Schrödinger equation is a univalent function while that via the WKB are multivalent functions. A difficulty arises when approximating a single-valued function by multivalent functions. So, matching the functions at the turning points is critical as we have to resolve the divergence of the wave functions. Hence, it is important that the solutions to be matched in the two regions should correspond to the same branch. So, the necessity to give an exact solution to the Schrödinger equation in a small region around the turning point. This permits to find the right matching prescription.

By assumption, the potential is smoothly or slowly varying at the boundaries between the **transition region** and the **classical accessible and inaccessible regions**. This implies that at the neighborhood of the turning point $q = q_0$, which is the domain of applicability of the quasi-classical approximation we assume the potential to be smoothly or slowly varying. So, based on this analysis, we thus obtain matching conditions where the wavefunction and its derivative must be continuous. It permits as well to Taylor series expand the potential energy $V(q)$ about the turning point $q = q_0$:

$$V(q) = V(q_0) + (q-q_0)V'(q_0) + \cdots \tag{8.48}$$

It is instructive to note that though we still need to solve the Schrödinger equation explicitly or numerically in the transition region, the WKB is still imperative in that it gives a simple form and much better intuition for what the wave function should look like outside the transition regions $\pm\delta$. Besides, a generic quantization condition can be derived without having details of the transition region, as long as the potential can be linearly approximated.

To find a more explicit solution for the WKB wave function and, consequently, the quantization condition on the energy, it is essential to model the transition region wave function explicitly. So, we assume a linear approximation to the potential is appropriate and is, of course, consistent with the essence of the quasi-classical approximation and so we consider

$$\lim_{q\to q_0}\frac{dV}{dq}<0 \tag{8.49}$$

and select q_0 to be the origin of the coordinate system and q measures the distance from the turning point. So, at the neighbourhood of the turning point $q = q_0$ then

$$V(q)\cong V(q_0)-(q-q_0)\mathrm{F}=E-(q-q_0)\mathrm{F} \tag{8.50}$$

and

$$p(q)=\sqrt{2m\left(E-V(q)\right)}=\sqrt{2(q-q_0)m\mathrm{F}} \tag{8.51}$$

So, from equation 7.65 we have

$$\frac{d^2}{dq^2}\psi+\frac{2(q-q_0)m\mathrm{F}}{\hbar^2}\psi=0 \tag{8.52}$$

The matching of the equations may then be done via the given turning points. For brevity, the one-dimensional motion should be bounded. We consider the wave function non-degenerate and real. This is applicable as well to the wave function $\psi_I(q)$ if:

$$C_1=\frac{C}{2}\exp\{i\phi\},C_2=\frac{C}{2}\exp\{-i\phi\} \tag{8.53}$$

So,

$$\psi_{\mathrm{I}}(q)=\frac{C}{2\sqrt{p}}\cos\left(\frac{1}{\hbar}\int_{q_0}^{q}pdq+\phi\right) \tag{8.54}$$

The expression for $\psi_{II}(q)$ should then be automatically rewritten:

$$\psi_{\mathrm{II}}(q)=\frac{C}{2\sqrt{|p|}}\exp\left\{-\frac{1}{\hbar}\int_{q}^{q_0}|p|dq\right\} \tag{8.55}$$

Matching the two wave functions gives $\phi=\frac{\pi}{4}$. The idea of matching involves the following: Without the time dependence, we examine the quasi-classical complex function on the real axis on which lies the turning point $q = q_0$. So, the quasi-classical approximation is applicable for all points on the plane except the turning point $q = q_0$ on the circle of radius ρ with center at $q = q_0$. In this case, for the motion described by the circle:

$$|p(q)| \to ip(q) \tag{8.56}$$

That is, we have the transition to the two terms in $\psi_1(q)$. So, since

$$|p(q)|^{\frac{1}{2}} \approx (q - q_0)^{\frac{1}{4}} \tag{8.57}$$

then

$$|p(q)|^{\frac{1}{2}} \to \rho^{\frac{1}{4}} \exp\left\{i\frac{\phi}{4}\right\} \tag{8.58}$$

From the excursion from the point $q = q_0$, the argument changes from 0 to π and so there arises the factor $\exp\left\{i\frac{\pi}{4}\right\}$. This implies that $\phi = \frac{\pi}{4}$.

So,

$$\psi_I(q) = \frac{C}{2\sqrt{p}}\cos\left(\frac{1}{\hbar}\int_{q_0}^{q} pdq + \frac{\pi}{4}\right) \tag{8.59}$$

If now we consider the case

$$\lim_{q \to q_0} \frac{dV}{dq} > 0 \tag{8.60}$$

then

$$V(q) = V(q_0) + (q - q_0)\mathrm{F} \tag{8.61}$$

and for $q < q_0 - \delta$, then

$$\psi(q) = \frac{C'}{\sqrt{p}}\cos\left(\frac{1}{\hbar}\int_{q}^{q_0} pdq - \frac{\pi}{4}\right) \tag{8.62}$$

and for $q > q_0 + \delta$, then

$$\psi(q) = \frac{C'}{2\sqrt{|p|}}\exp\left\{-\frac{1}{\hbar}\int_{q_0}^{q} |p|dq\right\} \tag{8.63}$$

8.1.3 Quasi-Classical Quantization

We examine Figure 8.2 and denote the classical turning points as a and b with the particle executing an oscillatory motion in the direction of the q-axis according to the laws of classical physics between these two turning points. This periodic motion is not necessarily simple harmonic. The domain between the

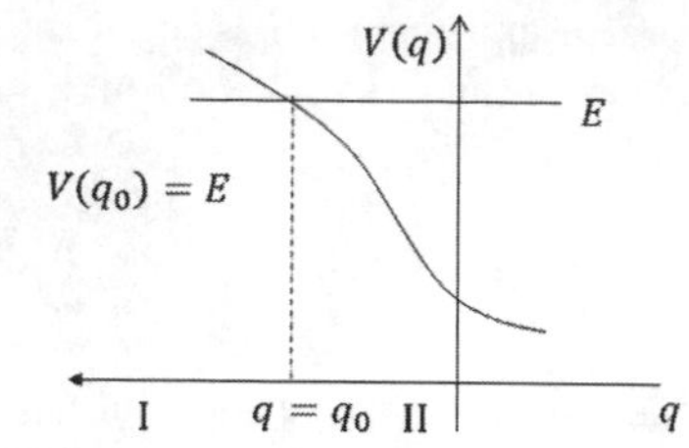

FIGURE 8.2 One turning point q_0 separating the classical accessible region from the inaccessible region. The classical accessible region $q > q_0$ lies left of the turning point while the classical inaccessible region $q < q_0$ is to the right of the turning point.

turning points a and b is classically accessible. We examine the problem quantum mechanically by using the quasi-classical approximation. The wave function can be rewritten in the classically accessible region $q < b$:

$$\psi(q) = \frac{C}{\sqrt{p}}\cos\left(\frac{1}{\hbar}\int_b^q p dq + \frac{\pi}{4}\right) = \frac{C}{\sqrt{p}}\cos\left(\frac{1}{\hbar}\int_q^b p dq - \frac{\pi}{4}\right) \tag{8.64}$$

We have considered that the cosine is an even function.

Consider the case of a bounded quasi-classical motion for $q > a$:

$$\psi(q) = \frac{C'}{\sqrt{p}}\cos\left(\frac{1}{\hbar}\int_a^q p dq - \frac{\pi}{4}\right) \tag{8.65}$$

The above two functions have to match since a bounded one-dimensional motion is nondegenerate. For this it is necessary that

$$C' = (-1)^n C \tag{8.66}$$

as well as

$$\cos\phi = \cos\phi' \tag{8.67}$$

This implies that

$$\frac{\phi + \phi'}{2} = n\frac{\pi}{2} \tag{8.68}$$

So,

$$\frac{1}{\hbar}\int_a^b p dq - \frac{\pi}{2} = n\pi, n = 0,1,2,\cdots \tag{8.69}$$

From here, considering the integral over the period of the classical oscillatory motion then we have the condition of **Bohr-Sommerfeld quantization**:

$$\frac{1}{\hbar}\oint p dq = 2\pi\left(n + \frac{1}{2}\right) \tag{8.70}$$

We find that this condition results in quantization of the energies as the only free parameter in $p(q)$ is the energy E and this condition puts a constraint on this energy E. Classically, the motion is seen to be bounded by the turning points and more so the motion must be periodic since the momentum can only take on two values for any given position, corresponding to right and left motion. If the motion were nonperiodic, then the momentum p should be allowed to take an infinite number of values for any coordinate q. That is why the integral is transformed into a line integral over one period of the motion, corresponding to an integral from a to b and back. This integral in 8.70 is the area enclosed in classical phase space (q, p) by the classical path which the particle would follow. In classical mechanics, the integral in 8.70 is literally the classical action along the classical path and is conserved over time for energy-conserving systems. The classical action for this case is for one period of the particle's motion in the potential well. **From relation 8.70, the Bohr-Sommerfeld quantization condition implies that the phase-space area enclosed by the path, or, equivalently, the classical action for one complete period, must be a multiple of $\hbar$.**

We find now the normalization constant C:

$$C^2\int_a^b \frac{1}{p}\cos^2\left(\frac{1}{\hbar}\int_a^q pdq-\frac{\pi}{4}\right)dq=1 \tag{8.71}$$

We suppose that n is sufficiently large and so

$$C^2\int_a^b \frac{1}{p}\overline{\cos^2\left(\frac{1}{\hbar}\int_a^q pdq-\frac{\pi}{4}\right)}dq=\frac{C^2}{2}\int_a^b \frac{1}{p}dq=\frac{C^2}{2}\frac{\pi}{m\omega}=1 \tag{8.72}$$

Here, ω is the classical oscillatory frequency.

8.1.4 Path Integral Link

We examine how the WKB method fits path integration where the wave function is expressed:

$$\psi(q,t)=\int dq'\mathrm{K}(q,t;q',t')\psi(q',t') \tag{8.73}$$

Here,

$$\mathrm{K}(q,t;q',t')=\int Dq(t)\exp\left\{\frac{i}{\hbar}\int_{t'}^t L(q,\dot{q},t)dt\right\}\equiv\int Dq(t)\exp\left\{\frac{i}{\hbar}\mathrm{S}[q]\right\} \tag{8.74}$$

In the quasi-classics $\hbar\to 0$, the action functional depends on the classical path $\bar{q}$:

$$\mathrm{S}=\mathrm{S}[\bar{q}]=\int_{t'}^t\left(\frac{p^2}{2m}-V(\bar{q})\right)dt''\equiv\int_{t'}^t\left(\mathrm{T}-V(\bar{q})\right)dt''\equiv\int_{t'}^t(2\mathrm{T}-E)dt'' \tag{8.75}$$

If E is the constant energy of the path under consideration then

$$\mathrm{S}=\mathrm{S}[\bar{q}]=\int_{t'}^t 2\mathrm{T}dt''+(t-t')E \tag{8.76}$$

From,

$$p=m\frac{dq}{dt} \tag{8.77}$$

then,

$$\int_{t'}^{t} 2\mathrm{T}dt'' = \frac{1}{m}\int_{t'}^{t} pm\frac{dq''}{dt''}dt'' = \int_{q'}^{q} pdq'' \tag{8.78}$$

and

$$\psi(q,t) = \int dq' \int Dq(t)\exp\left\{\frac{i}{\hbar}\int_{q'}^{q} pdq''\right\}\exp\left\{\frac{i}{\hbar}(t-t')E\right\}\psi(q',t') \tag{8.79}$$

Considering that

$$\psi(q,t) = \psi(q)\exp\left\{\frac{i}{\hbar}tE\right\} \tag{8.80}$$

The exponential factor with tE is irrelevant when incorporated in the wave function in 8.80 as it is just a time-evolving complex phase factor with no position dependence. This factor does not affect the probability density or the momentum and should therefore be dropped out in order for the result to have no explicit time dependence so that the function should depend only on the classical path and so

$$\psi(q) = \int dq' \int Dq\exp\left\{\frac{i}{\hbar}\int_{q'}^{q} pdq''\right\}\psi(q') \tag{8.81}$$

This is the classical limit of the solution in 8.64. The terms of the order of $\frac{1}{\sqrt{p}}$ can be obtained when we evaluate the transition amplitude keeping terms next to the leading order. We now examine some particular cases.

8.2 Potential Well

We examine the motion in a potential $V(q)$ satisfying $\lim_{q\to\pm\infty} V(q) = 0$ and so any solution ψ of the Schrödinger equation with the energy $E > 0$ behaves at $q \to \pm\infty$ as a linear combination of outgoing and incoming plane waves:

$$\lim_{q\to+\infty} \psi \approx \exp\left\{-\frac{i}{\hbar}pq\right\}, \lim_{q\to-\infty} \psi \approx \exp\left\{\frac{i}{\hbar}pq\right\} \tag{8.82}$$

Here,

$$p = \hbar\sqrt{2mE} \tag{8.83}$$

We select the solutions $\psi_{1,2}$ so that $\mathrm{K}(q, q', E)$ should have a physically correct behavior at $q \to \pm\infty$. Consider, the Fourier transform:

$$\mathrm{K}(q,t;q',0) = \int_{-\infty}^{\infty} \frac{dE}{2\pi\hbar}\mathrm{K}(q,q',E)\exp\left\{-\frac{iEt}{\hbar}\right\} \tag{8.84}$$

representing the wave function $\psi(q, t)$ at $t > 0$ that initially at $t = 0$ is a Dirac delta function at some finite position q'

$$\lim_{t\to 0}\psi(q,t) = \delta(q-q') \tag{8.85}$$

So, at any finite $t > 0$ and for $q \to +\infty$, the wave function $\psi(q, t)$ corresponds to a superposition of waves travelling out in the positive q direction and implies a travelling with positive momentum and positive energy:

$$\lim_{q\to+\infty} \mathrm{K}(q,q',E) \approx \exp\left\{\frac{i}{\hbar}pq\right\} \tag{8.86}$$

Similarly, for $q \to -\infty$, $\psi(q, t)$ is a superposition of waves travelling out in the negative q direction and implies a travelling with negative momentum:

$$\lim_{q\to-\infty} \mathrm{K}(q,q',E) \approx \exp\left\{-\frac{i}{\hbar}pq\right\} \tag{8.87}$$

So, we have all the constraints to fully fix $\psi_{1,2}$ and rewrite the general expression for $\mathrm{K}(q, q', E)$ as in 7.73. From the analyticity in the upper energy plane we do the continuation of $\sqrt{E}$ from $E > 0$ to $E < 0$ since the cut off $\sqrt{E}$ is chosen at $\mathrm{Im}E < 0$ to ensure analyticity of $\mathrm{K}(q, q', E)$ at $\mathrm{Im}E > 0$. So, for $E < 0$ or

$$E = \lim_{\delta\to+0}\left(-|E| + i\delta\right) \tag{8.88}$$

then K is in the space of normalized wave functions:

$$\lim_{q\to+\infty} \mathrm{K}(q,q',E) \approx \exp\left\{-\frac{1}{\hbar}|p|q\right\} \to 0 \tag{8.89}$$

$$\lim_{q\to-\infty} \mathrm{K}(q,q',E) \approx \exp\left\{\frac{1}{\hbar}|p|q\right\} \to 0 \tag{8.90}$$

Here,

$$|p| = \hbar\sqrt{2mE} \tag{8.91}$$

By analytic continuation of $\sqrt{E}$ from $E > 0$ via a path in the $\mathrm{Im}E > 0$ half-plane we get the outgoing wave boundary conditions with

$$\lim_{q\to+\infty} \psi_1 \approx \exp\left\{\frac{i}{\hbar}|p|q\right\},\ \lim_{q\to-\infty} \psi_2 \approx \exp\left\{-\frac{i}{\hbar}|p|q\right\} \tag{8.92}$$

and corresponds to the retarded propagator:

$$\mathrm{K}(q,q',E) = \frac{2mi}{\hbar W}\left[\theta(q-q')\psi_1(q)\psi_2(q') + \theta(q'-q)\psi_2(q)\psi_1(q')\right] \tag{8.93}$$

On the one hand, for the analytic continuation of $\sqrt{E}$ from $E > 0$ via a path in the $\mathrm{Im}E < 0$ half-plane we get incoming boundary conditions and corresponds to the advanced propagator being analytic in the lower energy half-plane.

For a free particle, $V(q) = 0$ then

$$\psi_1 = \exp\left\{\frac{i}{\hbar} pq\right\}, \psi_2 = \exp\left\{-\frac{i}{\hbar} pq\right\} \tag{8.94}$$

So, the fixed energy retarded propagator for the free particle:

$$K(q,q',E) = \frac{m}{p}\left[\theta(q-q')\exp\left\{\frac{i}{\hbar}p(q-q')\right\} + \theta(q'-q)\exp\left\{\frac{i}{\hbar}p(q'-q)\right\}\right] \tag{8.95}$$

or

$$K(q,q',E) = \frac{m}{p}\exp\left\{\frac{i}{\hbar}p|q-q'|\right\} \tag{8.96}$$

Switch on the potential, we find the following two solutions to the Schrödinger equation:

$$\lim_{q\to+\infty}\psi_1 = A_+\exp\left\{\frac{i}{\hbar}pq\right\}, \lim_{q\to-\infty}\psi_2 = A_-\exp\left\{-\frac{i}{\hbar}pq\right\} \tag{8.97}$$

The retarded propagator is then also

$$K(q,q',E) = \frac{m}{p}\left[\theta(q-q')\exp\left\{\frac{i}{\hbar}p(q-q')\right\} + \theta(q'-q)\exp\left\{\frac{i}{\hbar}p(q'-q)\right\}\right] \tag{8.98}$$

or

$$K(q,q',E) = \frac{m}{p}\exp\left\{\frac{i}{\hbar}p|q-q'|\right\} \tag{8.99}$$

8.3 Potential Barrier

We investigate the penetration of a barrier by a particle. Classically, if a particle possesses energy less than the height of the barrier then it will be completely reflected and, on the one hand, if the energy of the particle is greater than the potential barrier then the barrier behaves as if it were completely transparent. Quantum mechanically, when the energy of the particle is less than the height of the potential barrier $E < V(q)$, the particle will be reflected in addition to its transmission through the potential barrier. This implies, quantum mechanically, that the particle can tunnel through the potential barrier. Similarly, if the energy of the particle is greater than that of the potential barrier $E > V(q)$, the particle will be reflected as well as transmitted through the potential barrier. Note that the exact solution of the barrier problem exists for the square well potential. The WKB approach is very appropriate since all potentials are not ideal.

For a non-ideal problem, we switch on a slowly changing potential $V(q)$, assuming a barrier is located somewhere along the q-axis. The asymptotic solution $\psi_{1,2}$ of the Schrödinger equation gives

$$\psi_1 \equiv \begin{cases} \exp\left\{\frac{i}{\hbar}pq\right\} + B_+\exp\left\{-\frac{i}{\hbar}pq\right\}, q\to-\infty \\ A_+\exp\left\{\frac{i}{\hbar}pq\right\}, q\to+\infty \end{cases} \tag{8.100}$$

and

$$\psi_2 \equiv \begin{cases} A_{-}\exp\left\{-\frac{i}{\hbar}pq\right\}, q \to -\infty \\ \exp\left\{-\frac{i}{\hbar}pq\right\} + B_{-}\exp\left\{\frac{i}{\hbar}pq\right\}, q \to +\infty \end{cases} \tag{8.101}$$

From here, the transmission coefficient A_+ permit to find the transmission probability:

$$P = \left|A_+\right|^2 \tag{8.102}$$

The conservation of the probability current is guaranteed by the relation

$$\left|A_+\right|^2 + \left|B_+\right|^2 = 1 \tag{8.103}$$

The wave function ψ_1 in 8.100 describes an incoming wave travelling from $-\infty$:

$$\psi_{\text{in}} = \exp\left\{\frac{i}{\hbar}pq\right\} \tag{8.104}$$

which then is scattered at the barrier into an outgoing wave for $q \to -\infty$:

$$\psi_{\text{out}} = B_+ \exp\left\{-\frac{i}{\hbar}pq\right\} \tag{8.105}$$

and for $q \to +\infty$:

$$\psi_{\text{out}} = A_+ \exp\left\{\frac{i}{\hbar}pq\right\} \tag{8.106}$$

Similarly, the motion in the opposite direction is described by the wave function ψ_2 in 8.101. For $q \to -\infty$, we arrive at the Wronskian:

$$W = 2p\frac{i}{\hbar}A_{-} \tag{8.107}$$

So,

$$K(q,q',E) = \frac{m}{pA_{-}}\left[\theta(q-q')\psi_1(q)\psi_2(q') + \theta(q'-q)\psi_2(q)\psi_1(q')\right] \tag{8.108}$$

For the limits $q \to \infty$ and $q' \to -\infty$, then

$$\lim_{q \to +\infty, q' \to -\infty} K(q,q',E) = \frac{m}{p}A_+ \exp\left\{\frac{i}{\hbar}p(q-q')\right\} \equiv \frac{m}{p}\psi_{\text{out}}(q)\psi_{\text{in}}^*(q') \tag{8.109}$$

In the same manner we consider reflexion by a barrier by letting $q' \to -\infty$, $q \to -\infty$, and $q > q'$. Far away, with $V(q) \to 0$, we have:

$$K(q,q',E) \approx \frac{m}{p}\left(\psi_{\text{in}}(q) + \psi_{\text{out}}(q)\psi_{\text{in}}^*(q')\right) \tag{8.110}$$

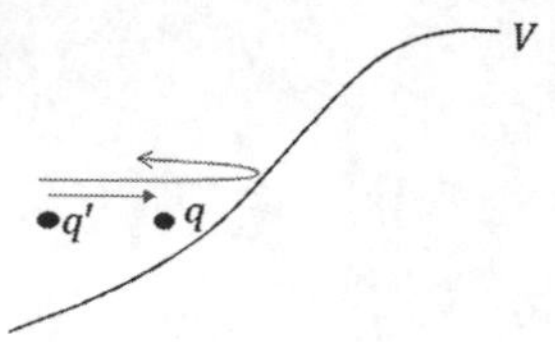

FIGURE 8.3 Depicts two possible classical paths from q' to q with the first one going directly from q' to q, while the second one is first reflected by the potential barrier before going to q.

or

$$K(q,q',E) = \frac{m}{p}\left(\exp\left\{\frac{i}{\hbar}p(q-q')\right\} + B_+ \exp\left\{-\frac{i}{\hbar}p(q+q')\right\}\right) \tag{8.111}$$

From the above equation we find the relics of two possible classical paths and involves a direct path from q' to q, and a path from q' to q where the particle is reflected by the barrier as depicted in Figure 8.3.

8.4 Quasi-Classical Derivation of the Propagator

We revisit the computation of $K(q, q', E)$ quasi-classically where neglecting higher derivatives of the potential $V(q)$ higher than cubic will give a good approximation.

Considering equations 5.2 and 5.41 we do the Fourier transform of the quasi-classical propagator:

$$K_{cl}(q,q',E) = \int_0^\infty dt \sqrt{-\frac{1}{2\pi i\hbar}\frac{\partial^2 S_{cl}}{\partial q'\partial q}} \exp\left\{\frac{i}{\hbar}(S_{cl} + Et)\right\} \tag{8.112}$$

Here, S_{cl} is the classical action and for $\hbar \to 0$, or equivalently for E large enough, the saddle point computation of the integral over t is a good approximation. We find the stationary point for the integral 8.112:

$$\frac{\partial S_{cl}}{\partial t} + E = 0 \tag{8.113}$$

or

$$-E_{cl}(\tau) + E = 0 \tag{8.114}$$

This implies that

$$\tau = t(E) \tag{8.115}$$

Here, $E_{cl}(\tau)$ is the energy of the classical trajectory for the motion from q' to q at a time t where τ is the stationary point. Considering the exponent in 8.112 then at the stationary point τ we have:

$$S_{cl} + E_{cl}\tau = \int_{q'}^{q} dq\, p(q) \tag{8.116}$$

where,

$$p(q) = \sqrt{2m(E - V(q))} \tag{8.117}$$

The second order derivative of S_{cl} around τ:

$$\frac{\partial^2}{\partial t^2} S_{cl} = -\frac{\partial}{\partial t} E_{cl} \tag{8.118}$$

From 8.117 then

$$\dot{q}_{cl} = v(q) = \sqrt{\frac{2\left(E_{cl} - V(q)\right)}{m}} \tag{8.119}$$

and the time dependence of the classical path q_{cl} is given:

$$t = \int_{q'}^{q} \frac{dq}{\dot{q}_{cl}} = \int_{q'}^{q} dq \sqrt{\frac{m}{2\left(E_{cl} - V(q)\right)}} \tag{8.120}$$

It follows from here that

$$1 = -\int_{q'}^{q} dq \frac{1}{m} \left(\frac{m}{2\left(E_{cl} - V(q)\right)} \right)^{\frac{3}{2}} \frac{\partial}{\partial t} E_{cl} \tag{8.121}$$

and

$$-\frac{\partial}{\partial t} E_{cl} = \left(\frac{1}{m} \int_{q'}^{q} \frac{dq}{v^3} \right)^{-1} = -v'v \frac{\partial^2}{\partial q' \partial q} S_{cl}, v' = v(q'), v = v(q) \tag{8.122}$$

We write the stationary phase integral:

$$\int d(t-\tau) \exp\left\{ \frac{i}{2\hbar} (t-\tau)^2 \frac{\partial^2}{\partial t^2} S_{cl} \right\} = \sqrt{\frac{2\pi i\hbar}{\frac{\partial^2}{\partial t^2} S_{cl}}} = \sqrt{2\pi i\hbar \frac{1}{m} \int \frac{dq}{v^3}} \tag{8.123}$$

From here considering 8.112 we have the quasi-classical propagator:

$$\mathrm{K}(q, q', E) = \sqrt{\frac{1}{v'v}} \exp\left\{ \frac{i}{\hbar} \int_{q'}^{q} dq p(q) \right\} \approx \psi_{\mathrm{WKB}}(q) \psi^{*}_{\mathrm{WKB}}(q') \tag{8.124}$$

Here, the wave function derived by applying the Wentzel-Kramer-Brillouin (WKB) approximation:

$$\psi_{\mathrm{WKB}}(q) \approx \sqrt{\frac{1}{v(q)}} \exp\left\{ \frac{i}{\hbar} \int_{q_0}^{q} dq p(q) \right\} \tag{8.125}$$

8.5 Reflection and Tunneling via a Barrier

We examine two basic applications of the WKB propagator formalism and, in particular, the process of reflection and tunneling via a barrier. We assume that the potential $V(q)$ vanishes as $q \to -\infty$, and the barrier is somewhere along the q-axis. Setting $q', q \to -\infty$ and $q > q'$ then the classical action from q'

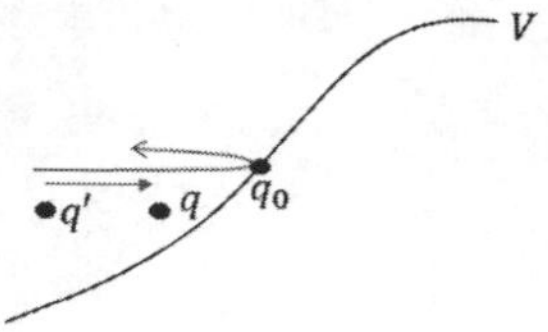

FIGURE 8.4 Two possible classical paths from q' to q with the first one moving from q' to q, while the second one moving first from q' to q_0, and then is reflected from q_0 to q.

to q has two solutions. As depicted in Figure 8.4, this includes one moving from q' to q, and one reflected first by the barrier at q_0 and then moving to q.

So, the Green's function is then the contribution of two classical paths:

$$K(q,q',E) = K_1(q,q',E) + K_2(q,q',E) \tag{8.126}$$

The contribution of the reflected path are two possible classical paths moving from q' to q with the first one moving from q' to q, while the second moving first from q' to q_0, and then is reflected from q_0 and moves to q. In the contribution $K_2(q, q', E)$ is inherent information on the phase shift induced by the reflexion and $K_1(q, q', E)$ is not influenced by the barrier:

$$K_2(q,q',E) = \sqrt{\frac{1}{v'v}} \exp\left\{-i\frac{\pi}{2}\right\} \exp\left\{\frac{i}{\hbar}\int_{q'}^{q_0} |p|dq + \frac{i}{\hbar}\int_{q}^{q_0} |p|dq\right\} \tag{8.127}$$

Here,

$$|p| = \sqrt{2m(E - V(q))} \tag{8.128}$$

Setting $q', q \to -\infty$ and $q > q'$ then

$$K(q,q',E) = K_1(q,q',E) + K_2(q,q',E) = \psi_{\text{out}}(q)\psi_0(q') \tag{8.129}$$

Here, $\psi_0(q')$ is the stationary solution of the Schrödinger equation. Equation 8.129 may also be represented as follows:

$$K(q,q',E) = \sqrt{\frac{1}{v'v}} \left[\exp\left\{\frac{i}{\hbar}\int_{q}^{q'} |p(\xi)|d\xi\right\} + \exp\left\{-i\frac{\pi}{2}\right\} \exp\left\{\frac{i}{\hbar}\int_{q}^{q_0} |p(\xi)|d\xi + \frac{i}{\hbar}\int_{q'}^{q_0} |p(\xi)|d\xi\right\}\right] \tag{8.130}$$

or

$$K(q,q',E) = \sqrt{\frac{1}{v'v}} \exp\left\{\frac{i}{\hbar}\int_{q}^{q_0} |p(\xi)|d\xi - i\frac{\pi}{4}\right\} \left[\exp\left\{-\frac{i}{\hbar}\int_{q'}^{q_0} |p(\xi)|d\xi + i\frac{\pi}{4}\right\} + \exp\left\{\frac{i}{\hbar}\int_{q'}^{q_0} |p(\xi)|d\xi - i\frac{\pi}{4}\right\}\right] \tag{8.131}$$

or

$$K(q,q',E) = \sqrt{\frac{1}{v}} \exp\left\{\frac{i}{\hbar}\int_{q}^{q_0} |p(\xi)|d\xi - i\frac{\pi}{4}\right\} \sqrt{\frac{1}{v'}} \cos\left(\frac{1}{\hbar}\int_{q'}^{q_0} |p(\xi)|d\xi - \frac{\pi}{4}\right) \tag{8.132}$$

So, considering 8.129 then

$$\psi_{\text{out}}(q)=\sqrt{\frac{1}{v}}\exp\left\{\frac{i}{\hbar}\int_{q}^{q_0}|p(\xi)|d\xi-i\frac{\pi}{4}\right\} \tag{8.133}$$

$$\psi_0(q')=\sqrt{\frac{1}{v'}}\cos\left(\frac{1}{\hbar}\int_{q'}^{q_0}|p(\xi)|d\xi-\frac{\pi}{4}\right) \tag{8.134}$$

Consider that the motion is bounded by the point q_b which is to the far left of the point q_0. In a similar manner, as the wave is reflected at the point q_b then we have the following solution:

$$\psi_{0b}(q')=\sqrt{\frac{1}{v'}}\cos\left(\frac{1}{\hbar}\int_{q_b}^{q'}|p(\xi)|d\xi-\frac{\pi}{4}\right) \tag{8.135}$$

Since the two stationary solutions 8.134 and 8.135 arise from the same problem then they must coincide up to a sign:

$$\psi_0(q')=\pm\psi_{0b}(q') \tag{8.136}$$

From here, follows the constraint for the integral over the closed contour resulting in the celebrated **Bohr-Sommerfeld quasi-classical quantization condition**:

$$\frac{1}{\hbar}\oint p(q)dq=\frac{2}{\hbar}\int_{q_b}^{q_0}|p(q)|dq=\frac{2}{\hbar}\int_{q_b}^{q'}|p(q)|dq+\frac{2}{\hbar}\int_{q'}^{q_0}|p(q)|dq=2\pi\left(n+\frac{1}{2}\right) \tag{8.137}$$

This condition results in quantization of the energies as the only free parameter in $p(q)$ is E. This condition puts a constraint on E.

8.6 Transparency of the Quasi-Classical Barrier

In classical mechanics, if a particle is moving from the left of the point q_0 then when it arrives point q_0 then it is reflected (Figure 8.5). There is no reason expecting the particle at point q_b. However, partitioning the abscissa into the accessible and inaccessible regions should be based on the potential and kinetic energies values. In quantum mechanics the operators of the kinetic and potential energies do not commute. The wave function in the classical inaccessible region in quantum mechanics can be different from zero due to the tunnel effect. In Figure 8.6 we consider the potential barrier $V(q)$ where a classical

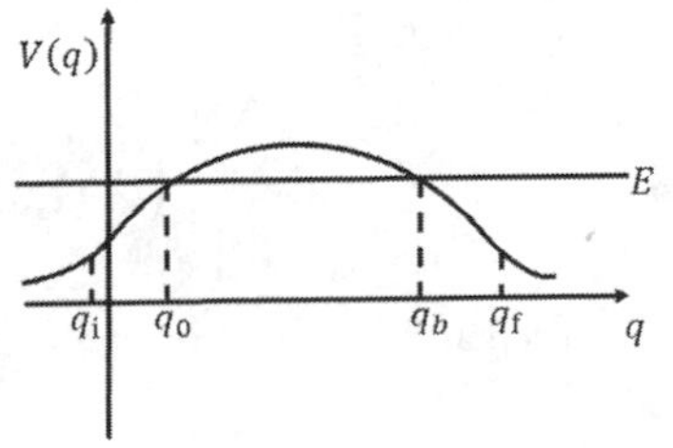

FIGURE 8.5 No classical trajectory from $\boldsymbol{q_i}$ to $\boldsymbol{q_f}$ with energy $\boldsymbol{E}$. The potential barrier is too high to permit a particle go past it and quantum physics in the quasi-classical limit predicts a non-vanishing probability for such a particle to tunnel through the barrier from $\boldsymbol{q_i}$ to $\boldsymbol{q_f}$.

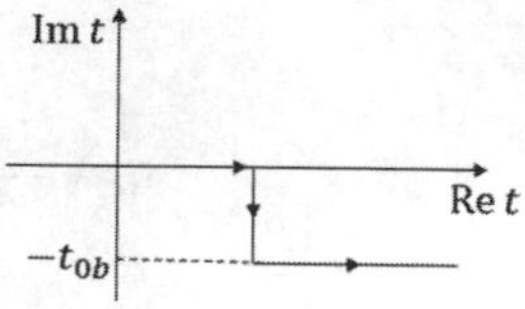

FIGURE 8.6 The evolution of the time along the trajectory in the complex plane where it acquires a negative imaginary part as the particle travels via the potential barrier where, $t_{0b} = \int_{q_0}^{q_b} \frac{dq}{|v(q)|}$.

particle with an energy E cannot go from $-\infty$ to $+\infty$. The wave function in the classical inaccessible region in the region right of the point q_0 in principle is different from zero due to the tunnel effect. This function is different from zero in the classical accessible region to the right of point q_b. The tail of the wave function from region $q_0 < q < q_b$ gets into region to the right of point q_b due to the tunnel effect.

We introduce the notion of transparency and suppose that for $q < q_0$, the particle has a free motion. We find a quasi-classical contribution to this process, that can be seen in the WKB approximation by examining the integral:

$$K(q_f,q_i,E) = \int_{-\infty}^{\infty} dt \int Dq \exp\left\{\frac{i}{\hbar}\left(S_{cl}(t) + Et\right)\right\} \tag{8.138}$$

We can find a stationary point provided t is complex and from the classical equation of motion, we have:

$$\frac{dq}{dt} = \sqrt{\frac{2(E - V(q))}{m}} \tag{8.139}$$

Considering the domain $q_0 < q < q_b$ while ensuring that the Green's function remain analytic then we have:

$$\frac{dq}{dt} = i\sqrt{\frac{2|E - V(q)|}{m}} \tag{8.140}$$

Doing integration over the full trajectory then the time dependence of the classical trajectory:

$$t = \int dt = \int_{q_i}^{q_f} dq \sqrt{\frac{m}{2(E - V(q))}} = \int_{q_i}^{q_0} \frac{dq}{v(q)} - i\int_{q_0}^{q_b} \frac{dq}{|v(q)|} + \int_{q_b}^{q_f} \frac{dq}{v(q)} \tag{8.141}$$

Figure 8.6 shows the evolution of time in the complex plane along the trajectory.
Regardless of the complex value of t, we still have:

$$\frac{i}{\hbar}\left(S_{cl}(t) + Et\right) = \frac{i}{\hbar}\int_{q_i}^{q_f} dq \sqrt{2m(E - V(q))} \tag{8.142}$$

where, $S_{cl}(t)$ is seen to be the classical eikonal and E, the energy for which the positions q_i and q_f are connected by a classical path.

So,

$$K(q_f,q_i,E) = \sqrt{\frac{1}{v(q_i)v(q_f)}} \exp\left\{\frac{i}{\hbar}\int_{q_i}^{q_f} dq \sqrt{2m(E - V(q))}\right\} \tag{8.143}$$

or

$$\mathrm{K}\left(q_f,q_i,E\right)=\psi_{\text{in}}^*\left(q_i\right)\psi_{\text{out}}\left(q_f\right) \tag{8.144}$$

where the first factor is the first de Broglie wave, the so-called incident wave

$$\psi_{\text{in}}^*\left(q_i\right)=\sqrt{\frac{1}{v\left(q_i\right)}}\exp\left\{\frac{i}{\hbar}\int_{q_i}^{q_0}\left|p\left(q\right)\right|dq\right\} \tag{8.145}$$

of the particle moving towards the barrier with the speed $v(q_i)$ and the transmitted wave moving with the speed $v(q_f)$:

$$\psi_{\text{out}}\left(q_f\right)=\mathrm{A}_+\sqrt{\frac{1}{v\left(q_f\right)}}\exp\left\{\frac{i}{\hbar}\int_{q_b}^{q_f}\left|p\left(q\right)\right|dq\right\} \tag{8.146}$$

and the transmission coefficient:

$$\mathrm{A}_+=\exp\left\{-\frac{1}{\hbar}\int_{q_0}^{q_b}\left|p\left(q\right)\right|dq\right\} \tag{8.147}$$

The tunneling probability is then:

$$\mathrm{P}=\left|\mathrm{A}_+\right|^2=\exp\left\{-\frac{2}{\hbar}\int_{q_0}^{q_b}\left|p\left(q\right)\right|dq\right\}=\exp\left\{-\frac{2}{\hbar}\int_{q_0}^{q_b}\sqrt{2m\left(V\left(q\right)-E\right)}dq\right\} \tag{8.148}$$

This leads to the weakening of the square of the amplitude of the wave function and as a result there is the passage of the particle through the barrier. It is obvious that this weakening of the amplitude of the wave function is as a result of the exponential function that for $q = q_0$ it is equal to unity and for $q = q_b$ we have:

$$\exp\left\{-\frac{1}{\hbar}\int_{q_0}^{q_b}\left|p\right|dq\right\}=\mathrm{A}_+ \tag{8.149}$$

The expression in 8.148 is true when the exponential function is very big. This implies that the points q_0 and q_b are sufficiently further apart. If these points are sufficiently closer to each other, then an exact solution is obtained when Figure 8.5 is approximated to a parabola and so we have:

$$\mathrm{P}=\left[1+\exp\left\{\frac{2}{\hbar}\int_{q_0}^{q_b}\left|p\right|dq\right\}\right]^{-1} \tag{8.150}$$

8.7 Homogenous Field

We consider now the potential energy

$$V\left(q\right)=\begin{cases}\mathrm{F}_0q, q>0, \mathrm{F}_0>0\\ \infty, q<0\end{cases} \tag{8.151}$$

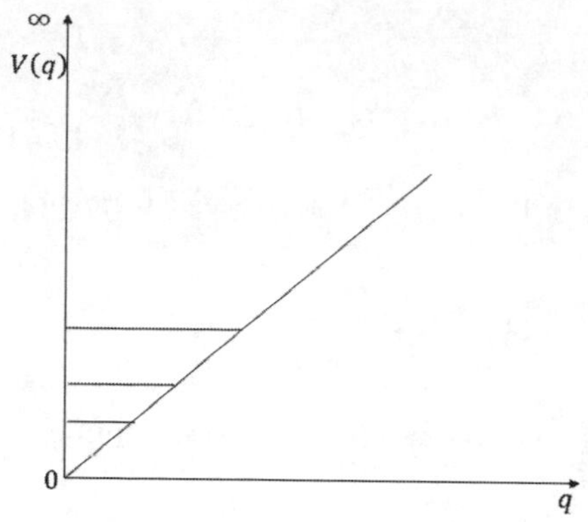

FIGURE 8.7 Representing a rectangular potential field $V(q)$.

which imitates that of a charged particle in a homogenous electric field. The momentum representation is convenient for the solution of the Schrödinger equation with such a potential energy. We represent the potential field in Figure 8.7.

The Schrödinger equation in the momentum representation:

$$\left(\frac{p^2}{2m}+i\hbar F_0\frac{\partial}{\partial p}-E\right)\langle p|E\rangle=0 \tag{8.152}$$

The solution of this equation yields the following wave function in the momentum representation:

$$\langle p|E\rangle=A(E)\exp\left\{\frac{i}{\hbar F}\left(\frac{p^3}{6m}-Ep\right)\right\} \tag{8.153}$$

We normalize the wave function $\langle p|E\rangle$ by the Dirac delta function with respect to the energy:

$$\int_{-\infty}^{\infty}\langle p|E\rangle\langle E'|p\rangle dp=\delta(E-E') \tag{8.154}$$

Thus

$$\delta(E-E')=|A(E)|^2\int_{-\infty}^{\infty}\exp\left\{\frac{i}{\hbar F}(E-E')p\right\}dp=|A(E)|^2 2\pi\delta\left(\frac{E-E'}{\hbar F_0}\right) \tag{8.155}$$

or

$$\delta(E-E')=|A(E)|^2 2\pi\hbar F_0\delta(E-E') \tag{8.156}$$

So,

$$|A(E)|^2 2\pi\hbar F_0=1 \tag{8.157}$$

and

$$\langle p|E\rangle=\frac{1}{(2\pi\hbar F_0)^{\frac{1}{2}}}\exp\left\{\frac{i}{\hbar F_0}\left(\frac{p^3}{6m}-Ep\right)\right\} \tag{8.158}$$

The boundary condition for the problem for the position space: The wave function in position space must vanish at $q = 0$. Transform $\langle p|E\rangle$ to the position space via the Fourier integral:

$$\psi_E(q) \equiv \langle q|E\rangle = \frac{1}{(2\pi\hbar)^{\frac{1}{2}}}\int_{-\infty}^{\infty}\exp\left\{\frac{ipq}{\hbar}\right\}\langle p|E\rangle dp \tag{8.159}$$

or

$$\psi_E(q) = \frac{1}{2\pi\hbar F_0^{\frac{1}{2}}}\int_{-\infty}^{\infty}\exp\left\{\frac{i}{\hbar F}\left(\frac{p^3}{6m} - Ep\right) + \frac{ipq}{\hbar}\right\}dp \tag{8.160}$$

or

$$\psi_E(q) = \frac{1}{2\pi\hbar F_0^{\frac{1}{2}}}\int_{-\infty}^{\infty}\exp\left\{\frac{i}{\hbar}\left[(q - q_0)p + \frac{p^3}{6mF_0}\right]\right\}dp, q_0 = \frac{E}{F_0} \tag{8.161}$$

Here, q_0 is the coordinate of the classical turning point where $V = E$. Letting,

$$\xi = \frac{p}{(2m\hbar F_0)^{\frac{1}{3}}}, z = \left(\frac{2mF_0}{\hbar^2}\right)^{\frac{1}{3}}(q - q_0), z\Big|_{q=0} \equiv z_0 = -\left(\frac{2mF_0}{\hbar^2}\right)^{\frac{1}{3}} q_0 \tag{8.162}$$

then

$$\psi_E(z) = \frac{1}{2}\left(\frac{2m}{\hbar^2 F_0^{\frac{1}{2}}}\right)^{\frac{1}{3}} A_i(z) \tag{8.163}$$

where, the Airy function $A_i(z)$ with argument z:

$$\int_{-\infty}^{\infty}\exp\left\{i\left(\frac{\xi^3}{3} + z\xi\right)\right\}d\xi \equiv \pi A_i(z) \tag{8.164}$$

From the condition $\psi_E(z_0) = 0$ then

$$A_i(z_0) = \int_{-\infty}^{\infty}\exp\left\{i\left(\frac{\xi^3}{3} + z_0\xi\right)\right\}d\xi = 0 \tag{8.165}$$

We obtain from here

$$A_i(z_0) = \int_{-\infty}^{\infty}\cos\left(\frac{\xi^3}{3} + z_0\xi\right)d\xi = 0 \tag{8.166}$$

We have considered the integral of the sin function (odd function) to be equal to zero. So,

$$\int_{-\infty}^{\infty}\cos\left(\frac{\xi^3}{3} + z\xi\right)d\xi = \pi A_i(z) \tag{8.167}$$

The boundary condition that $\psi_E(z)$ tends to zero as z tends to z_0 and gives discrete energy eigenvalues E_n (these are zeros of an Airy function):

$$A_i(z_0) = 0 \tag{8.168}$$

Let these zeros be

$$E'_n = -E_n\left(\frac{2m}{F_0^2\hbar^2}\right)^{\frac{1}{3}} \tag{8.169}$$

Then

$$E_n = -\left(\frac{F_0^2\hbar^2}{2m}\right)^{\frac{1}{3}} E'_n \tag{8.170}$$

The first three zeros of the Airy function E'_n:

$$E'_1 = -2.338, E'_2 = -4.088, E'_3 = -5.521 \tag{8.171}$$

In our case the position-space wave function is the Airy function $A_i(z)$.

We investigate the asymptotic behavior of the Airy function $A_i(z)$ for large z. For the classical inaccessible region $z > 0$ then

$$A_i(z) \approx \frac{1}{|z|^{\frac{1}{4}}}\exp\left\{\frac{2}{3}z^{\frac{3}{2}}\right\} \tag{8.172}$$

and for the classical accessible region $z < 0$ then

$$A_i(z) \approx \frac{1}{|z|^{\frac{1}{4}}}\exp\left\{i\left(\frac{2}{3}z^{\frac{3}{2}} + \frac{\pi}{4}\right)\right\} \tag{8.173}$$

But,

$$\frac{2}{3}z^{\frac{3}{2}} = \frac{2}{3}\left(\frac{2mF_0}{\hbar^2}\right)^{\frac{1}{2}}(q-q_0)^{\frac{3}{2}} = \left(\frac{2mF_0}{\hbar^2}\right)^{\frac{1}{2}}\int_{q_0}^{q}(q-q_0)^{\frac{1}{2}}\,dq \tag{8.174}$$

or

$$\left(\frac{2mF_0}{\hbar^2}\right)^{\frac{1}{2}}\int_{q_0}^{q}(q-q_0)^{\frac{1}{2}}\,dq = \frac{1}{\hbar}\int_{q_0}^{q}p\,dq \tag{8.175}$$

Here, the classical momentum $p(q)$ is written:

$$p(q) = \left[2m(E-V)\right]^{\frac{1}{2}} = \left[2(q-q_0)mF_0\right]^{\frac{1}{2}} \tag{8.176}$$

For the region $z > 0$ then

$$\frac{2}{3}z^{\frac{3}{2}} = \frac{1}{\hbar}\int_{q}^{q_0}|p|dq \tag{8.177}$$

Here,

$$p(q) = \sqrt{\left[2(q_0 - q)m\mathrm{F}_0\right]} \tag{8.178}$$

From here we then have

$$|z|^{\frac{1}{4}} = \frac{(2m\mathrm{F}_0\hbar)^{\frac{1}{6}}}{\sqrt{p}} \tag{8.179}$$

The above derivations permit us to find the asymptotic position space wave function $\langle p| E\rangle$ in the classical accessible region $q > q_0$:

$$\langle p|E\rangle = \psi_E(q) \approx \left(\frac{2m}{\pi\hbar^2}\right)^{\frac{1}{2}}\frac{1}{\sqrt{p}}\cos\left(\frac{1}{\hbar}\int_{q_0}^{q} pdq - \frac{\pi}{4}\right) \tag{8.180}$$

For the classical inaccessible region $q < q_0$:

$$\langle p|E\rangle = \psi_E(q) \approx \left(\frac{2m}{\pi\hbar^2}\right)^{\frac{1}{2}}\frac{1}{2\sqrt{|p|}}\exp\left\{-\frac{1}{\hbar}\int_{q}^{q_0}|p|dq\right\} \tag{8.181}$$

8.7.1 Motion in a Central Symmetric Field

We consider motion in a central symmetric field where the potential contains a centrifugal term

$$V_{\text{centrifugal}}(r) = \frac{\hbar^2 l(l+1)}{2mr^2} \tag{8.182}$$

in addition to the potential $V(r)$. For such problems, the singularity at $r = 0$ is so strong that the WKB approximation may not be application and besides, the factor $\hbar^2$ in 8.182 ruins the systematics of the Taylor series expansion terms in 8 when we have to consider

$$\frac{1}{2m}\left(\frac{d}{dr}S_0\right)^2 = E - V(r) - \frac{\hbar^2 l(l+1)}{2mr^2} \tag{8.183}$$

It is for this reason that for such a case, the quasi-classical treatment requires an amendment. In the above equations, r is the absolute value of the position radial vector.

8.7.1.1 Polar Equation

We consider the motion in a central symmetric field and begin by deriving the quantities found in the Hamiltonian where the polar angles are expressed via the square of the angular momentum operator $\hat{L}^2$:

$$\hat{L}^2 = -\hbar^2 \left\{ \frac{1}{\sin\theta} \frac{\partial}{\partial\theta} \left(\sin\theta \frac{\partial}{\partial\theta} \right) + \frac{1}{\sin^2\theta} \frac{\partial^2}{\partial\phi^2} \right\} = -\hbar^2 \Delta_{\theta,\phi} \quad (8.184)$$

Here the Laplacian, $\Delta_{\theta,\phi}$ acting only on the polar variables θ and ϕ is expressed as:

$$\Delta_{\theta,\phi} = \frac{1}{\sin\theta} \frac{\partial}{\partial\theta} \left(\sin\theta \frac{\partial}{\partial\theta} \right) + \frac{1}{\sin^2\theta} \frac{\partial^2}{\partial\phi^2} \quad (8.185)$$

The motion in a central symmetric field has the Schrödinger equation expressed in the form:

$$-\frac{\hbar^2}{2M} \frac{1}{r^2} \frac{\partial}{\partial r} \left(r^2 \frac{\partial}{\partial r} \Psi \right) + \frac{\hat{L}^2}{2Mr^2} \Psi + V(r)\Psi = E\Psi \quad (8.186)$$

Here, $V(r)$ is the potential and M the mass of the particle. In the central symmetric field, the potential $V(r)$ is independent of the distance between a given point and the center of the field where we place the origin of the coordinate system. The square of the angular momentum satisfies the eigenvalue equation:

$$\hat{L}^2 \Psi = L^2 \Psi \quad (8.187)$$

With

$$L^2 = \hbar^2 l(l+1), \quad l = 0,1,2,\cdots \quad (8.188)$$

The component of the operator $\hat{L}_z$ of the angular momentum in the z–axis direction satisfies the eigenvalue equation:

$$\hat{L}_z \Psi = m\hbar\Psi \quad (8.189)$$

$$l \geq m \geq -l \quad (8.190)$$

In the above relations, l is the azimuthal quantum number and m the magnetic quantum number. We substitute 8.40 into 8.43 considering 8.44:

$$\frac{1}{\sin\theta} \frac{\partial}{\partial\theta} \left(\sin\theta \frac{\partial\Psi}{\partial\theta} \right) + \frac{1}{\sin^2\theta} \frac{\partial^2\Psi}{\partial\phi^2} + l(l+1)\Psi = 0 \quad (8.191)$$

From here, for motion in a central symmetric field where we have a rotationally symmetric potential, the solution of the Schrödinger equation 8.186 can be obtained in the form of separation of variables with factors being the radial $R(r)$ and polar $\Theta_l(\theta)$ exp $\{im\phi\}$ parts respectively:

$$\Psi(r,\theta,\phi) = R(r)\Theta_l(\theta)\exp\{im\phi\} \quad (8.192)$$

We substitute 8.48 into 8.47, where the radial function $R(r)$ and the factor exp$\{im\phi\}$ will fall off:

$$\frac{1}{\sin\theta} \frac{d}{d\theta} \left(\sin\theta \frac{d\Theta_l}{d\theta} \right) + \left[l(l+1) - \frac{m^2}{\sin^2\theta} \right] \Theta_l = 0 \quad (8.193)$$

Let us do the change of variable $\cos\theta = \varsigma$ then we obtain the equation for the associated Legendre polynomials $P_l^m(\varsigma)$ of order l and rank m:

$$\Theta(\theta) = P_l^m(\cos\theta) = P_l^m(\varsigma) \tag{8.194}$$

and

$$P_l^m(\varsigma) = \left(1-\varsigma^2\right)^{\frac{m}{2}} \frac{d^m}{d\varsigma^m} P_l(\varsigma) \tag{8.195}$$

where

$$\mathrm{i}_l(\varsigma) = \frac{1}{2^l l!} \frac{d^l}{d\varsigma^l}\left(\varsigma^2 - 1\right)^l, \quad l \geq m \geq -l \tag{8.196}$$

is the so-called **Rodriguez recurrence formula** for the Legendre polynomial $P_l(\varsigma)$. Thus, the eigenfunction $\Psi(r,\theta,\phi)$:

$$\Psi(r,\theta,\phi) = R(r) Y_{lm}(\theta,\phi) \tag{8.197}$$

where

$$Y_{lm}(\theta,\phi) = N_{lm} P_l^m(\cos\theta) \exp\{im\phi\} \tag{8.198}$$

is a spherical function called the **spherical harmonic** that is an eigenfunction of the operator of the square of the angular momentum $\hat{\vec{L}}^2$ and N_{lm} is a normalization constant for the associated Legendre polynomials.

For brevity, let $m = 0$ then equation 8.193 becomes

$$\frac{1}{\sin\theta}\frac{d}{d\theta}\left(\sin\theta \frac{d}{d\theta}\Theta_l\right) + \left[l(l+1) - \frac{m^2}{\sin^2\theta}\right]\Theta_l = 0 \tag{8.199}$$

$$\frac{d^2}{d\theta^2} P_l + \cot\theta \frac{d}{d\theta} P_l + l(l+1) P_l = 0 \tag{8.200}$$

Letting,

$$P_l(\cos\theta) = \frac{X(\theta)}{\sin^{\frac{1}{2}}\theta} \tag{8.201}$$

then

$$\frac{d}{d\theta}\frac{X(\theta)}{\sin^{\frac{1}{2}}\theta} = \frac{X'(\theta)}{\sin^{\frac{1}{2}}\theta} - \frac{1}{2}\frac{X(\theta)}{\sin^{\frac{3}{2}}\theta}\cos\theta \tag{8.202}$$

$$\frac{d^2}{d\theta^2}\frac{X(\theta)}{\sin^{\frac{1}{2}}\theta} = \frac{X''(\theta)}{\sin^{\frac{1}{2}}\theta} - \frac{X'(\theta)}{\sin^{\frac{3}{2}}\theta}\cos\theta + \frac{3}{4}\frac{X(\theta)}{\sin^{\frac{5}{2}}\theta}\cos^2\theta + \frac{1}{2}\frac{X(\theta)}{\sin^{\frac{1}{2}}\theta} \tag{8.203}$$

and

$$X''(\theta)+\left[\left(l+\frac{1}{2}\right)^2+\frac{1}{4\sin^2\theta}\right]X(\theta)=0 \tag{8.204}$$

For quasi-classical approximation $l \gg 1$ then $l\theta \gg 1$ and $l(\pi-\theta) \gg 1$ is everywhere fulfilled except at the small neighborhood of $\theta = 0$ and $\theta = \pi$.

So, for

$$\frac{1}{l\theta} \ll 1 \tag{8.205}$$

then

$$\left(l+\frac{1}{2}\right)^2+\frac{1}{4\sin^2\theta}=\left(l+\frac{1}{2}\right)^2\left(1+\frac{1}{4\left(l+\frac{1}{2}\right)^2\sin^2\theta}\right)\approx\left(l+\frac{1}{2}\right)^2 \tag{8.206}$$

and

$$X''(\theta)+\left(l+\frac{1}{2}\right)^2X(\theta)=0 \tag{8.207}$$

This is an equation of a harmonic oscillator with solution:

$$X(\theta)=C\sin\left(\left(l+\frac{1}{2}\right)\theta+\phi\right) \tag{8.208}$$

If now,

$$\theta \ll 1 \tag{8.209}$$

then considering

$$\cot\theta \approx \frac{1}{\theta}, l(l+1)\to\left(l+\frac{1}{2}\right)^2 \tag{8.210}$$

Equation 8.200 becomes

$$\frac{d^2}{d\theta^2}P_l+\frac{1}{\theta}\frac{d}{d\theta}P_l+\left(l+\frac{1}{2}\right)^2P_l=0 \tag{8.211}$$

with solution

$$P_l(\theta)=J_0\left(\left(l+\frac{1}{2}\right)\theta\right) \tag{8.212}$$

Here, $J_0(\theta)$ is the Bessel function of the first kind and of order 0 and argument θ. So, the asymptotic solution for 8.204 takes the form

$$\mathrm{P}_l(\theta) \approx \sqrt{\frac{2}{\pi l}} \frac{1}{\sqrt{\theta}} \sin\left(\left(l+\frac{1}{2}\right)\theta + \frac{\pi}{4}\right) \tag{8.213}$$

Hence, the quasi-classical expression for $\mathrm{P}_l(\theta)$:

$$\mathrm{P}_l(\theta) \approx \sqrt{\frac{2}{\pi l}} \frac{1}{\sin^{\frac{1}{2}}\theta} \sin\left(\left(l+\frac{1}{2}\right)\theta + \frac{\pi}{4}\right) \tag{8.214}$$

and

$$\Theta_l = i^l \left(l+\frac{1}{2}\right)^{\frac{1}{2}} \mathrm{P}_l(\cos\theta) \approx i^l \sqrt{\frac{2}{\pi}} \frac{1}{\sin^{\frac{1}{2}}\theta} \sin\left(\left(l+\frac{1}{2}\right)\theta + \frac{\pi}{4}\right) \tag{8.215}$$

8.7.1.2 Radial Equation for a Spherically Symmetric Potential in Three Dimensions

We substitute 8.200 into 8.186 and find the solution of the Schrödinger equation, which is the eigenfunction of the operators $\hat{L}^2$ and $\hat{L}_z$, i.e., the eigenfunction of such a state where E, $L^2$and $L_z$have a defined value:

$$-\frac{\hbar^2}{2mq^2}\frac{d}{dq}\left(q^2\frac{dR}{dq}\right) + \frac{\hbar^2 l(l+1)}{2mq^2}R + V(q)R = ER \tag{8.216}$$

Here, for consistency in the entire heading,

$$r \equiv q, V(r) \equiv V(q), R(r) \equiv R(q) \tag{8.217}$$

In equation 8.216, it is convenient to introduce the following function to ease solution:

$$R(q) = \frac{X(q)}{q} \tag{8.218}$$

and we have the simplified radial equation:

$$-\frac{\hbar^2}{2m}\frac{d^2}{dq^2}X + \frac{\hbar^2 l(l+1)}{2mq^2}X + V(q)X = EX \tag{8.219}$$

So,

$$V_{\text{centrifugal}}(q) \equiv V_{\text{eff}}(q) = \frac{\hbar^2 l(l+1)}{2mq^2} \geq 0 \tag{8.220}$$

The corresponding force to this term (which is equal to minus the gradient of the given term) tends always to repel the particle from the force center and so the name of the term **centrifugal potential (or**

centrifugal barrier) and similar to that in classical mechanics for the Kepler problem. In addition to other boundary conditions (three conditions: wave function should be **finite, continuous and single-valued**), X should satisfy the condition:

$$X(0)=0 \tag{8.221}$$

The effective potential $V_{\text{eff}}(q)$ is the sum of the true potential $V(q)$ and the rotational energy $\frac{L^2}{2mq^2}$ as in classical mechanics. It should be noted that $q \geq 0$ and for small q then the de Broglie wavelength:

$$\lambda = \frac{\hbar}{p} \approx \frac{q}{l} \tag{8.222}$$

and so, for the quasi-classical approximation $l \gg 1$ then

$$\left|\frac{d\lambda}{dq}\right| \ll 1 \tag{8.223}$$

Hence, if l is small, the quasi-classical condition is violated by the centrifugal forces.

8.7.2 Motion in a Coulombic Field

8.7.2.1 Hydrogen Atom

The hydrogen atom is the most important example of the motion of a particle in a Coulomb field. This problem is solved exactly in the analytic form. We suppose that the nucleus of the hydrogen atom has a charge Ze located at the origin of the coordinate axis. The electrons and protons attract each other with the force $\frac{Ze^2}{q^2}$. This corresponds to the potential

$$V(q) = -\frac{Ze^2}{q} \equiv -\frac{\alpha}{q}, \alpha = Ze^2 \tag{8.224}$$

with the following conditions satisfied:

$$|V| \approx |E|, q \approx q_0 \equiv \frac{\alpha}{|E|} \tag{8.225}$$

$$\lambda = \frac{\hbar}{p} = \frac{\hbar}{\sqrt{2m|E|}} \ll q_0 \tag{8.226}$$

Here q is the coordinate of the relative motion and q_0 is one of the turning points. We consider the proton to be the center of our coordinate system and so equation 8.219 now becomes:

$$\left(-\frac{\hbar^2}{2m}\frac{d^2}{dq^2} + V_{\text{eff}}(q)\right)X = EX \tag{8.227}$$

where the effective potential:

$$V_{\text{eff}}(q) = \frac{\hbar^2 l(l+1)}{2mq^2} - \frac{\alpha}{q} \tag{8.228}$$

The first term is always positive or zero. Equation 8.227 can be solved quasi-classically. So, considering the initial condition

$$X(0)=0 \tag{8.229}$$

We examine the s state where $l = 0$. So, in this case,

$$V_{\text{eff}}(q)=-\frac{\alpha}{q} \tag{8.230}$$

and from equation 8.227 we have

$$\frac{d^2X}{dq^2}+\frac{2m}{\hbar^2}\left(-|E|+\frac{\alpha}{q}\right)X=0 \tag{8.231}$$

The turning points for this problem are now

$$q=a=0, q=b=q_0 \tag{8.232}$$

The quasi-classical solution at the right-hand side of the turning point has the form:

$$X_{\text{qs}}(q)=\frac{C}{2\sqrt{|p|}}\exp\left\{-\frac{1}{\hbar}\int_b^q|p|dq\right\}, q>b, p=\sqrt{2m\left(-|E|+\frac{\alpha}{q}\right)} \tag{8.233}$$

Such a solution for $q < b$, corresponds to the wave function:

$$X_{\text{qs}}(q)=\frac{C}{\sqrt{p}}\sin\left(\frac{1}{\hbar}\int_q^b pdq+\frac{\pi}{4}\right) \tag{8.234}$$

Such a solution is true when the quasi-classical approximation condition is satisfied. It is interesting that for small distances $q \to 0$, such a solution seems not to match the quasi-classical approximation for the Coulombic field. So, in order that the boundary condition defined in 8.229 be satisfied, this should indeed correspond to the energy levels for small distances. The quasi-classical solution 8.234 for small distances then matches with the exact solution of the Schrödinger equation. This is in case if we simply consider that

$$X_{\text{qs}}(0)=0 \tag{8.235}$$

For this, the energy levels will not be all that exact.

Consider small distances when

$$q \ll q_0=\frac{\alpha}{|E|} \tag{8.236}$$

then the Schrödinger equation has the form:

$$\left(\frac{d^2}{dq^2}+\frac{\tilde{\alpha}}{q}\right)X=0, \tilde{\alpha}=\frac{2m\alpha}{\hbar^2} \tag{8.237}$$

The solution of this equation should satisfy condition 8.229:

$$X(q)=\tilde{C}\sqrt{q}J_1\left(2\sqrt{\tilde{\alpha}q}\right) \tag{8.238}$$

Here, $J_1(z)$ is the Bessel function of the first kind and of order 1 and argument z. For $\sqrt{\alpha q} \gg 1$ then we have the asymptotic wave function:

$$X(q)\cong\tilde{C}\sqrt[4]{\frac{q}{\pi^2\tilde{\alpha}}}\sin\left(2\sqrt{\tilde{\alpha}q}-\frac{\pi}{4}\right) \tag{8.239}$$

We find the quasi-classical solution for $q < b$, considering 8.234 in the form:

$$X_{qs}(q)=\frac{C'}{\sqrt{p}}\sin\left(\frac{1}{\hbar}\int_0^q pdq+\beta\right) \tag{8.240}$$

We examine the domain of q satisfying the condition:

$$q \ll q_0, \frac{1}{\sqrt{\tilde{\alpha}}} \ll \sqrt{q} \tag{8.241}$$

So,

$$\left|\frac{d\lambda}{dq}\right| \approx (\tilde{\alpha}q)^{-\frac{1}{2}} \ll 1 \tag{8.242}$$

This is the condition of applicability of the quasi-classical approximation. From

$$p \approx \sqrt{\frac{2m\alpha}{q}}=\hbar\sqrt{\frac{\tilde{\alpha}}{q}} \tag{8.243}$$

$$X_{qs}(q)=C\sqrt{\frac{q}{\hbar^2\tilde{\alpha}}}\sin\left(2\sqrt{\tilde{\alpha}q}+\beta\right) \tag{8.244}$$

Within the framework of the above conditions, the exact solution of the Schrödinger equation has the form in 8.239. Comparing 8.239 with 8.244 then we have

$$\beta=-\frac{\pi}{4} \tag{8.245}$$

We consider 8.234 and 8.240 and the fact that in the Coulombic field, the solution is expected to be true when the boundary condition in 8.229 should be satisfied and permits to find the energy levels (Figure 8.8).

We have proven that solving the quantization problem for such a motion then the turning points are expected to be:

$$a=0, b=q_0 \tag{8.246}$$

So, the initial condition now becomes

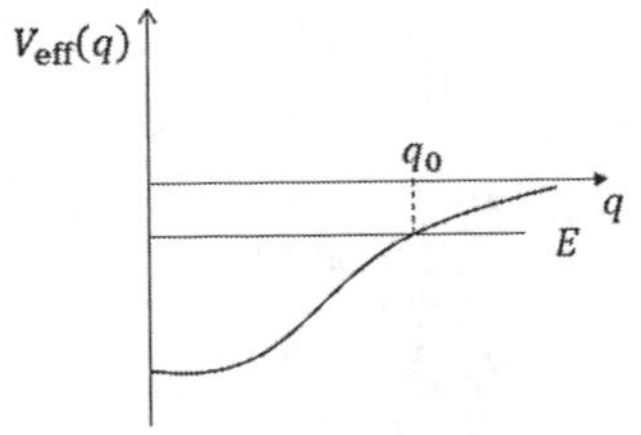

FIGURE 8.8 Representing the variation of an effective potential $V_{eff}(q)$ with the coordinate q.

$$X(q_0)=0 \tag{8.247}$$

Hence, for the internal accessible classical motion then the Bohr-Sommerfeld quantization corresponds to the s state:

$$\frac{1}{\hbar}\int_0^q pdq+\beta+\frac{1}{\hbar}\int_q^b pdq+\frac{\pi}{4}\equiv\frac{1}{\hbar}\int_0^b\sqrt{2m\left(-\left|E_{n_q}\right|+\frac{\alpha}{q}\right)}dq=\pi\left(n_q+1\right) \tag{8.248}$$

or

$$\frac{1}{\hbar}\int_0^{q_0}\sqrt{2m\left(-\left|E_{n_q}\right|+\frac{\alpha}{q}\right)}dq=\frac{1}{\hbar}\oint pdq=\pi\left(n_q+1\right) \tag{8.249}$$

The radial quantum number is $n_q = 0, 1, \cdots$. We calculate the integral in the Bohr-Sommerfeld quantization:

$$\mathcal{K}\equiv\frac{1}{\hbar}\int_0^{q_0}\sqrt{2m\left(-\left|E_{n_q}\right|+\frac{\alpha}{q}\right)}dq=\int_0^{q_0}\sqrt{\left(-\frac{2m\left|E_{n_q}\right|}{\hbar^2}+\frac{2m\alpha}{q\hbar^2}\right)}dq \tag{8.250}$$

Letting,

$$\kappa^2=\frac{2m\left|E_{n_q}\right|}{\hbar^2} \tag{8.251}$$

then

$$\mathcal{K}=\int_0^{q_0}\sqrt{-\frac{2m\left|E_{n_q}\right|}{\hbar^2}+\frac{2m\alpha}{q\hbar^2}}dq=\sqrt{\tilde{\alpha}}\int_0^{q_0}\sqrt{\frac{1}{q}-\frac{\kappa^2}{\tilde{\alpha}}}dq \tag{8.252}$$

and letting, also

$$\frac{1}{q}-\frac{\kappa^2}{\tilde{\alpha}}=z^2,\ \frac{\kappa^2}{\tilde{\alpha}}=\epsilon^2=\frac{2mE}{\hbar^2\tilde{\alpha}}=\frac{E}{\alpha}=\frac{1}{q_0} \tag{8.253}$$

$$q=\frac{1}{z^2+\epsilon^2},dq=-\frac{2zdz}{\left(z^2+\epsilon^2\right)^2} \tag{8.254}$$

then

$$\mathcal{K} = -\sqrt{\tilde{\alpha}}\int_{\infty}^{0} \frac{2z^2 dz}{\left(z^2+\epsilon^2\right)^2} = 2\sqrt{\tilde{\alpha}}\int_{0}^{\infty} \frac{z^2 dz}{\left(z^2+\epsilon^2\right)^2} = 2\sqrt{\tilde{\alpha}}\int_{0}^{\infty} \frac{\left(z^2+\epsilon^2-\epsilon^2\right)dz}{\left(z^2+\epsilon^2\right)^2} \tag{8.255}$$

or

$$\mathcal{K} = 2\sqrt{\tilde{\alpha}}\left[\int_{0}^{\infty} \frac{dz}{z^2+\epsilon^2} - \epsilon^2\int_{0}^{\infty} \frac{dz}{\left(z^2+\epsilon^2\right)^2}\right] \tag{8.256}$$

or

$$\mathcal{K} = 2\sqrt{\tilde{\alpha}}\left(1+\frac{\epsilon}{2}\frac{d}{d\epsilon}\right)\int_{0}^{\infty} \frac{dz}{z^2+\epsilon^2} \tag{8.257}$$

or

$$\mathcal{K} = 2\sqrt{\tilde{\alpha}}\left(1+\frac{\epsilon}{2}\frac{d}{d\epsilon}\right)\frac{1}{\epsilon}\tan^{-1}\frac{z}{\epsilon}\bigg|_{0}^{\infty} = 2\sqrt{\tilde{\alpha}}\left(1+\frac{\epsilon}{2}\frac{d}{d\epsilon}\right)\frac{\pi}{2\epsilon} = \frac{\pi\sqrt{\tilde{\alpha}}}{2\epsilon} \tag{8.258}$$

From here we have

$$\frac{\pi\sqrt{\tilde{\alpha}}}{2\epsilon} = \pi\left(n_q+1\right) \tag{8.259}$$

or

$$\frac{\tilde{\alpha}}{4\epsilon^2} = \left(n_q+1\right)^2 \tag{8.260}$$

and the energy levels:

$$E = -\frac{mZ^2e^4}{2\hbar^2\left(n_q+1\right)^2} \tag{8.261}$$

9

Free Particle and Harmonic Oscillator

9.1 Eigenfunction and Eigenvalue

We find the eigenfunctions and eigenenergy values via the Feynman path integration with the help of the transition amplitude K for the free particle as well as the harmonic oscillator. This will be done without solving the Schrödinger equation and implies not solving the wave equation directly but everything determined from the transition amplitude K .

9.1.1 Free Particle

We examine an important step towards the path integral formulation of quantum mechanics by considering the time-dependent transition amplitude for a free particle of mass m and satisfies the Schrödinger equation with respect to the variables q and t:

$$\mathrm{K}\left(q,t;q',t'\right)=\int dp\exp\left\{-\frac{ip^2T}{2m\hbar}\right\}C_p\left(q',t'\right)\psi_p\left(q\right),T=t-t' \tag{9.1}$$

Here, the eigenfunction of the momentum operator as well as the Hamiltonian is:

$$\psi_p\left(q\right)=\frac{1}{\left(2\pi\hbar\right)^{\frac{1}{2}}}\exp\left\{\frac{i}{\hbar}pq\right\} \tag{9.2}$$

with a momentum eigenvalue p out of a continuous spectrum. The coefficient $C_p(q',t')$ in 9.1 can be defined from the condition when the transition amplitude achieves the expression of the Dirac delta function $\delta(q-q')$ when $t=t'$:

$$\int dp\psi_p^*\left(q'\right)\psi_p\left(q\right)=\delta\left(q-q'\right) \tag{9.3}$$

This implies,

$$C_p=\psi_p^*\left(q'\right) \tag{9.4}$$

We show the validity of the result by recovering the transition amplitude via 9.1 and 9.4:

$$\mathrm{K}\left(\mathrm{q,t;q',t'}\right)=\int \mathrm{dp}\exp\left\{-\frac{\mathrm{ip^2T}}{2m\hbar}\right\}\psi_{\mathrm{p}}^*\left(\mathrm{q'}\right)\psi_{\mathrm{p}}\left(\mathrm{q}\right) \tag{9.5}$$

or

$$K(q,t;q',t') = \frac{1}{2\pi\hbar}\int dp \exp\left\{-\frac{i}{\hbar}\left(\frac{ip^2T}{2m\hbar} + pq' - pq\right)\right\} \tag{9.6}$$

or

$$K(q,t;q',t') = \left(\frac{m}{2i\pi\hbar T}\right)^{\frac{1}{2}} \exp\left\{\frac{im(q-q')^2}{2\hbar T}\right\} \tag{9.7}$$

From transition amplitude in 9.5 the energy of the free particle is:

$$E_p = \frac{p^2}{2m} \tag{9.8}$$

The transition amplitude 9.7 can be rewritten in the 3D form:

$$K(r,t;r',t') = \left(\frac{m}{2i\pi\hbar T}\right)^{\frac{3}{2}} \exp\left\{\frac{im(r-r')^2}{2\hbar T}\right\} \tag{9.9}$$

It should be noted that the dependence of the transition amplitude on the difference of the coordinate $(q - q')$ and time T is related to the homogeneity of space and time for the free particle.

From relation 9.9, the quantity $\frac{m(r-r')^2}{2T}$ at the exponent is the classical action as seen in 5.72 for the 1D problem:

$$S = \frac{m}{2}\int_{t'}^{t} v^2 dt \tag{9.10}$$

We examine some particular cases where the transition amplitude can again be expressed via the classical action. For brevity, we consider the 1D wave function that is the solution of the time-dependent Schrödinger equation:

$$\Psi(q,t) = A(t)\exp\left\{\frac{i}{\hbar}S(q,t)\right\} \tag{9.11}$$

Substituting this in the time-dependent Schrödinger equation while separating the imaginary and real parts then we have:

$$\frac{\partial S}{\partial t} + \frac{1}{2m}\left(\frac{\partial S}{\partial q}\right)^2 + V = 0 \tag{9.12}$$

that has the form of the Hamilton-Jacobi equation with S coinciding with the classical action and then

$$\frac{1}{A}\frac{dA}{dt} + \frac{1}{2m}\frac{\partial^2 S}{\partial q^2} = 0 \tag{9.13}$$

Here, the first summand from our condition is dependent only on t. It follows that the second summand is also only dependent on t. This can only be feasible when

$$V(q,t) = \beta(t) + \gamma(t)q + \delta(t)q^2 \tag{9.14}$$

So, expressions 8.11 as well as 8.21, where, S is the classical action are valid for the potential in 9.14 describing a time-dependent homogenous arbitrary electric and magnetic fields as well as a time-dependent driven harmonic oscillator with a variable frequency.

9.1.2 Transition Amplitude for a Particle in a Homogenous Field

We apply this to an example for the transition amplitude for a particle in a homogenous field. We find the transition amplitude for this case by first evaluating the action:

$$S = \frac{m}{2}\int_{t'}^{t} L(t)dt \tag{9.15}$$

Here,

$$L = \frac{m\dot{q}^2}{2} + F_0 q(t) \tag{9.16}$$

From the equation of motion

$$\frac{d}{dt}\frac{\partial}{\partial \dot{q}}L - \frac{\partial}{\partial q}L = 0 \tag{9.17}$$

then

$$\frac{d}{dt}\frac{\partial}{\partial \dot{q}}L = \frac{d}{dt}(m\dot{q}) = m\ddot{q}, \quad \frac{\partial L}{\partial q} = F_0 \tag{9.18}$$

or

$$m\ddot{q} - F_0 = 0 \tag{9.19}$$

So,

$$\int_{t'}^{t} m\ddot{q}dt = \int_{t'}^{t} F_0 dt \tag{9.20}$$

or

$$\dot{q} - \dot{q}' = \frac{F_0}{m}(t - t') \tag{9.21}$$

From where,

$$q - q' = \dot{q}'(t - t') + \frac{F_0}{2m}(t - t')^2 \tag{9.22}$$

or

$$q - q' = v'(t - t') + \frac{F_0}{2m}(t - t')^2 \tag{9.23}$$

Here, q' is the initial coordinate and

$$v' = \frac{q-q'}{t-t'} - \frac{F_0}{2m}(t-t') \tag{9.24}$$

the velocity at the initial time moment t'. From here, the action is obtained as follows:

$$S = \frac{m(q-q')^2}{2T} + \frac{1}{2}F_0(q+q')T - \frac{F_0^2 T^3}{24m}, T = t - t' \tag{9.25}$$

We substitute this in 9.13 to obtain

$$A = \frac{F}{T^{\frac{1}{2}}} \tag{9.26}$$

For the case of a free particle then

$$F_0 \to 0 \tag{9.27}$$

and we find F from the transition amplitude for a free particle:

$$F = \left(\frac{m}{2i\pi\hbar}\right)^{\frac{1}{2}} \tag{9.28}$$

So, the transition amplitude for a free particle subjected to a homogenous field is:

$$K(q,t;q',t') = \left(\frac{m}{2i\pi\hbar T}\right)^{\frac{1}{2}} \exp\left\{\frac{i}{\hbar}\left[\frac{m(q-q')^2}{2T} + \frac{1}{2}F_0(q+q')T - \frac{F_0^2}{24m}T^3\right]\right\} \tag{9.29}$$

From the spectral representation of the transition amplitude in 7.45 considering 7.29, the eigenfunctions (expressed via Airy functions) and eigenvalues (roots of the Airy functions) are obtained respectively in 8.163 and 8.169.

From knowledge of the path integral of the harmonic oscillator we study another harmonic system which involves a particle of charge e in a homogenous electric field with electric field strength $\vec{\varepsilon}$. The transition amplitude for the 3D case:

$$K(r,t;r',t') = \left(\frac{m}{2i\pi\hbar T}\right)^{\frac{3}{2}} \exp\left\{\frac{i}{\hbar}\left[\frac{m(r-r')^2}{2T} + \frac{1}{2}e\varepsilon(r+r')T - \frac{e^2\varepsilon^2}{24m}T^3\right]\right\} \tag{9.30}$$

9.2 Harmonic Oscillator

For the harmonic oscillator, the Feynman transition amplitude is written:

$$K(q_b,t_b;q_a,t_a) = \left(\frac{m\omega}{2\pi i\hbar\sin\omega T}\right)^{\frac{1}{2}} \exp\left\{\frac{im\omega}{2\hbar\sin\omega T}\left[(q_a^2+q_b^2)\cos\omega T - 2q_a q_b\right]\right\} \tag{9.31}$$

Though the eigenvalues and eigenfunctions could be obtained in a straightforward manner [1], this time we apply not a straightforward procedure but consider the generating function for the Hermite polynomials by the use of the Mehler formula [17–19]:

$$G(x,y,\xi)=\frac{1}{\sqrt{1-\xi^2}}\exp\left\{\frac{2\xi xy-\xi^2\left(x^2+y^2\right)}{1-\xi^2}\right\} \tag{9.32}$$

that is analytic over the ξ plane with cuts going from $-\infty$ to -1 and from $+1$ to $+\infty$. For $|\xi|<1$ then follows the expansion for the generating function:

$$G(x,y,\xi)=\sum_{n=0}^{\infty}\frac{1}{n!}\left(\frac{\xi}{2}\right)^n H_n(x)H_n(y) \tag{9.33}$$

So, the product of the Hermite polynomials:

$$H_n(x)H_n(y)=2^n\frac{d^n}{d\xi^n}G(x,y,\xi)\Big|_{\xi=0} \tag{9.34}$$

where $H_n(x)$ are the Hermite polynomials with argument x. Letting

$$x\equiv\left(\frac{m\omega}{\hbar}\right)^{\frac{1}{2}}q_a, y\equiv\left(\frac{m\omega}{\hbar}\right)^{\frac{1}{2}}q_b, \xi=\exp\{-i\omega T\}, T=(t_b-t_a) \tag{9.35}$$

$$H_0(x)=1, H_1(x)=2x, H_2(x)=4x^2, H_n(x)=(-1)^n\exp\{x^2\}\frac{d^n}{dx^n}\exp\{-x^2\} \tag{9.36}$$

and

$$\frac{\xi}{1-\xi^2}=\frac{1}{2i\sin\omega T},\frac{1+\xi^2}{1-\xi^2}=\frac{1+\exp\{-2i\omega T\}}{1-\exp\{-2i\omega T\}}=\frac{\cos\omega T}{i\sin\omega T} \tag{9.37}$$

From here we arrive at the spectral representation

$$K(q_b,t_b;q_a,t_a)=\sum_n\Phi_n(q_b)\Phi_n^*(q_a)\exp\left\{-i\left(n+\frac{1}{2}\right)\right\} \tag{9.38}$$

Matching this with formula 9.33, the eigenenergy spectrum and eigenwave functions are, respectively:

$$E_n=\hbar\omega\left(n+\frac{1}{2}\right) \tag{9.39}$$

$$\Phi_n(q)=C_nH_n\left(\frac{q}{q_0}\right)\exp\left\{-\frac{q^2}{2q_0^2}\right\} \tag{9.40}$$

Here, q_0 is the natural length scale of the oscillator:

$$q_0 = \sqrt{\frac{\hbar}{m\omega}} \tag{9.41}$$

and the normalization constant

$$C_n = \frac{1}{\sqrt{q_0 2^n n! \sqrt{\pi}}} \tag{9.42}$$

The oscillator has a discrete eigenenergy spectrum and brings us to the physical result of the quantization of the energy of the oscillator.

The **energy of the zero oscillation** (**zero-point** or **ground state energy**) of the oscillator:

$$E_0 = \frac{\hbar\omega}{2} \tag{9.43}$$

Examine the graphs of $V(q)$, E and Φ_n^2 (Figure 9.1). The existence of the ground state energy is a quantum mechanical result that differs not only from the classical solution but also from Bohr solution. If the problem on the oscillator is solved by the Bohr method: condition of Bohr's quantization:

$$\oint p dq = n\hbar\omega \tag{9.44}$$

then we have:

$$E_n = n\hbar\omega \tag{9.45}$$

It follows from here that it is not advisable to neglect completely the kinetic energy of an oscillator. Obviously for the origin of the energy is selected the potential energy of the oscillator at the equilibrium position. The oscillator has some sought of a **zero-point energy (ground state energy)** that is a characteristic peculiarity of some problems in quantum mechanics. It can be shown that the appearance of E_0 is as a result of the Heisenberg uncertainty. At the absolute zero a system has the possible minimum energy (for the oscillator it is E_0). The motion does not cease, but is conserved in the form of zero oscillations and so follows that the material is not destroyed.

The example of a macroscopic system of harmonic oscillators is a solid state. The thermal motion leads to small oscillations of atoms (ions) about the lattice sites of crystalline substances. Introducing normal coordinates, the system of small oscillations may lead to the totality of independent harmonic oscillators. The eigenwave functions of the oscillator have the behaviour under space reflection:

$$\Phi_n(-q) = (-1)^n \Phi_n(q) \tag{9.46}$$

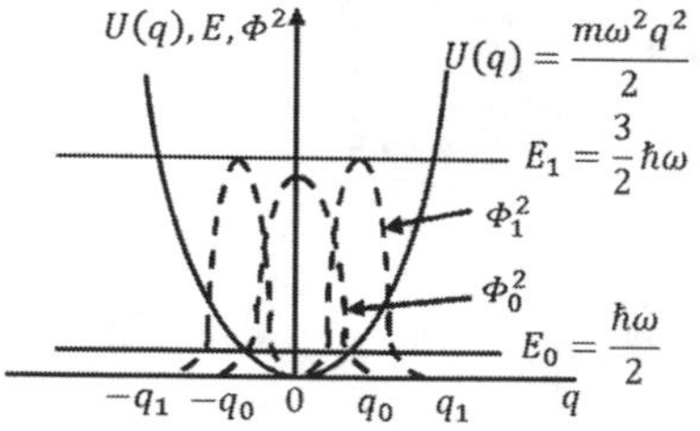

FIGURE 9.1 The ground state and the first excited state of a harmonic oscillator together with their respective eigenstates and eigenenergies.

This is implied from the Hermite polynomials. The quantity $(-1)^n$ is the parity.

It can be shown that the wave functions of two different states are orthogonal:

$$\int_{-\infty}^{\infty} \Phi_m(q)\Phi_n(q)dq = 0, \quad m \neq n \tag{9.47}$$

The motion is bounded if the potential energy at infinity is infinite (though from one side). It is obvious that in such a domain of space where the potential energy is infinite, the wave function should tend to zero as the average of the total energy

$$\bar{E} = \bar{T} + \bar{U} \tag{9.48}$$

and

$$\bar{T} = -\frac{\hbar^2}{2m}\int \Phi^* \Delta \Phi dq \tag{9.49}$$

It is necessary that

$$\Phi\Big|_{\to \pm\infty} \to 0 \tag{9.50}$$

It follows that we are concerned with a bounded one-dimensional motion. It is bounded in the sense that the wave function is different from zero for a point infinitely far removed from it.

1. The finite motion corresponds to a discrete energy spectrum.
2. The energy level is non-degenerate: each energy level corresponds to one wave function.
3. The wave function is real-valued.

Property 9.3 follows from 9.2: The non-degenerate level corresponds to a real-valued function. The functions Φ and Φ^* correspond to one and the same energy level for a non-degenerate case. It follows that

$$\Phi = \Phi^* \tag{9.51}$$

are real-valued functions.

The ground state wave function $\Phi_0(q)$:

$$\Phi_0(q) = C_0 \exp\left\{-\frac{q^2}{q_0^2}\right\} \tag{9.52}$$

and $\Phi_0^2(q)$ is a Gaussian function that is the maximum probability about the equilibrium position. The wave function for the level $n = 1$ is:

$$\Phi_1(q) = C_1 q \exp\left\{-\frac{q^2}{q_0^2}\right\} \tag{9.53}$$

and the probability density of the level ($n = 1$):

$$\Phi_1^2(q) = C_1^2 q^2 \exp\left\{-2\frac{q^2}{q_0^2}\right\} \tag{9.54}$$

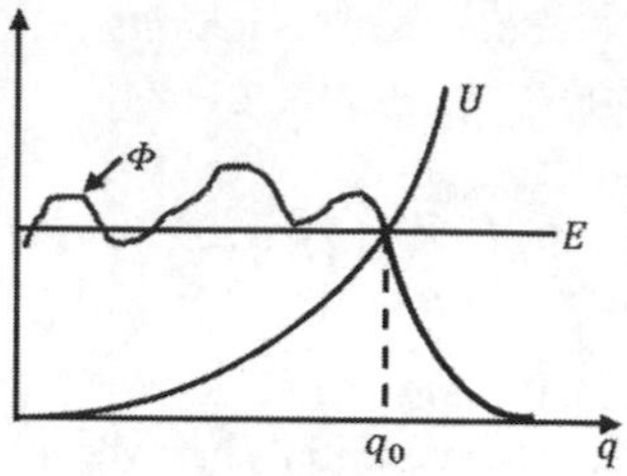

FIGURE 9.2 The tail of the wave function (right of the point q_0) going into the classical inaccessible zone (Tunnel effect).

There exist two maxima for $q = \pm q_0$. Suppose $E = \frac{3}{2}\hbar\omega$, then the oscillator should move from $-q_1$ to $+q_1$. It may not be found on the LHS of $-q_1$ or on the RHS of $+q_1$ which are classically inaccessible domains ($V > E$). In quantum mechanics, there is not such an inaccessible domain as there exists in that domain a probability different from zero that we may find an oscillator. This is obvious from the wave function. It does not tend to zero but asymptotically tend to zero (Figure 9.2).

In quantum mechanics, there is no sense in the partition of the total energy E into the kinetic T and the potential V energy. There is no sense at the point say q for $T \lesseqgtr V$. There is a sense only in E. As the operators of the kinetic $\hat{T}$ and potential energies $\hat{V}$ are not mutually commutable as $\hat{q}$ and $\hat{p}$ are not mutually commutable then each of them do not commute with the Hamiltonian $\hat{\mathrm{H}}$ and we have to exercise some caution. It should be noted that $\overline{E} = \overline{T} + \overline{V}$ is independent of q and $\overline{T}$ is always greater than zero. If we consider some cases for which V is dependent on q then we have a motion in a central symmetric field.

The Feynman transition amplitude 9.31 is only valid when:

$$0 < \omega T < \pi \tag{9.55}$$

We investigate it for larger time moments by supposing

$$T = \frac{n\pi}{\omega} + \tau, 0 < \tau < \frac{\pi}{\omega}, n \in \mathbb{N} \tag{9.56}$$

then from

$$\sin\omega T = \exp\{in\pi\}\sin\omega\tau, \cos\omega T = \exp\{in\pi\}\cos\omega\tau \tag{9.57}$$

equation 9.31 can now be rewritten:

$$\mathrm{K}(q_b, t_b; q_a, t_a) = \left(\frac{m\omega}{2\pi\hbar\sin\omega\tau}\right)^{\frac{1}{2}} \exp\left\{-\frac{i\pi}{2}\left(\frac{1}{2}+n\right) + \frac{im\omega}{2\hbar\sin\omega T}\left[\left(q_a^2+q_b^2\right)\cos\omega T - 2q_a q_b\right]\right\} \tag{9.58}$$

which is the formula for the propagator with the Maslov correction. It is instructive to note that:

$$\left|\sin\omega T\right| = \sin\omega\tau \tag{9.59}$$

For $\tau \to 0$, this implies we consider the propagator at caustics and so

$$\mathrm{K}\left(q_b, q_a, \frac{n\pi}{\omega}\right) = \lim_{\tau\to 0}\left(\frac{m}{2\pi i\hbar\tau}\right)^{\frac{1}{2}} \exp\left\{-\frac{in\pi}{2} + \frac{im}{2\hbar\tau}\left[\left(q_a^2+q_b^2\right) - 2\exp\{-in\pi\}q_a q_b\right]\right\} \tag{9.60}$$

or

$$K\left(q_b,q_a,\frac{n\pi}{\omega}\right)=\exp\left\{-\frac{in\pi}{2}\right\}\delta\left(q_a-(-1)^n q_b\right) \tag{9.61}$$

9.3 Transition Amplitude Hermiticity

We show that the transition amplitude $K(q_b, t_b; q_a, t_a)$ is a Hermitian function by considering the normalization of the function $\Psi(q_b, t_b)$.

From equation 7.20, then

$$\int\Psi(q_b,t_b)\Psi^*(q_b,t_b)dq_b=\iiint dq_b dq_a dq'_a K(q_b,t_b;q_a,t_a)\Psi(q_a,t_a)K^*(q_b,t_b;q'_a,t_a)\Psi^*(q'_a,t_a) \tag{9.62}$$

or

$$\int\Psi(q_b,t_b)\Psi^*(q_b,t_b)q_b=\int dq_a dq'_a\Psi(q_a,t_a)\Psi^*(q'_a,t_a)\int K(q_b,t_b;q_a,t_a)K^*(q_b,t_b;q'_a,t_a)dq_b \tag{9.63}$$

For the orthogonality of the function $\Psi(q_b, t_b)$ then

$$\int K(q_b,t_b;q_a,t_a)K^*(q_b,t_b;q'_a,t_a)dq_b=\delta(q_a-q'_a) \tag{9.64}$$

So,

$$\int\Psi(q_b,t_b)\Psi^*(q_b,t_b)q_b=\int dq_a dq'_a\Psi(q_a,t_a)\Psi^*(q'_a,t_a)\delta(q_a-q'_a) \tag{9.65}$$

Thus,

$$\int\Psi(q_b,t_b)\Psi^*(q_b,t_b)q_b=\int dq_a\Psi(q_a,t_a)\Psi^*(q_a,t_a) \tag{9.66}$$

Equation 9.64 gives the property of orthogonality of the transition amplitude $K(q_b, t_b; q_a, t_a)$. If we multiply 9.64 from left and right by $K^*(q_a, t_a; q_c, t_c)$ and then take the integral over q'_a then we have

$$\int dq'_a\int K(q_b,t_b;q_a,t_a)K^*(q_b,t_b;q'_a,t_a)K^*(q_a,t_a;q_c,t_c)dq_b=\int dq'_a K^*(q_a,t_a;q_c,t_c)\delta(q_a-q'_a) \tag{9.67}$$

From here,

$$K^*(q_a,t_a;q_c,t_c)=\int K(q_b,t_b;q_a,t_a)K^*(q_b;t_b;q_c,t_c)dq_b \tag{9.68}$$

$$K^*(q_a,t_a;q_c,t_c)=\int K^*(q_a,t_a;q_b,t_b)K^*(q_b,t_b;q_c,t_c)dq_b \tag{9.69}$$

So, from the definition, it follows that

$$K(q_b,t_b;q_a,t_a)=K^*(q_a,t_a;q_b,t_b) \tag{9.70}$$

This shows the **property of Hermiticity of the transition amplitude** $K(q_b, t_b; q_a, t_a)$.

10

Matrix Element of a Physical Operator via Functional Integral

We limit ourselves for convenience to a system with one degree of freedom and having the Lagrangian of a particle in a potential field, as in 5.107. For the transition amplitude of this system we have expression 5.97. We have the following questions to ask:

1. Is it possible on the basis of functional integration to obtain the formula for the evaluation of the matrix element of a physical operator of the type 4.82?
2. How can we avoid the difficulty related to the normalization factor in the evaluation of these matrix elements?

In order to answer these questions, we define the matrix element of a physical operator using the quantum mechanical definition. We limit ourselves to the quantum mechanical evaluation of the matrix element of the operator of a physical quantity F. We consider the Heisenberg localized state $|q_i, t_i\rangle$. Suppose the initial time moment t_i corresponds to the initial state $|q_i, t_i\rangle$ and the final time moment t_f corresponds to the final state $|q_f, t_f\rangle$. Consider the Heisenberg operator $\hat{F}(t)$ of a physical quantity F and $t_i \leq t \leq t_f$. By definition, considering the normalized state, the matrix element in the q representation of the operator $\hat{F}(t)$ at the given time moment t is given by the relation:

$$F_{if} = \frac{\left\langle q_f, t_f | \hat{F}(t) | q_i, t_i \right\rangle}{\left\langle q_f, t_f | q_i, t_i \right\rangle} \tag{10.1}$$

Moving to the Schrödinger representation and considering the functional integration from the transition amplitude, then we have in the numerator of 10.1 the following:

$$\left\langle q_f, t_f | \hat{F}(t) | q_i, t_i \right\rangle = q_f \left| \hat{U}(t_f - t) \hat{F} \hat{U}(t - t_i) \right| q_i \rangle \tag{10.2}$$

So, considering the closure relation in equation 3.11 then

$$\left\langle q_f, t_f | \hat{F}(t) | q_i, t_i \right\rangle = \int dq' dq'' \left\langle q_f | \hat{U}(t_f - t) | q'' \right\rangle \left\langle q'' | \hat{F} | q' \right\rangle \left\langle q' | \hat{U}(t - t_i) | q_i \right\rangle \tag{10.3}$$

or

$$\left\langle q_f, t_f | \hat{F}(t) | q_i, t_i \right\rangle = \int dq' dq'' \int Dq(t) \exp\left\{ -\frac{i}{\hbar} S\left[q_f, t_f; q'', t \right] \right\} \left\langle q'' | \hat{F} | q' \right\rangle \exp\left\{ \frac{i}{\hbar} S\left[q', t; q_i, t_i \right] \right\} \tag{10.4}$$

For brevity, we suppose the operator $\hat{F}$ to be diagonalized in the q representation:

$$\left\langle q''|\hat{F}|q'\right\rangle = F(q)\delta(q''-q') \tag{10.5}$$

Then considering

$$\int dq' F\left[q'(t)\right]\int Dq(t)\{\ldots\} = \int Dq(t)F\left[q(t)\right]\{\ldots\} \tag{10.6}$$

we have

$$\left\langle q_f,t_f|\hat{F}(t)|q_i,t_i\right\rangle = \int Dq(t)F\left[q(t)\right]\exp\left\{\frac{i}{\hbar}S\left[q_f,t_f;q_i,t_i\right]\right\} \tag{10.7}$$

In the same fashion the denominator in 10.1 is as follows:

$$\left\langle q_f,t_f|q_i,t_i\right\rangle = \int Dq(t)\exp\left\{\frac{i}{\hbar}S\left[q_f,t_f;q_i,t_i\right]\right\} \tag{10.8}$$

So, the matrix element via functional integration:

$$F_{if} = \int Dq(t)P(q)F\left[q(t)\right] \tag{10.9}$$

where

$$P(q) = \frac{\exp\left\{\frac{i}{\hbar}S\left[q_f,t_f;q_i,t_i\right]\right\}}{\int Dq(t)\exp\left\{\frac{i}{\hbar}S\left[q_f,t_f;q_i,t_i\right]\right\}} \tag{10.10}$$

In the quantum theory, the so-called matrix element 10.9 is defined from the so-called time-ordered product $\hat{T}$ of operators:

$$\hat{T}\prod_{i=1}^{N}\hat{F}\left[q(t_i)\right] \equiv \hat{T}\hat{F}(t) \tag{10.11}$$

The time-ordering operator orders all the operators in the product per time:

$$\hat{T}\left[\hat{F}(t_1)\hat{F}(t_2)\right] = \theta(t_1-t_2)\hat{F}(t_1)\hat{F}(t_2)+\theta(t_2-t_1)\hat{F}(t_2)\hat{F}(t_1) \tag{10.12}$$

Here, $\theta(t)$ is the Heaviside step function.

The formula 10.9 may now be generalized for the case of a time-ordered product of some Heisenberg operators as follows:

$$\frac{\left\langle q_f,t_f|\hat{T}\hat{F}(t)|q_i,t_i\right\rangle}{\left\langle q_f,t_f|q_i,t_i\right\rangle} = \int Dq(t)P(q)F\left[q(t)\right] \tag{10.13}$$

where

$$F[q(t)] \equiv \prod_{i=1}^{N} F[q(t_i)] \tag{10.14}$$

10.1 Matrix Representation of the Transition Amplitude of a Forced Harmonic Oscillator

We analyze a single harmonic oscillator linearly coupled to some external force $\gamma(t)$ described by the Lagrangian:

$$L = \frac{m\dot{q}^2}{2} - \frac{m\omega^2 q^2}{2} + \gamma(t)q \tag{10.15}$$

We assume for brevity that an oscillator with the force $\gamma(t)$ in state $|\Phi_n(q,t)\rangle$ is turned on from time moment $t_a = 0$ and later on at the state $|\Phi_m(q,t)\rangle$ is turned off at time moment $t_b = T$. The transition amplitude G_{mn} describes the oscillator initially in the state with $|\Phi_n(q,t)\rangle$ at time t_a and now found in the state with $|\Phi_m(q,t)\rangle$ at time t_b:

$$G_{mn} = \langle \Phi_m | \hat{U}(t_b, t_a) | \Phi_n \rangle = \langle \Phi_m | \exp\left\{-\frac{i(t_b - t_a)}{\hbar}\right\} | \Phi_n \rangle \tag{10.16}$$

Inserting identity operators over position states then

$$G_{mn} = \int dq_b \int dq_a \langle \Phi_m | q_b \rangle \langle q_b | \exp\left\{-\frac{it_b \hat{H}}{\hbar}\right\} \exp\left\{\frac{it_a \hat{H}}{\hbar}\right\} | q_a \rangle \langle q_a | \Phi_n \rangle \tag{10.17}$$

$$G_{mn} = \int \Phi_m^*(q_b) K_{FHO}(q_b, t_b; q_a, t_a) \Phi_n(q_a) dq_a dq_b \tag{10.18}$$

Supposing $|\Phi_n(t_a)\rangle$ and $|\Phi_m(t_b)\rangle$ are Schrödinger state vectors at the indicated times, then from 7.36 we obtain

$$G_{mn} = \exp\left\{\frac{i}{\hbar}E_n t_b - \frac{i}{\hbar}E_m t_a\right\} \int \Phi_m^*(q_b, t_b) K_{FHO}(q_b, t_b; q_a, t_a) \Phi_n(q_a, t_a) dq_a dq_b \tag{10.19}$$

From 5.134, 5.143 and 5.166 then for easy computation, we find the expression of G_{mn} by considering the situation $m = n = 0$, where G_{00} is the amplitude for which an external perturbation does not change, the oscillator's ground state:

$$G_{00} = \exp\left\{-\frac{i\omega T}{2}\right\} \int \Phi_0^*(q_b, t_b) K_{FHO}(q_b, t_b; q_a, t_a) \Phi_0(q_a) dq_a dq_b \tag{10.20}$$

Here,

$$K_{FHO}(q_b, t_b; q_a, t_a) = F(T) \exp\left\{\frac{i}{\hbar} S_{Hcl}[0, T]\right\} \tag{10.21}$$

and considering equations 5.120 and 5.121 we have:

$$S_{\mathrm{Hcl}}\left[0,T\right]=\frac{m\omega}{2\sin\omega T}\left[\left(q_a^2+q_b^2\right)\cos\omega T-2q_aq_b+2\tilde{\mathcal{F}}\left(T\right)\right] \tag{10.22}$$

with

$$\tilde{\mathcal{F}}\left(T\right)=\int_0^T q_{\mathrm{H}}\left(t\right)\gamma\left(t\right)dt \tag{10.23}$$

$$q_{\mathrm{H}}\left(t\right)=q_a\Gamma_a\left(T-t\right)+q_b\Gamma\left(t\right) \tag{10.24}$$

$$\Gamma_a\left(T-t\right)\equiv\frac{\sin\omega\left(T-t\right)}{\sin\omega T},\Gamma\left(t\right)\equiv\frac{\sin\omega t}{\sin\omega T},T=t_b-t_a \tag{10.25}$$

Computing the integral in 10.20 we find the amplitude to remain in the ground state,

$$\mathrm{G}_{00}=\exp\left\{-\frac{1}{4m\hbar\omega}\int_0^T\int_0^\tau\gamma\left(t\right)\gamma\left(t'\right)\exp\left\{-i\omega\left|t-t'\right|\right\}dtdt'\right\} \tag{10.26}$$

Let us now calculate a general case. We make use of the **generating function for the Hermite polynomials**:

$$\exp\left\{-q^2+2q\xi\right\}=\sum_{n=0}^{\infty}\frac{q^n}{n!}\mathrm{H}_n\left(\xi\right) \tag{10.27}$$

On this basis we can construct a generating function for the matrix elements G_{mn}. Let us introduce the generating function from definition:

$$\mathcal{G}\left(x,y\right)=\sum_{n,m}\mathrm{G}_{mn}\frac{x^my^n}{\sqrt{n!m!}}=\sum_{n,m}\exp\left\{-\frac{i}{\hbar}E_mT\right\}\int\frac{x^m}{\sqrt{m!}}\Phi_m^*\left(q_b,t_b\right)\mathrm{K}_{\mathrm{FHO}}\left(q_b,t_b;q_a,0\right)\frac{y^n}{\sqrt{n!}}\Phi_n\left(q_a,0\right)dq_adq_b \tag{10.28}$$

Considering

$$\Phi_m^*\left(q_b\right)=\exp\left\{-\frac{m\omega}{2\hbar}q_b^2\right\}\mathrm{H}_m\left(q_b\right) \tag{10.29}$$

therefore, it follows from 10.27 and 10.28 that

$$\sum_m\frac{x^m}{\sqrt{m!}}\exp\left\{-\frac{i}{\hbar}E_mT\right\}\Phi_m\left(q_b\right)=\exp\left\{-\frac{i\omega T}{2}\right\}\left(\frac{m\omega}{\pi\hbar}\right)^{\frac{1}{4}}\exp\left\{\frac{m\omega}{2\hbar}q_b^2-\left[\exp\left\{i\omega T\right\}\frac{x}{\sqrt{2}}-q_b\sqrt{\frac{m\omega}{\hbar}}\right]^2\right\} \tag{10.30}$$

$$\sum_n\frac{y^n}{\sqrt{n!}}\Phi_n\left(q_a\right)=\exp\left\{\frac{i\omega t_a}{2}\right\}\left(\frac{m\omega}{\pi\hbar}\right)^{\frac{1}{4}}\exp\left\{\frac{m\omega}{2\hbar}q_a^2-\left[\frac{y}{\sqrt{2}}-q_a\sqrt{\frac{m\omega}{\hbar}}\right]^2\right\} \tag{10.31}$$

Further, we suppose that $t_b = T$, $t_a = 0$. From these expressions we have the following:

$$\mathcal{G}(x,y) = \mathrm{G}_{00} \exp\left\{xy + ixb_\omega^* + iyb_\omega\right\} \tag{10.32}$$

where

$$b_\omega = \frac{1}{\sqrt{2m\hbar\omega}} \int_0^T \gamma(\tau)\exp\{i\omega\tau\}d\tau \tag{10.33}$$

Considering,

$$xy + ixb_\omega^* + iyb_\omega = \left(x + ib_\omega\right)\left(y + ib_\omega^*\right) + b_\omega b_\omega^* \tag{10.34}$$

then

$$\mathcal{G}(x,y) = \mathrm{G}_{00} \exp\left\{b_\omega b_\omega^*\right\} \exp\left\{\left(x + ib_\omega\right)\left(y + ib_\omega^*\right)\right\} \tag{10.35}$$

or

$$\mathcal{G}(x,y) = \mathrm{G}_{00} \exp\left\{b_\omega b_\omega^*\right\} \sum_{p=0}^{\infty} \frac{\left(x + ib_\omega\right)^p \left(y + ib_\omega^*\right)^p}{p!} \tag{10.36}$$

Considering the sum as a Newton binomial then

$$\mathcal{G}(x,y) = \mathrm{G}_{00} \exp\left\{b_\omega b_\omega^*\right\} \sum_{n,m} \sum_{p>n,m} \frac{p!x^m\left(ib_\omega\right)^{p-m}}{m!(p-m)!p!} \frac{p!y^n\left(ib_\omega^*\right)^{p-n}}{n!(p-n)!} \tag{10.37}$$

or

$$\mathcal{G}(x,y) = \mathrm{G}_{00} \exp\left\{b_\omega b_\omega^*\right\} \sum_{n,m} \sum_{p>n,m} \frac{p!\left(ib_\omega\right)^{p-m}\left(ib_\omega^*\right)^{p-n} x^m y^n}{m!n!(p-m)!(p-n)!} \tag{10.38}$$

or

$$\mathcal{G}(x,y) = \mathrm{G}_{00} \exp\left\{b_\omega b_\omega^*\right\} \sum_{n,m} \frac{1}{\sqrt{m!n!}} \sum_{p>n,m} \frac{p!\left(ib_\omega\right)^{p-m}\left(ib_\omega^*\right)^{p-n} x^m y^n}{\sqrt{m!n!}(p-m)!(p-n)!} \tag{10.39}$$

Since from definition

$$\mathcal{G}(x,y) = \sum_{n,m} \mathrm{G}_{mn} \frac{x^m y^n}{\sqrt{m!n!}} \tag{10.40}$$

Then

$$G_{mn} = G_{00} \exp\left\{b_\omega b_\omega^*\right\} \frac{1}{\sqrt{m!n!}} \sum_{p>n,m} \frac{p!\left(ib_\omega\right)^{p-m}\left(ib_\omega^*\right)^{p-n}}{(p-m)!(p-n)!} \tag{10.41}$$

Now that all matrix elements between the energy eigenstates are determined, we can say that these represent the complete solution of the problem of a forced harmonic oscillator. Depending on the particular driving function $\gamma(t)$, the amplitudes for any particular transitions, can, in principle, be found.

10.1.1 Charged Particle Interaction with Phonons

We apply the knowledge of the transition amplitude matrix to the case of an electron interacting with lattice vibrations (phonons) in a polar medium. The electron in its motion carries with itself, the self-created polarization and this constitutes a many-body problem. More details on a charge particle interacting with lattice vibrations in a polar medium will be seen later with the polaron theorywhere we will find the polaron ground state energy (variational upper-bound ground state energy approximation) applying the Feynman variational principle. .

If the wave function at an initial time moment t_1 is $\Psi\left(\vec{r}_1, q_1\right)$ then K is also provided and the wave function $\Psi\left(\vec{r}_2, q_2\right)$ at an arbitrary time moment t_2 can be written as:

$$\Psi\left(\vec{r}_2, q_2\right) = \iint d\vec{r}_1 dq_1 K\left(\vec{r}_2, q_2, t_2 \middle| \vec{r}_1, q_1, t_1\right) \Psi\left(\vec{r}_1, q_1\right) \tag{10.42}$$

with the transition amplitude K being

$$K\left(\vec{r}_2, q_2, t_2 \middle| \vec{r}_1, q_1, t_1\right) = \int_{\vec{r}_1}^{\vec{r}_2} D\vec{r} \int_{q_1}^{q_2} Dq \exp\left\{\frac{i}{\hbar} S\left[\vec{r}, \vec{r}', q, q'\right]\right\} \tag{10.43}$$

Here, the action functional for the composite particle $S\left[\vec{r}, \vec{r}', q, q'\right]$:

$$S\left[\vec{r}, \vec{r}', q, q'\right] = S_L[q] - S_L\left[q'\right] + S_{e-L}\left[\vec{r}, q\right] - S_{e-L}\left[\vec{r}', q'\right] \tag{10.44}$$

is considered linear in the electron-phonon-coupling-amplitude and quadratic in the lattice vibration coordinates where $S_e\left[\vec{r}\right]$, the electronic action functional; $S_L[q']$ and $S_{e-L}\left[\vec{r}, q\right]$ are respectively, the lattice and electron-lattice action functionals. The transition amplitude considering 7.42 or 7.43 is rewritten

$$K\left(\vec{r}_2, q_2, t_2 \middle| \vec{r}_1, q_1, t_1\right) = \sum_n \exp\left\{-i \frac{E_n\left(t_2 - t_1\right)}{\hbar}\right\} \Psi_n^*\left(\vec{r}_1, q_1\right) \Psi_n\left(\vec{r}_2, q_2\right) \tag{10.45}$$

Here, $\vec{r}$ and q are, respectively, the electronic and the polarization field oscillator coordinates, $\Psi\left(\vec{r}, q\right)$ a complete orthonormal set of eigenfunctions of the full Hamiltonian of the composite particle:

$$\hat{H}\left(\vec{r}, q\right) = \hat{H}_e\left(\vec{r}\right) + \hat{H}_L(q) + \hat{H}_{e-L}\left(\vec{r}, q\right) \tag{10.46}$$

and E_n represents the corresponding energy eigenvalues.

The ground-state energy, E_0, can be conveniently obtained from equation 10.45 by multiplying both sides of 10.45 by $\psi_0^*(q_2)\psi_0(q_1)$ and then integrate over the variables q_1 and q_2, giving

$$\iint dq_1 dq_2 \psi_0^*(q_2) K\left(\vec{r}_2, q_2, t_2 \middle| \vec{r}_1, q_1, t_1\right) \psi_0(q_1) = \sum_n \exp\left\{-i \frac{E_n\left(t_2 - t_1\right)}{\hbar}\right\} \Phi_{n0}^*\left(\vec{r}_1\right) \Phi_{n0}\left(\vec{r}_2\right) \tag{10.47}$$

where,

$$\Phi_{n0}(\vec{r}) = \int dq\psi_0^*(q)\Psi_n(\vec{r},q) \tag{10.48}$$

and $\psi_0(q)$ is a function of q that is not orthogonal to the full ground state eigenfunction $\Psi_0(\vec{r},q)$.

From 10.17 to 10.21 and 10.47 then the amplitude for which an external perturbation does not change, the oscillator's ground state, is:

$$\text{G}_{00} = \iint dq_1 dq_2 \psi_0^*(q_2)\text{K}(\vec{r}_2,q_2,t_2|\vec{r}_1,q_1,t_1)\psi_0(q_1) \tag{10.49}$$

Evaluating equation 10.47, by considering the analytic continuation $t = -i\hbar\tau$ (where τ is a real number):

$$0 \le \tau \le \beta \tag{10.50}$$

then

$$t_2 - t_1 = -i\hbar\beta \tag{10.51}$$

and letting

$$\vec{r}_2 = \vec{r}_1 = 0 \tag{10.52}$$

so, from 10.47 we have

$$\text{G}_{00} = \sum_n \exp\{-\beta E_n\}\Phi_{n0}^*(0)\Phi_{n0}(0) \to \exp\{-\beta E_n\}\Phi_{00}^*(0)\Phi_{00}(0) \tag{10.53}$$

So, the exact expression for the composite's particle ground-state energy in terms of the transition amplitude:

$$E_0 = -\lim_{\beta\to 0}\frac{1}{\beta}\ln \text{G}_{00} \tag{10.54}$$

Here,

$$\text{G}_{00} = \int D\vec{r}\exp\{S[\vec{r}]\} \tag{10.55}$$

and the initial bonafide many-body problem via the Feynman approach is reduced to a one-body problem (because the phonon coordinates are adequately eliminated) and described by the action functional:

$$S[\vec{r}] = -\int_0^\beta \frac{m\dot{\vec{r}}^2}{2\hbar^2}d\tau + \Phi_\omega[\vec{r},\vec{r}\,'] \tag{10.56}$$

with the first term being the kinetic energy of the electron, $\Phi_\omega[\vec{r},\vec{r}\,']$, considering 10.26 describes the **full functional of the electron-phonon interaction (influence phase)**:

$$\Phi_\omega[\vec{r},\vec{r}\,'] = \frac{\hbar}{4m\omega}\int_0^\beta d\tau\int_0^\tau d\tau'\gamma(\vec{r}(\tau))\gamma(\vec{r}(\tau'))\exp\{-\omega|\tau-\tau'|\} \tag{10.57}$$

This functional is responsible for the phonon-mediated attractive interaction of the electron at time τ with itself at an earlier time τ' and also depend on the quantity $|\tau - \tau'|$ indicative of a retarded function depending on the past, thereby signifying interaction with the past where the perturbative motion of the electron takes "**time**" to propagate in the crystal lattice. This suggests the self-interaction to be stronger in the near past than it is in the distant past.

The ground state energy in 10.54 could not be possibly evaluated with ease via path integral as it is non-Gaussian. The Feynman solution to the composite particle problem provides a technical mathematical tool on how to calculate this non-Gaussian path integral. This tool allows the building of a class of exactly solvable models corresponding to quadratic functionals on whose bases is built a variational technique that gives a good upper bound to the composite particle's ground state energy. For this, Feynman introduced the notion of a model system with a fictitious particle that we examine together with the coupling amplitude $\gamma(t)$ in chapter 18 on the polaron theory.

11

Path Integral Perturbation Theory

The importance of the path integral formalism should be clear by now and we have seen how it can be useful to obtain matrix elements and expectation values. However, its usefulness may not be very obvious when it is used for interacting particles. For such an interaction, when the non-Gaussian part of the action is weak compared to the quadratic part, then one expects perturbation expansion to be applicable. This is really much easier than in the operator formalism since we deal only with c-numbers in the path integration.

Often in quantum mechanics we may be confronted with systems that are very difficult to solve exactly. This implies that for such systems the Hamiltonian may not be diagonalized exactly and so we resolve to approximate solutions that may sometimes give results close to true experimental values. In quantum mechanics as in classical mechanics there are relatively few systems of physical interest for which the equations of motion may be solved exactly. Approximate methods are thus expected to play an important role in virtually all applications of the theory. For quantum mechanics, an exact solution of the Schrödinger equation exists only for few idealized problems. The perturbation theory is applied to those cases in which the real system can be described by a small change in an easily solvable idealized system.

In the evaluation of path integration, when the exponential functions are quadratic in q and $\dot{q}$ then the integral is exactly solvable. But the alternative method, the so-called perturbation theory, can be employed as mentioned. For that reason, we rewrite the transition amplitude 4.82:

$$K\left(q_b,t_b;q_a,t_a\right)=\int_{q_a}^{q_b}Dq(t)\exp\left\{\frac{i}{\hbar}\int_{t_a}^{t_b}L(q,\dot{q},t)dt\right\}\equiv\int_{q_a}^{q_b}Dq(t)\exp\left\{\frac{i}{\hbar}S[q]\right\} \tag{11.1}$$

where the Lagrangian $L(q,\dot{q},t)$:

$$L(q,\dot{q},t)=\frac{m}{2}\dot{q}^2-U(q) \tag{11.2}$$

From the Lagrangian 11.2, we show that the transition amplitude 4.82, in which the potential $U(q)$ is inherent, can be viewed as a sequence of interactions with the potential separated by free propagations.

It is useful further to represent the perturbation series pictorially by denoting the full transition amplitude with a double undirected straight line:

the unperturbed transition amplitude by undirected straight line:

and insertion of the perturbation by a wiggly line:

Let us select the potential $V_0(q)$ then we write the following transition amplitude for that potential:

$$\mathrm{K}_0\left(q_b,t_b;q_a,t_a\right)=\int_{q_a}^{q_b} Dq(t)\exp\left\{\frac{i}{\hbar}\int_{t_a}^{t_b}\left[\frac{m}{2}\dot{q}^2-V_0(q)\right]dt\right\} \tag{11.3}$$

and diagrammatically represented:

To find the transition amplitude for subsequent potentials then we introduce the so-called perturbation $v(q)$

$$v(q)\equiv U(q)-V_0(q) \tag{11.4}$$

then follows the transition amplitude:

$$\mathrm{K}\left(q_b,t_b;q_a,t_a\right)=\int_{q_a}^{q_b} Dq(t)\exp\left\{\frac{i}{\hbar}\int_{t_a}^{t_b}\left[\frac{m}{2}\dot{q}^2-v(q)-V_0(q)\right]dt\right\} \tag{11.5}$$

or

$$\mathrm{K}\left(q_b,t_b;q_a,t_a\right)=\int_{q_a}^{q_b} Dq(t)\exp\left\{\frac{i}{\hbar}\int_{t_a}^{t_b}\left[\frac{m}{2}\dot{q}^2-V_0(q)\right]dt\right\}\exp\left\{-\frac{i}{\hbar}\int_{t_a}^{t_b}v(q)dt\right\} \tag{11.6}$$

that we represent pictorially by a double line:

We expand the second exponential in series of the perturbation $v(q)$ (**Neumann-Liouville expansion or Dyson series**):

$$\mathrm{K}\left(q_b,t_b;q_a,t_a\right)=\int_{q_a}^{q_b} Dq(t)\exp\left\{\frac{i}{\hbar}\int_{t_a}^{t_b}\left[\frac{m}{2}\dot{q}^2-V_0(q)\right]dt\right\}$$
$$\left(1+\left(-\frac{i}{\hbar}\right)\int_{t_a}^{t_b}v(s)ds+\frac{1}{2}\left(-\frac{i}{\hbar}\right)^2\int_{t_a}^{t_b}\int_{t_a}^{t_b}v(s)v(s')dsds'+\cdots\right) \tag{11.7}$$

or

$$\mathrm{K}\left(q_b,t_b;q_a,t_a\right)=\mathrm{K}_0\left(q_b,t_b;q_a,t_a\right)+\mathrm{K}_1\left(q_b,t_b;q_a,t_a\right)+\mathrm{K}_2\left(q_b,t_b;q_a,t_a\right)+\cdots \tag{11.8}$$

that we represent pictorially (Figure 11.1):

So, the full amplitude can be written as sum of partial amplitudes in which the particle is not scattered plus scattered once plus scattered twice and so on. Here, for the zero order of the perturbation theory

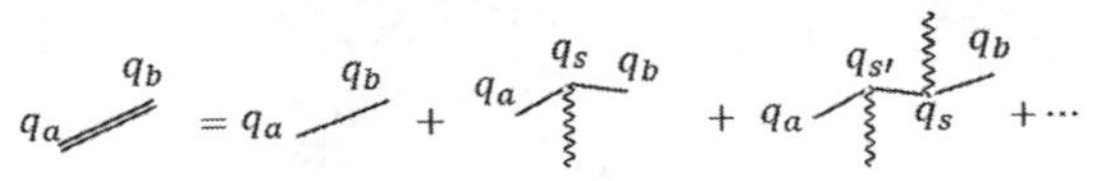

FIGURE 11.1 Pictorial representation of the full transition amplitude as sum of partial amplitudes in which the particle is not scattered plus scattered once plus scattered twice and so on.

$$K_0(q_b,t_b;q_a,t_a) = \int_{q_a}^{q_b} Dq(t) \exp\left\{\frac{i}{\hbar}\int_{t_a}^{t_b}\left[\frac{m}{2}\dot{q}^2 - V_0(q)\right]dt\right\} \tag{11.9}$$

is the contribution to the transition amplitude resulting from the particle going from q_a to q_b in the time $(t_b - t_a)$, without being affected by the perturbation $v(q)$. For the first order of the perturbation theory then

$$K_1(q_b,t_b;q_a,t_a) = \left(-\frac{i}{\hbar}\right)\int_{q_a}^{q_b} Dq(t) \exp\left\{\frac{i}{\hbar}\int_{t_a}^{t_b}\left[\frac{m}{2}\dot{q}^2 - V_0(q)\right]dt\right\}\int_{t_a}^{t_b} v(s)\,ds \tag{11.10}$$

which is the contribution to the transition amplitude when the particle is scattered once due to the perturbation $v(q)$ (Figure 11.2).

The last factor of 11.10 is in the discretized form:

$$\int_{t_a}^{t_b} v(s)\,ds = \varepsilon \sum_i v(q_i) \tag{11.11}$$

where i indicates the sum over time slices and then the transition amplitude

$$K(q_b,t_b;q_a,t_a) = \frac{1}{A}\int_{-\infty}^{+\infty}\cdots\int_{-\infty}^{+\infty}\prod_{i=1}^{N-1}\frac{dq_i}{A}\exp\left\{\frac{i\varepsilon}{\hbar}\sum_{i=0}^{N-1}\left[\frac{m}{2}\left(\frac{q_{i+1}-q_i}{\varepsilon}\right)^2 - U(q_i)\right]\right\} \tag{11.12}$$

where the normalization factor is

$$\frac{1}{A} = \left(\frac{m}{2\pi i\hbar\varepsilon}\right)^{\frac{1}{2}} \tag{11.13}$$

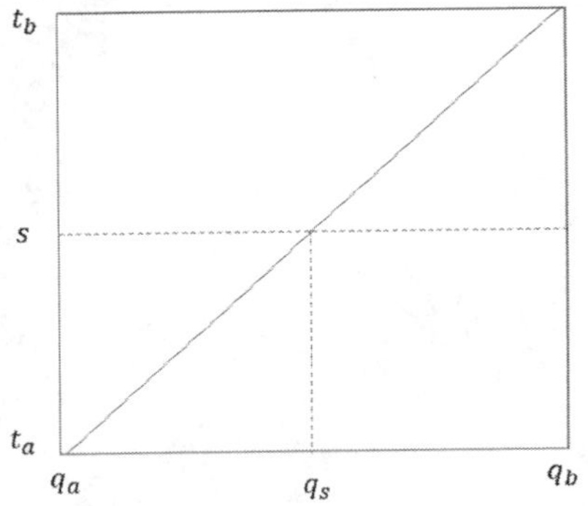

FIGURE 11.2 Illustrating the time-ordering procedure in 11.7 or 11.8 and showing the initial, final and intermediate time moments together with the respective coordinates during the evolution of the system.

Consider that the particles travel from point q_a at the time moment t_a via point q_s with time moment s to point q_b with time moment t_b:

$$t_a \le s \le t_b \tag{11.14}$$

From here and 11.12, the first order perturbation may be written as an integral equation for the transition amplitude:

$$K_1(q_b,t_b;q_a,t_a) = \int_{t_a}^{t_b} ds \int_{-\infty}^{\infty} dq_s K_0(q_b,t_b;q_s,t_s)\left(-\frac{i}{\hbar}\right)v(q_s)K_0(q_s,t_s;q_a,t_a) \tag{11.15}$$

or

$$K_1(q_b,t_b;q_a,t_a) = \int K_0(q_b,t_b;q_s,t_s)\left(-\frac{i}{\hbar}\right)v(q_s)K_0(q_s,t_s;q_a,t_a)d\tau_s \tag{11.16}$$

where

$$d\tau_s = dq_s ds \tag{11.17}$$

The pictorial representation is as follows (Figure 11.3)

For the second order of the perturbation theory we consider first the following Figure 11.4 (shaded area).

From Figure 11.4, we have

$$s \le s' \le t_b, t_a \le s \le t_b, t_a \le s \le s', t_a \le s' \le t_b \tag{11.18}$$

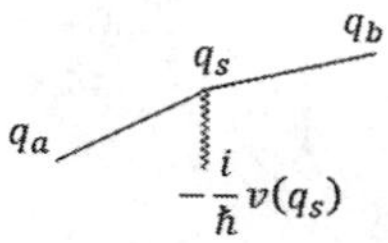

FIGURE 11.3 Pictorial representation of the first order perturbation theory.

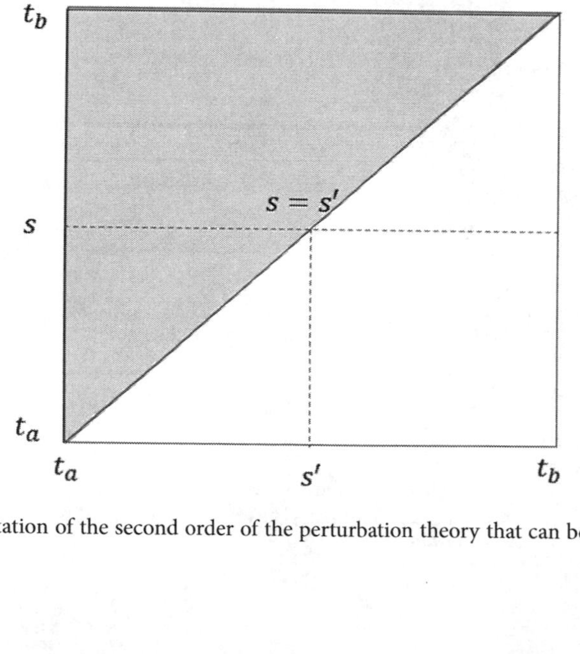

FIGURE 11.4 Representation of the second order of the perturbation theory that can be evaluated by the shaded area.

From these time intervals then

$$\left(\int_{t_a}^{t_b} v(q_s)ds\right)^2 = \int_{t_a}^{t_b} v(q_s)ds \int_{t_a}^{t_b} v(q_{s'})ds' \tag{11.19}$$

or

$$\left(\int_{t_a}^{t_b} v(q_s)ds\right)^2 = \int_{t_a}^{t_b} v(q_s)ds \int_{t_a}^{t_b} v(q_{s'})ds' = \iint_{t_a}^{t_b} v(q_s)v(q_{s'})dsds' \tag{11.20}$$

or

$$\left(\int_{t_a}^{t_b} v(q_s)ds\right)^2 = \int_{t_a}^{t_b}\int_{t_a}^{s} v(q_s)v(q_{s'})dsds'_{s'<s} + \int_{t_a}^{t_b}\int_{t_a}^{s} v(q_s)v(q_{s'})dsds'_{s'>s} \tag{11.21}$$

From here and Figure 11.4 we observe that the integrals covering the triangle above the diagonal of the square and the triangle below the diagonal of the square are equal:

$$\left(\int_{t_a}^{t_b} v(q_s)ds\right)^2 = 2\int_{t_a}^{t_b}\int_{t_a}^{s} v(q_s)v(q_{s'})dsds' \tag{11.22}$$

We can therefore write the second order of the perturbation theory as follows:

$$K_2(q_b,t_b;q_a,t_a) = \int K_0(q_b,t_b;q_s,s)\left(-\frac{i}{\hbar}\right)v(q_s)K_0(q_s,s;q_{s'},s')\left(-\frac{i}{\hbar}\right)v(q_{s'})K_0(q_{s'},s';q_a,t_a)d\tau_s d\tau_{s'} \tag{11.23}$$

The pictorial representation is as in Figure 11.5.

This procedure may be continued for higher orders of the perturbation theory and we have

$$\begin{aligned} K_v(q_b,t_b;q_a,t_a) &= K_0(q_b,t_b;q_s,s) + \int K_0(q_b,t_b;q_s,t_s)\left(-\frac{i}{\hbar}\right)v(q_s)K_0(q_s,t_s;q_a,t_a)d\tau_s \\ &\quad + \int K_0(q_b,t_b;q_s,s)\left(-\frac{i}{\hbar}\right)v(q_s)K_0(q_s,s;q_{s'},s') \\ &\quad \left(-\frac{i}{\hbar}\right)v(q_{s'})K_0(q_{s'},s';q_a,t_a)d\tau_s d\tau_{s'} + \cdots \end{aligned} \tag{11.24}$$

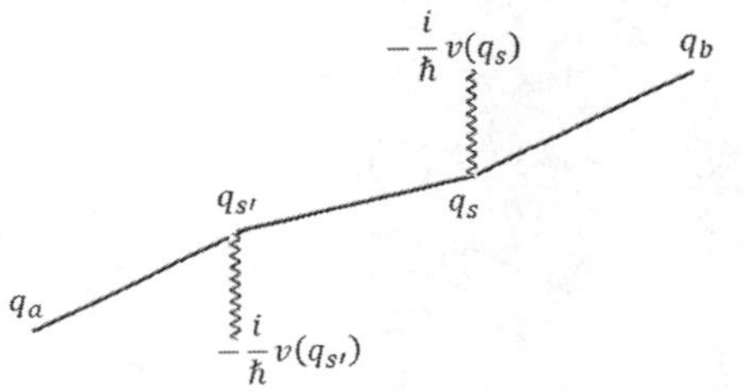

FIGURE 11.5 Pictorial representation of the second order of the perturbation theory.

We rewrite this as follows:

$$K_v(q_b,t_b;q_a,t_a) = K_0(q_b,t_b;q_s,s) + \int K_0(q_b,t_b;q_s,t_s)\left(-\frac{i}{\hbar}\right)v(q_s)d\tau_s$$
$$\left[K_0(q_s,t_s;q_a,t_a) + \int K_0(q_s,s;q_{s'},s')\left(-\frac{i}{\hbar}\right)v(q_{s'})K_0(q_{s'},s';q_a,t_a)d\tau_{s'}\cdots\right] \quad (11.25)$$

This may be rewritten in the following form by iteration:

$$K_v(q_b,t_b;q_a,t_a) = K_0(q_b,t_b;q_a,t_a) + \int K_0(q_b,t_b;q_s,t_s)\left(-\frac{i}{\hbar}\right)v(q_s)K_v(q_s,t_s;q_a,t_a)d\tau_s \quad (11.26)$$

So, for K_v we have the integral form

$$K_v(q_b,t_b;q_a,t_a) = K_0(q_b,t_b;q_a,t_a) + \left(-\frac{i}{\hbar}\right)\int K_0(q_b,t_b;q_c t_c)v(q_c)K_v(q_c t_c;q_a,t_a)d\tau_c \quad (11.27)$$

This mimics the usual equation for the time-evolution operator in the **interaction picture** if one writes equation 11.27 in operator form. The pictorial representation that is the **full transition amplitude equals unscattered transition amplitude plus sum of all processes with the last scattering at time t_c (Figure 11.6)**:

This integral equation is linked with the transition amplitude via a given potential. If v is the quantity then the perturbation series is limited to only a few terms, and we can use 11.27 to obtain a similar equation for wave function:

$$\psi(q_b,t_b) = \int K_v(q_b,t_b;q_a,t_a)\psi(q_a,t_a)dq_a \quad (11.28)$$

or

$$\psi(q_b,t_b) = \int K_0(q_b,t_b;q_a,t_a)\psi(q_a,t_a)dq_a + \left(-\frac{i}{\hbar}\right)\int K_0(q_b,t_b;q_c t_c)v(q_c)K_v(q_c t_c;q_a,t_a)\psi(q_a,t_a)d\tau_c dq_a \quad (11.29)$$

The first summand is an unchanging interaction (incident wave) and the second a scattering wave. Due to the Bloch scattering theory we swap K_v for K_0 and

$$\psi(q_b,t_b) = \int K_0(q_b,t_b;q_a,t_a)\psi(q_a,t_a)dq_a\left(-\frac{i}{\hbar}\right)\int K_0(q_b,t_b;q_c t_c)v(q_c)K_v(q_c t_c;q_a,t_a)\psi(q_a,t_a)d\tau_c dq_a \quad (11.30)$$

Considering

$$\Psi(q_b,t_b) = \int dq_c K(q_b,t_b;q_c,t_c)K(q_c,t_c;q_a,t_a) \quad (11.31)$$

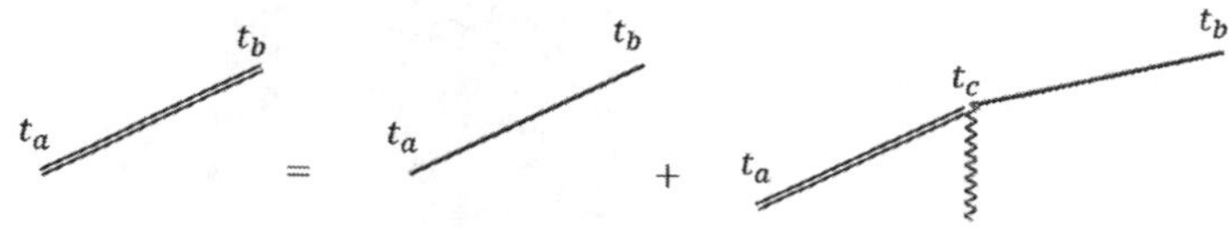

FIGURE 11.6 Pictorial representation of the full transition amplitude equals unscattered transition amplitude plus sum of all processes with the last scattering at time t_c.

and the fact that states have delta-normalization as seen earlier, then

$$\Psi(q_b,t_b) = \int dq\delta(q_b - q)\mathrm{K}(q,t_b;q_a,t_a) = \mathrm{K}(q_b,t_b;q_a,t_a) \tag{11.32}$$

and so

$$\Psi(q_b,t_b) = \int dq_c\mathrm{K}(q_b,t_b;q_c,t_c)\Psi(q_c,t_c) \tag{11.33}$$

then from

$$d\tau_c = dq_c dt_c \tag{11.34}$$

we have

$$\psi(q_b,t_b) = \psi_0(q_b,t_b) - \frac{i}{\hbar}\int \mathrm{K}_0(q_b,t_b;q_c,t_c)v(q_c)\psi(q_c,t_c)d\tau_c \tag{11.35}$$

Here, $\psi_0(q_b, t_b)$ is the unperturbed wave function satisfying the unperturbed Schrödinger equation

$$\left(i\hbar\frac{\partial}{\partial t_b} - \hat{\mathrm{H}}_0\right)\psi_0(q_b,t_b) = 0 \tag{11.36}$$

and the transition amplitude satisfying the relation

$$i\hbar\frac{\partial}{\partial t_b}\left[\theta\mathrm{K}_0\right] - \hat{\mathrm{H}}\left[\theta\mathrm{K}_0\right] = i\hbar\delta(q_b - q_c)\delta(t_b - t_c) \tag{11.37}$$

Here, θ is the **Heaviside step function**

$$\theta(t_b - t_c) = \begin{cases} 1, & t_b > t_c \\ 0, & t_b < t_c \end{cases} \tag{11.38}$$

which permits equation 11.35 to be rewritten as follows:

$$\psi(q_b,t_b) = \psi_0(q_b,t_b) - \frac{i}{\hbar}\int \theta(t_b - t_c)\mathrm{K}_0(q_b,t_b;q_c,t_c)v(q_c)\psi(q_c,t_c)d\tau_c \tag{11.39}$$

From here, considering 11.37 then

$$\left(i\hbar\frac{\partial}{\partial t_b} - \hat{\mathrm{H}}_0\right)\psi(q_b,t_b) = 0 - i\hbar\frac{i}{\hbar}\int \delta(q_b - q_c)\delta(t_b - t_c)v(q_c)\psi(q_c,t_c)d\tau_c \tag{11.40}$$

or

$$\left(i\hbar\frac{\partial}{\partial t_b} - \hat{\mathrm{H}}_0\right)\psi(q_b,t_b) = v(q_b)\psi(q_b,t_b) \tag{11.41}$$

or

$$i\hbar\frac{\partial}{\partial t_b}\psi(q_b,t_b) = \left(\hat{\mathrm{H}}_0 + v(q_b)\right)\psi(q_b,t_b) = \hat{\mathrm{H}}\psi(q_b,t_b) \tag{11.42}$$

So, equation 11.39 is an integral equation for $\psi(q_b, t_b)$ and imitates the Schrödinger equation.

11.1 Time-Dependent Perturbation

One of the principal tasks in quantum mechanics is the **evaluation of transition probabilities** from say state Φ_n to another state Φ_m. This takes place under the influence of a time-dependent perturbation $v(q,t)$ that tailors the transition. When does this transition of a system from one state to another makes sense? It may make sense if the reason for the transition, i.e., $v(q,t)$ acts only within a finite time period, say from $t = t_0$ to $t = T$. With the exception of this time period, the total energy is a **constant of motion** that can be measured. For the case when the Hamiltonian is dependent on time t, there are no stationary solutions of the Schrödinger equation. Hence the identification of bound states with discrete energy levels and stationary eigenfunctions must be modified. The **time-dependent perturbation theory** or so-called **method of variation of constants** assures that.

The Hamiltonian $\hat{H}$ of the system:

$$\hat{H} = \hat{H}_0 + \hat{v}(t) \tag{11.43}$$

where $\hat{H}_0$ is the unperturbed Hamiltonian and describes a conservative system i.e., that which is independent of time t. The index 0 stands for the time independence. The perturbation $v(t)$ is dependent on time t and has the effect of causing transitions between eigenstates of $\hat{H}_0$ that will be stationary in the absence of $v(t)$.

Instead of decomposing the Hamiltonian in 11.43, the adiabatic approximation assures that $\hat{H}$ has parameters that change very slowly with time. This follows that the system can be described approximately by means of stationary eigenfunctions of the instantaneous Hamiltonian. We finally consider the sudden approximation in which $\hat{H}$ is constant in time except for very short time intervals when it changes from one form to another.

If

$$\hat{H}(t) = f(t) \tag{11.44}$$

then the state of a system should be described by the time Schrödinger equation:

$$i\hbar \frac{\partial}{\partial t}\Psi = \hat{H}\Psi \tag{11.45}$$

Example

Solve equation 11.45 considering that

$$\hat{v}(t) = f(t) \tag{11.46}$$

Thus

$$\Psi_n^{(0)}(t) = \Psi_n^{(0)} \exp\left\{-i\frac{E_n t}{\hbar}\right\} \tag{11.47}$$

where $\Psi_n^{(0)}(t)$is the eigenfunction and E_n eigenvalue of the Hamiltonian $\hat{H}_0$. Equation 11.47 is the solution of equation 11.45 for $\hat{H}_0$. The functions 11.47 form a **complete set of basis functions**. The solution of 11.45 may be expanded in terms of these functions:

$$\Psi(t)=\sum_{n} C_n(t)\Psi_n^{(0)}(t) \tag{11.48}$$

If the Hamiltonian $\hat{H}$ should have not been dependent on time t then $C_m$should also not have been dependent on time t. If we substitute 11.48 and 11.43 into 11.45 then

$$i\hbar\frac{\partial}{\partial t}\Psi=\left(\hat{H}_0+\hat{v}(t)\right)\Psi \tag{11.49}$$

We consider this time-dependent perturbation $v(q, t)$ described by the following transition amplitude K_u:

$$K_{u+v}(q_b,t_b;q_a,t_a)=K_u(q_b,t_b;q_a,t_a)-\frac{i}{\hbar}\int K_u(q_b,t_b;q_c t_c)v(q_c,t_c)K_u(q_c t_c;q_a,t_a)d\tau_c \tag{11.50}$$

We show that this formula mimics the generalized formula of the perturbation theory. We write the propagator K_u via the time-independent Schrödinger eigenstates $|\Phi_n\rangle$ [1]:

$$K_u(q_b,t_b;q_c,t_c)=\sum_{n}\exp\left\{-\frac{iE_n(t_b-t_c)}{\hbar}\right\}\Phi_n(q_b)\Phi_n^*(q_c)\theta(t_b-t_c)\equiv\int Dq(t)\exp\left\{\frac{i}{\hbar}S[q(t)]\right\} \tag{11.51}$$

Here, the **Heaviside step function**:

$$\theta(t_b-t_c)=\begin{cases}1 & , \quad t_b>t_c\\ 0 & , \quad t_b<t_c\end{cases} \tag{11.52}$$

entails compelling the system at a starting point to be evolving towards the future and

$$\hat{H}|\Phi_n\rangle=E_n|\Phi_n\rangle \tag{11.53}$$

with E_n being the eigenenergies for the eigenstates n. Expression 7.43 might be viewed as if the particle moves around in all possible locations in the energy space (acquiring different phase factors) as it propagates from the selected initial position to the final position.

We substitute 11.51 into 11.50 so that in the second summand of 11.50 we have

$$-\frac{i}{\hbar}\int K_u(q_b,t_b;q_c t_c)v(q_c,t)K_u(q_c t_c;q_a,t_a)d\tau_c=-\frac{i}{\hbar}\int\sum_{n}\exp\left\{-\frac{iE_n(t_b-t_c)}{\hbar}\right\}\Phi_n(q_b)\Phi_n^*(q_c)v(q_c,t_c)$$
$$\sum_{m}\exp\left\{-\frac{iE_m(t_c-t_a)}{\hbar}\right\}\Phi_m(q_c)\Phi_m^*(q_a)d\tau_c \tag{11.54}$$

The perturbed amplitude in the energy eigenbasis is then:

$$K_{u+v}(q_b,t_b;q_a,t_a)=K_u(q_b,t_b;q_a,t_a)-\frac{i}{\hbar}\int\sum_{n}\exp\left\{-\frac{iE_n(t_b-t_c)}{\hbar}\right\}\Phi_n(q_b)\Phi_n^*(q_c)v(q_c,t_c)$$
$$\sum_{m}\exp\left\{-\frac{iE_m(t_c-t_a)}{\hbar}\right\}\Phi_m(q_c)\Phi_m^*(q_a)d\tau_c \tag{11.55}$$

Letting the matrix element of the perturbation operator constructed on the eigenfunctions $\Phi_n^*(q_c)$ and $\Phi_m(q_c)$:

$$v_{nm}(t_c) = \int \Phi_n^*(q_c) v(q_c, t_c) \Phi_m(q_c) dq_c \tag{11.56}$$

The quantity $v_{nm}(t_c)$ is the matrix element of the potential and is time dependent as the perturbation itself is dependent explicitly on time. The quantity λ_{nm}, representing the expansion coefficient or the probability amplitude that at time moment t_b the particle will be found in the state with m if at the initial time moment t_a the particle was found in the state with n, is expressed as follows:

$$\lambda_{nm}(t_b, t_a) = -\frac{i}{\hbar} \exp\left\{-\frac{i}{\hbar}(E_n t_b + E_m t_a)\right\} \int_{t_a}^{t_b} v_{nm}(t_c) \exp\{-i\omega_{mn} t_c\} dt_c \tag{11.57}$$

where,

$$\omega_{mn} = \frac{E_m - E_n}{\hbar} \tag{11.58}$$

is the **Bohr angular frequency** for the transition:

$$E_m \rightarrow E_n \tag{11.59}$$

Then

$$K_{u+v}(q_b, t_b; q_a, t_a) = K_u(q_b, t_b; q_a, t_a) + \sum_{nm} \lambda_{nm}(t_b, t_a) \Phi_n(q_b) \Phi_m^*(q_a) \tag{11.60}$$

Multiplying this relation by $\Phi_{m'}(q_b)$ and then integrate over t_b we represent the perturbation as small series representation via the expansion coefficient $\lambda_{nm}(t_b, t_a)$ as function of the perturbation v:

$$\lambda_{nm} = \delta_{nm} \exp\left\{-\frac{iE_n(t_b - t_a)}{\hbar}\right\} + \delta_{nm}^{(1)} + \delta_{nm}^{(2)} + \cdots \tag{11.61}$$

For the zero order correction, we assume that before the introduction of the perturbation the system is found at the n state. For the first order correction we have

$$\delta_{nm}^{(1)} = -\frac{i}{\hbar} \exp\left\{-\frac{i}{\hbar}(E_n t_b + E_m t_a)\right\} \int_{t_a}^{t_b} v_{nm}(t_c) \exp\{-i\omega_{mn} t_c\} dt_c \tag{11.62}$$

Suppose that the wave function at the time moment t_1 is $\Phi_n(q_1)$ then the wave function at the time moment t_2 is:

$$\begin{aligned} \Phi_n(q_2) = \int_{-\infty}^{\infty} K_v(q_2, t_2; q_1, t_1) \Phi_n(q_1) dq_1 &= \sum_{kl} \lambda_{kl}(t_2, t_1) \Phi_k(q_2) \int_{-\infty}^{\infty} \Phi_l^*(q_1) \Phi_n(q_1) dq_1 \\ &= \sum_k \lambda_{kn}(t_2, t_1) \Phi_k(q_2) \end{aligned} \tag{11.63}$$

This implies that the wave function at the time moment t_2 may have the form

$$\Phi_n(q_2) = \sum_k \lambda_{kn}(t_2, t_1) \Phi_k(q_2) \tag{11.64}$$

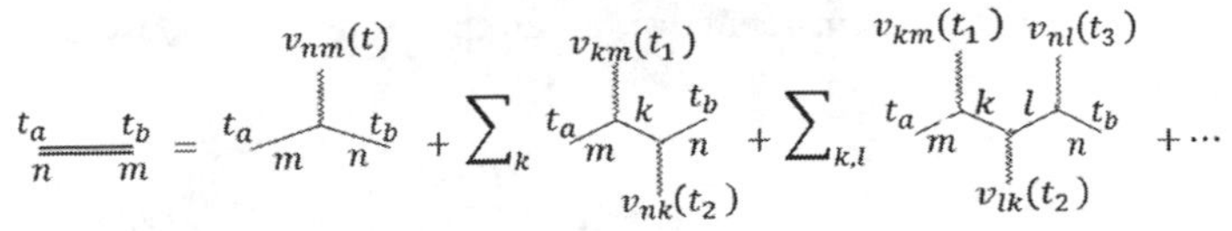

FIGURE 11.7 Transition amplitude represented pictorially in higher orders of the perturbation theory.

This is the first order term with interest being the scattering probability from the state with quantum number n to one with m at the time interval dt expressed via

$$\left(-\frac{i}{\hbar}\right)v_{nm}(t)dt \tag{11.65}$$

The transition amplitude in 11.60 can be represented pictorially in higher orders of the perturbation theory (Figure 11.7):

The first order describes the direct transition from the state with m to one with n and the second order together with higher orders transitions via intermediate states where all the intermediate **virtual** states $k, l, \cdots$ are summed over, and the **interaction times**

$$t_1 < t_2 < \cdots < t_n \tag{11.66}$$

are integrated over. The evaluation of the transition amplitude via the sequence of the first order is valid for the case when the perturbation is sufficiently small, as indicated earlier. However, there exist cases when the velocity of variation of the perturbation $v(t)$ is insignificant that v may not be small. This case may be observed from slow atomic collisions where the adiabatic approximation based on the eigenfunctions and eigenvalues of the full Hamiltonian is very appropriate, as shown at the start of this section.

11.2 Transition Probability

We examine the transition between two discrete levels where, if before the introduction of the perturbation the system was at rest in the stationary state $E_n$and after the perturbation the change in energy gave E_m, then the transition probability of the system from the initial state n to the final state m in the first order of the perturbation remains invariant and has the form:

$$\mathrm{P}_{n\to m} = \frac{1}{\hbar^2}\left|\int_{-\infty}^{\infty} v_{nm}(t)\exp\{-i\omega_{mn}t\}dt\right|^2 \tag{11.67}$$

We assume the absence of degeneracy in this case in addition to the fact that the perturbation is small. From 11.61, the probability of the system to remain in the state n for $t \to \infty$ when the perturbation is switched on earlier is rewritten:

$$\mathrm{P}_n = \left|\lambda_{nn}(t\to\infty)\right|^2 \tag{11.68}$$

Letting,

$$\tilde{v}_{nm} = \int_{-\infty}^{\infty} v_{nm}(t)\exp\{-i\omega_{mn}t\}dt \tag{11.69}$$

When the perturbation is time-independent, we can do the t integrals and for the first order correction we have

$$\delta_{nm}^{(1)} = \frac{1}{\hbar\omega_{mn}}\left(\exp\left\{-\frac{iE_nT}{\hbar}\right\} - \exp\left\{-\frac{iE_mT}{\hbar}\right\}\right)v_{nm}, T = t_b - t_a \tag{11.70}$$

The first order contribution to the transition probability of the system from the initial state n to the final state m (after switching on the perturbation) for $n \neq m$:

$$\mathrm{P}_{n\to m} = \frac{1}{\hbar^2\omega_{mn}^2}\left|\exp\left\{-\frac{iE_mT}{\hbar}\right\}\left(\exp\{-iT\omega_{mn}\}-1\right)\right|^2 |v_{nm}|^2 \tag{11.71}$$

or

$$\mathrm{P}_{n\to m} = \frac{1}{\hbar^2}|v_{nm}|^2 \frac{4\sin^2\frac{\omega_{mn}T}{2}}{\omega_{mn}^2} = \frac{f(\mathrm{T},\omega_{mn})}{\hbar^2}|v_{nm}|^2 \tag{11.72}$$

This quantity oscillates in time between zero and $\frac{4|v_{nm}|^2}{\hbar^2\omega_{mn}^2}$.

11.3 Time-Energy Uncertainty Relation

To understand the main features of the result for the transition probability we examine the behavior of the function $f(t,\omega)$ for different values of ω:

$$f(t,\omega) = \frac{4\sin^2\frac{\omega t}{2}}{\omega^2} \tag{11.73}$$

The function $f(t,\omega)$ has basically a large peak (resonant peak), centered at $\omega = 0$, with height

$$\lim_{\omega\to 0} f(t,\omega) = \lim_{\omega\to 0}\frac{4\sin^2\frac{\omega t}{2}}{\omega^2} = \lim_{\omega\to 0}\frac{\sin^2\frac{\omega t}{2}}{\left(\frac{\omega t}{2}\right)^2}t^2 = t^2 \tag{11.74}$$

and width $\cong \frac{2\pi}{t}$ (Figure 11.8). This resonance width is the distance between the first zeros on either side of the resonant frequency. The resonance width shows that the longer the perturbation acts, the narrower the resonance would be. Away from the value of the resonance frequency $=0$, the function f oscillates with damped amplitude in much the same way as a diffraction pattern. So, for larger t, there is only an appreciable transition probability for those states whose energy lies in a band-width ΔE of about the initial energy E_n:

$$t \cong \frac{2\pi}{\omega} = \frac{2\pi\hbar}{|E_m - E_n|} = \frac{2\pi\hbar}{\Delta E} \tag{11.75}$$

or

$$\Delta E \cong \frac{2\pi\hbar}{t} \tag{11.76}$$

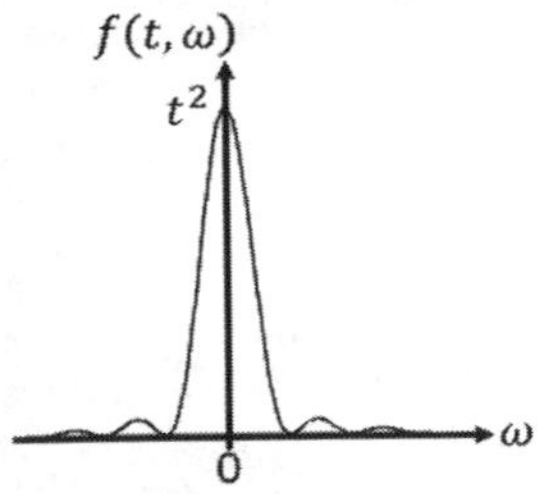

FIGURE 11.8 The variation of the function $f(t, \omega)$ on the frequency ω for fixed t with the large peak (resonance peak), centered at $\omega = 0$, with height t^2 and width inversely proportional to t.

So, from here

$$t\Delta E \geq \hbar \tag{11.77}$$

This is the time-energy uncertainty relation and was obtained for the first time by Bohr. For the conservative system, the greater the energy uncertainty, the more rapid the time evolution. For the time t, the perturbation is tuned on. The time here is a parameter and not an observable.

From the integral

$$\int_{-\infty}^{\infty} \frac{\sin^2 x}{x^2} dx = \pi \tag{11.78}$$

and then

$$\int_{-\infty}^{\infty} f(t,\omega) d\omega = 2\pi t \tag{11.79}$$

and $f(t, \omega)$ achieves an infinite peak of area $2\pi t$ as $t \to \infty$:

$$\lim_{t\to\infty} f(t,\omega) \cong 2\pi t\delta(\omega) \tag{11.80}$$

So, for small t, the peak is broader and implies large energy change. For large t, the peak is narrow.

We already know that

$$\Delta p \Delta q \geq \hbar \tag{11.81}$$

is the Heisenberg uncertainty relation. This shows a limitation on the exactness of measurement of the coordinate q and momentum p of a particle simultaneously. In 11.77 such a limitation on the measurement of the energy E and time t may seem not to be apparent. It says that if we do two measurements of the energy E of a system that is prepared at time moment $t = 0$ and at $t \neq 0$, then the minimum of the dispersion of these two values of the energy is defined by 11.77:

$$\Delta E \geq \frac{\hbar}{t} \tag{11.82}$$

The energy may be measured with an instrument. Relation 11.77 is satisfied as if there were no small perturbation.

Consider the disintegration of a nucleus of an atom so that

$$E \to E_1 + E_2 \tag{11.83}$$

Here, E is the initial energy and E_1 and E_2 are the energies of the systems as a result of disintegration. Suppose E_1 and E_2 are known with great accuracy, then all the uncertainty in ΔE is related to the initial state. What should be the significance of the time t in 77? The time $t = \tau$ is the **lifetime** due to the disintegration:

$$\Delta E \geq \frac{\hbar}{\tau} = \hbar W \tag{11.84}$$

If a system is subjected to disintegration, then the energy of the corresponding state is a quantity which is totally undefined and this is described by the equality 11.84 and is called the **widening** of the energy level that is an observable physical effect. This appears in spectroscopy in the fine structure. In place of spectral lines, we have stripes. It may be shown that the width of the stripes is equal to W. Quantum mechanics gives us the possibility to evaluate this width. So, the energy of an unstable state cannot be determined with arbitrary accuracy. It can only be within an uncertainty of the order ΔE, which is the **natural width** of the given state.

11.4 Density of Final State

11.4.1 Transition Rate

We examine the case of a continuous spectrum. We consider one of the cases associated with the operator $\hat{H}_0$ where there exists the discrete spectrum together with the continuous spectrum. So, for this we suppose a transition not to a single final state but to be a range ν of final states about the energy E_m. This situation arises in a potential well where the continuous spectrum is situated above the discrete spectrum in the energy scale.

We examine a many-particle system with a good example being the helium atom. The helium nucleus has two protons and two neutrons. Around the nucleus we have two electrons. The two electrons have an attractive force for the nucleus in addition to the repulsion between the two electrons. We assume that the two electrons are distinguishable, though in reality this is not the case. If we consider the spin property of the two electrons, then they are indeed distinguishable. For brevity, we neglect the spin coupling interaction. The helium atom is only slightly more complex than the hydrogen atom and may not be solved analytically. Variational techniques are usually applied to the helium atom that may aid in solving more complex problems.

For brevity, we ignore the mutual repulsion between the two electrons then it is possible to talk of the state of each electron separately in the field of the Helium nucleus. The system will then be equivalent to two hydrogen atoms. The Hamiltonian for such a problem will then be that of two hydrogenic atoms. We can then find the ground state energy for this system. This ground state energy is the same as the binding energy (or ionization energy) which can be measured easily. Comparing the ground state energy of the hydrogenic atom with that of the helium atom we observe that the mutual repulsion between the two electrons in the helium atom contributes significantly to the ground state energy. This comparison permits us to place on the same energy scale, the discrete spectrum of the helium atom, the excited states of the two electrons, the continuous spectrum of two-electron system, and the ionized helium atom with electron continuous spectrum. From the interaction between electrons we can show the nonexistence of the one-electronic state. For the many-body problem, the discrete spectrum is below the continuous spectrum on the energy scale. The energy of the excited state of the helium atom is the sum of the ground state energy of the ion and the free electron.

We consider the case associated with the Hamiltonian operator $\hat{H}_0$ where, after switching on the perturbation at $t = 0$, there exists a discrete spectrum together with the continuous spectrum and the transition $dP_{n\to\nu}$ is not to a single final state but to a range ν of final states about the energy E_n of the discrete spectrum. This should be a type of approximation mentioned above that permits the solution of the problem.

Since the final state is a continuous spectrum, then we examine the probability that the final state will have energy in the interval:

$$E \to E + dE \tag{11.85}$$

In this problem, the total angular moment of the system equals the angular moment of the excited atom. For convenience, we will not write the expressions for the angular moments without which the material will still be understood. We begin with the transition for the first order perturbation theory. We suppose that the energy E of the continuous spectrum is the unique parameter within the range 11.85 and so

$$dP_{n\to\nu} = \rho(E_\nu)dE_\nu \tag{11.86}$$

Then further we perform integration over the parameter E. This implies evaluating the transition probability:

$$P_{n\to\nu} = \int P_{n\to E}\varrho(E)dE \tag{11.87}$$

In the above relations, $\varrho(E)dE$ is the **number of states within the range** $E \to E + dE$ while the function $\varrho(E)$ is the so-called **density of final states.** Assuming the range ν to be small enough so that $\varrho(E)$ and $v_{n\nu}$ are constant within the given range. For exceedingly large T then

$$\lim_{T\to\infty} \frac{4\sin^2 \frac{\omega_{\nu n}T}{2}}{\omega_{\nu n}^2} = 2\pi T\delta(\omega_{\nu n}) \tag{11.88}$$

and the only dominant contribution to the integral arises from the energy range that corresponds to the narrow central peak of the function $f(t,\omega)$. So, the integration limits can safely be extended to infinity without affecting the given result:

$$P_{n\to\nu} = \int_{-\infty}^{\infty} \varrho(E)\frac{4\sin^2 \frac{\omega_{\nu n}T}{2}}{\omega_{\nu n}^2}|v_{n\nu}|^2 dE \tag{11.89}$$

or

$$P_{n\to\nu} = \frac{|v_{n\nu}|^2 \varrho(E_\nu)}{\hbar^2}\int_{-\infty}^{\infty} \frac{4\sin^2 \frac{\omega_{\nu n}T}{2}}{\omega_{\nu n}^2}\hbar d\omega_{\nu n} \tag{11.90}$$

or

$$P_{n\to\nu} = \frac{2\pi}{\hbar}|v_{n\nu}|^2 \varrho(E_\nu)T \tag{11.91}$$

Here, the transition probability $P_{n\to\nu}$ is proportional to the time T and so this permits the introduction of the **transition rate** (or **transition probability per unit time**):

$$W_\nu = \frac{2\pi}{\hbar}|v_{n\nu}|^2 \varrho(E_\nu) \tag{11.92}$$

This is the so-called **Fermi Golden Rule**. It is instructive to note that in the matrix element v_{nv}, one of the parameters v from the wave function is due to the continuous spectrum and the normalization by the Dirac delta function $\delta(E - E')$ has the dimension $(\text{energy})^{\frac{1}{2}}$. The above derivations have been possible by considering the condition:

$$\frac{2\pi}{\hbar}\left|v_{nv}\right|^2 T \ll 1 \tag{11.93}$$

So, to know what happens within the following condition

$$\frac{2\pi}{\hbar}\left|v_{nv}\right|^2 T \geq 1 \tag{11.94}$$

is impossible. However, the most exact solution of the system is still based on equations 11.50 or 11.55, where the summation over intermediate times is replaced by integration over intermediate times.

11.5 Continuous Spectrum due to a Constant Perturbation

We consider the transition in a continuous spectrum due to a constant perturbation. In this case, the perturbation is brought about by a quantum transition independent of time t. The transition probability of the states within the interval $\nu \to \nu + d\nu$:

$$dP_v(t) = \frac{\left|v_{vv_0}\right|^2}{\hbar^2}\frac{4\sin^2\frac{\omega_{vv_0}t}{2}}{\omega_{vv_0}^2}dv \tag{11.95}$$

If we do the change $t = T \to \infty$, then

$$dW_v = \frac{2\pi}{\hbar}\left|v_{vv_0}\right|^2 \delta\left(E_v - E_{v_0}\right)dv \tag{11.96}$$

This is the transition probability per unit time.

It is necessary to differentiate the transition probability from the transition rate. The transition rate is a quantity with dimension per second. It has the units of frequency. Consider dW to relate the optical transition in an atom and to be of the order 10^7sec^{-1}. It is the mean transition frequency and its inverse is called the mean lifetime τ. Suppose there exists a transition where at time moment $t = 0$ with N_0 atoms. We find the number of atoms $N(t)$ at time moment $t \neq 0$ from the equation:

$$dN = -NWdt \tag{11.97}$$

So,

$$N(t) = N_0 \exp\{-Wt\} \tag{11.98}$$

where W is the full transition probability per unit time t:

$$W = \frac{1}{\tau} \tag{11.99}$$

Here, τ is the mean lifetime relative to the given transition.

11.6 Harmonic Perturbation

For the case of $v = f(t)$ being a periodic dependence then we have

$$\hat{v}(t) = \hat{u}\exp\{i\omega t\} + \hat{w}\exp\{-i\omega t\} \tag{11.100}$$

Here, ω is a constant angular frequency; $\hat{u}$ and $\hat{w}$ are operators that explicitly do not depend on time t as $\hat{v}$ should be Hermitian, i.e., $\hat{v}^{\dagger} = \hat{v}$ which is easily seen after the substitution in 11.100:

$$\hat{u}\exp\{i\omega t\} + \hat{w}\exp\{-i\omega t\} = \hat{u}^{\dagger}\exp\{-i\omega t\} + \hat{w}^{\dagger}\exp\{i\omega t\} \tag{11.101}$$

From where

$$\hat{u}^{\dagger} = \hat{w} \tag{11.102}$$

This follows that

$$u_{nm} = w_{mn}^{*} \tag{11.103}$$

Here we set to introduce the perturbation at $t = 0$ and cut it off at $t = T$. Thus, if we consider 11.100 and 11.95 then

$$dP_{n\to v} = \frac{1}{\hbar^2}\left| u_{vn}\frac{\exp\{i(\omega_{vn}+\omega)T\}-1}{i(\omega_{vn}+\omega)} + u_{vn}^{*}\frac{\exp\{i(\omega_{vn}-\omega)T\}-1}{i(\omega_{vn}-\omega)}\right|^2 dv \tag{11.104}$$

For fixed t i.e., T, the transition probability dP_{nv} is a function of only the variable ω. For a resonance phenomenon we have for any of the two denominators in 11.104 the following frequencies

$$\omega_{vn} \approx \pm\omega \tag{11.105}$$

or

$$E_v \approx E_n \pm \hbar\omega \tag{11.106}$$

These two cases may correspond, respectively, to the absorption and emission of a quantum of radiation when electromagnetic interactions are involved.

Consider $\omega \geq 0$. Then $\omega_{vn} = \omega(-\omega)$ gives the resonance conditions corresponding to $\omega_{vn} > 0$ $(\omega_{vn} < 0)$. For $\omega_{vn} > 0$, the system goes from the lower energy level E_n to the higher-level E_v by the resonant absorption of an energy quantum $\hbar\omega$ (Figure 11.9).

For $\omega_{vn} < 0$, the resonant perturbation stimulates the passage of a system from the higher-level E_v to the lower level E_n. This is accompanied by the induced emission of an energy quantum $\hbar\omega$. Further, we

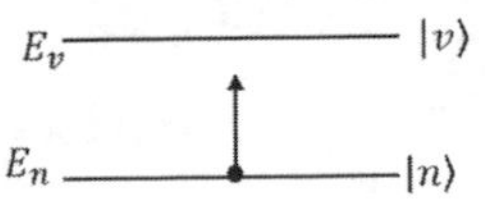

FIGURE 11.9 The resonance condition corresponding to $\omega_{vn} > 0$ $(\omega_{vn} < 0)$. For $\omega_{vn} > 0$ the system goes from the lower energy level E_n to the higher-level E_v by the resonant absorption of an energy quantum $\hbar\omega$.

consider the case for which $\omega_{\nu n} > 0$ and $\omega > 0$ as the case for which $\omega_{\nu n} < 0$ is considered analogously. Consider again 11.104. We do the denotation

$$\mathcal{P}^{+}_{\nu\to n} = \frac{\exp\{i(\omega_{\nu n}+\omega)T\}-1}{i(\omega_{\nu n}+\omega)} = \frac{\exp\left\{i(\omega_{\nu n}+\omega)\frac{T}{2}\right\}\sin\left[(\omega_{\nu n}+\omega)\frac{T}{2}\right]}{(\omega_{\nu n}+\omega)\frac{1}{2}} \quad (11.107)$$

$$\mathcal{P}^{-}_{\nu\to n} = \frac{\exp\{i(\omega_{\nu n}-\omega)T\}-1}{i(\omega_{\nu n}-\omega)} = \frac{\exp\left\{i(\omega_{\nu n}-\omega)\frac{T}{2}\right\}\sin\left[(\omega_{\nu n}-\omega)\frac{T}{2}\right]}{(\omega_{\nu n}-\omega)\frac{1}{2}} \quad (11.108)$$

The denominator of $\mathcal{P}^{-}_{\nu\to n}$ is zero for $\omega = \omega_{\nu n}$ and that of $\mathcal{P}^{+}_{\nu\to n}$ is zero for $\omega = -\omega_{\nu n}$. Thus, if we consider ω close to $\omega_{\nu n}$then $\mathcal{P}^{-}_{\nu\to n}$ should be the dominant term compared to $\mathcal{P}^{+}_{\nu\to n}$. This follows that $\mathcal{P}^{-}_{\nu\to n}$ is the **resonant term** while $\mathcal{P}^{+}_{\nu n}$ is the **anti-resonant term**. It should be noted that $\mathcal{P}^{+}_{\nu\to n}$ becomes resonant for negative $\omega_{\nu n}$ close to ω and as well close to $-\omega_{\nu n}$.

It should be noted that we deal with a situation in which the initial state $|n\rangle$ is a discrete bound state and final state $|\nu\rangle$ is one of a continuous set of dissociated states. Then

$$E_\nu > E_n$$

and we need consider only $\mathcal{P}^{-}_{\nu\to n}$. Thus 11.104 becomes:

$$d\mathrm{P}_{\nu\to n} = \mathrm{i}_{\nu\to n}(\omega,T)d\nu \ , \ \mathrm{P}_{\nu\to n}(\omega,T) = \frac{1}{\hbar^2}|u_{\nu n}|^2 \frac{4\sin^2(\omega_{\nu n}-\omega)\frac{T}{2}}{(\omega_{\nu n}-\omega)^2} \quad (11.109)$$

We consider the moment $T\to\infty$. It should be noted that

$$\lim_{T\to\infty}\frac{\sin^2\alpha T}{\alpha^2} = \pi\delta(\alpha) \quad (11.110)$$

The resonance width $\Delta\alpha = \frac{2\pi}{T}$ can be obtained at the zeros on both sides of the resonant point and for the transition probability 11.109 then

$$(\omega_{\nu n}-\omega)\frac{T}{2} = \pm\pi \quad (11.111)$$

Then from here the time T is larger if the width is smaller. It should be noted that during the time interval $[0, T]$ the perturbation performs numerous oscillations that are sinusoidal perturbations (Figure 11.10).

If T were small, then the perturbation would not have time to oscillate and this will be equivalent to a perturbation linear in time.

For $T\to\infty$, the maximum of the function $\frac{|u_{\nu n}|^2 T^2}{\hbar^2}$ tends to infinity and the width of the curve becomes smaller and the curve increases in height. This follows that the curve tends to a Dirac δ-function and we have

$$\pi T\delta\left(\frac{\omega_{\nu n}-\omega}{2}\right) = 2\pi\hbar T\delta(E_\nu - E_n - \hbar\omega) \quad (11.112)$$

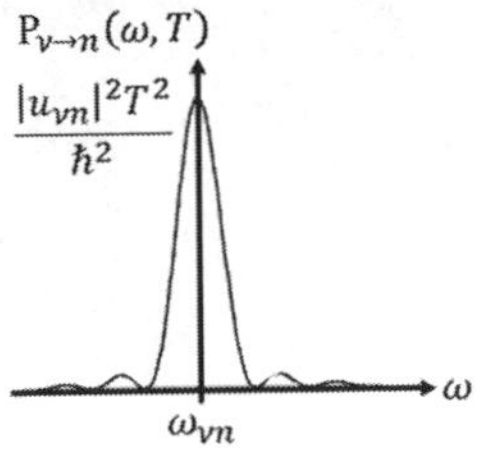

FIGURE 11.10 The variation of the transition probability $P_{v\to n}(\omega, T)$ on the frequency ω for fixed T with the large peak (resonant peak), centered at $\omega_{vn} = \omega$, with height $\frac{|u_{vn}|^2 T^2}{\hbar^2}$ and width inversely proportional to T.

So,

$$dW_v = \frac{2\pi}{\hbar}|u_{vn}|^2 \delta\left(E_v - E_n - \hbar\omega\right)dv \tag{11.113}$$

which is the transition rate. This formula cannot yield a numerical value since it contains a δ -function. In order to compare 11.112 with the experiment it is necessary to do away with the δ -function using integration.

Suppose that the energy $E$is the unique parameter, then

$$dv = \rho\left(E_v\right)dE_v \tag{11.114}$$

We now integrate 11.113 over the parameter $E$and this results in the transition rate for the continuous spectrum :

$$W_v = \frac{2\pi}{\hbar}|v_{vn}|^2 \rho\left(E_v\right) \tag{11.115}$$

Here, $\rho(E_v)$ is the density of state of the continuous spectrum and

$$E_v = E_n + \hbar\omega \tag{11.116}$$

From where it shows that if in the perturbation, there is absorption of light, then $\hbar\omega$ is the quantum of light and 11.116 is the equation for the Einstein photo effect.

12

Transition Matrix Element

Suppose a given state is described by the wave function $\chi(q, t)$. What, then, should be the probability that in the process of evolution from the state $\psi(q_a, t_a)$ at a point q_a to a state $\psi(q_b, t_b)$ at a point $q_b$that given state appears? This should be

$$K\left(q_b,t_b;q_a,t_a\right)=\int K\left(q_b,t_b;q_c,t_c\right)K\left(q_c,t_c;q_a,t_a\right)dq_c \tag{12.1}$$

Considering that the probability transition amplitude mimics the wave function, then

$$K\left(q_b,t_b;q_a,t_a\right)=\psi\left(q_b,t_b\right) \tag{12.2}$$

So, for the initial function, $\psi(q_a, t_a)$ then the final wave function is as follows:

$$\psi\left(q_b,t_b\right)=\int K\left(q_b,t_b;q_c,t_c\right)\psi\left(q_c,t_c\right)dq_c \tag{12.3}$$

We find the relation between the wave functions $\psi(q_b, t_b)$ and $\chi(q, t)$:

$$\psi\left(q_b,t_b\right)=\int K^*\left(q,t;q_a,t_a\right)\psi\left(q_a,t_a\right)dq_a\equiv\int\chi^*\left(q,t\right)\psi\left(q_a,t_a\right)dq_a \tag{12.4}$$

What should be the transition probability that at the state $\psi(q_b, t_b)$ we have the state $\chi(q, t)$:

$$\left|\psi\left(q_b,t_b\right)\right|^2=\int\chi^*\left(q_c,t_c\right)K\left(q_c,t_c;q_a,t_a\right)\psi\left(q_a,t_a\right)dq_adq_c \tag{12.5}$$

or

$$\left|\psi\left(q_c,t_c\right)\right|^2=\iint_{q_a}^{q_c}Dq\left(t\right)\chi^*\left(q_c,t_c\right)\exp\left\{\frac{i}{\hbar}S\left[q_c,q_a\right]\right\}\psi\left(q_a,t_a\right)dq_adq_c=\left\langle\chi\middle|1\middle|\psi\right\rangle_S \tag{12.6}$$

So, we generalize this formula

$$\left\langle\chi\middle|F\left[q\right]\middle|\psi\right\rangle_S=\int\chi^*\left(q_b,t_b\right)\psi\left(q_a,t_a\right)dq_adq_b\int_{q_a}^{q_b}Dq\left(t\right)F\left[q\left(t\right)\right]\exp\left\{\frac{i}{\hbar}S\left[q_b,q_a\right]\right\} \tag{12.7}$$

We take an example in the perturbation theory and we consider the path from the initial point q_a to the final point q_b. So, for brevity,

$$\left\langle F_S\right\rangle=\int_{q_a}^{q_b}Dq\left(t\right)F\left[q\left(t\right)\right]\exp\left\{\frac{i}{\hbar}S\left[q_b,q_a\right]\right\} \tag{12.8}$$

This definition permits an evolution of the Feynman apparatus that, for the case of a simple functional, mimics that of quantum mechanics.

We suppose that the action S may have two summands:

$$S = S_0 + \sigma \tag{12.9}$$

We do series expansion with respect to σ:

$$\langle 1_S \rangle = \int_{q_a}^{q_b} Dq(t) \exp\left\{\frac{i}{\hbar}(S_0 + \sigma)\right\} \cong \int_{q_a}^{q_b} Dq(t) \exp\left\{\frac{i}{\hbar} S_0\right\}\left[1 + \frac{i}{\hbar}\sigma + \frac{1}{2}\left(\frac{i}{\hbar}\sigma\right)^2 + \cdots\right] \tag{12.10}$$

or

$$\langle 1_S \rangle \cong \langle 1_{S_0} \rangle + \frac{i}{\hbar}\sigma_{S_0} + \frac{1}{2}\left(\frac{i}{\hbar}\right)^2 \sigma^2{}_{S_0} + \cdots \tag{12.11}$$

So, for

$$\langle \chi | 1 | \psi \rangle_S = \int \chi^*(q_b, t_b) \psi(q_a, t_a) dq_a dq_b \int_{q_a}^{q_b} Dq(t) \exp\left\{\frac{i}{\hbar} S[q_b, q_a]\right\} \tag{12.12}$$

then

$$\langle \chi | 1 | \psi \rangle_S \cong \frac{i}{\hbar} \langle \chi | \sigma | \psi \rangle_{S_0} + \langle \chi | \sigma^2 | \psi \rangle_{S_0} + \cdots \tag{12.13}$$

We suppose the perturbation to be time-independent then we can do t integrals. Letting,

$$\mathrm{F} = \int_{q_a}^{b} v(s) ds \tag{12.14}$$

then

$$\langle \chi | \mathrm{F}[q] | \psi \rangle_S = \int_{q_a}^{b} ds \int_{q_a}^{q_b} Dq(t) \chi^*(q_b, t_b) v(s) \exp\left\{\frac{i}{\hbar} S_0[q_b, t_b, q_a, t_a]\right\} \psi(q_a, t_a) dq_a dq_b \tag{12.15}$$

We time slice the interval:

$$t_a \le s \le t_b \tag{12.16}$$

This transforms the action functional to

$$S_0[q_b, t_b, q_a, t_a] = S_0[q_b, t_b, q_c, t_c] + S_0[q_c, t_c, q_a, t_a] \tag{12.17}$$

and so

$$\int_{q_a}^{q_b} Dq(t) \exp\left\{\frac{i}{\hbar} S_0[q_b, t_b, q_a, t_a]\right\} = \int\left(\int_{q_s}^{q_b} Dq(t) \exp\left\{\frac{i}{\hbar} S_0[q_b, t_b, q_s, s]\right\} \int_{q_a}^{q_s} Dq(t) \exp\left\{\frac{i}{\hbar} S_0[q_s, s, q_a, t_a]\right\}\right) dq_s \tag{12.18}$$

or

$$\int_{q_a}^{q_b} Dq(t)\exp\left\{\frac{i}{\hbar}S_0\left[q_b,t_b,q_a,t_a\right]\right\} = \int K(q_b,t_b,q_s,s)K(q_s,s,q_a,t_a)dq_s \tag{12.19}$$

So,

$$\langle \chi|F[q]|\psi\rangle_S = \int_{t_a}^{t_b} ds \int Dq(t)\chi^*(q_c,t_c)K(q_c,t_c,q_s,s)v(s)K(q_s,s,q_a,t_a)\psi(q_a,t_a)dq_a dq_c dq_s \tag{12.20}$$

Usually, the connection with the ordinary quantum mechanics is done via the transition amplitude:

$$\psi(q_s,s) = \int K(q_s,s,q_a,t_a)\psi(q_a,t_a)dq_a \tag{12.21}$$

and considering the fact that K is Hermitian then

$$\int K^*(q_s,s,q_c,t_c)\chi^*(q_c,t_c)dq_c = \chi^*(q_s,s) \equiv K^*(q_s,s;q,s') \tag{12.22}$$

Considering the identity

$$K(q_b,t_b,q_s,s) = \sum_n \exp\left\{-\frac{iE_n(t_b-s)}{\hbar}\right\}\Phi_n(q_s)\Phi_n^*(q_b) \tag{12.23}$$

and

$$\psi(q_b,t_b) = K(q_b,t_b,q_s,s) \tag{12.24}$$

then

$$\langle \chi|F[q]|\psi\rangle_S = \int_{t_a}^{t_b} ds \int \sum_n \exp\left\{-\frac{iE_n(t_b-s)}{\hbar}\right\}\Phi_n(q_b)\Phi_n^*(q_s)v(s) \sum_m \exp\left\{-\frac{iE_m(s-t_a)}{\hbar}\right\}\Phi_m(q_a)\Phi_m^*(q_s)dq_s \tag{12.25}$$

We evaluate the matrix element to the first order of K :

$$v_{nm}(s) = \int \Phi_n^*(q_s)v(s)\Phi_m(q_s)dq_s \tag{12.26}$$

then

$$\langle \chi|F[q]|\psi\rangle_S = \int_{t_a}^{t_b} ds \sum_{nm} \exp\left\{\frac{i(E_n-E_m)s}{\hbar}\right\}v_{nm}(s)\exp\left\{-\frac{i(E_n t_b - E_m t_a)}{\hbar}\right\}\Phi_n(q_b)\Phi_m(q_a) \tag{12.27}$$

We evaluate the matrix element in the second order

$$v(s)v(s')_{S_0}\rangle = \int_{t_a}^{t_b} ds \int_{t_a}^{t_b} ds' \int_{q_a}^{q_b} Dq(t)v(s)v(s')\exp\left\{\frac{i}{\hbar}S_0\right\} \tag{12.28}$$

or

$$\left\langle v(s)v(s')_{S_0}\right\rangle = \int_{t_a}^{t_b} ds \int_{t_a}^{t_b} ds' \int_{-\infty}^{\infty} dq_s \int_{-\infty}^{\infty} dq_{s'} \mathrm{K}\left(q_b,t_b,q_{s'},s'\right)\mathrm{K}\left(q_{s'},s',q_s,s\right)\mathrm{K}\left(q_s,s,q_a,t_a\right)v(s)v(s') \quad (12.29)$$

or

$$\left\langle v(s)v(s')_{S_0}\right\rangle = \int_{t_a}^{t_b} ds \int_{t_a}^{t_b} ds' \int_{-\infty}^{\infty} dq_s \int_{-\infty}^{\infty} dq_{s'} \sum_n \exp\left\{-\frac{iE_n\left(t_b-s'\right)}{\hbar}\right\}\Phi_n\left(q_{s'}\right)\Phi_n^*\left(q_b\right)v(s')$$
$$\times\sum_m \exp\left\{-\frac{iE_m\left(s'-s\right)}{\hbar}\right\}\Phi_m\left(q_s\right)\Phi_m^*\left(q_{s'}\right)v(s)$$
$$\sum_k \exp\left\{-\frac{iE_k\left(s-t_a\right)}{\hbar}\right\}\Phi_k\left(q_a\right)\Phi_k^*\left(q_s\right) \quad (12.30)$$

or

$$\left\langle v(s)v(s')_{S_0}\right\rangle = \int_{t_a}^{t_b} ds \int_{t_a}^{t_b} ds' \sum_{nmk} \Phi_n^*\left(q_b\right)\Phi_k\left(q_a\right)v_{km}v_{mn}\exp\left\{-\frac{iE_n\left(t_b-s'\right)}{\hbar}-\frac{iE_m\left(s'-s\right)}{\hbar}-\frac{iE_k\left(s-t_a\right)}{\hbar}\right\} \quad (12.31)$$

But,

$$\int_{t_a}^{t_b} ds \int_{t_a}^{t_b} ds' \exp\left\{-\frac{iE_n\left(t_b-s'\right)}{\hbar}-\frac{iE_m\left(s'-s\right)}{\hbar}-\frac{iE_k\left(s-t_a\right)}{\hbar}\right\} = \int_{t_a}^{t_b} ds \int_{t_a}^{t_b} ds' \exp\left\{-i\omega_{mn}s'-i\omega_{km}s-\frac{iE_nt_b-iE_kt_a}{\hbar}\right\} \quad (12.32)$$

Then for $k = n$ we have

$$\int_{t_a}^{t_b} ds \int_{t_a}^{t_b} ds' \exp\left\{-i\omega_{mn}\left(s'-s\right)-\frac{iE_n\left(t_b-t_a\right)}{\hbar}\right\} = \exp\left\{-\frac{iE_n\left(t_b-t_a\right)}{\hbar}\right\}\frac{i}{\omega_{mn}}\left[\exp\left\{-i\omega_{mn}t_b\right\}\right.$$
$$\left.-\exp\left\{-i\omega_{mn}t_a\right\}\right]\frac{i}{-\omega_{mn}}\left[\exp\left\{i\omega_{mn}t_b\right\}-\exp\left\{i\omega_{mn}t_a\right\}\right] \quad (12.33)$$

and

$$\int_{t_a}^{t_b} ds \int_{t_a}^{t_b} ds' \exp\left\{-i\omega_{mn}\left(s'-s\right)-\frac{iE_nT}{\hbar}\right\} = \frac{4\sin^2\left(\frac{\omega_{mn}T}{2}\right)}{\left|\omega_{mn}\right|^2}\exp\left\{-\frac{iE_nT}{\hbar}\right\} \quad (12.34)$$

So,

$$\left\langle v(s)v(s')_{S_0}\right\rangle = \sum_{nm}\Phi_n^*\left(q_b\right)\Phi_n\left(q_a\right)v_{nm}v_{mn}\frac{4\sin^2\left(\frac{\omega_{mn}T}{2}\right)}{\left|\omega_{mn}\right|^2}\exp\left\{-\frac{iE_nT}{\hbar}\right\} \quad (12.35)$$

The quantity

$$v_{n\to m}^{(2)} = v_{nm}v_{mn}\frac{4\sin^2\left(\frac{\omega_{mn}T}{2}\right)}{\left|\omega_{mn}\right|^2}\exp\left\{-\frac{iE_nT}{\hbar}\right\} \quad (12.36)$$

is the transition matrix element through a virtual state. With this result based on

$$\lambda_{nm} = \delta_{nm} \exp\left\{-\frac{iE_n(t_b - t_a)}{\hbar}\right\} + \delta_{nm}^{(1)} + \delta_{nm}^{(2)} \tag{12.37}$$

to the second order of the perturbation and

$$\omega_{mn} = -\omega_{nm} \tag{12.38}$$

as well as

$$v_{nm}^{*} = v_{mn} \tag{12.39}$$

and due to the Hermiticity of v, then considering the following property applied above:

$$\int_{t_a}^{t_b} ds \int_{t_a}^{s} ds' f(s,s') = \int_{t_a}^{t_b} ds' \int_{s'}^{t_b} ds f(s,s') \tag{12.40}$$

we have the transition rate

$$\mathrm{W}_{n\to m}^{(2)} = 1 + \frac{1}{\hbar^2}\left[\int_{-\infty}^{\infty} v_{nn}(t)dt\right]^2 - \frac{1}{\hbar^2}\sum_m \left|\int_{-\infty}^{\infty} v_{nm}(t)\exp\{-i\omega_{mn}t\}dt\right|^2 \tag{12.41}$$

or

$$\mathrm{W}_{n\to m}^{(2)} = 1 - \frac{1}{\hbar^2}\sum_m{}' \left|\int_{-\infty}^{\infty} v_{mn}(t)\exp\{i\omega_{nm}t\}dt\right|^2 = 1 - \sum_m{}' \mathrm{W}^{(1)} \tag{12.42}$$

or

$$\mathrm{W}_{n\to m}^{(2)} = 1 - \frac{1}{\hbar^2}\sum_m{}' \left|\int_{-\infty}^{\infty} v_{mn}(t)\exp\{i\omega_{nm}t\}dt\right|^2 = 1 - \sum_m{}' \mathrm{W}^{(1)} \tag{12.43}$$

The apostrophe on the summation sign implies the absence of the term with $n = m$. The expression in (12.43) obviously also expresses the normalization of the wave function of the system with the approximation to the second order before the switching on of the perturbation. The contribution from the second order of the perturbation of the transition probability can therefore be written as:

$$\mathrm{P}_{n\to m} = \frac{1}{\hbar^2}|v_{nm}|^2 \frac{4\sin^2\left(\frac{\omega_{mn}T}{2}\right)}{|\omega_{mn}|^2} \exp\left\{-\frac{iE_nT}{\hbar}\right\} \tag{12.44}$$

13

Functional Derivative

As in path integration one deals with classical functions and functionals rather than with operators then fairly simple **classical** manipulations of path integration can lead to non-trivial quantum mechanical identities. We consider a simple example of the functional derivative that relates some functional, say, F[q] where the function $q(t)$ changes by $y(t)$ and the transition amplitude K remains invariant:

$$\delta F = F[q+y] - F[q] = \int \frac{\delta F(s)}{\delta q(s)} y(s) ds \equiv \int A[q,s] y(s) ds \tag{13.1}$$

and

$$dF = Adq \tag{13.2}$$

Here the functional derivative

$$A = \frac{\delta F(s)}{\delta q(s)} \tag{13.3}$$

is also a functional. The action S is also a functional of $y(s)$.

Suppose we have the function $F(q,\dot{q},t)$ then

$$\delta F = \int_{t_a}^{t_b} F(q+y,\dot{q}+\dot{y},t)dt - \int_{t_a}^{t_b} F(q,\dot{q},t)dt \tag{13.4}$$

or

$$\delta F = \int_{t_a}^{t_b} \left[\frac{\partial F}{\partial \dot{q}} \dot{y} + \frac{\partial F}{\partial q} y \right] dt = \int_{t_a}^{t_b} \left[\frac{d}{dt} \left(\frac{\partial F}{\partial \dot{q}} y \right) - \frac{d}{dt} \frac{\partial F}{\partial \dot{q}} y + \frac{\partial F}{\partial q} y \right] dt \tag{13.5}$$

or

$$\delta F = \frac{\partial F}{\partial \dot{q}} y \Big|_{t_a}^{t_b} - \int_{t_a}^{t_b} \left[-\frac{d}{dt} \frac{\partial F}{\partial \dot{q}} + \frac{\partial F}{\partial q} \right] y dt \tag{13.6}$$

But at the end points:

$$y(t_a) = y(t_b) = 0 \tag{13.7}$$

So,

$$\frac{\delta F}{\delta q} = -\frac{d}{dt}\frac{\partial F}{\partial \dot{q}} + \frac{\partial F}{\partial q} \tag{13.8}$$

We consider the functional F[q] and examine it as a function of an infinite number of paths $\left\{F\left[q_i\right]\right\}_{i=1}^{N}$:

$$F\left[q\right] = \sum_{i=1}^{N} F\left[q_i\right] \tag{13.9}$$

So,

$$F\left[q_1 + y_1, q_2 + y_2, \cdots\right] - F\left[q_1, q_2, \cdots\right] = \sum_{i=1}^{N} \frac{\delta F}{\delta q_i} y_i \tag{13.10}$$

We examine an infinitesimal time slice:

$$\Delta s = \varepsilon = s_{i+1} - s_i \tag{13.11}$$

then

$$\int \frac{1}{\varepsilon}\frac{\delta F}{\delta q_i} y ds = \delta F \tag{13.12}$$

and

$$\frac{1}{\varepsilon}\frac{\delta F}{\delta q_i} = \frac{\delta F(s)}{\delta q(s)} \tag{13.13}$$

We take the following example

$$F(s) = q(s) \tag{13.14}$$

then

$$F(s) = \int q(s')\delta(s' - s) ds' \tag{13.15}$$

where the derivative of F(s) is simply given by the Dirac delta distribution:

$$\frac{\delta F(s)}{\delta q(s)} = \delta(s' - s) \tag{13.16}$$

From

$$F\left[q + y\right] - F\left[q\right] = \int \left(q(s') + y(s')\right)\delta(s' - s) ds' - \int q(s')\delta(s' - s) ds' \tag{13.17}$$

or

$$F\left[q + y\right] - F\left[q\right] = \int y(s')\delta(s' - s) ds' = \int \frac{\delta F(s)}{\delta q(s)} y(s) ds \tag{13.18}$$

then again

$$\frac{\delta F(s)}{\delta q(s)} = \delta(s' - s) \tag{13.19}$$

13.1 Functional Derivative of the Action Functional

We find the functional derivative $\frac{\delta S}{\delta q(s)}$ of the action functional by choosing $q(t)$ as the **classical path.** The nearby paths, which are not chosen by the system, we denote by $q(t)$. The classical path extremizes the action S in comparison with all neighboring paths,

$$q(t) = \bar{q}(t) + y(t) \tag{13.20}$$

with the same end points $q(t_a) \equiv q_a$ and $q(t_b) \equiv q_b$. The mathematical problem (**calculus of variations)** finds the curve for which the action is an extremum (Figure 13.1)

$$\delta S[q] = S[q + y] - S[q] \tag{13.21}$$

The variation in the action is zero (the action is stationary) near the extremum for all variations of the path about the classical path such that the quantity $y(t)$ vanishes at the end points:

$$y(t_a) = y(t_b) = 0 \tag{13.22}$$

This may be done as

$$\delta S[q] = S[q + y] - S[q] = \int_{t_a}^{t_b} L(q + y, \dot{q} + \dot{y}, t)\,dt - \int_{t_a}^{t_b} L(q, \dot{q}, t)\,dt \tag{13.23}$$

or

$$\delta S[q] = \int_{t_a}^{t_b} \left[\frac{\partial L}{\partial \dot{q}} \dot{y} + \frac{\partial L}{\partial q} y \right] dt = \int_{t_a}^{t_b} \left[\frac{d}{dt}\left(\frac{\partial L}{\partial \dot{q}} y \right) - \frac{d}{dt}\frac{\partial L}{\partial \dot{q}} y + \frac{\partial L}{\partial q} y \right] dt \tag{13.24}$$

or

$$\delta S[q] = \frac{\partial L}{\partial \dot{q}} y \Big|_{t_a}^{t_b} - \int_{t_a}^{t_b} \left[-\frac{d}{dt}\frac{\partial L}{\partial \dot{q}} + \frac{\partial L}{\partial q} \right] y\,dt \tag{13.25}$$

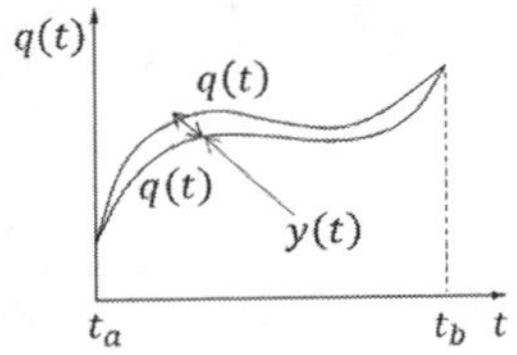

FIGURE 13.1 Illustrates the path of a particle moving from a starting coordinate $q(t_a) \equiv q_a$ to an ending coordinate $q(t_b) \equiv q_b$ of a given path.

From the condition in 13.22, then the first summand is zero and

$$\delta S[q] = -\int_{t_a}^{t_b}\left[-\frac{d}{dt}\frac{\partial L}{\partial \dot{q}} + \frac{\partial L}{\partial q}\right] y dt \tag{13.26}$$

So, the functional derivative of the action functional S is understood as the functional:

$$\frac{\delta S}{\delta q(s)} = \frac{d}{dt}\frac{\partial L}{\partial \dot{q}} - \frac{\partial L}{\partial q} \tag{13.27}$$

The Euler-Lagrange equation of motion for the classical path $q(t)$ is found from the condition that

$$\frac{\delta S}{\delta q(s)} = 0 \tag{13.28}$$

and

$$\frac{d}{dt}\frac{\partial L}{\partial \dot{q}} - \frac{\partial L}{\partial q} = 0 \tag{13.29}$$

We vary now the end point

$$(q_b, t_b) \to (q_b, t_b) + (\delta q_b, \delta t_b = 0) \tag{13.30}$$

but keeping (q_a, t_a) fixed while using the definition of the canonical momentum p conjugate to q:

$$p = \frac{\delta L}{\delta \dot{q}} \tag{13.31}$$

then

$$\delta S = \frac{\delta L}{\delta \dot{q}}\delta q\Big|_{t_a}^{t_b} = p\delta q\Big|_{t_a}^{t_b} = p_b\delta q_b - p_a\delta q_a = p_b\delta q_b \tag{13.32}$$

So,

$$\frac{\delta S}{\delta q_b} = p_b \tag{13.33}$$

We find now the energy E_b (Energy Function or Hamiltonian) at b:

$$\frac{dS}{dt_b} = L = \frac{\delta S}{\delta t_b} + \frac{\delta S}{\delta q_b}\dot{q}_b \tag{13.34}$$

From here,

$$\frac{\delta S}{\delta t_b} = L - p_b\dot{q}_b = -E_b \tag{13.35}$$

or

$$E_b = -\frac{\delta S}{\delta t_b} \tag{13.36}$$

Equations 13.33 and 13.36 are the so-called **Hamilton-Jacobi equations**.

Example 1

We consider the example of a free particle with Lagrange function:

$$L = \frac{m\dot{q}^2}{2} \tag{13.37}$$

The classical action has been found in 5.72 as

$$S_{cl}\left[q_b, t_b; q_a, t_a\right] = \frac{m}{2T}\left(q_b - q_a\right)^2, T = t_b - t_a \tag{13.38}$$

From the Hamilton-Jacobi equations:

$$\frac{\delta S_{cl}}{\delta q_b} = p_b = mv \tag{13.39}$$

$$E_b = -\frac{\delta S_{cl}}{\delta t_b} = \frac{m}{2}v^2 \tag{13.40}$$

Example 2

We examine the harmonic oscillator where the classical path $\bar{q}(t)$ and classical action $S_{cl}[0, T]$ have been obtained respectively in 5.121 and 5.125:

$$\bar{q}(t) = \frac{1}{\sin\omega T}\left[q_a \sin\omega(T - t) + q_b \sin\omega t\right] \tag{13.41}$$

$$S_{cl}\left[0, T\right] = \frac{m\omega}{2\sin\omega T}\left[\left(q_a^2 + q_b^2\right)\cos\omega T - 2q_a q_b\right] \tag{13.42}$$

From the Hamilton-Jacobi equations:

$$p_b = \dot{\bar{q}}\Big|_{t=t_b}, \quad E_b = \frac{m}{2}\left(\dot{\bar{q}}^2 + \omega^2\bar{q}^2\right)\Big|_{t=t_b} \tag{13.43}$$

13.2 Functional Derivative and Matrix Element

We find a general formula via which we can find the link between the transition matrix element and the differential of the matrix element. For that, we introduce the expectation value of the functional:

$$\langle \mathrm{F} \rangle_S = \int Dq\mathrm{F}\left[q\right]\exp\left\{\frac{i}{\hbar}S\right\} \tag{13.44}$$

and

$$\left\langle F[q+y]\right\rangle_S = \int DqF[q+y]\exp\left\{\frac{i}{\hbar}S[q+y]\right\} \tag{13.45}$$

Here, S is the action functional.
But,

$$\left\langle F[q+y]\right\rangle_S = \langle F\rangle_S \tag{13.46}$$

then we do Taylor series expansion over y:

$$\left\langle F[q+y]\right\rangle_S = \int Dq\left(\exp\left\{\frac{i}{\hbar}S[q]\right\}F[q] + \int\exp\left\{\frac{i}{\hbar}S[q]\right\}\frac{\delta F[q]}{\delta q(s)}y(s)ds + \frac{i}{\hbar}\int F[q]\frac{\delta S[q]}{\delta q(s)}y(s)ds\right) \tag{13.47}$$

Here, s is the time moment and from 13.47 then

$$\begin{aligned}\left\langle F[q+y]\right\rangle_S &= \int Dq\exp\left\{\frac{i}{\hbar}S[q]\right\}F[q] + \int Dq\int\frac{\delta F[q]}{\delta q(s)}\exp\left\{\frac{i}{\hbar}S[q]\right\}y(s)ds\\ &+\frac{i}{\hbar}\int Dq\int F[q]\frac{\delta S[q]}{\delta q(s)}\exp\left\{\frac{i}{\hbar}S[q]\right\}y(s)ds = \int Dq\exp\left\{\frac{i}{\hbar}S[q]\right\}F[q]\end{aligned} \tag{13.48}$$

From here, the path integral of a total derivative is zero:

$$\int Dq\exp\left\{\frac{i}{\hbar}S[q]\right\}\int y(s)ds\left(\frac{\delta F[q]}{\delta q(s)} + \frac{i}{\hbar}F[q]\frac{\delta S[q]}{\delta q(s)}\right) = 0 \tag{13.49}$$

The consequence derived from such identities are known as the **Schwinger-Dyson equations** in the quantum field theory [8]. From expression 13.49 then the **Schwinger action principle** is:

$$\left\langle\frac{\delta F[q]}{\delta q(s)}\right\rangle_S = -\frac{i}{\hbar}\left\langle F[q]\frac{\delta S[q]}{\delta q(s)}\right\rangle_S \tag{13.50}$$

This may be rewritten

$$\left\langle \delta F[q]\right\rangle_S = -\frac{i}{\hbar}\left\langle F[q]\delta S[q]\right\rangle_S \tag{13.51}$$

Some applications of this formula may be done for the variation of the action functional:

$$S[q] = \int_{t_a}^{t_b} L\left(q,\dot{q},t\right)dt \tag{13.52}$$

with the variation being

$$\delta S = \int_{t_a}^{t_b}\left[\frac{\partial L}{\partial\dot{q}}\dot{y} + \frac{\partial L}{\partial q}y\right]dt = \int_{t_a}^{t_b}\left[\frac{d}{dt}\left(\frac{\partial L}{\partial\dot{q}}y\right) - \frac{d}{dt}\frac{\partial L}{\partial\dot{q}}y + \frac{\partial L}{\partial q}y\right]dt \tag{13.53}$$

The first summand in the last equation equals zero and we have

$$\delta S = \int_{t_a}^{t_b} \left[-\frac{d}{dt}\frac{\partial L}{\partial \dot{q}} y + \frac{\partial L}{\partial q} y \right] dt \tag{13.54}$$

From here

$$\frac{\delta S[q]}{\delta q} = -\frac{d}{dt}\frac{\partial L}{\partial \dot{q}} + \frac{\partial L}{\partial q} \tag{13.55}$$

For the extremal value of the action functional then the Euler-Lagrange equations are obtained from

$$\frac{\delta S[q]}{\delta q} = 0 \tag{13.56}$$

We use the above formulae to find different results in quantum mechanics such as the equation of motion when we have

1. Suppose F = 1 then δF = 0 and

$$\langle \delta S[q] \rangle_S = \left\langle \int_{t_a}^{t_b} \left[-\frac{d}{dt}\frac{\partial L}{\partial \dot{q}} + \frac{\partial L}{\partial q} \right] \right\rangle_S y dt = 0 \tag{13.57}$$

Since y is an arbitrary function then

$$\left\langle \left[-\frac{d}{dt}\frac{\partial L}{\partial \dot{q}} + \frac{\partial L}{\partial q} \right] \right\rangle_S = 0 \tag{13.58}$$

This is the Euler-Lagrange equation of motion for q.

2. Suppose

$$L = \frac{m\dot{q}^2}{2} - V(q) \tag{13.59}$$

then we have the following classical equation of motion which is the standard classical result for Newton's second law in the presence of a potential $V(q)$:

$$m\langle \ddot{q} \rangle = -\left\langle \frac{\delta V}{\delta q} \right\rangle \tag{13.60}$$

However, in classical mechanics there is only one path and the average disappears. In our case, equation 13.60 is the quantum version of the classical equation of motion which is the path integral version of the **Ehrenfest theorem**, showing how the expectation values obey classical laws. We see that equation 13.60 also shows the connection with Newtonian equations of motion derived by Ehrenfest. Generally, the expression for the average integral is taken over all paths.

3. Consider also

$$\langle \delta F[q] \rangle_S = -\frac{i}{\hbar} \langle F[q] \delta S[q] \rangle_S \tag{13.61}$$

We rewrite the action functional in the form:

$$S=\sum_i\left[\frac{m\left(q_i-q_{i-1}\right)^2\varepsilon}{2\varepsilon^2}-\varepsilon V\left(q_i\right)\right]=\varepsilon\sum_i L\left(\frac{\left(q_i-q_{i-1}\right)}{\varepsilon}\right) \tag{13.62}$$

Consider relation 13.50 holds for an arbitrary value of the parameters. It is possible then to rewrite it at a discrete point q_k corresponding to the value of $q(s)$ at time $s = s_k$. So, taking the special form of the function F[q]:

$$\mathrm{F}\left[q\right]=q_k \tag{13.63}$$

and differentiating over q_k then then the **Schwinger action principle is rewritten**:

$$\left\langle\frac{\delta}{\delta q_k}\mathrm{F}\right\rangle_S=-\frac{i}{\hbar}\left\langle\mathrm{F}\frac{\delta S}{\delta q_k}\right\rangle_S \tag{13.64}$$

So,

$$\langle 1\rangle_S=-\frac{i}{\hbar}\left\langle\left[\frac{m\left(q_k-q_{k-1}\right)}{\varepsilon}-\frac{m\left(q_{k+1}-q_k\right)}{\varepsilon}-\varepsilon\frac{\delta V\left(q_k\right)}{\delta q_k}\right]q_k\right\rangle_S \tag{13.65}$$

The Hamilton-Jacobi relation may also be obtained:

$$p_k=\frac{\delta S}{\delta q_k} \tag{13.66}$$

If $V(q_k)$ is a smooth function then for the limit $\varepsilon\to 0$ we have

$$\varepsilon q_k\frac{\delta V\left(q_k\right)}{\delta q_k}\to 0 \tag{13.67}$$

and

$$\left\langle\left[\frac{m\left(q_{k+1}-q_k\right)q_k}{\varepsilon}-\frac{q_k m\left(q_k-q_{k-1}\right)}{\varepsilon}\right]\right\rangle_S=\frac{\hbar}{i}\langle 1\rangle_S \tag{13.68}$$

Letting, the momentum evaluated at time $t+\frac{\varepsilon}{2}$:

$$\frac{m\left(q_{k+1}-q_k\right)}{\varepsilon}\equiv p_{t+\frac{\varepsilon}{2}} \tag{13.69}$$

while the second term on the left-hand side of 13.68 is the momentum $p_{t-\frac{\varepsilon}{2}}$ evaluated at time $t-\frac{\varepsilon}{2}$ then

$$\langle 1\rangle_S=\frac{i}{\hbar}\left\langle\left[p_{t+\frac{\varepsilon}{2}}-p_{t-\frac{\varepsilon}{2}}\right]q_k\right\rangle_S \tag{13.70}$$

From here we have the path integral analogue of the commutation relation for the momentum and the coordinate:

$$\left\langle \left[p_{t+\frac{\varepsilon}{2}} q_k - q_k p_{t-\frac{\varepsilon}{2}} \right] \right\rangle_S = -i\hbar \tag{13.71}$$

This corresponds to different time intervals. If we take the momentum and coordinate at a time moment t, then the commutation disappears:

$$q_k \frac{q_k - q_{k-1}}{\varepsilon} = q_{k+1} \frac{q_{k+1} - q_k}{\varepsilon} + o(\varepsilon) \tag{13.72}$$

Since, the terms represent one and the same value calculated at the moment different from ε then

$$-q_k \frac{m(q_{k+1} - q_k)}{\varepsilon} + q_{k+1} \frac{m(q_{k+1} - q_k)}{\varepsilon} = -\frac{m(q_{k+1} - q_k)^2}{\varepsilon} \tag{13.73}$$

$$\left\langle \frac{(q_{k+1} - q_k)^2}{\varepsilon^2} \right\rangle = \frac{\hbar}{im\varepsilon} \langle 1 \rangle \tag{13.74}$$

For $\varepsilon \to 0$, the square of the velocity is an infinite value also implying an infinite square value of the momentum. This implies the absence of a path. The matrix element of the square of the velocity is of the order $\frac{1}{\varepsilon}$ and unboundedly increases when $\varepsilon \to 0$. Thus, the path of the quantum mechanical particle is not a smooth curve with a defined geometry. This implies, with a defined velocity and defined by a curve with very small chaotic description. For the given definition, the average quadratic speed for a small final time interval will be large for small ε. However, we will average over different time slices Δt, the chaotic paths that will lead to different **average drift velocities**. In summary, for the average of the square of the velocity, the paths are not distinguished.

From the above-mentioned remarks, we deduce that the expectation value of the square of the velocity is not expected to be evaluated as in 13.74 and so generally we calculate expectation value or matrix element of the change of the coordinate with time:

$$\left\langle \frac{m(q_{k+1} - q_k)}{\varepsilon} \right\rangle_S = \int Dq \chi^* \frac{m(q_{k+1} - q_k)}{\varepsilon} \exp\left\{ \frac{i}{\hbar} S \right\} \psi \, dq_b dq_a \tag{13.75}$$

But,

$$q_{k+1} \to q(t+\varepsilon), q_k \to q(t) \tag{13.76}$$

$$\chi^*(q_b) \mathrm{K}(q_b, q_{t+\varepsilon}) = \mathrm{K}^*(q_{t+\varepsilon}, q_b) \chi^*(q_b) \to \chi^*(q_{t+\varepsilon}) \tag{13.77}$$

We examine the following difference:

$$\int \chi^*(t+\varepsilon) q \psi(t+\varepsilon) dq - \int \chi^*(t) q \psi(t) dq \tag{13.78}$$

We Taylor series expand the wave functions in the above expression in ε:

$$\chi^*(t+\varepsilon) = \chi^*(t) + \frac{\partial \chi^*(t)}{\partial t} \varepsilon \tag{13.79}$$

$$\psi(t+\varepsilon)=\psi(t)+\frac{\partial\psi(t)}{\partial t}\varepsilon \tag{13.80}$$

From these expansions we have

$$\left\langle\frac{m(q_{k+1}-q_k)}{\varepsilon}\right\rangle_S=\frac{m\varepsilon}{\varepsilon}\int\frac{\chi^*\mathrm{H}^*q\psi}{-i\hbar}dq+\frac{m\varepsilon}{\varepsilon}\int\frac{\chi q\mathrm{H}\psi}{i\hbar}dq \tag{13.81}$$

Considering

$$\frac{\partial\chi^*}{\partial t}=\frac{\mathrm{H}^*\chi^*}{-i\hbar} \tag{13.82}$$

then

$$\left\langle\frac{m(q_{k+1}-q_k)}{\varepsilon}\right\rangle_S=-\frac{m}{i\hbar}\int\chi^*\left(\mathrm{H}^*q-q\mathrm{H}\right)\psi\,dq \tag{13.83}$$

We find the average momentum and shows the transition from ordinary quantum mechanics to the quantum mechanics of our time (Feynman approach).

We calculate the transition matrix element for arbitrary coordinates for the quadratic actions. We examine

$$\exp\left\{\frac{i}{\hbar}\int f(t)q(t)dt\right\} \tag{13.84}$$

and take its average value

$$\left\langle\exp\left\{\frac{i}{\hbar}\int f(t)q(t)dt\right\}\right\rangle_S=\int Dq\exp\left\{\frac{i}{\hbar}S\right\}\exp\left\{\frac{i}{\hbar}\int f(t)q(t)dt\right\} \tag{13.85}$$

or

$$\left\langle\exp\left\{\frac{i}{\hbar}\int f(t)q(t)dt\right\}\right\rangle_S=\exp\left\{\frac{i}{\hbar}S'\right\}\int_0^0 Dy\exp\left\{\frac{i}{\hbar}S[y]\right\} \tag{13.86}$$

where

$$S'=\exp\left\{\frac{i}{\hbar}S+\frac{i}{\hbar}\int f(t)q(t)dt\right\} \tag{13.87}$$

with $S[y]$ being the quadratic part of the functional S.

So,

$$\exp\left\{\frac{i}{\hbar}(S'_{cl}-S_{cl})\right\}\int Dy\exp\left\{\frac{i}{\hbar}\left(S[q]+S[y]\right)\right\}=\exp\left\{\frac{i}{\hbar}(S'_{cl}-S_{cl})\right\}\int Dq\exp\left\{\frac{i}{\hbar}S[q]\right\} \tag{13.88}$$

or

$$\exp\left\{\frac{i}{\hbar}(S'_{cl}-S_{cl})\right\}\int Dy\exp\left\{\frac{i}{\hbar}\left(S[q]+S[y]\right)\right\}=\exp\left\{\frac{i}{\hbar}(S'_{cl}-S_{cl})\right\}\langle 1\rangle_S \tag{13.89}$$

So,

$$\left\langle \exp\left\{\frac{i}{\hbar}\int f(t)q(t)dt\right\}\right\rangle_S=\exp\left\{\frac{i}{\hbar}(S'_{cl}-S_{cl})\right\}\langle 1\rangle_S \tag{13.90}$$

This is a useful formula that we use further. If we take the functional derivative from left and right over the function f then

$$\left\langle \frac{i}{\hbar}q_i\exp\left\{\frac{i}{\hbar}\int f_i(t)q_i(t)dt\right\}\right\rangle_S=\frac{i}{\hbar}\frac{\delta S'_{cl}}{\delta f_i}\exp\left\{\frac{i}{\hbar}(S'_{cl}-S_{cl})\right\}\langle 1\rangle_S \tag{13.91}$$

Taking the limit $f\to 0$ then

$$\langle q_i\rangle_S=\frac{i}{\hbar}\frac{\delta S'_{cl}}{\delta f_i}\langle 1\rangle_S \tag{13.92}$$

Since we have evaluated the matrix element of the coordinate, we then do the second derivative to find $\langle q_iq_k\rangle_S$:

$$\left\langle q_i\frac{i}{\hbar}q_k\exp\left\{\frac{i}{\hbar}\int f_i(t)q_i(t)dt\right\}\right\rangle_S=\left(\frac{\delta^2S'_{cl}}{\delta f_i\delta f_k}+\frac{i}{\hbar}\frac{\delta S'_{cl}}{\delta f_i}\frac{\delta S'_{cl}}{\delta f_k}\right)\exp\left\{\frac{i}{\hbar}(S'_{cl}-S_{cl})\right\}\langle 1\rangle_S \tag{13.93}$$

For $f\to 0$ then

$$\langle q_iq_k\rangle_S=\left(\frac{\hbar}{i}\frac{\delta^2S'_{cl}}{\delta f_i\delta f_k}+\frac{\delta S'_{cl}}{\delta f_i}\frac{\delta S'_{cl}}{\delta f_k}\right)\langle 1\rangle_S \tag{13.94}$$

This constitutes a useful trick which allows us to determine the average values $\langle q_iq_k\rangle_S$.

14

Quantum Statistical Mechanics Functional Integral Approach

14.1 Introduction

We establish a rather deep relation between statistical mechanics and quantum mechanics at an imaginary time. For the quantum field theory, this implies a relation between finite temperature quantum field theory in Minkowski space and quantum field theory in Euclidean space with one compact **Euclidean time** direction. The quantum mechanical language being a proper formulation of statistical physics as a molecular theory is basically the study of quantum equilibrium as well as non-equilibrium systems. A quantum system found in a thermodynamic equilibrium state with an external classical system (thermostat or thermal reservoir) may be characterized by fixed values of thermodynamic parameters of the system, say, for example, temperature T, volume V and number of particles N. In this way the total energy of the thermostat is assumed to be much greater than the energy of the system. The behavior of the quantum system may be assumed to be independent of the nature of the thermostat. The most convenient object of the quantum theory to describe such non-isolated systems is the density matrix since the partition function does not determine any local thermodynamic quantities. It is for this reason that we introduced quantum mechanical and quantum statistical density matrices which are necessary to do the investigation via path integration. This enables us to make a transition to the quasi-classical approximation of quantum statistical mechanics, which, in turn, enables us to evaluate the first quantum correction to the classical Boltzmann distribution function.

14.2 Density Matrix

14.2.1 Partition Function

In this heading, we consider two systems that can interact with each other. However, we are particularly interested to study the properties of one of them. We distinguish between them, by referring to one of them as the system and to the other as the environment. One direct consequence of quantum mechanics is that, in principle, we require the general wave function of the entire system (the **system** and the **environment**) that may seem difficult. Usually, in practice, it is convenient to identify the entire system as **system** and the **environment**. This will permit us solve and investigate the properties of the isolated system as if it constitutes the entire system. When the isolated system is very small compared to the entire system, a statistical physics approach appears to be particularly appropriate.

Path integration turns out to provide an elegant manner of studying statistical mechanics. Here the central object in statistical mechanics is the partition function that can be written as a path integral.

When a system in thermal equilibrium with a bath of temperature T can be in one of the N states, then the statistical probability of the system with discrete energy spectrum $\{E_n\}$ is given by

$$W_n = Z^{-1}\exp\{-\beta E_n\} \tag{14.1}$$

Here, $\beta = \frac{1}{T}$ is the inverse temperature in units of the energy, and the normalization factor,

$$Z = Z(\beta) = \sum_n \exp\{-\beta E_n\} \tag{14.2}$$

is called the **canonical partition function** (central physical quantity in thermodynamics) that encodes probabilistic information about the system. All the standard thermodynamic quantities for a given system, e.g., free energy, internal energy, entropy, pressure, etc., can be derived from it. In equation 14.2, the lowest energy level E_0, the so-called, ground state energy or zero-point energy, is generally the dominant energy level contributing to the partition function. By virtue of the Heisenberg uncertainty principle, this ground state energy is always larger than the minimum value of potential energy. This is because a particle can never be at rest anywhere in a given potential or a particle with a particular momentum can be everywhere in a given potential.

14.3 Expectation Value of a Physical Observable

Consider a physical observable F defined by its quantum mechanical operator $\hat{F}$, and the evolution of the system is described by a Hamiltonian $\hat{H}$ with the basis vectors $|\Phi_n\rangle$, defined by the eigenvalue equation:

$$\hat{H}|\Phi_n\rangle = E_n|\Phi_n\rangle \tag{14.3}$$

From quantum mechanics, the diagonal matrix element F_{nn} is described by the following equation

$$F_{nn} = \langle\Phi_n|\hat{F}|\Phi_n\rangle = \int dq\Phi_n^*(q)\hat{F}(q)\Phi_n(q) \tag{14.4}$$

with $\Phi_n(q) = \langle q|\Phi_n\rangle$, and $\Phi_n^*(q) = \langle\Phi_n|q\rangle$. The matrix element F_{nn} in 14.4 permits to evaluate the expectation value of the physical observable F quantized as an operator $\hat{F}$:

$$\bar{F} = \sum_n F_{nn}W_n = \sum_n F_{nn}Z^{-1}\exp\{-\beta E_n\} = \sum_n W_n\int dq\Phi_n^*(q)\hat{F}(q)\Phi_n(q) \tag{14.5}$$

or

$$\bar{F} = \int dq'dq\delta(q'-q)\hat{F}(q)\sum_n W_n\Phi_n^*(q')\Phi_n(q) \tag{14.6}$$

So, for a mixed state the expectation value of the physical observable F is written as:

$$\bar{F} = \int dq'dq\delta(q'-q)\hat{F}(q)\rho(q,q') = \mathrm{Tr}\left[\hat{F}\hat{\rho}\right] \tag{14.7}$$

We see that it has two averaging procedures, one is the usual quantum mechanical average procedure in 14.4 and the other is the classical averaging of multiplying the probability of being in a state by the value of being in that state as in 14.5 as well as in 14.6.

14.4 Density Matrix

In relation 14.7, it is obvious that the quantity

$$\rho(q,q') = \sum_n W_n \Phi_n^*(q')\Phi_n(q) \tag{14.8}$$

Since the coefficients W_n are real then $\rho(q,q')$ is obviously a Hermitian operator and so can be diagonalized. Here, $\rho(q,q')$ is the so-called **density matrix** in the coordinate representation – a fundamental quantity that is the summit in quantum statistical mechanics from where all concepts are derived as well as the concepts of thermal equilibrium and temperature T clarified. This requires the definition of the weighted function of eigenstates for any operator,

$$\rho(q,q') = \sum_n W_n \Psi_n(q)\Psi_n^*(q')) = \sum_n \langle q|\Psi_n\rangle W_n \langle\Psi_n|q'\rangle \tag{14.9}$$

This is the so-called **statistical density matrix** (that is positive definite) in the coordinate representation that is the matrix element of the density operator and also determines the thermal average of the particle density of a quantum-statistical system:

$$\hat{\rho} \equiv \sum_i W_i |\Psi_i\rangle\langle\Psi_i| \tag{14.10}$$

In relation 14.9, the quantities $\Psi_n(q)$ and W_n are the eigenfunctions and eigenvalues of $\hat{\rho}$ indicative that $\rho(q,q')$ corresponds to a mixture of pure states of the wave functions $\Psi_n(q)$ with the respective weights W_n. In 14.10, $|\Psi_n\rangle\langle\Psi_n|$ can be interpreted as the probability distribution of the system in the eigenstate $|\Psi_n\rangle$, while W_n, the normalized probability to encounter the system in the state $|\Psi_n\rangle$. So, $\rho(q,q')$ should be the normalized average particle density in space. The advantage of working with $\rho(q,q')$ rather than the wave functions $\Psi_n(q)$ is that we can more easily treat an infinite volume that avoids the complications of the boundaries of the system. For the evaluation of the partition function Z by explicit calculation of E_n we must take a large, but finite volume. It is instructive to note that at low temperatures, only the lowest energy state survives and $\rho(q,q')$ achieves the particle distribution at the ground state. At high temperatures, quantum effects are expected to be irrelevant and we therefore expect matrix density $\rho(q,q')$ to imitate that of a classical particle distribution.

The expression 14.9 applies generally to a mixed state. The physical sense of the density operator in 14.10 entails that for any Hermitian operator $\hat{\rho}$ we may always find such a representation, say $|\Psi_i\rangle$, for which it is diagonalized. From 14.10, we may reformulate quantum mechanics: Any system is described by the density matrix in 14.10 with

(a) $|\Psi_i\rangle$ being some complete orthonormal set of vectors and the probability W_i has the following properties:

$$W_i \geq 0 \tag{14.11}$$

$$\sum_i W_i = 1 \tag{14.12}$$

(b) For a given operator, $\hat{F}$ the quantum mechanical-statistical expectation value can be found via a trace, for any representation,

$$\langle F\rangle \equiv \bar{F} = \int dq'dq\delta(q'-q)\hat{F}(q)\rho(q,q') = \mathrm{Tr}\left[\hat{F}\hat{\rho}\right] \tag{14.13}$$

Since $\langle \Psi_i | \hat{F} | \Psi_i \rangle$ is the expectation value of $\hat{F}$ in the state $|\Psi_i\rangle$ then from a) to d) the density matrix ρ_{ii} in the diagonal $|\Psi_i\rangle$ representation is, simply, interpreted as the probability of the system to be found in the state $|\Psi_i\rangle$. Generally, deriving the density matrix from 14.10 as well as in 14.13 we do not define concretely the exact diagonal representation and at least suppose they exist. If all but one of the W_i is zero, then the system is in the **pure state** $|\Psi_{ipure}\rangle$:

$$W_i = \delta_{ii}\delta_{ii_{pure}} \tag{14.14}$$

$$\rho_{ij} = \delta_{ij}\delta_{ii_{pure}} \tag{14.15}$$

and otherwise it is in the **mixed state**. So, the density matrix handles both pure and mixed states.

14.5 Density Matrix in the Energy Representation

Solving a quantum mechanical problem, there is a necessity to partition the full system into two, i.e., the system of interest and the environment that is generally very large compared to the system of interest. So, partitioning the full system into two, we arrive at the fact that the pure states cannot generally describe a quantum mechanical system that is devoid of the environment. So, we do not have knowledge on whether the full system is either in the pure or mixed state. Then the necessity to reformulate quantum mechanics via a more general density matrix ρ satisfying its equation of motion.

We find the wave function with i labelling the state of the ensemble, in the energy representation, via the time-dependent Schrödinger equation:

$$i\hbar\frac{\partial}{\partial t}\left|\Psi^{(i)}\right\rangle = \hat{H}\left|\Psi^{(i)}\right\rangle \tag{14.16}$$

with solution

$$\left|\Psi_n^{(i)}(q,t)\right\rangle = \exp\left\{-i\frac{E_n t}{\hbar}\right\}\left|\Phi_n(q)\right\rangle \tag{14.17}$$

Here, $|\Phi_n(q)\rangle$ is the nth eigenfunction and E_n the corresponding eigenvalue of the Hamiltonian $\hat{H}$ satisfying the stationary Schrödinger equation:

$$\hat{H}\left|\Phi_n(q)\right\rangle = E_n\left|\Phi_n(q)\right\rangle \tag{14.18}$$

The complete set of normalized eigenfunctions acts as an orthonormal basis set:

$$\left|\Psi^{(i)}(q,t)\right\rangle = \sum_n C_n^{(i)}(t)\left|\Phi_n(q)\right\rangle \tag{14.19}$$

The coefficients $C_n^{(i)}(t)$ describe the wave function in the energy representation and the expectation values via this representation:

$$\bar{F} = \sum_{n,m} F_{mn}\rho_{nm} = \sum_n \left(\hat{F}\hat{\rho}\right)_{nn} = \mathrm{Tr}\left[\hat{F}\hat{\rho}\right] \tag{14.20}$$

where

$$F_{mn} = \left\langle \Phi_m \right|\hat{F}\left|\Phi_n\right\rangle \tag{14.21}$$

and the matrix element of the density matrix operator:

$$\rho_{nm} = \sum_i W_i C_n^{(i)}(t) C_m^{(i)*}(t) \tag{14.22}$$

In order to find ρ_{nm} it is useful to differentiate 14.22 with respect to time and we have the evolution equation of the matrix elements according to Heisenberg, which is the quantum dynamics of ρ_{nm} given by the differential equation [1]:

$$i\hbar\frac{\partial}{\partial t}\rho_{nm} = \left[H\rho - \rho H\right]_{nm} \tag{14.23}$$

and in the operator representation the following evokes the Heisenberg equation of motion for the density matrix operator $\hat{\rho}$:

$$i\hbar\frac{\partial}{\partial t}\hat{\rho} = \left[\left(\hat{H}\hat{\rho}\right) - \left(\hat{\rho}\hat{H}\right)\right] = \left[\hat{H},\hat{\rho}\right] \tag{14.24}$$

The formal solution of 14.24 yields

$$\hat{\rho}(t) = \hat{\rho}\left(t,t_0,\hat{\rho}(t_0)\right) \equiv \hat{U}(t,t_0)\hat{\rho}(t_0)\hat{U}^{\dagger}(t,t_0), t \geq t_0 \tag{14.25}$$

Here, $\hat{U}^{\dagger}$ and $\hat{U}$ are unitary operators:

$$\hat{U}^{\dagger}(t,t_0) = \hat{T}^{-}\exp\left\{\frac{i}{\hbar}\int_{t_0}^{t}\hat{H}(t')dt'\right\}, \hat{U}(t,t_0) = \hat{T}^{+}\exp\left\{-\frac{i}{\hbar}\int_{t_0}^{t}\hat{H}(t')dt'\right\} \tag{14.26}$$

Here, $\hat{T}^{+}$ is the operator of the chronological-ordered product that orders the times chronologically with the latest time to the left while $\hat{T}^{-}$, is the anti-chronological-ordered product.

For the given initial conditions,

$$\hat{\rho}(t_0) = \hat{\rho}_0 \tag{14.27}$$

the description of the quantum system by the density matrix becomes closed. So, this is completely equivalent to the description by the wave function for which its evolution is described by the Schrödinger equation.

Solving equation 14.24 permits us to have 14.25 enabling us to find the expectation value of the operator $\hat{F}(t)$ of the physical quantity:

$$\bar{F} = \mathrm{Tr}\left\{\hat{F}(t)\hat{\rho}(0)\right\} \tag{14.28}$$

Here the observables have time dependence with

$$\hat{F}(t) = \hat{U}(t)F(0)\hat{U}^{\dagger}(t) \tag{14.29}$$

According to Heisenberg, the quantum dynamics of $\hat{F}(t)$ is given by the differential equation, the so-called Heisenberg equation of motion:

$$i\hbar\frac{\partial}{\partial t}\hat{F}(t)=\left[\hat{H},\hat{F}(t)\right] \tag{14.30}$$

For the case of the density matrix $\hat{\rho}$ this Heisenberg equation of motion is rewritten as:

$$i\hbar\frac{\partial}{\partial t}\hat{\rho}(t)=\left[\hat{H},\hat{\rho}(t)\right] \tag{14.31}$$

In the case of an equilibrium state the distribution is time-independent and $\hat{\rho}$ commutes with $\hat{H}$ i.e.,

$$\left[\hat{\rho},\hat{H}\right]=\left[\hat{\rho}\hat{H}-\hat{H}\hat{\rho}\right]=0 \tag{14.32}$$

Thus, $\hat{\rho}$ is an integral of motion in the equilibrium state.

We describe the quantum sub-system found in thermodynamic equilibrium with a thermostat [1] via the matrix density for fixed thermodynamic parameters *T, V* and *N*. By a thermostat is understood a classical (or quasi-classical) system with energy greater than the quantum sub-system. For this reason, we ignore its physical nature. Physically, such an equilibrium quantum sub-system should be described by Gibbs distribution and, in particular, suppose that the density matrix 14.22 is diagonalized in the energy representation, the eigenstates and eigenvalues of which satisfy the following equation:

$$\hat{H}|\Phi_n\rangle=E_n|\Phi_n\rangle \tag{14.33}$$

So, from here and considering 14.25 and 14.26 then

$$\hat{\rho}(t)=\sum_n\sum_m\langle\Phi_n|\rho(0)|\Phi_m\rangle\exp\left\{-\frac{i}{\hbar}(E_n-E_m)t\right\}|\Phi_n\rangle\langle\Phi_m| \tag{14.34}$$

This implies that the system is in a stationary state when off-diagonal terms vanish, i.e.,

$$\rho_{nm}=\rho_{nn}\delta_{nm} \tag{14.35}$$

For the equilibrium case with a diagonal matrix density, then we find the expectation value:

$$\bar{F}=\sum_{n,m}\rho_{nm}F_{mn}=\sum_{n,m}\rho_{nn}\delta_{nm}F_{mn}=\sum_n\rho_{nn}F_{nn} \tag{14.36}$$

For the equilibrium case, the density matrix is defined by one index i.e., $\rho_{nn}=W_n$ and

$$\bar{F}=\sum_n W_nF_{nn} \tag{14.37}$$

where W_n is the probability (population) of the nth quantum state and due to

$$\bar{F}=\bar{F}^* \tag{14.38}$$

it follows that

$$\rho_{nm}=\rho^*_{mn} \tag{14.39}$$

Consider again 14.10 then the probability of finding the system in the state $|\Psi_i\rangle$ is dependent only on the energy of the given state in the following way:

$$W_i = Z^{-1}\exp\{-\beta E_i\} \tag{14.40}$$

From 14.10, the density matrix of the statistical system is:

$$\hat{\rho} = \sum_n W_n |\Phi_n\rangle\langle\Phi_n| \tag{14.41}$$

Considering 14.3, we can rewrite the density matrix in the form

$$\hat{\rho} = Z^{-1}\exp\left\{-\beta\hat{H}\right\} \tag{14.42}$$

The partition function is expressed via the matrix density 14.42 since

$$\sum_n W_n = Z^{-1}\sum_n \exp\{-\beta E_n\} = Z^{-1}\mathrm{Tr}\left\{\exp\left\{-\beta\hat{H}\right\}\right\} = 1 \tag{14.43}$$

from where the **von Neumann, quantum-statistical partition function**:

$$Z = \mathrm{Tr}\left\{\exp\left\{-\beta\hat{H}\right\}\right\} \tag{14.44}$$

or as a sum over the Boltzmann factors of all eigenstates $|n\rangle$ of the Hamiltonian:

$$Z = \sum_n \exp\{-\beta E_n\} \tag{14.45}$$

In 14.44, Tr $\hat{A}$ denotes the trace of the operator $\hat{A}$. If the quantum Hamiltonian $\hat{H}$ has a discrete spectrum $\{E_n\}$, then the formula in 14.44 allows one to calculate the energy levels E_n from the partition function $Z(T, V, N)$. The partition function $Z(T, V, N)$ is a univalent function of thermodynamic parameters and contains all thermodynamic information [1]. In particular, the thermodynamic potential for fixed T, V and N is the free energy of the system F [1] that may be obtained from relation 14.44 via the following formula:

$$F(T,V,N) = -\frac{1}{\beta}\ln Z \tag{14.46}$$

From 14.13, we can compute the average energy [1] $\bar{E}$:

$$\bar{E} \equiv \mathrm{Tr}\left(\hat{H}\hat{\rho}\right) = Z^{-1}\mathrm{Tr}\left\{\hat{H}\exp\left\{-\beta\hat{H}\right\}\right\} = -Z^{-1}\frac{\partial}{\partial\beta}Z = -\frac{\partial}{\partial\beta}\ln Z \tag{14.47}$$

The entropy S [1] may be obtained from the thermodynamic relation

$$F = \bar{E} - TS \tag{14.48}$$

15

Partition Function and Density Matrix Path Integral Representation

15.1 Density Matrix Path Integral Representation

For now, let us proceed in the evaluation of the density matrix via the functional integration in a similar manner as we did for the transition amplitude.

15.1.1 Density Matrix Operator Average Value in Phase Space

First we write the density matrix operator average value in phase space:

$$\langle q|\hat{\rho}(t)|q'\rangle = \int \mathrm{K}\left(q,q',t|q_0,q'_0,t_0\right)\langle q_0|\hat{\rho}(t_0)|q'_0\rangle dq_0 dq'_0 \tag{15.1}$$

Here,

$$\mathrm{K}\left(q,q',t|q_0,q'_0,t_0\right) \equiv \langle q|\hat{\mathrm{U}}(t,t_0)|q_0\rangle\langle q'_0|\hat{\mathrm{U}}^{\dagger}(t,t_0)|q'\rangle \tag{15.2}$$

This gives the basis for path integral application in non-equilibrium statistical physics as well as the kinetic theory. We consider a statistical ensemble of a quantum system at the absolute temperature T and described by the Hamiltonian of a stationary system:

$$\hat{\mathrm{H}} = \hat{\mathrm{H}}(\hat{p},\hat{q}) \tag{15.3}$$

For brevity, the unnormalized density matrix can be written as:

$$\hat{\rho}(\beta) = \exp\left\{-\beta\hat{\mathrm{H}}\right\}, \beta = \frac{1}{T} \tag{15.4}$$

and the normalized equilibrium density matrix as:

$$\hat{\rho}_{\mathrm{eq}}(\beta) = Z^{-1}(\beta)\hat{\rho}(\beta) \tag{15.5}$$

Here, the statistical partition function $Z(\beta)$ is:

$$Z(\beta) = \mathrm{Tr}\left\{\hat{\rho}(\beta)\right\} \tag{15.6}$$

From the above, equation 15.4 can equally be obtained from the solution of the following Cauchy problem:

$$-\frac{\partial}{\partial\beta}\rho_{nm}(\beta)=E_n\rho_{nm}(\beta),\rho_{nm}(0)=\delta_{nm} \tag{15.7}$$

which can be rewritten in the operator form referred to as the Bloch equation [1,20] for the density matrix of a canonical ensemble:

$$-\frac{\partial}{\partial\beta}\hat{\rho}(\beta)=\hat{\mathrm{H}}\,\hat{\rho}(\beta),\hat{\rho}(0)=\hat{\mathbb{I}} \tag{15.8}$$

and the Bloch equation in the coordinate q-representation:

$$-\frac{\partial}{\partial\beta}\rho(q,q',\beta)=\hat{\mathrm{H}}\,\rho(q,q',\beta),\rho(q,q',0)=\delta(q-q') \tag{15.9}$$

The second equation in 15.9 is the boundary condition at $\beta=0$ and expresses the completeness condition on the eigenfunctions.

In the formulation of statistical physics, we pass from the real time path integral formulation to the Euclidean time formulation where the pure imaginary time is introduced by the analytic continuation of the time interval to the negative imaginary value:

$$t=-i\hbar\tau \tag{15.10}$$

where τ is a real number [2] on the negative imaginary axis (Figure 15.1):

$$0\le\tau\le\beta \tag{15.11}$$

and

$$t-t_0=-i\hbar\beta \tag{15.12}$$

So, similarly, considering the Cauchy problems 4.7 and 4.9 then the solution of 4.8 imitates 4.11:

$$\hat{\rho}(\beta)=\hat{\mathrm{T}}\exp\left\{-\int_0^\beta\hat{\mathrm{H}}(\tau)d\tau\right\} \tag{15.13}$$

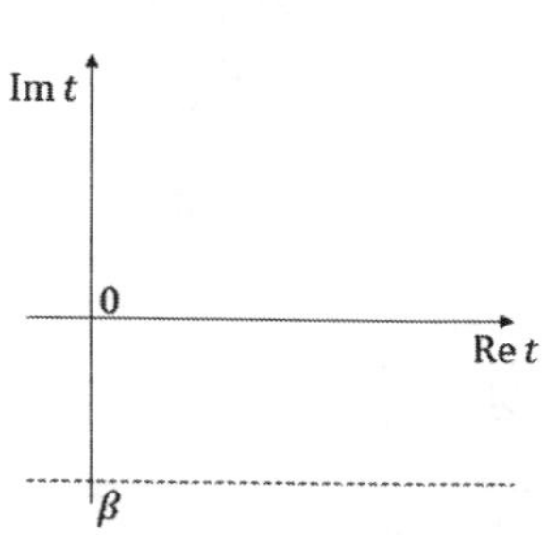

FIGURE 15.1 The analytic continuation of the time t on the negative imaginary axis.

Here $\hat{T}$ is the operator of the chronological-ordered product that orders the times chronologically with the latest time to the left and the ordering parameter τ has the dimension of inverse energy.

We can now find the matrix element of the equilibrium statistical density matrix operator in the path integral representation:

$$\langle q|\hat{\rho}(\beta)|q_0\rangle = \int_{q_0}^{q} Dp(\tau)Dq(\tau)\exp\{S[p(\tau),q(\tau),-i\hbar\beta]\} \tag{15.14}$$

Here,

$$S[p(\tau),q(\tau),-i\hbar\beta] = \int_0^{\beta}\left[\frac{i}{\hbar}p(\tau)\dot{q}(\tau) - H(p(\tau),q(\tau))\right]d\tau \tag{15.15}$$

is the dimensionless and real quantum-statistical action functional (**Euclidean action**) and τ, the **Euclidean time**.

We can now rewrite the statistical partition function in 6:

$$Z(\beta) = \int \delta(q-q')\langle q|\hat{\rho}(\beta)|q'\rangle dq'dq \tag{15.16}$$

or

$$Z(\beta) = \int \delta(q-q')\delta(q_0-q'_0)dq'dqdq_0dq'_0\int_{q'_0}^{q'} Dp(\tau)Dq(\tau)\exp\{S[p(\tau),q(\tau),-i\hbar\beta]\} \tag{15.17}$$

This involves path integral in the sub-space of generalized coordinates. We find how 15.14 relates 15.17 by calculating the matrix element of the normalized equilibrium statistical density matrix defined in 5:

$$\langle q|\hat{\rho}_{eq}(\beta)|q_0\rangle = Z^{-1}(\beta)\langle q|\hat{\rho}(\beta)|q_0\rangle \tag{15.18}$$

15.1.1.1 Generalized Gaussian Functional Path Integral in Phase Space

For brevity, we consider harmonic oscillation with the Lagrangian:

$$L(q,\dot{q}) = -\frac{1}{2\hbar^2}\dot{q}m\dot{q} - \frac{i}{\hbar}qb\dot{q} - \frac{1}{2}qkq \tag{15.19}$$

The boundary condition for our problem is, respectively, the starting and ending coordinates of the path:

$$q(0) \equiv q_0, q(\beta) \equiv q \tag{15.20}$$

We write the Euler-Lagrange equation described by the relation in 3.30:

$$\frac{1}{\hbar^2}\mathbf{m}\ddot{q} - \frac{2i}{\hbar}\mathbf{b}\dot{q} + \mathbf{k}q = 0 \tag{15.21}$$

with solution being the classical path q_{cl}.

From quasi-classical approximation, we select a fluctuating path $q'(\tau)$ interpreted as the quantum fluctuations around the classical path $q_{cl}(t')$:

$$q(\tau)=q_{cl}(\tau)+q'(\tau) \tag{15.22}$$

The classical action is obtained:

$$S(q_{cl}(\tau),\beta)=-\frac{1}{2\hbar^2}\left[q_{cl}(\beta)\mathbf{m}\dot{q}_{cl}(\beta)-q_{cl}(0)\mathbf{m}\dot{q}_{cl}(0)\right] \tag{15.23}$$

So,

$$\langle q|\hat{\rho}(\beta)|q_0\rangle=F(\beta)\exp\{S(q_{cl}(\tau),\beta)\} \tag{15.24}$$

where the reduced propagator is:

$$F(\beta)=\left[\frac{1}{\left(2\pi\hbar^2\beta\right)^s}\det\mathbf{m}\times\det\left[\frac{\beta\hbar\left(\mathbf{k}'-\mathbf{b}'^2\right)^{\frac{1}{2}}}{\sinh\left(\beta\hbar\left(\mathbf{k}'-\mathbf{b}'^2\right)^{\frac{1}{2}}\right)}\right]\right]^{\frac{1}{2}} \tag{15.25}$$

We apply this to the case of a charged particle in a uniform magnetic field in the z-direction where the action functional is:

$$S(q_{cl}(\tau),\beta)=-\frac{m\omega_c}{4\hbar}\left[-2i(yx_0-xy_0)+\left[(x-x_0)^2+(y-y_0)^2\right]\coth\left(\frac{\omega_c\beta\hbar}{2}\right)\right]-\frac{m}{2\hbar^2\beta}(z-z_0)^2 \tag{15.26}$$

and the reduced propagator:

$$F(\beta)=\left[\left(\frac{m}{2\pi\hbar^2\beta}\right)^3\left(\frac{\omega_c\beta\hbar}{2\sinh\frac{\omega_c\beta\hbar}{2}}\right)^2\right]^{\frac{1}{2}} \tag{15.27}$$

The partition function can now be written as:

$$Z(\beta)=VF(\beta) \tag{15.28}$$

where V is the volume.

15.1.2 Density Matrix via Transition Amplitude

We consider the Hamiltonian describing our system to be

$$\hat{H}=\frac{\hat{p}^2}{2m}+U(q) \tag{15.29}$$

and calculate the density matrix via a sequence of many time steps:

$$\rho(q,q',\beta) \cong \lim_{\substack{\varepsilon\to 0\\ N\to\infty}} \left(\frac{m}{2\pi i\hbar(-i\varepsilon)}\right)^{\frac{N}{2}} \int_{-\infty}^{+\infty}\cdots\int_{-\infty}^{+\infty}\prod_{i=1}^{N-1} dq_i \exp\left\{\frac{i}{\hbar}\sum_{i=0}^{N-1} S[q_i,q_{i+1}]\right\} \tag{15.30}$$

where,

$$S[q_i,q_{i+1}] = \left[\frac{m}{2}\left(\frac{q_{i+1}-q_i}{-i\varepsilon}\right)^2 - U(q_i)\right](-i\varepsilon) \tag{15.31}$$

The integral 15.30 may also be written as

$$\rho(q,q',\beta) = \int_{q'}^{q} \exp\left\{\underline{S}[q,q',-i\hbar\beta]\right\} Dq(\tau) \tag{15.32}$$

Here

$$\underline{S} = \underline{S}[q(\tau),-i\hbar\beta] \equiv \int_0^{\beta}\left(-\frac{m}{\hbar^2}\dot{q}^2 - U(q)\right)d\tau \tag{15.33}$$

is the dimensionless and real quantum-statistical **Euclidean action functional** and τ the **Euclidean time**, as seen earlier.

We observe that in quantum mechanics, the contribution of each path is weighted by $\exp\left\{\frac{i}{\hbar}S[q]\right\}$ and in quantum statistical physics weighted by $\exp\left\{\underline{S}[q,q',-i\hbar\beta]\right\}$. In quantum mechanics, the action $S[q]$ is evaluated by the method of the stationary phase while $\underline{S}$ in quantum statistical physics is evaluated by the turning point method. The dominant contribution to the functional integral weighted by $\exp\{S_{cl}\}$ is given by the classical action that achieves its extremal value for the classical path $\bar{q}$. For the evaluation of the density matrix, it is important to consider the fluctuation about this classical path.

We evaluate the partition function that encodes probabilistic information about the system:

$$Z(\beta) = \mathrm{Tr}\left\{\rho(q,q',\beta)\right\} \tag{15.34}$$

via the method of the stationary point. For this we have to consider

$$\delta S[q] = 0 \tag{15.35}$$

that yields the Euler-Lagrange equation for the classical path:

$$m\ddot{\bar{q}} + \left.\frac{dV(q)}{dq}\right|_{q=\bar{q}(\tau)} = 0, \quad q(0) = q(\beta) = q_1 \tag{15.36}$$

We select the fluctuation $y(\tau)$ of the classical path:

$$q(\tau) = \bar{q}(\tau) + y(\tau) \tag{15.37}$$

$$y(0) = y(\beta), Dq(\tau) = Dy(\tau)$$

The potential $V(q)$ should be expanded in a functional Taylor series:

$$V(q) \cong V(\bar{q}) + V'(\bar{q})y + \frac{1}{2!}V''(\bar{q})y^2 + \cdots \tag{15.38}$$

The next terms in this functional Taylor series:

$$\frac{1}{3!}U'''(\bar{q})y^3 + \cdots \cong o(y^3) \tag{15.39}$$

So, the partition function is

$$Z \cong \mathrm{F}(\beta)\int_{-\infty}^{\infty} dq_1 \exp\left\{-\frac{1}{\hbar}S[\bar{q}(\tau), -i\hbar\beta]\right\} \tag{15.40}$$

where

$$\mathrm{F}(\beta) = \int_0^0 Dy(\tau)\exp\left\{-\frac{1}{\hbar}\int_0^\beta \left(\frac{m}{2}\dot{y}^2 + \frac{1}{2!}U''(\bar{q})y^2\right)d\tau\right\} \tag{15.41}$$

and

$$\bar{q}(0) = \bar{q}(\beta) = q_1, \quad y(0) = y(\beta) \tag{15.42}$$

We apply this approximation to the case of a quantum harmonic oscillator that yields exact results:

$$V(q) = \frac{m}{2}\omega^2 q^2, \quad V''(q) = m\omega^2 \tag{15.43}$$

$$V^{(n)}(q) = 0, \quad n \geq 3 \tag{15.44}$$

So, from the equation of motion 15.36 we find the classical path

$$q(\tau) = \frac{1}{\sinh\hbar\omega\beta}\left[q'\sinh\hbar\omega(\tau'' - \tau) - q''\sinh\hbar\omega(\tau' - \tau)\right] \tag{15.45}$$

By setting,

$$q' = q = q_1, \tau' = 0, \tau'' = \beta \tag{15.46}$$

then the classical path can be rewritten

$$q(\tau) = \frac{q_1}{2\sinh\hbar\omega\beta}\left[\left(\exp\{\hbar\omega\beta\} - 1\right)\exp\{-\hbar\omega\tau\} + \left(1 - \exp\{-\hbar\omega\beta\}\right)\exp\{\hbar\omega\tau\}\right] \tag{15.47}$$

and the classical action

$$S_{cl}[0, -i\hbar\beta] = m\omega\tanh\frac{\beta\hbar\omega}{2}q_1^2 \tag{15.48}$$

The function $\mathrm{F}(\beta)$ can be obtained in the same manner as for the case seen earlier:

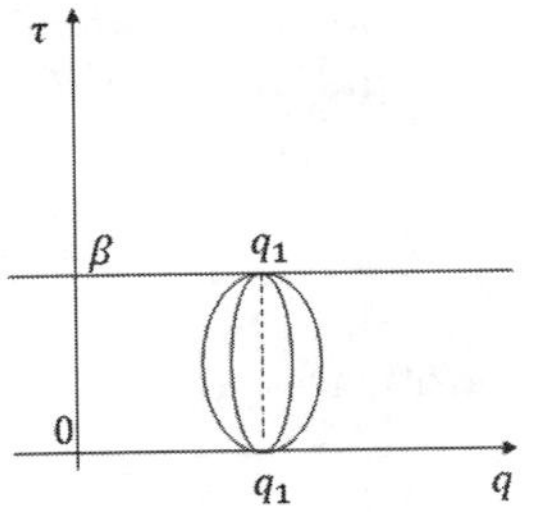

FIGURE 15.2 The closed path over which path integration is taken.

$$F(\beta)=\left(\frac{m\omega}{2\pi\hbar\sinh\beta\hbar\omega}\right)^{\frac{1}{2}} \tag{15.49}$$

So, the partition function in 15.40 can now be evaluated from considering the closed path in Figure 15.2:

$$Z\cong F(\beta)\int_{-\infty}^{\infty}dq_1\exp\left\{-\frac{m\omega}{\hbar}\tanh\frac{\beta\hbar\omega}{2}q_1^2\right\}=\frac{1}{2\sinh\frac{\beta\hbar\omega}{2}} \tag{15.50}$$

From 15.50 the matrix density can be written as:

$$\rho(q_1,q_1,-i\hbar\beta)=F(\beta)\exp\left\{-\frac{m\omega}{\hbar}\tanh\frac{\beta\hbar\omega}{2}q_1^2\right\} \tag{15.51}$$

which is the probability for finding the system at point q_1 .

For the free particle $\omega\to 0$ and

$$\rho(q_1,q_1,-i\hbar\beta)=\left(\frac{m}{2\pi\hbar^2\beta}\right)^{\frac{1}{2}}\equiv\frac{1}{l_e(\hbar\beta)} \tag{15.52}$$

where

$$l_e(\hbar\beta)\equiv\left(\frac{2\pi\hbar^2\beta}{m}\right)^{\frac{1}{2}} \tag{15.53}$$

is called the **de Broglie wavelength associated with the temperature** T or, in short, the **thermal de Broglie wavelength**. These results could equally be obtained from the equation of motion of the density matrix.

15.2 Partition Function in the Path integral Representation

In this heading we examine the partition function as well as the density matrix. We extrapolate them to examples of physical interest such as a system of free particles (ideal gas) as well as a continuum of oscillators (black body radiation of electromagnetic eigenmodes, phonon gas in a crystal lattice and so on). Finding the result for one free particle or oscillator, this may be extrapolated to a large number of particles. In the formulation of statistical physics, we pass from the real time path integral formulation to the Euclidean time formulation where the pure imaginary time is obtained after the analytic continuation $t = -i\hbar\tau$ (where τ is a real number) [2]:

$$0 \le \tau \le \beta \tag{15.54}$$

So, in the imaginary timeframe, the transition amplitude K in quantum mechanics is transformed to $\rho(q,q')$ by doing the change of variable

$$\frac{i\left(t_b - t_a\right)}{\hbar} = \beta \tag{15.55}$$

where the density matrix is $\rho(q,q')$ [1,9]. So, all formulae obtained in path integration in quantum mechanics may be transformed into statistical physics formulae via the change of variables

$$t_b - t_a = -i\hbar\beta \tag{15.56}$$

We consider the Hamiltonian describing the system to be

$$\hat{\mathrm{H}} = \frac{\hat{p}^2}{2m} + U\left(q\right) \tag{15.57}$$

The potential $U(q)$ can be Taylor series expanded about the classical path $\bar{q}\left(t\right)$:

$$U\left(q\right) = U\left(\ddot{q} + y\right) \cong U\left(\bar{q}\right) + U'\left(\bar{q}\right)y + \frac{1}{2!}U''\left(\bar{q}\right)y^2 + \frac{1}{3!}U'''\left(\bar{q}\right)y^3 + \cdots \tag{15.58}$$

The following terms in this functional Taylor series are infinitesimally small:

$$\frac{1}{3!}U'''\left(\bar{q}\right)y^3 + \cdots \cong o\left(y^3\right) \tag{15.59}$$

For brevity, we conveniently apply the stationary phase method to a simple case of an oscillator when

$$U\left(q\right) = \frac{m}{2}\omega^2 q^2, U'' = m\omega^2 \tag{15.60}$$

$$U^{(n)}\left(q\right) = 0, \quad n \ge 3 \tag{15.61}$$

So,

$$\mathrm{F}\left(T\right) = \int_0^0 Dy\left(t\right)\exp\left\{\frac{i}{\hbar}\int_0^T \frac{m}{2}\left(\dot{y}^2 - \omega^2 y^2\right)dt\right\} \tag{15.62}$$

The procedure for its evaluation is seen earlier and we have

$$\mathrm{F}\left(T\right) = \left(\frac{m\omega}{2\pi i\hbar \sin\omega T}\right)^{\frac{1}{2}} \tag{15.63}$$

that is indeed just the transition amplitude from $y(t_a) = 0$ to $y(t_b) = 0$.

For now, we proceed in the evaluation of the density matrix via the transition amplitude.

We consider the propagator K rewritten in an orthonormal basis set of eigenvectors $\{|\Phi_n\rangle\}$ of the time-independent Schrödinger equation:

$$K(q_b,t_b;q_a,t_a) = \sum_n \exp\left\{-\frac{iE_n(t_b - t_a)}{\hbar}\right\}\Phi_n(q_b)\Phi_n^*(q_a)\theta(t_b - t_a) \quad (15.64)$$

where the **Heaviside step function** is:

$$\theta(t_b - t_a) = \begin{cases} 1 & , \quad t_b > t_a \\ 0 & , \quad t_b < t_a \end{cases} \quad (15.65)$$

and

$$\hat{H}|\Phi_n\rangle = E_n|\Phi_n\rangle \quad (15.66)$$

with E_n being the eigenenergies for the eigenstates n. Hence, we have the following identity linking the Schrödinger eigenfunctions with eigenenergies to Feynman path integral:

$$K(q_b,t_b;q_a,t_a) \equiv \int Dq(t)\exp\left\{\frac{i}{\hbar}S[q(t)]\right\} \quad (15.67)$$

The identity 15.67 is still valid even when we pass from the real time path integral formulation to the Euclidean time formulation where the pure imaginary time is obtained after the analytic continuation $t = -i\hbar\tau$ (where τ is a real number) [1,2].

From

$$\text{Tr}\{\hat{\rho}\} = \sum_n \langle\Psi_n|\hat{\rho}|\Psi_n\rangle = \sum_n \langle\Psi_n|\left(\sum_i W_i|\Psi_i\rangle\langle\Psi_i|\right)|\Psi_n\rangle = \sum_n\sum_i \delta_{ni}W_i\delta_{in} = 1 \quad (15.68)$$

It follows that the probabilities must sum up to 1 and so, for example, the trace of the transformed transition amplitude will give the statistical mechanical definition of the thermal partition function Z for a system at temperature T:

$$\text{Tr}\{\hat{K}(-i\hbar\beta)\} = Z(\beta) \quad (15.69)$$

This is a very important quantity in statistical physics, as it can be used to derive many other important statistical quantities, e.g. the free energy:

$$F(T,V) = -T\ln Z \quad (15.70)$$

and the internal energy:

$$E(S,V) = T^2\frac{\partial}{\partial T}\ln Z \quad (15.71)$$

with V and S being the volume and entropy. It is instructive to note that the time coordinate $-i\hbar\beta$ is imaginary and so equation 15.69 assumes that the transition amplitude has been analytically extended into the imaginary time plane. So, for the transition amplitude

$$K(q,t_b;q',t_a)=\int_{q'}^{q}\exp\left\{\frac{i}{\hbar}\int_{t_a}^{t_b}L(\dot{q},q)dt\right\}Dq(t) \tag{15.72}$$

of a particle defined by the Lagrangian $L(\dot{q},q)$:

$$L(\dot{q},q)=\frac{m\dot{q}^2}{2}-U(q) \tag{15.73}$$

where

$$\dot{q}=\frac{dq}{dt}=\frac{i}{\hbar}\frac{dq}{d\tau} \tag{15.74}$$

then the action functional:

$$S=\int_{t_a}^{t_b}\left(\frac{m\dot{q}^2}{2}-U(q)\right)dt=\int_0^{\beta}\left(-\frac{m}{\hbar^2}\left(\frac{dq}{d\tau}\right)^2-U(q)\right)(-i\hbar)d\tau \tag{15.75}$$

and

$$\frac{i}{\hbar}S=\int_0^{\beta}\left(-\frac{m}{\hbar^2}\dot{q}^2-U(q)\right)d\tau\equiv\underline{S}=\underline{S}\left[q(\tau),-i\hbar\beta\right] \tag{15.76}$$

Here, $\underline{S}\left[q(\tau),-i\hbar\beta\right]$ is dimensionless and real quantum-statistical action functional (**Euclidean action**) obtained from the usual **Minkowski action** and τ, the **Euclidean time**. The quantum-statistical action in equation 15.76 effectively relates the Hamiltonian instead of the Lagrangian for the paths running along the imaginary time axis. So, we may obtain the density matrix expressed via the path-integral formalism:

$$\rho(q,q')=K(q,q',-i\hbar\beta)=\int_{q'}^{q}\exp\left\{\underline{S}\left[q,q',-i\hbar\beta\right]\right\}Dq(\tau) \tag{15.77}$$

This is the quantum-statistical path integral for the density matrix $\rho(q,q')$**.** It is instructive to note that doing the change of variable $t=-i\hbar\tau$ for the differential dt and for the kinetic energy (or $\dot{q}(t)$) in equation 15.73, the quantum-statistical action functional in equation 15.76 effectively relates the Hamiltonian instead of the Lagrangian for the paths running along the imaginary time axis. The measure $Dq(\tau)$ denotes the summation (functional integration) over all closed paths (function) $q(\tau)$ in the time interval β. So, we consider **all possible paths**, by which the system travels between the initial and final configurations in the **time** β with the density matrix ρ being the sum of contribution from each path.

The partition function via the transition amplitude is

$$Z=\sum_n\int dq\left\langle\Phi_n(q)\middle|\Phi_n(q)\right\rangle\exp\{-\beta E_n\} \tag{15.78}$$

or

$$Z=\int\delta(q-q')dq'dq\sum_n\left\langle\Phi_n(q)\middle|\Phi_n(q')\right\rangle\exp\{-\beta E_n\} \tag{15.79}$$

This is convergent provided the spectrum is known and the n-sum converges. Of course, the sum should converge since $\exp\{-\beta E_n\}$ decreases with n. From

$$\sum_n \langle \Phi_n(q) | \Phi_n(q') \rangle \exp\{-\beta E_n\} = Z\rho(q,q') \tag{15.80}$$

Then, in coordinate representation the partition function is the trace of the density matrix:

$$Z = \int \delta(q-q')dqdq'Z\rho(q,q') = \mathrm{Tr}\{Z\rho(q,q')\} \tag{15.81}$$

This is the sum of the diagonal element $\rho(q,q)$ which is an invariant quantity with the representing basis functions for the matrix. This **provides the central physical quantity in quantum statistics, the so-called, the quantum canonical partition function** Z:

$$Z = \int \delta(q-q')dqdq'Z\rho(q,q') = \int \delta(q-q')dqdq'\mathrm{K}(q,q',-i\hbar\beta) \tag{15.82}$$

So, the quantum-mechanical-statistical partition function is related in a simple manner to the trace of the quantum-mechanical transition amplitude for time-independent Hamiltonians:

$$Z = \mathrm{Tr}\{\mathrm{K}(q,q',-i\hbar\beta)\} \tag{15.83}$$

Hence, the thermal partition function is equivalent to a functional integral over a compact Euclidean time $\tau \in [0,\beta]$ with boundaries identified. Therefore, the integration variables $q(\tau)$ are periodic in τ with period β. So, the partition function is derived considering only those cases when the final configuration mimics the initial configuration, and we sum over all possible initial configurations. In $\mathrm{K}(q,q',-i\hbar\beta)$ for large β the ground state energy dominates provided the spectrum is not continuous, but **gapped (discrete)**. In this manner the spectrum/masses of the system numerically can be determined where it is advantageous to consider positions q_a and q_b or profiles around them in a fashion that the prefactor $\langle \Phi_n(q_b) | \Phi_n(q_a) \rangle$ is large.

15.3 Particle Interaction with a Driven or Forced Harmonic Oscillator: Partition Function

We have seen so far that quadratic actions give Gaussian integrals that are easily solvable via Feynman path integration with the strength of this method seen when solving interacting multi-particle problems. As the number of degrees of freedom concerned is very large then solving the problem via differential equations is extremely difficult, since the equations can barely be made tractable. So, it is very important to apply the Feynman path integral technique to resolve this problem. If the action functional has a non-quadratic part that is **small,** then we employ approximation methods with one of such methods being the perturbation theory that in many-particle physics and quantum field theory is most easily evaluated via path integrals. We examine an example of such a problem which is the particle interaction with a driven or forced harmonic oscillator. This is a problem that involves two subsystems, i.e. a particle with position vector $\vec{r}$ and mass m subjected to some potential $U(\vec{r})$ and interacting with an oscillator with mass M and the coordinate being q. The coupling between the oscillator and the particle is enforced by the driving term, $\gamma(\vec{r}(t))$. The example may be the interaction of the electron with phonons in a crystal lattice. The entire system is described by the Lagrangian:

$$L(\vec{r},q) = \frac{m\dot{\vec{r}}^2}{2} - U(\vec{r}) + \frac{M\dot{q}^2}{2} - \frac{M\omega^2 q^2}{2} + \gamma(\vec{r})q \tag{15.84}$$

Considering the imaginary time, $t = -i\hbar\tau$ then

$$L(\vec{r},q) = L_e(\vec{r}) + L_q(q) + L_{\text{int}}(\vec{r},q) \tag{15.85}$$

where

$$L_e(\vec{r}) = -\frac{m\dot{\vec{r}}^2}{2\hbar^2} - U(\vec{r}), L_q(q) = -\frac{M\dot{q}^2}{2\hbar^2} - \frac{M\omega^2 q^2}{2}, L_{\text{int}}(\vec{r},q) = \gamma(\vec{r})q \tag{15.86}$$

The partition function for this system is the trace of the transition amplitude:

$$Z = \int d\vec{r}d\vec{r}'\delta(\vec{r}-\vec{r}')\int_{r'}^{r} Dr\exp\left\{\int_0^\beta L_e(\vec{r}(\tau))d\tau\right\}\mathrm{F}[\vec{r},\vec{r}'] \tag{15.87}$$

where the partition function of the forced harmonic oscillator is

$$Z_{\text{FHO}} = \int dqdq'\delta(q-q')\int_{q'}^{q} Dq\exp\left\{\int_0^\beta L_{\text{FHO}}(\vec{r},q)d\tau\right\} \tag{15.88}$$

To evaluate this path integral, we apply the quasi-classical approximation for the new path for which we select a small variation of the path

$$\delta q(t) \equiv y(t) \tag{15.89}$$

so that

$$q(t) = \bar{q}(t) + y(t), \quad y(t_a) = y(t_b) = 0, \quad Dq(t) = Dy(t) \tag{15.90}$$

So,

$$\mathrm{F}[\vec{r},\vec{r}'] = \int_{q'}^{q} Dq\exp\left\{\int_0^\beta\left[-\frac{M\dot{q}^2}{2\hbar^2} - \frac{M\omega^2 q^2}{2} + \gamma(\vec{r}(\tau))q\right]\right\} \tag{15.91}$$

or

$$Z_{\text{FHO}} = \exp\{S[q(\tau), -i\hbar\beta]\}\int_0^0 Dy(t)\exp\left\{\int_0^\beta \frac{M}{2}\left(-\frac{\dot{y}^2}{\hbar^2} - \omega^2 y^2\right)d\tau\right\} \tag{15.92}$$

or

$$Z_{\text{FHO}} = \mathrm{F}(T)\exp\{S[q(\tau), -i\hbar\beta]\} \tag{15.93}$$

where

$$\mathrm{F}(T) = \left(\frac{m\omega}{2\pi i\hbar\sin\omega T}\right)^{\frac{1}{2}} \tag{15.94}$$

and

$$S\left[q(\tau),-i\hbar\beta\right]=\frac{m\omega}{2\hbar\sinh\beta\hbar\omega}\left[\left(q_a^2+q_b^2\right)\cosh\left(\beta\hbar\omega\right)-2q_aq_b+2\tilde{\mathcal{F}}\left(T\right)-2\tilde{\mathcal{F}}_{\mathrm{NH}}\left(T\right)\right] \tag{15.95}$$

with

$$\tilde{\mathcal{F}}_{\mathrm{NH}}\left(T\right)=\frac{1}{m\omega}\int_0^{\beta}\int_0^{s}\gamma\left(s\right)\mathcal{F}\left(s,t\right)\gamma\left(t\right)dtds \tag{15.96}$$

and

$$\mathcal{F}\left(s,t\right)=\frac{\sinh\hbar\omega\left(\beta-s\right)\sinh\hbar\omega\left(t-t_a\right)}{m\omega\sinh\beta\hbar\omega} \tag{15.97}$$

We may exclude first the variable of the oscillator q. To do this we use the fact that the delta function enforces $q=q'$ or $q_a=q_b$ and evaluate a Gaussian integral in 15.92 via the formula

$$\int_{-\infty}^{+\infty}dq\exp\left\{-aq^2+bq\right\}=\left(\frac{\pi}{a}\right)^{\frac{1}{2}}\exp\left\{\frac{b^2}{4a}\right\},\quad a>0 \tag{15.98}$$

then from

$$Z_{\mathrm{FHO}}=\mathrm{F}\left(-i\beta\hbar\right)\int_{-\infty}^{\infty}dq\exp\left\{S\left[q(\tau),-i\hbar\beta\right]\right\} \tag{15.99}$$

the Gaussian integral in 15.99 yields

$$Z_{\mathrm{FHO}}\equiv Z_0\mathrm{F}\left[\vec{r},\vec{r}'\right] \tag{15.100}$$

with the influence functional

$$\mathrm{F}\left[\vec{r},\vec{r}'\right]=\exp\left\{-\Phi_{\omega}\left[\vec{r},\vec{r}'\right]\right\} \tag{15.101}$$

and the influence phase:

$$\Phi_{\omega}\left[\vec{r},\vec{r}'\right]=\frac{\hbar}{4M\omega}\int_0^{\beta}d\tau\int_0^{\tau}d\tau'\gamma\left(\vec{r}\left(\tau\right)\right)\gamma\left(\vec{r}\left(\tau'\right)\right)\mathrm{F}_{\omega}\left(\left|\tau-\tau'\right|\right) \tag{15.102}$$

$$\mathrm{F}_{\omega}\left(\left|\tau-\tau'\right|\right)=\frac{\cosh\hbar\omega\left(\left|\tau-\tau'\right|-\frac{\beta}{2}\right)}{\sinh\left(\frac{\beta\hbar\omega}{2}\right)},\quad Z_L=\left[2\sinh\left(\frac{\beta\hbar\omega}{2}\right)\right]^{-1} \tag{15.103}$$

Then in our case we have

$$Z=Z_0\int d\vec{r}d\vec{r}'\delta\left(\vec{r}-\vec{r}'\right)\int_{r'}^{r}D\mathrm{r}\mathrm{F}\left[\vec{r},\vec{r}'\right]\exp\left\{\int_0^{\beta}\left[-\frac{m\dot{\vec{r}}^{2}}{2\hbar^2}-U\left(\vec{r}\right)\right]d\tau\right\} \tag{15.104}$$

We observe that to evaluate the partition function in 15.91, the coordinates of the oscillator should first be eliminated by path integration, then the result is the path integral over the coordinate $\vec{r}$ of the particle.

15.4 Free Particle Density Matrix and Partition Function

This is a methodical approach with the physics being the system of free particles, for example say, ideal gas. The result for the free particle may be applied to a large number of free particles. Applying the Cauchy problem in the coordinate q representation for the free particle we have:

$$\frac{\partial\rho(q,q',\beta)}{\partial\beta}=\frac{\hbar^2}{2m}\frac{\partial^2}{\partial q^2}\rho(q,q',\beta),\rho(q,q',0)=\delta(q-q') \tag{15.105}$$

This is a system of equations of a diffusive motion with standard solution for given initial conditions:

$$\rho(q,q',\beta)=\left(\frac{m}{2\pi\beta\hbar^2}\right)^{\frac{1}{2}}\int_{-\infty}^{\infty}d\xi\delta(q'-\xi)\exp\left\{-\frac{m}{2\beta\hbar^2}(q-\xi)^2\right\} \tag{15.106}$$

or

$$\rho(q,q',\beta)=\left(\frac{m}{2\pi\beta\hbar^2}\right)^{\frac{1}{2}}\exp\left\{-\frac{m}{2\beta\hbar^2}(q-q')^2\right\} \tag{15.107}$$

For the free particle transition amplitude (see details in [1] when $U(q) = 0$ in 4.39 of chapter 4), we expect a motion with constant velocity,

$$v=\frac{q_b-q_a}{t_b-t_a}=\dot{q} \tag{15.108}$$

Then following the procedure from 4.47 to 4.79 of chapter 4, the transition amplitude

$$\mathrm{K}(q_b,t_b;q_a,t_a)\equiv\langle q_b,t_b|q_a,t_a\rangle\equiv q_b\left|\exp\left\{-\frac{i\hat{\mathrm{H}}t}{\hbar}\right\}\right|q_a \tag{15.109}$$

and in terms of the classical action

$$\mathrm{K}(q_b,t_b;q_a,t_a)=\left(\frac{m}{2\pi i\hbar(t_b-t_a)}\right)^{\frac{1}{2}}\exp\left\{\frac{im(q_b-q_a)^2}{2\hbar(t_b-t_a)}\right\}=\left(\frac{m}{2\pi i\hbar(t_b-t_a)}\right)^{\frac{1}{2}}\exp\left\{\frac{i}{\hbar}S_{\mathrm{cl}}\left[q_b,q_a\right]\right\} \tag{15.110}$$

where the classical action

$$S_{\mathrm{cl}}\left[q_a,q_b\right]=\frac{m(q_b-q_a)^2}{2(t_b-t_a)} \tag{15.111}$$

was derived in [1] via the classical equation of motion. The quantity 15.110 simply is a diffusive motion or spreading of a localized wave packet that started as a Dirac delta-function centered at q.

We apply statistical physics by doing the analytic continuation in 15.55, which transforms the transition amplitude K in quantum mechanics to the density matrix $\rho(q_b, q_a)$:

where, the density matrix is $\rho(q_b, q_a)$ [1,5,9]. So, all formulae obtained in path integration in quantum mechanics may be transformed into statistical physics formulae via the change of variables (analytical prolongation)

$$t_b - t_a = -i\hbar\beta \tag{15.112}$$

and the quantum-statistical density matrix for the free particles corresponds formally to

$$K(q_b, q_a, -i\hbar\beta) = \rho(q_b, q_a, \beta) = \left(\frac{m}{2\pi i\hbar(-i\hbar\beta)}\right)^{\frac{1}{2}} \exp\left\{\frac{im(q_b - q_a)^2}{2\hbar(-i\hbar\beta)}\right\} \tag{15.113}$$

or

$$\rho(q_b, q_a, \beta) = \left(\frac{m}{2\pi\beta\hbar^2}\right)^{\frac{1}{2}} \exp\left\{-\frac{m(q_b - q_a)^2}{2\hbar\beta\hbar^2}\right\} \tag{15.114}$$

This corresponds to the solution of 15.105 for the one-dimensional case and for $\beta = 0$ we have

$$\rho(q_b, q_a, 0) = \delta(q_b - q_a) \tag{15.115}$$

If L is the length of the one-dimensional system then

$$\exp\{-\beta F\} = Z = \int \rho(q_b, q_a, \beta)\, dq = \left(\frac{m}{2\pi\beta\hbar^2}\right)^{\frac{1}{2}} L \tag{15.116}$$

For the case of the three-dimensional system with volume V then

$$\rho(q_b, q_a, \beta) = \left(\frac{m}{2\pi\beta\hbar^2}\right)^{\frac{3}{2}} \exp\left\{-\frac{m(q_b - q_a)^2}{2\hbar\beta\hbar^2}\right\} \tag{15.117}$$

$$\exp\{-\beta F\} = Z = \left(\frac{m}{2\pi\beta\hbar^2}\right)^{\frac{3}{2}} V \tag{15.118}$$

For the case of N particles then

$$Z(T, V, N) = \left(\frac{m}{2\pi\beta\hbar^2}\right)^{\frac{3N}{2}} V^N \tag{15.119}$$

This permits us to derive all the thermodynamic quantities.

15.5 Quantum Harmonic Oscillator Density Matrix and Partition Function

Detail derivation of the action functional and transition amplitude of the harmonic oscillator may be found in the reference [1]. The Lagrangian of a harmonic oscillator has the form:

$$L = \frac{m\dot{q}^2}{2} - \frac{m\omega^2 q^2}{2} \tag{15.120}$$

and the action for the classical path of the harmonic oscillator is found to be

$$S_{\text{Hcl}}\left[0,T\right] = \frac{m\omega}{2\sin\omega T}\left[\left(q_a^2 + q_b^2\right)\cos\omega T - 2q_a q_b\right], \quad T = t_b - t_a \tag{15.121}$$

and the transition amplitude

$$\mathrm{K}\left(q_b, T; q_a, 0\right) = \mathrm{F}\left(T\right)\exp\left\{\frac{i}{\hbar} S_{\text{Hcl}}\left[0, T\right]\right\} \tag{15.122}$$

with

$$\mathrm{F}\left(T\right) = \left(\frac{m\omega}{2\pi i\hbar \sin\omega T}\right)^{\frac{1}{2}} \tag{15.123}$$

From the procedure 15.74 to 15.77, then the quantum-statistical density matrix $\rho(q_b, q_a, -i\hbar\beta)$ for the harmonic oscillator is obtained:

$$\mathrm{K}\left(q_b, q_a, -i\hbar\beta\right) = \rho\left(q_b, q_a, -i\hbar\beta\right) = \mathrm{F}\left(-i\beta\hbar\right)\exp\left\{\underline{S}_{\text{Hcl}}\left[0, -i\hbar\beta\right]\right\} \tag{15.124}$$

with

$$\underline{S}_{\text{Hcl}}\left[0, -i\hbar\beta\right] = -\frac{m\omega}{2\hbar\sinh\left(\beta\hbar\omega\right)}\left[\left(q_a^2 + q_b^2\right)\cosh\left(\beta\hbar\omega\right) - 2q_a q_b\right] \tag{15.125}$$

We find the partition function from the quantum-statistical density matrix via the trace:

$$Z = \mathrm{Tr}\left\{\mathrm{K}\left(q_b, q_a, -i\hbar\beta\right)\right\} \tag{15.126}$$

This involves setting $q_b = q_a = q$. This implies integrating over periodic paths starting and ending at q as in Figure 100. Integrating over q is equivalent to integrating over all periodic paths with period $\hbar\beta$. This yields the Gaussian integral

$$Z = \int dq \mathrm{F}\left(-i\beta\hbar\right)\exp\left\{-\frac{m\omega}{2\hbar\sinh\left(\beta\hbar\omega\right)}\left[\cosh\left(\beta\hbar\omega\right) - 1\right]2q^2\right\} \tag{15.127}$$

or

$$Z \equiv Z_{\text{HO}} = \mathrm{F}(-i\beta\hbar)\left(\frac{\pi\hbar\sinh(\beta\hbar\omega)}{m\omega 2\sinh^2\left(\frac{\beta\hbar\omega}{2}\right)}\right)^{\frac{1}{2}} = \frac{1}{2\sinh\left(\frac{\beta\hbar\omega}{2}\right)} \tag{15.128}$$

This agrees with the partition function for the one-dimensional oscillator evaluated from

$$Z = \sum_n \exp\{-\beta E_n\} = \sum_{n=0}^{\infty} \exp\left\{-\beta\hbar\omega\left(n+\frac{1}{2}\right)\right\}$$
$$= \exp\left\{-\frac{\beta\hbar\omega}{2}\right\}\sum_{n=0}^{\infty}\exp\{-\beta\hbar\omega n\} \tag{15.129}$$

or

$$Z = \exp\left\{-\frac{\beta\hbar\omega}{2}\right\}\sum_{n=0}^{\infty}\exp\{-\beta\hbar\omega n\} = \frac{\exp\left\{-\frac{\beta\hbar\omega}{2}\right\}}{1-\exp\{-\beta\hbar\omega\}} \tag{15.130}$$

or

$$Z = \frac{1}{2\frac{\left(\exp\left\{\frac{\beta\hbar\omega}{2}\right\}-\exp\left\{-\frac{\beta\hbar\omega}{2}\right\}\right)}{2}} = \frac{1}{2\sinh\left(\frac{\beta\hbar\omega}{2}\right)} \tag{15.131}$$

This equation may be obtained via the Cauchy problem in the coordinate q-representation for the harmonic oscillator:

$$\frac{\partial\rho(q,q',\beta)}{\partial\beta} = \frac{\hbar^2}{2m}\frac{\partial^2}{\partial q^2}\rho(q,q',\beta) - \frac{m\omega^2 q^2}{2}\rho(q,q',\beta), \rho(q,q',0) = \delta(q-q') \tag{15.132}$$

The exact solution for this problem is written exactly in 15.124 and 15.125.

The free energy F of the harmonic oscillator is evaluated:

$$F = -T\ln Z = \frac{\hbar\omega}{2} + T\ln\left(1-\exp\{-\beta\hbar\omega\}\right) \tag{15.133}$$

The average over the ensemble of the energy $\bar{E}$ in thermodynamics is also called internal energy and may be denoted by $\bar{E}$:

$$\bar{E} = T^2\frac{\partial}{\partial T}\ln Z \tag{15.134}$$

or

$$\bar{E} = T^2\frac{\partial}{\partial T}\ln Z = -T^2\frac{\partial}{\partial T}\ln\left[2\sinh\left(\frac{\beta\hbar\omega}{2}\right)\right] = \frac{\hbar\omega}{2T^2}T^2F_{\omega}(0) \tag{15.135}$$

or

$$\bar{E} = \frac{\hbar\omega}{2} F_\omega(0) = \hbar\omega\left(\bar{n} + \frac{1}{2}\right) \tag{15.136}$$

where the average occupation of statistical oscillators (Bose-Einstein distribution $\bar{n}$):

$$\bar{n} = \frac{1}{\exp\{-\beta\hbar\omega\} - 1}, F_\omega(0) = \frac{\cosh\frac{\beta\hbar\omega}{2}}{\sinh\left(\frac{\beta\hbar\omega}{2}\right)} \tag{15.137}$$

The model of real physical systems are not the oscillators themselves but the selection of infinite number of free or forced oscillators. Suppose we have an infinite number of selections of free oscillators with the spectrum $\{\omega_i\}$. From the property of additivity of the free energy F, and the internal energy $\bar{E}$ of the system, and in the absence of these expressions in the terms describing an interaction of oscillators then we have, respectively, the free energy F and the internal energy $\bar{E}$:

$$F = \sum_i F_i = \sum_i \left(\frac{\hbar\omega_i}{2} + T\ln\left(1 - \exp\{-\beta\hbar\omega_i\}\right)\right) \tag{15.138}$$

$$\bar{E} = \sum_i U_i = \sum_i \hbar\omega_i\left(\bar{n}_i + \frac{1}{2}\right) \tag{15.139}$$

Here, the average selection of occupation numbers $\bar{n}_i$ is given by:

$$\bar{n}_i = \frac{1}{\exp\{\beta\hbar\omega_i\} - 1} \tag{15.140}$$

Except the $\frac{\hbar\omega_i}{2}$ term, the contribution to F of the ith oscillator is negligible when $\beta\hbar\omega_i \gg 1$. For low temperatures, the high-frequency modes are frozen out and so do not contribute to the specific heat.

For the case of the Lagrangian for an oscillator subjected to a time-dependent force $\gamma(t)$ then

$$L = \frac{m\dot{q}^2}{2} - \frac{m\omega^2 q^2}{2} + \gamma(t)q \tag{15.141}$$

and the classical action is found to be

$$S_{\mathrm{FHO}} = \frac{m\omega}{2\sin\omega T}\left[\begin{matrix}\left(q_a^2 + q_b^2\right)\cos\omega T - 2q_a q_b + 2\tilde{\mathcal{F}}(T) - \\ -2\tilde{\mathcal{F}}_{\mathrm{NH}}(T)\end{matrix}\right] \tag{15.142}$$

with

$$\tilde{\mathcal{F}}_{\mathrm{NH}}(T) = \frac{1}{m\omega}\int_0^T\int_0^s \gamma(s)\mathcal{F}(s,t)\gamma(t)\,dt\,ds \tag{15.143}$$

$$\mathcal{F}(s,t) = \frac{\sin\omega(T-s)\sin\omega(t-t_a)}{m\omega\sin\omega T} \tag{15.144}$$

$$\tilde{\mathcal{F}}(T)=\int_0^T q_{\mathrm{H}}(t)\gamma(t)dt \tag{15.145}$$

$$q_{\mathrm{H}}(t)=q_a\Gamma_a(T-t)+q_b\Gamma(t) \tag{15.146}$$

$$\Gamma_a(T-t)\equiv\frac{\sin\omega(T-t)}{\sin\omega T},\Gamma(t)\equiv\frac{\sin\omega t}{\sin\omega T},T=t_b-t_a \tag{15.147}$$

If $\gamma(t)\equiv 0$ then we achieve the action of a free harmonic oscillator. The transition amplitude of a forced harmonic oscillator corresponds to a purely quantum mechanical result is found to be

$$\mathrm{K}_{\mathrm{FHO}}(q_b,t_b;q_a,t_a)=\mathrm{F}(T)\exp\left\{\frac{i}{\hbar}S_{\mathrm{FHO}}\right\} \tag{15.148}$$

From the procedure 15.74 to 15.77, then the quantum-statistical density matrix $\rho(q_b,q_a,-i\hbar\beta)$ for the forced harmonic oscillator

$$\rho(q_b,q_a,-i\hbar\beta)=\mathrm{F}(-i\beta\hbar)\exp\left\{S_{\mathrm{FHO}}\left[0,-i\hbar\beta\right]\right\} \tag{15.149}$$

Here, the classical action is found to be

$$S_{\mathrm{FHO}}\left[q(\tau),-i\hbar\beta\right]=\frac{m\omega}{2\hbar\sinh\beta\hbar\omega}\left[\left(q_a^2+q_b^2\right)\cosh(\beta\hbar\omega)-2q_aq_b+2\tilde{\mathcal{F}}(T)-2\tilde{\mathcal{F}}_{\mathrm{NH}}(T)\right] \tag{15.150}$$

Considering the standard Gaussian integral:

$$\int_{-\infty}^{+\infty}dq\exp\left\{-aq^2+bq\right\}=\left(\frac{\pi}{a}\right)^{\frac{1}{2}}\exp\left\{\frac{b^2}{4a}\right\},\quad a>0 \tag{15.151}$$

Then the partition function for the forced harmonic oscillator becomes

$$Z_{\mathrm{FHO}}=Z_{\mathrm{HO}}\exp\left\{-\frac{\hbar}{4m\omega}\int_0^\beta\int_0^\beta dsds'\gamma(s)\gamma(s')F_\omega(|s-s'|)\right\} \tag{15.152}$$

where

$$\mathrm{F}_\omega(|s-s'|)=\frac{\cosh\hbar\omega\left(|s-s'|-\frac{\beta}{2}\right)}{\sinh\left(\frac{\hbar\omega\beta}{2}\right)} \tag{15.153}$$

Here, the driving force γ is not yet defined. We can equally rewrite the partition function in terms of occupation numbers, say, for example, phonons via

$$\mathrm{F}_\omega(|s-s'|)=\frac{\exp\left\{\hbar\omega\left(|s-s'|-\frac{\beta}{2}\right)\right\}+\exp\left\{-\hbar\omega\left(|s-s'|-\frac{\beta}{2}\right)\right\}}{2\left(\frac{\exp\left\{\frac{\hbar\omega\beta}{2}\right\}-\exp\left\{-\frac{\hbar\omega\beta}{2}\right\}}{2}\right)} \tag{15.154}$$

or

$$F_{\omega}\left(\left|s-s'\right|\right)=\frac{\exp\left\{\hbar\omega\left|s-s'\right|\right\}+\exp\left\{-\hbar\omega\left(\left|s-s'\right|-\beta\right)\right\}}{\exp\left\{\hbar\omega\beta\right\}-1} \tag{15.155}$$

But the mean occupation $\bar{n}\left(\hbar\omega\beta\right)$ is

$$\bar{n}\left(\hbar\omega\beta\right)\equiv\frac{1}{\exp\left\{\hbar\omega\beta\right\}-1} \tag{15.156}$$

and

$$\frac{\exp\left\{\hbar\omega\beta\right\}}{\exp\left\{\hbar\omega\beta\right\}-1}=1+\frac{1}{\exp\left\{\hbar\omega\beta\right\}-1}=1+\bar{n}\left(\hbar\omega\beta\right) \tag{15.157}$$

So,

$$F_{\omega}\left(\left|s-s'\right|\right)=\bar{n}\left(\hbar\omega\beta\right)\exp\left\{\hbar\omega\left|s-s'\right|\right\}+\left(1+\bar{n}\left(\hbar\omega\beta\right)\right)\exp\left\{-\hbar\omega\left|s-s'\right|\right\} \tag{15.158}$$

or

$$F_{\omega}\left(\left|s-s'\right|\right)=\left(\exp\left\{\hbar\omega\left|s-s'\right|\right\}+\exp\left\{-\hbar\omega\left(\left|s-s'\right|-\beta\right)\right\}\right)\bar{n}\left(\hbar\omega\beta\right) \tag{15.159}$$

All these formulae will be useful in the evaluation of the polaron and bipolaron characteristics in the quasi-0D quantum dot and quasi-1D quantum wire.

16

Quasi-Classical Approximation in Quantum Statistical Mechanics

We begin by first showing how, with the Feynman functional integration in the representation of the matrix density, we can make a transition to classical distribution. The Feynman density matrix as seen earlier has the form in 16.77:

$$\rho(q,q') = \int_{q'}^{q} \exp\left\{\underline{S}\left[q,q',-i\hbar\beta\right]\right\} Dq(\tau) \tag{16.1}$$

where

$$\underline{S} = \underline{S}\left[q(\tau),-i\hbar\beta\right] \equiv \int_0^\beta \left(-\frac{m}{2\hbar^2}\dot{q}^2 - U(q)\right) d\tau \tag{16.2}$$

We find which paths dominate for the transition to the classical case where the characteristic parameter is small, $\beta \to 0$. This implies very high temperatures T. It is instructive to note that in the stationary phase approximation, only a small region about the extremum contributes to the integral. For the given integral 16.2 those paths not far from $q \to q'$ are dominant since the **time** β is that for which the system should return to point q. Outside of this region, the integrand rapidly oscillates and so gives to the leading order a negligible contribution. This is linked to the fact that for very high temperatures T or very small β, a very high velocity $\left|\frac{dq}{d\tau}\right|$ is needed in order to traverse a large distance within a small time interval. But, in the integrand 16.1 we find the quantity

$$\exp\left\{\int_0^\beta \left(-\frac{m}{2\hbar^2}\dot{q}^2\right) d\tau\right\} \tag{16.3}$$

that is very small due to a very high velocity $\left|\frac{dq}{d\tau}\right|$. So, we assume that q and q' are so close that we bring out $\exp\{-\beta U(q)\}$ of the integral in 16.1 though an over approximation:

$$\rho(q,q') \cong \exp\left\{-\beta U(q)\right\} \int_{q'}^{q} \exp\left\{-\frac{m}{2\hbar^2}\int_0^\beta \dot{q}^2 d\tau\right\} Dq(\tau) = \exp\left\{-\beta U(q)\right\} \rho_{\text{free}}(q,q') \tag{16.4}$$

Here, the free particle density matrix has the form:

$$\rho_{\text{free}}(q,q') = \left(\frac{m}{2\pi\beta\hbar^2}\right)^{\frac{1}{2}} \exp\left\{-\frac{m(q-q')^2}{2\hbar\beta\hbar^2}\right\} \tag{16.5}$$

The inverse of the factor $\left(\frac{m}{2\pi\beta\hbar^2}\right)^{\frac{1}{2}}$ in the above equation is the thermal de Broglie wavelength. So, for the classical particle when $q = q' = q_1$, the density matrix has the form

$$\rho(q,q') \cong \left(\frac{m}{2\pi\beta\hbar^2}\right)^{\frac{1}{2}} \exp\{-\beta U(q)\} \tag{16.6}$$

So, the partition function for the system is the trace:

$$Z \cong \int_{-\infty}^{\infty} dq\rho(q,q) = \left(\frac{m}{2\pi\beta\hbar^2}\right)^{\frac{1}{2}} \int_{-\infty}^{\infty} dq \exp\{-\beta U(q)\} \tag{16.7}$$

Hence the over approximation gives, of course, the classical approximation since 16.6 gives the Boltzmann distribution over the coordinates for a one-dimensional motion of a particle in an external field $U(q)$. The inverse of the factor $\left(\frac{m}{2\pi\beta\hbar^2}\right)^{\frac{1}{2}}$ in equation 16.7 mimics the thermal de Broglie wavelength. Relation 16.7 holds only for sufficiently smooth potentials.

16.1 Centroid Effective Potential

It is interesting to approximate the quantum partition function in a classical fashion without explicitly solving the Schrödinger equation. We consider the classical result and show if we could obtain the classical distribution or the first quantum mechanical correction. Bringing out $U(q)$ of the expression in the action 16.2 while considering it as a lower or upper bound is immaterial. This is a very over approximation that supposes points very close to each other. So $U(q)$ achieves the same value when the points q and q' are so close.

We investigate first the condition for which the classical approximation is good. We consider the potential $U(q)$ to be invariant for the motion from point q' to point q. This is due to the smallness of the **characteristic time** β. So, from 16.1 we have

$$\rho(q,q') \cong \exp\{-\beta U(q')\} \int_{q'}^{q} \exp\left\{-\frac{m}{2\hbar^2}\int_0^\beta \dot{q}^2 d\tau\right\} Dq(\tau) \tag{16.8}$$

or

$$\rho(q,q') \cong \left(\frac{m}{2\pi\beta\hbar^2}\right)^{\frac{1}{2}} \exp\{-\beta U(q')\} \exp\left\{-\frac{m}{2\hbar^2\beta}(q-q')^2\right\} \tag{16.9}$$

From here, we can rewrite the following exponential function

$$\exp\left\{-\frac{m}{2\hbar^2\beta}(q-q')^2\right\} = \exp\left\{-\frac{1}{2}\left(\left(\frac{m}{\hbar^2\beta}\right)^{\frac{1}{2}} \Delta q\right)^2\right\} \tag{16.10}$$

This shows that the contribution to the functional integral are from paths that differ by

$$\Delta q > \left(\frac{\beta \hbar^2}{m} \right)^{\frac{1}{2}} \tag{16.11}$$

This renders the exponential function in 16.9 or 16.10 very small. So, the characteristic distance for which the path gives a dominant contribution to the density matrix should be of the order

$$\Delta q \approx \left(\frac{\beta \hbar^2}{m} \right)^{\frac{1}{2}} \tag{16.12}$$

It follows from 16.8 to 16.10 we have a classical approximation true only when for the characteristic length 16.12, the potential $U(q)$ is invariant.

We find the partition function

$$Z = \int_{-\infty}^{\infty} dq \rho(q,q) \tag{16.13}$$

in the quasi-classical approximation. This implies evaluating the correction $o(\hbar)$ to the classical expression 16.7 that involves the **centroid path integrals.** Instead of integrating over all possible initial/final positions of paths to obtain the partition function, we may group the closed paths in accordance with their **time-average positions** $\bar{q}$, the **so-called centroids** that helps in expansion about those positions:

$$\bar{q} = \frac{1}{\beta} \int_0^{\beta} q(\tau) d\tau \tag{16.14}$$

We evaluate the partition function from the following conditions over the centroid positions:

1. **Fix the arbitrary value of $\bar{q}$,**
2. **For given values of end points of the path $q = q'$, we sum for those paths for which the $\bar{q}$ is the time-average position or centroid is $\bar{q}$,**
3. **We select all possible values of q and sum over them,**
4. **In the resultant expression, we select all possible values of $\bar{q}$ and sum over them.**

Since the quasi-classical approximation implies a sufficiently small change of the potential $U(q)$ corresponding to points above the average point $\bar{q}$, the **so-called centroid** of the path, then we proceed by Taylor series expansion for $U(q)$ locally around $\bar{q}$ to obtain the **centroid potential** $U(\bar{q})$ in addition to some order corrections to it:

$$U(q(\tau)) = U(\bar{q} + (q(\tau) - \bar{q})) \cong U(\bar{q}) + \left. \frac{dU}{dq(\tau)} \right|_{q(\tau)=\bar{q}} (q(\tau) - \bar{q}) + \frac{1}{2!} \left. \frac{d^2U}{dq^2(\tau)} \right|_{q(\tau)=\bar{q}} (q(\tau) - \bar{q})^2 + \cdots \tag{16.15}$$

Neglecting higher order terms and considering the time integral associated with the first order derivative $U'(\bar{q})$ that is zero due to equation 16.14, i.e.,

$$U'(\bar{q}) \int_0^{\beta} (q(\tau) - \bar{q}) d\tau = 0 \tag{16.16}$$

So, we are left with the quadratic integral

$$\int_0^\beta U(q(\tau))d\tau = \beta U(\bar{q}) + \int_0^\beta \frac{1}{2}\frac{d^2U}{dq^2(\tau)}\bigg|_{q(\tau)=\bar{q}} (q(\tau)-\bar{q})^2 d\tau \tag{16.17}$$

or

$$\int_0^\beta U(q(\tau))d\tau = \beta U(\bar{q}) + \int_0^\beta \frac{1}{2}U''(\bar{q})(q(\tau)-\bar{q})^2 d\tau \tag{16.18}$$

Note that the effective quasi-classical potential based on the centroid density of path integrals should be less sensitive to the curvature or the variation of the original **classical** potential U and, so, the centroid effective potential should be more classical-like. Then substituting 16.15 into density matrix relation (or temperature Green's function) we carry out integration of the Gaussian integral over the centroid positions:

$$K = \exp\{-\beta U(\bar{q})\}\int_{q'}^{q} Dq(\tau)\exp\left\{\int_0^\beta\left[-\frac{m}{2\hbar^2}\dot{q}^2(\tau) - \frac{1}{2}U''(\bar{q})(q(\tau)-\bar{q})^2\right]d\tau\right\} \tag{16.19}$$

This integral is taken considering an additional auxiliary condition: The point $\bar{q}$ is the same for all paths when the centroid is fixed. Since the action is stationary at classical paths, we are obliged to express the general path that resolves this Feynman difficulty by introducing new variables:

$$q(\tau) = \bar{q} + y(\tau) \tag{16.20}$$

Here $y(\tau)$ represents the fluctuations around the classical path. From here,

$$\frac{1}{\beta}\int_0^\beta q(\tau)d\tau = \bar{q} + \frac{1}{\beta}\int_0^\beta y(\tau)d\tau \tag{16.21}$$

and the functional integration measure $Dy(\tau)$ is expressed now via the Dirac-delta-functional picking or constraining a closed path with the centroid position $\bar{q}$ at some given value:

$$Dy(\tau) \to \delta\left(\frac{1}{\beta}\int_0^\beta y(\tau)d\tau\right)Dy(\tau) \tag{16.22}$$

This approach is appropriate in the quantum theory when we examine coupled quantum systems as well as gauged fields. Applying this approach to the given integral then for all paths we obtain the required paths for the centroid transition amplitude:

$$K = \exp\{-\beta U(\bar{q})\}\int_y^y Dy(\tau)\delta\left(\frac{1}{\beta}\int_0^\beta y(\tau)d\tau\right)\exp\left\{\int_0^\beta\left(-\frac{m}{2\hbar^2}\dot{y}^2 - \frac{1}{2}U''(\bar{q})y^2\right)d\tau\right\} \tag{16.23}$$

From

$$\delta\left(\frac{1}{\beta}\int_0^\beta y(\tau)d\tau\right) = \int_{-\infty}^{\infty}\frac{d\kappa}{2\pi}\exp\left\{\frac{i\kappa}{\beta}\int_0^\beta y(\tau)d\tau\right\} \tag{16.24}$$

The quantity $\frac{1}{\beta}$ is introduced in order to render the exponential function dimensionless. So,

$$K = \exp\{-\beta U(\bar{q})\}\int_{-\infty}^{\infty}\frac{d\kappa}{2\pi}\int_{y}^{y}Dy(\tau)\exp\left\{\int_{0}^{\beta}\left(\frac{i\kappa}{\beta}y(\tau)-\frac{m}{2\hbar^2}\dot{y}^2-\frac{1}{2}U''(\bar{q})y^2\right)d\tau\right\} \tag{16.25}$$

We do another change of variables:

$$\xi(\tau)=y(\tau)-\frac{i\kappa}{\beta U''(\bar{q})},\ D\xi(\tau)=Dy(\tau),\dot{\xi}(\tau)=\dot{y}(\tau) \tag{16.26}$$

$$y^2=\xi^2(\tau)-\left(\frac{\kappa}{\beta U''(\bar{q})}\right)^2+\frac{2i\kappa}{\beta U''(\bar{q})}\xi(\tau) \tag{16.27}$$

$$\dot{y}^2=\dot{\xi}^2(\tau) \tag{16.28}$$

$$\begin{aligned}\frac{i\kappa}{\beta}y(\tau)-\frac{m}{2\hbar^2}\dot{y}^2-\frac{1}{2}U''(\bar{q})y^2 &= \frac{i\kappa}{\beta}\left(\xi(\tau)+\frac{i\kappa}{\beta U''(\bar{q})}\right)-\frac{m}{2\hbar^2}\dot{\xi}^2(\tau)\\ &-\frac{1}{2}U''(\bar{q})\left(\xi^2(\tau)-\left(\frac{\kappa}{\beta U''(\bar{q})}\right)^2+\frac{2i\kappa}{\beta U''(\bar{q})}\xi(\tau)\right)\end{aligned} \tag{16.29}$$

or

$$\begin{aligned}\frac{i\kappa}{\beta}y(\tau)-\frac{m}{2\hbar^2}\dot{y}^2-\frac{1}{2}U''(\bar{q})y^2 &= \xi(\tau)\left(-\frac{1}{2}U''(\bar{q})\frac{2i\kappa}{\beta U''(\bar{q})}+\frac{i\kappa}{\beta}\right)-\frac{m}{2\hbar^2}\dot{\xi}^2(\tau)\\ &-\frac{1}{2}U''(\bar{q})\xi^2(\tau)+\frac{1}{2}U''(\bar{q})\left(\frac{\kappa}{\beta U''(\bar{q})}\right)^2+\frac{i\kappa}{\beta}\frac{i\kappa}{\beta U''(\bar{q})}\end{aligned} \tag{16.30}$$

or

$$\frac{i\kappa}{\beta}y(\tau)-\frac{m}{2\hbar^2}\dot{y}^2-\frac{1}{2}U''(\bar{q})y^2=-\frac{m}{2\hbar^2}\dot{\xi}^2(\tau)-\frac{1}{2}U''(\bar{q})\xi^2(\tau) \tag{16.31}$$

From here relation 16.25 becomes

$$K = \exp\{-\beta U(\bar{q})\}\int_{-\infty}^{\infty}\frac{d\kappa}{2\pi}\exp\left\{-\frac{\kappa^2}{2\beta U''(\bar{q})}\right\}\int_{\xi}^{\xi}D\xi(\tau)\exp\left\{\int_{0}^{\beta}\left(-\frac{m}{2\hbar^2}\dot{\xi}^2(\tau)-\frac{1}{2}U''(\bar{q})\xi^2(\tau)\right)d\tau\right\} \tag{16.32}$$

We discuss on the regime of validity of this approximation of deriving the transition amplitude in quasi-classical approximation. The factor $\exp\{-\beta U(\bar{q})\}$ may be responsible for only the global phase factor. However, the rest of the factor determines the magnitude of the quantum fluctuations. In the integrand, we have the expression for the density matrix of a harmonic oscillator. It is obvious this density matrix is due to contributions from the leading order term in the fluctuations. This implies that higher order terms in the fluctuations were neglected earlier by not considering the dots in 16.15. The quasi-classical approximation consists in neglecting these higher order terms. So, we need just write the result of the density matrix:

$$\rho(q,q',\beta)=\int_{\xi}^{\xi} D\xi(\tau)\exp\left\{\int_0^{\beta}\left(-\frac{m}{2\hbar^2}\dot{\xi}^2(\tau)-\frac{1}{2}U''(\bar{q})\xi^2(\tau)\right)d\tau\right\} \tag{16.33}$$

We apply the validity of our result to the case of harmonic centroid potential characterized by the frequency ω by letting,

$$U''(\bar{q})=m\omega^2 \tag{16.34}$$

then

$$\rho(q,q',\beta)=\int_{\xi}^{\xi} D\xi(\tau)\exp\left\{\int_0^{\beta}\left(-\frac{m}{2\hbar^2}\dot{\xi}^2(\tau)-\frac{m\omega^2}{2}\xi^2(\tau)\right)d\tau\right\} \tag{16.35}$$

with the special result for $\xi = \xi'$:

$$\rho(q,q,\beta)=\left(\frac{m\omega}{2\pi\sinh(\hbar\omega\beta)}\right)^{\frac{1}{2}}\exp\left\{-\frac{m\omega}{\hbar}\tanh\left(\frac{\hbar\omega\beta}{2}\right)\xi^2\right\} \tag{16.36}$$

Considering 16.32, we do integration over κ as well as ξ. This Gaussian integral is evaluated as the trace of the partition function:

$$Z=\left(\frac{m\omega^2\beta}{2\pi}\right)^{\frac{1}{2}}\int_{-\infty}^{\infty}d\bar{q}\frac{1}{2\sinh\left(\frac{\hbar\omega\beta}{2}\right)}\exp\{-\beta U(\bar{q})\} \tag{16.37}$$

In the quasi-classical approximation (limit of high temperatures) $\hbar\omega\beta \ll 1$, we expand 16.37 via this small parameter $\hbar\omega\beta$ up to the term $o[(\hbar\omega\beta)^2]$:

$$Z\cong\left(\frac{m\omega^2\beta}{2\pi}\right)^{\frac{1}{2}}\int_{-\infty}^{\infty}d\bar{q}\left[\hbar\omega\beta\left(1+\frac{1}{24}(\hbar\omega\beta)^2\right)\right]^{-1}\exp\{-\beta U(\bar{q})\} \tag{16.38}$$

or

$$Z\cong\left(\frac{m}{2\pi\beta\hbar^2}\right)^{\frac{1}{2}}\int_{-\infty}^{\infty}d\bar{q}\left(1-\frac{1}{24}(\hbar\omega\beta)^2\right)\exp\{-\beta U(\bar{q})\} \tag{16.39}$$

or

$$Z\cong\left(\frac{m}{2\pi\beta\hbar^2}\right)^{\frac{1}{2}}\int_{-\infty}^{\infty}d\bar{q}\exp\left\{-\beta U(\bar{q})-\frac{(\hbar\omega\beta)^2}{24}\right\} \tag{16.40}$$

or, generally,

$$Z \cong \left(\frac{m}{2\pi\beta\hbar^2}\right)^{\frac{1}{2}} \int_{-\infty}^{\infty} d\bar{q}\, \exp\left\{-\beta\left(U(\bar{q}) + \frac{\beta\hbar^2}{24m}U''(\bar{q})\right)\right\} \tag{16.41}$$

The integrand is some distribution. So, given the centroid potential $U(\bar{q})$, thermodynamic and quantum dynamic quantities can be accurately determined.

16.2 Expectation Value

Later in the book, we calculate the full expression of the transition amplitude K via Feynman path integral. This results in the full action functional that is not easily calculated due to retardation effects, as seen earlier under chapter 10. To proceed, Feynman introduced a variational principle that could resolve this problem by first considering the path average of any function of path, say, F[q] defined as:

$$\left\langle \mathrm{F}[q]\right\rangle_{S_0} = \int Dq\, \mathrm{P}(q)\mathrm{F}[q]s \tag{16.42}$$

due to the convexity of the following exponential functional:

$$\mathrm{P}(q) = \frac{\exp\{S_0[q]\}}{\int Dq \exp\{S_0[q]\}} \tag{16.43}$$

where S_0 is the action functional of the model system. Considering equation 10.54 we expect to find

$$E_0 \leq E_F \tag{16.44}$$

Here, E_F is the upper bound of the ground state energy we expect to find. The basis of the inequality 16.44 is that for a real variable q, the curve

$$\Psi = \exp\{q\} \tag{16.45}$$

is always concaved away from the q-axis. We write the tangent to the curve Ψ at point $\langle q\rangle$ on the q-axis:

$$\Psi_1 = \exp\{q\}\left(q - \langle q\rangle + 1\right) \tag{16.46}$$

It may be seen that the functions Ψ_1 and Ψ coincide at the point $q = \langle q\rangle \equiv \bar{q}$:

$$\exp\{q\} \geq \Psi_1 = \exp\{q\}\left(q - \langle q\rangle + 1\right) \tag{16.47}$$

So, considering the real function of the path, F[q] then

$$\exp\{\mathrm{F}[q]\} \geq \exp\left\langle\{\mathrm{F}[q]\}\right\rangle\left(q - \langle q\rangle + 1\right) \tag{16.48}$$

and, consequently,

$$\left\langle \exp\{\mathrm{F}[q]\}\right\rangle_{S_0} = \int Dq\mathrm{P}(q)\exp\{\mathrm{F}[q]\} \geq \exp\{\mathrm{F}[q]\}\int Dq\mathrm{P}(q)\left(q - \langle q\rangle + 1\right) \tag{16.49}$$

So, follows the **Feynman-Jensen-Peierls inequality** and is a consequence of the convexity of the exponential function:

$$\left\langle \exp\{F[q]\}\right\rangle_{S_0} \geq \exp\left\{\left\langle F[q]\right\rangle_{S_0}\right\} \tag{16.50}$$

So, from

$$\rho(q,q') = \int_{q'}^{q} Dq(\tau)\exp\left\{\int_0^\beta \left[-\frac{m}{2\hbar^2}\dot{q}^2(\tau) - U(q(\tau))\right]d\tau\right\} \tag{16.51}$$

or

$$\rho(q,q') = \int_{q'}^{q} Dq(\tau)\exp\left\{\int_0^\beta \left[-\frac{m}{2\hbar^2}\dot{q}^2(\tau) - U(\bar{q}) + U(\bar{q}) - U(q(\tau))\right]d\tau\right\} \tag{16.52}$$

or

$$\rho(q,q') = \exp\{-\beta U(\bar{q})\}\int Dq \exp\{F[q]\}\exp\{S_0[q]\} \tag{16.53}$$

where,

$$F[q] = \int_0^\beta \left[U(\bar{q}) - U(q(\tau))\right]d\tau \tag{16.54}$$

then

$$\rho(q,q') = \exp\{-\beta U(\bar{q})\}\left\langle \exp\{F[q]\}\right\rangle_{S_0} \geq \exp\left\{-\beta U(\bar{q}) + \left\langle F[q]\right\rangle_{S_0}\right\} \tag{16.55}$$

We examine

$$\left\langle F[q]\right\rangle_{S_0} = \int Dq \exp\{S_0[q]\}\int_0^\beta \left[U(\bar{q}) - U(q(\tau))\right]d\tau \tag{16.56}$$

or, equivalently,

$$\left\langle F[q]\right\rangle_{S_0} = \int_0^\beta d\tau \int Dq \exp\{S_0[q]\}\left[U(\bar{q}) - U(q(\tau))\right] \tag{16.57}$$

From here, it follows that the functional integral is independent of τ. We show that the functional integral is such of a structure that it should be independent of τ. The example for such a path that is a function of τ. It is necessary to start from a point and end at the same point as seen in Figure 29 that shows some twice periodicity.

We examine a family of paths obtained from the given paths. The sum implies that summing over one of the family of paths:

$$\left\langle F[q]\right\rangle_{S_0} = \int_0^\beta d\tau \int Dq \exp\{S_0[q]\}\left[U(\bar{q}) - U(q(\tau))\right] \tag{16.58}$$

or

$$\langle F[q]\rangle_{S_0} = \beta \int Dq \exp\{S_0[q]\}\left[U(\bar{q}) - U(q(\tau))\right] \tag{16.59}$$

or

$$\langle F[q]\rangle_{S_0} = \beta U(\bar{q}) \int Dq \exp\{S_0[q]\} - \beta U(q(\tau)) \int Dq \exp\{S_0[q]\} \tag{16.60}$$

Here, we suppose that the lower bound is $U(q(\tau))$.

17

Feynman Variational Method

In most cases solving a quantum mechanical problem exactly is not always possible. In such a case, an analytic solution to the Schrödinger equation is appropriate. This results to numerical solutions that may not necessarily be the best approach. Interest sometimes may not necessarily be in the detailed eigenfunctions, but rather in the energy levels and the qualitative features of the eigenfunctions. So, the numerical solutions are usually less intuitively reasonable. It is possible to find the values of the energy levels to be slightly sensitive to the deviation of the wave function from its true form. So, the expectation value of the energy for an approximate wave function can be a very good estimate of the corresponding energy eigenvalue. **Using an approximate wave function dependent on some set of variational parameters and minimizing its energy by these variational parameters yield such energy estimates. This technique is the so-called variational method due to this minimization procedure.**

The technique is very effective when determining the ground state energies and efficiently complements the WKB approximation that works well with relatively highly excited states where the De Broglie wavelength is short compared to the distance scale on which the wavelength changes. It is possible to formulate a variational principle based on Feynman's functional integral method, as indicated in chapter 16. It applies to quantum mechanical as well as statistical systems at finite temperatures. In quantum mechanics one uses the direct variational method known as the **Ritz method** that we show below under this heading. For $T = 0$, limit of the Feynman variational method leads to the Ritz variational method.

Let us analyse Feynman's variational method. Consider the partition function in terms of the dimensionless scaled action $\underline{S}\left[q(\tau)\right]$:

$$Z = \int dq \int_q^q \exp\left\{\underline{S}\left[q\right]\right\} Dq(\tau) = \int \delta(q - q') dq dq' \int_{q'}^q \exp\left\{\underline{S}\left[q\right]\right\} Dq(\tau) \tag{17.1}$$

We can introduce the trial or model action functional $\underline{S}_0\left[q(\tau)\right]$ that is hoped to be close to the true action, $\underline{S}\left[q(\tau)\right]$, and suppose we can evaluate both of these. Imagine that we cannot evaluate the true partition function, due to whatever mathematical difficulties, but we are able to evaluate a partition function for this trial action. Then the true partition function is

$$Z = \int \delta(q - q') dq dq' \int_{q'}^q \exp\left\{\underline{S} - \underline{S}_0\right\} \exp\left\{\underline{S}_0\right\} Dq(\tau) \tag{17.2}$$

or

$$Z = \int \delta(q - q') dq dq' \int_{q'}^q \exp\left\{\underline{S} - \underline{S}_0\right\} \exp\left\{\underline{S}_0\right\} Dq(\tau) \frac{\int \delta(q - q') dq dq' \int_{q'}^q \exp\left\{\underline{S}_0\right\} Dq(\tau)}{\int \delta(q - q') dq dq' \int_{q'}^q \exp\left\{\underline{S}_0\right\} Dq(\tau)} \tag{17.3}$$

From here it follows that

$$Z = Z_0 \left\langle \exp\{\underline{S} - \underline{S_0}\} \right\rangle_{S_0} \tag{17.4}$$

where the trial partition function is

$$Z_0 = \int \delta(q - q') dq dq' \int_{q'}^{q} \exp\{\underline{S_0}\} Dq(\tau) \tag{17.5}$$

and the correction term is

$$\left\langle \exp\{\underline{S} - \underline{S_0}\} \right\rangle_{S_0} = \frac{\int \delta(q - q') dq dq' \int_{q'}^{q} \exp\{\underline{S} - \underline{S_0}\} \exp\{\underline{S_0}\} Dq(\tau)}{\int \delta(q - q') dq dq' \int_{q'}^{q} \exp\{\underline{S_0}\} Dq(\tau)} \tag{17.6}$$

A thermodynamic quantity directly related to the partition function Z is the Helmholtz free energy F that can be obtained via the partition function Z from equation 17.4:

$$F = -T \ln Z \tag{17.7}$$

and also

$$F = E - T\tilde{S} \tag{17.8}$$

where $\tilde{S}$ is the entropy in this case. Thus, the internal energy is

$$E = -T \ln Z + T\tilde{S} \tag{17.9}$$

We apply the Feynman-Jensen-Peierls inequality 17.50 to estimate the partition function, 17.4:

$$\left\langle \exp\{\underline{S} - \underline{S_0}\} \right\rangle_{S_0} \geq \exp\left\{ \left\langle \underline{S} - \underline{S_0} \right\rangle_{S_0} \right\} \tag{17.10}$$

Then

$$Z = Z_0 \left\langle \exp\{\underline{S} - \underline{S_0}\} \right\rangle_{S_0} \geq Z_0 \exp\left\{ \left\langle \underline{S} - \underline{S_0} \right\rangle_{S_0} \right\} \tag{17.11}$$

and from

$$F = -T \ln Z \tag{17.12}$$

then

$$F = -T \ln Z_0 - T \ln \left\langle \exp\{\underline{S} - \underline{S_0}\} \right\rangle_{S_0} \tag{17.13}$$

But

$$\ln \left\langle \exp\{\underline{S} - \underline{S_0}\} \right\rangle_{S_0} \geq \ln \exp\left\{ \left\langle \underline{S} - \underline{S_0} \right\rangle_{S_0} \right\} = \left\langle \underline{S} - \underline{S_0} \right\rangle_{S_0} \tag{17.14}$$

Thus,

$$F \leq -T \ln Z_0 - T \left\langle \underline{S} - \underline{S_0} \right\rangle_{S_0} \equiv F_F \tag{17.15}$$

which defines F_F, the Feynman variational free energy and mimics the perturbation theory. This relates the true Helmholtz free energy:

$$F_F \geq F = E - T\tilde{S} \tag{17.16}$$

In the limit of zero temperature $T \to 0$, the free energy becomes the ground state energy $F = E$:

$$E \leq F_F = -T \ln Z_0 - T\left\langle \underline{S} - \underline{S}_0 \right\rangle_{\underline{S}_0} \tag{17.17}$$

or

$$E \leq F_F = -\frac{1}{\beta} \ln Z_0 - \frac{1}{\beta}\left\langle \underline{S} - \underline{S}_0 \right\rangle_{\underline{S}_0} \tag{17.18}$$

To find an optimal variational action $\underline{S}_0$ the right-hand-side (RHS) of this equation has to be minimized. Relation 17.18 should be the **generalization of the upper limit of energy from Hamiltonians** (i.e. the well-known **Ritz variational principle** shown below) to general actions. So, the variational energy (F_F) is an upper bound of the true internal energy of the system at low temperatures.

For a particle in a magnetic field, the Lagrangian has the form:

$$L = \frac{m\dot{q}^2}{2} + \frac{e}{c}\left(\vec{v}, \vec{\mathrm{A}}\right) + U(q) \tag{17.19}$$

Here, e, c, $\vec{v}$ and $\vec{\mathrm{A}}$ are, respectively, the electronic charge, speed of light, velocity of the particle and vector potential. Applying statistical physics, we do the analytical prolongation:

$$t = -i\hbar\tau \;,\; 0 \leq \tau \leq \beta \tag{17.20}$$

So,

$$L(-i\hbar\tau) = -\frac{m}{2\hbar^2}\left(\frac{dq}{d\tau}\right)^2 + \frac{ie}{\hbar c}\left(\frac{dq}{d\tau}, \vec{\mathrm{A}}\right) + U(q), \;\; \vec{v} = \frac{dq}{dt} = \frac{i}{\hbar}\frac{dq}{d\tau} \tag{17.21}$$

From here, we observe that with the magnetic field, the action functional has an imaginary part. However, it is still possible to employ formulae 17.13 to 17.18.

Now we show that for low temperatures, the Feynman variational principle coincides with that of Ritz. For any trial wave function $|\Psi\rangle$, we may calculate the expectation value of the energy,

$$E[\Psi] \equiv E_\Psi = \left\langle \Psi | \hat{\mathrm{H}} | \Psi \right\rangle \tag{17.22}$$

and is called the functional of the wave function $|\Psi\rangle$ since expression 17.22 maps a wave function $|\Psi\rangle$ onto a number E_Ψ. We can obtain the energy eigenvalues by requiring that E_Ψ should be stationary with respect to $|\Psi\rangle$. This implies, if there exist a function $|\Psi\rangle$ such that, for small variations $|\delta\Psi\rangle$ away from $|\Psi\rangle$, the corresponding variation δE in E_Ψ vanishes, then $|\Psi\rangle$ is an eigenstate of the Hamiltonian with energy E_Ψ. It is the same kind of requirement one places on the classical action in Lagrangian mechanics that yields the differential equation for a classical path and is basically the calculus of variations with the E_Ψ functional instead of the $\underline{S}\left[q(\tau)\right]$ functional.

Suppose $|\Phi_n\rangle$ is an eigenfunction of the Hamiltonian $\hat{\mathrm{H}}$ with eigenvalue E_n:

$$\hat{\mathrm{H}}\left|\Phi_n\right\rangle = E_n\left|\Phi_n\right\rangle \tag{17.23}$$

then it is possible that the energy functional E_Ψ should be stationary with respect to variations away from $|\Phi_n\rangle$. We suppose that, for whatever reason, it is difficult to solve this eigenvalue problem exactly. However, in the Ritz variational method, we want to find the lowest energy eigenvalue E_0. To do this, one formulates a normalized trial eigenstate function, $|\Psi\rangle$, which could (in principle) be expressed in terms of the energy eigenstates:

$$|\Psi\rangle = \sum_n C_n |\Phi_n\rangle, \quad \langle\Psi|\Psi\rangle = \sum_n |C_n|^2 = 1 \tag{17.24}$$

The trial state is an estimate of the true ground state. The expectation value of the energy in this trial state is obtained by averaging the Hamiltonian $\hat{H}$:

$$E_\Psi = \left\langle \Psi|\hat{H}|\Psi \right\rangle = \left(\sum_n C_n |\Phi_n\rangle \right)^* \hat{H} \left(\sum_{n'} C_{n'} |\Phi_n\rangle \right) = \sum_{nn'} E_{n'} C_n^* C_{n'} \left\langle \Phi_n | \Phi_n \right\rangle \tag{17.25}$$

or

$$E_\Psi = \sum_{nn'} E_{n'} C_n^* C_{n'} \delta_{nn'} = \sum_n E_n |C_n|^2 \geq \sum_n E_0 |C_n|^2 \geq E_0 \tag{17.26}$$

where E_0 is the actual ground state energy of the system and implies that the energy obtained from the trial state is an upper limit estimate of the ground state energy:

$$E_0 \leq E_\Psi \tag{17.27}$$

As is well known, any trial state can be used to estimate E_0 this way; however, trial wave functions with the correct symmetry (for example, no nodes) will give estimates much closer to the true ground state energy.

We show that for $T \to 0$, only the lowest energy state survives and we get the direct Ritz variational principle. It is necessary to make a transition from the functional integral representation to the matrix element representation. Consider the Lagrangian

$$L(-i\hbar\tau) = -\frac{m\dot{q}^2}{2\hbar^2} - U(q(\tau)) \tag{17.28}$$

then the trial Lagrangian that imitates the system can be written in the form

$$L_0(-i\hbar\tau) = -\frac{m\dot{q}^2}{2\hbar^2} - U'(q(\tau)) \tag{17.29}$$

From here we write the difference in the action functional

$$S - S_0 = \int_0^\beta \left(L(\tau) - L'(\tau)\right) d\tau = -\int_0^\beta \left(U(\tau) - U'(\tau)\right) d\tau = -\int_0^\beta \Delta U(\tau) d\tau \tag{17.30}$$

We find again the Feynman variational principle for the case $\beta \to \infty$ via the partition function:

$$Z' = \sum_n \exp\{-\beta E'_n\} \tag{17.31}$$

Here, E'_n is the energy level of the trial system. For $\beta \to \infty$, all levels lying above E'_0 faster disappears and in the sum 17.31, the dominant role is played by the ground state with energy E_0:

$$\exp\{-\beta E'_0\} \gg \exp\{-\beta E'_n\}\ ,\ n \neq 0 \tag{17.32}$$

Thus,

$$\lim_{\beta\to\infty} Z' \cong \exp\{-\beta E'_0\} \tag{17.33}$$

It remains to evaluate

$$E \le F_F = E'_0 - \frac{1}{\beta}\left\langle \underline{S} - \underline{S}_0 \right\rangle_{\underline{S}_0}\ ,\ \beta \to \infty \tag{17.34}$$

In this case, the free energy for $\beta \to \infty$ becomes simply the energy of the system:

$$E \le F_F = E'_0 - \frac{1}{\beta}\left\langle \underline{S} - \underline{S}_0 \right\rangle_{\underline{S}_0} \ge E_0 \tag{17.35}$$

This confirms again the Feynman variational principle where in our case

$$\left\langle \Phi_0 \middle| \hat{H}' \middle| \Phi_0 \right\rangle = E'_0 \tag{17.36}$$

Here, Φ_0 is the wave function of the trial system with Hamiltonian $\hat{H}'$ that relates the trial action functional $\underline{S}_0$.

We examine now the second term $\left\langle \underline{S} - \underline{S}_0 \right\rangle_{\underline{S}_0}$ that, by definition:

$$\left\langle \underline{S} - \underline{S}_0 \right\rangle_{\underline{S}_0} = \frac{\int \delta(q-q')dqdq' \int_{q'}^{q} (\underline{S} - \underline{S}_0)\exp\{\underline{S}_0\} Dq(\tau)}{\int \delta(q-q')dqdq' \int_{q'}^{q} \exp\{\underline{S}_0\} Dq(\tau)} \tag{17.37}$$

or

$$\left\langle \underline{S} - \underline{S}_0 \right\rangle_{\underline{S}_0} = \frac{\int \delta(q-q')dqdq' \int_{q'}^{q} \left(-\int_{0}^{\beta} \Delta U(\tau) d\tau \right) \exp\{\underline{S}_0\} Dq(\tau)}{\int \delta(q-q')dqdq' \int_{q'}^{q} \exp\{\underline{S}_0\} Dq(\tau)} \tag{17.38}$$

We move from functional integral representation to matrix representation:

$$\rho(q,q') = \mathrm{K}(q,q',-i\hbar\beta) = \int_{q'}^{q} \exp\{\underline{S}_0[q,q',-i\hbar\beta]\} Dq(\tau) \tag{17.39}$$

We find how a particle found at (q, t), arriving from the past in some situation relates the transition amplitude $\mathrm{K}(q, q', -i\hbar\beta)$ describing a particle coming from q' and arriving at q via the intermediate point q_τ:

$$K(q,q',-i\hbar\beta) = \int dq_\tau K(q,-i\hbar\beta;q_\tau,\tau)K(q_\tau,\tau;q',0) \tag{17.40}$$

But,

$$K(q_\tau,\tau;q',\sigma) = \sum_n \Phi_n^*(q_\tau)\Phi_n(q')\exp\{-E'_n(\tau-\sigma)\} \tag{17.41}$$

Here Φ_n is the eigenfunction of the Hamiltonian $\hat{H}'$ corresponding to the action functional $\underline{S}_0$. So, considering $M_{\Delta U}(\beta)$ to be the numerator of 17.38 then

$$M_{\Delta U}(\beta) = \int_0^\beta d\tau \int_{-\infty}^{\infty} dq' \int_{-\infty}^{\infty} dq_\tau \Delta U(\tau) \sum_{n,m} \Phi_n^*(q')\Phi_n(q_\tau)\Phi_m^*(q_\tau)\Phi_m(q')\exp\{-E_n(\beta-\tau)\}\exp\{-E_m\tau\} \tag{17.42}$$

or

$$M_{\Delta U}(\beta) = \int_0^\beta d\tau \sum_n \Delta U_{nn}\exp\{-E_n(\beta-\tau)\}\exp\{-E_n\tau\} \tag{17.43}$$

or

$$M_{\Delta U}(\beta) = \int_0^\beta d\tau \sum_n \Delta U_{nn}\exp\{-\beta E_n\} \tag{17.44}$$

or

$$M_{\Delta U}(\beta) = \beta \sum_n \Delta U_{nn}\exp\{-\beta E_n\} \tag{17.45}$$

For $\beta \to \infty$, then

$$M_{\Delta U}(\beta) \cong \beta \Delta U_{00}\exp\{-\beta E'_0\} \tag{17.46}$$

Hence,

$$\langle \underline{S} - \underline{S}_0 \rangle_{\underline{S}_0} = -\beta \Delta U_{00} \tag{17.47}$$

and so

$$E \le F_F = E_0' - \frac{1}{\beta} - \langle \beta \Delta U_{00} \rangle_{\underline{S}_0} = E_0' + \Delta U_{00} \tag{17.48}$$

This corresponds to

$$E = E_0' + \Delta U_{00} = \left\langle \Phi_0 \left| \left(\hat{H}_e + U' \right) \right| \Phi_0 \right\rangle + \langle \Phi_0 | (U - U') | \Phi_0 \rangle \tag{17.49}$$

or

$$E = \left\langle \Phi_0 \middle| \left(\hat{H}_e + U \right) \middle| \Phi_0 \right\rangle \tag{17.50}$$

Here,

$$\hat{H} = \hat{H}_e + U \tag{17.51}$$

is the exact Hamiltonian of the system that we average by the trial wave function Φ_0. This will give the exact expression of the direct variational method. This implies the evaluation of the Feynman functional variational method leads to the direct variational method. This is exactly shown as we modelled the field. So, the Feynman variational principle can be applied to those cases where the Ritz variational principle is applicable.

18

Polaron Theory

18.1 Introduction

In the following, we invoke the application of the path integral to explore the capacity of the solution of a quantum mechanical problem when a particle or system of particles is coupled to degrees of freedom of an external environment. As an application of the path integral, let us consider the polaron problem as an example illustrating how the theory discussed above works in practice. The polaron problem arises due to the motion of an electron in the conduction band of a polar semiconductor or ionic crystal.

The Bloch approximation is the usual initial approach to the description of a conduction electron. It considers the lattice ions to be rigidly attached to their lattice sites. In the Bloch approximation an electron with energy near the bottom of the conduction band moves as a free particle with a band mass which is an oversimplification for an ionic crystal. This is justified by the fact that in such crystals the lattice ions are unscreened charges and so respond significantly to the presence of a conduction electron which, in turn, induces a significant lattice ion displacement from their static configurations assumed in the Bloch approximation. The electron experiences, in turn, not only the static field assumed in the Bloch approximation but also the incremental electric field due to lattice ion displacements. Assuming that the electronic wave function varies negligibly over several lattice spacings then the electron can be regarded essentially as moving under the influence of a continuous macroscopic polarization field arising from the lattice ion displacements.

So, in this manner, the motion of an electron in a polar semiconductor or ionic crystal may be regarded as generating quantized vibrations or phonons and so the Bloch electron will move with its self-created polarization in the phonon cloud. In this scenario when the characteristic phonon frequencies are sufficiently low, the local deformation of ions, caused by the electronic motion, creates a potential well that traps the electron even in a perfect crystal lattice where Landau predicted this **self-trapping phenomenon** [21]. This results in a polaron state: **A polaron is a quasiparticle in a polar semiconductor or an ionic crystal that results from a conduction electron (or hole) interacting together with its self-induced polarization in a phonon cloud (bosons)** [21,22]. This quasi-particle is heavier and less mobile than the bare electron.

The polaron concept is one of the simplest examples of a particle interacting with a quantized field that dramatically modifies the properties of the particle. It has served as a testing ground for various techniques in quantum field theory and, in particular, the Feynman path integral theory. Recently, the polaron concept has experienced a renewed interest due to theoretical advances in ultracold quantum gases, quantum information, and high energy physics that will serve as an impetus for future developments in the polaron theory [23,24]. These polaronic effects reported in ultracold atomic gases where an impurity immersed in the quantum gas is dressed by the gas' excitations are motivated by the fact that generally quantum gases possess a large experimental tunability of microscopic parameters in the ensemble compared to solids. So it allows us to check many theoretical predictions that give new insights into different polaronic interaction mechanisms [23,24]. In this case an impurity interacting with a

Bose-Einstein condensate is no longer accurately described by the Fröhlich Hamiltonian as the coupling between the impurity and the boson bath gets stronger [24].

Interaction between fermions and bosons is one of the cornerstones of many-particle physics and continues to attract researchers' unrelenting attention. This electron-phonon interaction is ubiquitous in condensed matter and materials physics. Examples include [25–28]:

- **underpinning the temperature dependence of the electrical resistivity in metals and the carrier mobility in semiconductors;**
- **contributing to conventional superconductivity;**
- **contributing to optical absorption in indirect gap semiconductors;**
- **enabling the thermalization of hot carriers;**
- **determining the temperature dependence of electron energy bands in solids;**
- **distorting band structures and phonon dispersion relations of metals, resulting to characteristic kinks and Kohn anomalies in photoemission and Raman/neutron spectra, respectively;**
- **playing a role in the areas of spintronics and quantum information.**

Notwithstanding the fundamental and practical importance of electron-phonon interactions, there is yet appropriate theoretical studies to be done due to the complexity of the electron-phonon interaction.

In the present polaron study, the interaction between electrons will be described by 3D coulomb potentials of corresponding symmetries and the Fröhlich 3D Hamiltonian chosen to describe electron-phonon interaction. Consequently, interface phonons will not be considered. This approach seems to be adequate since the integral polaron effects resulting from the summation over all phonon modes appear to be weakly dependent on the details of the phonon spectra. In [22,29,30] interface-type longitudinal polar optical phonons have no contribution to polaron effects. In [22,31,32] bulk-type phonons play the dominant role in the polaron energy shift.

Under this chapter we consider a parabolic confinement potential where rigid interface boundaries are absent and interface-like phonon modes are smoothly distributed in space rather than localized near a sharp boundary and so we examine the electron interaction only with 3D longitudinal polar optical phonons (3D-phonon approximation) [2,22]. So, interface phonons will not be considered and this approach seems adequate since integral polaron effects result from the summation over all phonon spectra. The model with parabolic confinement potential is preferable since

- **it examines polaron states covering all values of Fröhlich electron-phonon-coupling constant, α_F;**
- **it is introduced for technological reasons;**
- **at small values of the radius of the quantum wire or quantum dot, the parabolic confinement potential can be regarded as a model for the real potential.**

We limit our investigation only to strong coupling polarons.

In a polar semiconductor or an ionic crystal, when two electrons (or two holes) together with their self-induced polarization in a phonon cloud (bosons) interact with each other simultaneously via a Coulomb force and via the electron-phonon-electron interaction either two independent polarons can occur or the phonon-mediated attraction dominates the direct Coulomb-mediated repulsion so that the net effect is the attraction between the fermionic particles, forming a composite bosonic quasi particle called a bipolaron [2,22,33]. This implies a bound state of two polarons. So, the birth of bipolarons depends on the competition between the direct Coulomb-mediated repulsion and phonon-mediated attraction via the electron-phonon interaction [2,22].

Bipolarons can be free and characterized by translational invariance or localized. A many-electron system on a lattice coupled with any bosonic field results in a charged Bose-liquid constituting small bipolarons in the strong coupling regime. Likewise, the bipolaron, an electron and a hole interacting with each other simultaneously in a polarizable medium through the Coulomb force and via the electron-phonon-hole interaction, form a quasi-particle called, polaronic exciton. Bipolarons are suspected to be the possible particles participating in high-T_C superconductivity.

As seen earlier, applying the Feynman variational principle for path integrals results in an upper bound for the polaron self-energy at all α_F, that at weak and strong coupling gives accurate limits. Feynman obtained the smooth interpolation for the ground state energy, between weak and strong coupling polarons. In our problems we evaluate the asymptotic expansions of Feynman's polaron energy for better understanding of the physics. The Feynman variational principle is one of the most effective superior method when investigating the polaron problem for arbitrary values of the electron-phonon-coupling constant, α_F [34].

18.2 Polaron Energy and Effective Mass

One of the celebrated applications of the Feynman path integral approach in statistical mechanics or solid-state physics is the motion of electrons in an ionic crystal resulting in a polaron where the Feynman path integral approach gives results that are visibly superior over those obtained via conventional approaches. Interest in the polaron concept is not only due to its description of particular physical properties of charge carriers in polarizable (or deformable) solids but because it constitutes also an interesting field theoretical model consisting of a fermion interacting with a scalar boson field [2,21].

The polaron concept was introduced by Landau [35] as an auto localization of an electron in a polar crystal and, subsequently, Landau and Pekar [27,36,37] investigated the self-energy and the effective mass of the polaron and was shown by Fröhlich [38] to correspond to the adiabatic or strong-coupling regime [39]. From the Fröhlich polaron Hamiltonian which is a sum of three operators (kinetic energy of the electron, energy of the phonon modes–treated as harmonic oscillators and the electron-phonon interaction energy) considers the electron to interact only with the longitudinal optical modes and the interaction being Coulombic. Further, in 1950 Fröhlich and collaborators considered his 1937 polaron model two modifications where they treated the crystal as a dielectric continuum, requiring the spatial extension of the polaron to be much larger than the lattice spacing - the so-called large polaron model [40]. In this model they introduced a dimensionless constant, $\sqrt{\alpha_F}$, in front of the interaction term of the Hamiltonian with determining the strength of the electron-phonon coupling. Due to this extended interest, the dimensionless **Fröhlich electron-phonon coupling constant,** α_F, is regarded as an independent parameter. For weak coupling polaron, the dimensionless Fröhlich electron-phonon coupling constant is $\alpha_F < 1$, strong coupling $\alpha_F > 7$, then the intermediate coupling is between these ranges. The strong coupling condition, for bulk crystals is not even achieved for strong ionic crystals such as alkali halides. One of the goals of theoretical research is to develop suitable mathematical techniques that are accurate for all coupling regimes that can also be appropriate in dealing with similar many-body problems like a nucleon interacting with its meson field.

It is possible to reduce the lower bound of the electron-phonon coupling constant's threshold value in nanocrystals to within weak or intermediate-coupling range when the electron and the phonon confinements potentials are considered and tailors the polaron characteristics [2,21]. The properties (binding or (self-)energy, effective mass, response to external electric or magnetic fields such as mobility and impedance) characterizing this quasi-particle differ from those of the band-electron. The polaron effective mass has been obtained experimentally by cyclotron resonance measurements made on different materials where the dimensionless polaron coupling constant, α_F, lies within the intermediate coupling range. So, it will be necessary, to develop suitable accurate formulae for the polaron characteristics and, in particular, the effective mass and the energy expressed through the dimensionless Fröhlich electron-phonon coupling constant, α_F.

The above path integral method permits the evaluation of not only the partition function but also the ground state energy of the system as well as the dependence of the energy on the total momentum p of the system for an absolute zero temperature, $\beta \to \infty$. Consequently, the interaction of the electron with lattice vibrations should not change the form of the energy spectrum of the system. In the absence of interaction, the total momentum p of the system should be a unique continuous quantum number. The total momentum p is the full integral of motion and as a result of interaction the dependence of the energy on the momentum p in the case of an isotropic polarized crystal, should have the form [2]:

$$E = E_0(\nu) + p^2 E_2(\nu) + p^4 E_4(\nu) + \cdots \tag{18.1}$$

The quantities $E_0(\nu)$, $E_2(\nu)$, $E_4(\nu)$, $\cdots$ are understood as expansion coefficients for the given totality of discrete quantum numbers ν of the system. We now evaluate the partition function of the system with the energy spectrum 18.1. We limit ourselves to the first two terms at low temperatures, $\beta \to \infty$, since other terms are infinitesimally small for a given momentum.

We calculate the polaron effective mass from the expression of the polaron energy in 18.1 via

$$m^* = \frac{1}{2E_2(\nu)} \tag{18.2}$$

Considering only the first two terms of 18.1 for low temperatures, we find the partition function of the system:

$$Z = \frac{V}{(2\pi\hbar)^3} \int d\vec{p} \exp\left\{-\beta\left(E_0(\nu) + p^2 E_2(\nu)\right)\right\} \tag{18.3}$$

or

$$Z = \frac{V}{(2\pi\hbar)^3} \exp\{-\beta E_0(\nu)\} \left(\frac{\pi}{\beta E_2(\nu)}\right)^{\frac{3}{2}} \tag{18.4}$$

From here, considering 18.2 then

$$Z = \frac{V}{(2\pi\hbar)^3} \exp\{-\beta E_0(\nu)\} \left(\frac{2m\pi}{\beta}\right)^{\frac{3}{2}} \left(\frac{m^*}{m}\right)^{\frac{3}{2}} \tag{18.5}$$

and

$$\ln Z = \ln\left[V l_e^{-3}(\hbar\beta)\right] - \beta E_0(\nu) + \frac{3}{2}\ln\left(\frac{m^*}{m}\right) \tag{18.6}$$

where,

$$l_e(\hbar\beta) \equiv \left(\frac{2\pi\hbar^2\beta}{m}\right)^{\frac{1}{2}} \tag{18.7}$$

is the **thermal de Broglie wavelength**; m is the electron band-mass and V is a normalization volume which is to be taken infinitely large. From 18.6, the coefficient of β gives the polaron ground state energy $E_0(\nu)$ while the term independent of β in the argument of the logarithmic function, the polaron dimensionless effective mass $\frac{m^*}{m}$ and is useful in relation with the experimentally determined polaron effective mass to compute the electron band-mass for the sake of comparison with independent theoretical calculations.

18.3 Functional Influence Phase

The problem linked with the functional influence phase arises due to three types of situations:

1. **Classical fluctuation problem**
2. **Quantum mechanics problem with statistical fields (fluctuating statistical field)**
3. **The quantum statistical problem where one of the subsystems has to be eliminated.**

Often, we may have a quantum mechanical problem where the potential $U(r)$ is not a rigorously given function but bound to be chosen. For example, say, a crystal with impurities chaotically distributed in the bulk giving rise to lattice vibrations that are responsible for the decoherence in the system. One can say this results from the so-called **system-plus-bath model** where the full system is partitioned into the relevant system consisting of a few degrees of freedom and a thermal bath represented by a large or infinite number of degrees of freedom. The action functional for this model is written as:

$$S = S_e\left[r\right] + S\left[q\right] + S_{\text{int}}\left[r,q\right] \tag{18.8}$$

where, the first summand,

$$S_e\left[\vec{r}\right] = \frac{m}{2}\int \dot{\vec{r}}^{\,2}\left(t\right)dt \tag{18.9}$$

is the electronic action functional due to the band electron, the second:

$$S\left[q\right] = \frac{1}{2}\sum_{\nu_{\vec{\kappa}j}}\int\left(\dot{q}^2_{\nu_{\vec{\kappa}j}} - \omega^2_{\nu_{\vec{\kappa}j}} q^2_{\nu_{\vec{\kappa}j}}\right)dt \tag{18.10}$$

the phonon contribution due to the polarization field. This consist of a large number of atoms interacting with each other and oscillating around their equilibrium states with small deviations that performing the normal mode transformation, can describe the bath by a number of independent harmonic oscillators [1]. The third summand is

$$S_{\text{int}}\left[r,q\right] = \sum_{\nu_{\vec{\kappa}j}}\int \gamma_{\nu_{\vec{\kappa}_j}}\left(t\right) q_{\nu_{\vec{\kappa}_j}}\left(t\right)dt \tag{18.11}$$

It describes the interaction of the particle with the phonon system and assumed to be bilinear in the system and bath coordinates. Here, we denote the electron coordinate by $\vec{r}$ to facilitate the distinction from the environmental coordinates (polarization-field oscillator coordinates or lattice coordinates) $q_{\nu_{\vec{\kappa}_j}}$ for a mode of frequency $\omega_{\nu_{\vec{\kappa}_j}}$ and $\nu_{\vec{\kappa}_j}$, the oscillator quantum number with $\vec{\kappa}_j$ being the phonon wave vector numbered by the polarization j.

Certainly, the system's degree of freedom does not have to be related to a real particle but may be quite abstract. We show below this form of the system-bath interaction permits to take into account the influence of the environment on the system exactly using the path integral technique. The transition amplitude for such a system is written:

$$\mathrm{K} = \int DrDr' \exp\left\{\frac{i}{\hbar}\left(S_e\left[r\right] - S_e\left[r'\right]\right)\right\}\int DqDq' \exp\left\{\frac{i}{\hbar}\left(S\left[q\right] + S_{\text{int}}\left[r,q\right] - S\left[q'\right] - S_{\text{int}}\left[r,q'\right]\right)\right\} \quad (18.12)$$

Taking path integration over q and q' (tracing out the environment) yields the influence functional $\mathrm{F}[r, r']$ expressed via the influence phase $\Phi[r, r']$:

$$\mathrm{F}\left[r,r'\right] \equiv \int DqDq' \exp\left\{\frac{i}{\hbar}\left(S\left[q\right] + S_{\text{int}}\left[r,q\right] - S\left[q'\right] - S_{\text{int}}\left[r,q'\right]\right)\right\} = \exp\left\{\frac{i}{\hbar}\Phi\left[r,r'\right]\right\} \quad (18.13)$$

So,

$$\mathrm{K} = \int DrDr' \exp\left\{\frac{i}{\hbar}\left(S_e\left[r\right] - S_e\left[r'\right]\right)\right\}\mathrm{F}\left[r,r'\right] \quad (18.14)$$

This functional is analogous to that of forces in classical mechanics. In classical mechanics we can also not examine the external body and take only the force that acts on the element of the system. Every change in, say, system A provokes a change in, say, system B. This provokes the change in the force that acts on system A in return. This change in the force is completely considered by the influence functional $\mathrm{F}[r, r']$ with the influence phase $\Phi[r, r']$. It is important to evaluate the influence phase $\Phi[r, r']$. For this, there should be concrete applicable problems with an excellent example of such a problem being a lattice in which is found an electron (or hole) in an ionic or polar medium that polarizes its neighborhood. Qualitatively, this is understood as a moving electron (hole) "dressing" itself with this polarization, which not only renormalizes the effective mass of the electron (hole) but makes its propagation (partially) incoherent. Without the lattice vibrations, a periodic potential act on the electron (hole) that will approximate the effect of this potential by supposing the electron (hole) to behave as a free particle but with a mass being that of the electron (hole) band-mass. Typically, the electron is coupled to longitudinal optical phonons, say, for example, in an ionic or polar crystal, having some long-wavelength frequency $\omega_{\nu_{\vec{\kappa}j}}$ numbered by the quantum number $\nu_{\vec{\kappa}_j}$ at wave vector $\vec{\kappa}_j$ with polarization j. The polaron moves with greater difficulty through the lattice than an undressed electron would do.

The polaron concept is of interest because it describes particular physical properties of charge carriers in polarizable media and the idea constitutes an interesting field theoretical model consisting of a fermion interacting with a scalar boson (phonon) field. Landau introduced the polaron concept as the self-localized state of a charge carrier in a homogeneous polar medium [41] then subsequently Landau and Pekar investigated the self-energy and the effective mass of the polaron [27,42–47]. Fröhlich showed this to correspond to the adiabatic or strong-coupling regime [27,48,49] as well as showing the typical behavior for the weak-coupling polaron for the first time. With the help of Fröhlich's work, Feynman found higher orders of the weak-coupling expansion for the polaron energy [34] and this Feynman model has remained the most successful approach to the polaron problem over the years.

This complex problem has been rendered simple by Feynman when the crystal vibrations are modelled to solve the polaron problem with less effort. Thus, Feynman introduces a procedure that imitates the perturbation theory and better imitates the physical situation described by the Lagrangian L:

$$L = -\frac{m\dot{r}^2}{2\hbar^2} - \frac{1}{2}\sum_{\nu_{\vec{\kappa}j}}\left(\frac{\dot{q}^2_{\nu_{\vec{\kappa}j}}}{\hbar^2} + \omega^2_{\nu_{\vec{\kappa}j}} q^2_{\nu_{\vec{\kappa}j}}\right) + \sum_{\nu_{\vec{\kappa}j}} \gamma_{\nu_{\vec{\kappa}j}}\left(\tau\right) q_{\nu_{\vec{\kappa}j}}\left(\tau\right) \quad (18.15)$$

This is the full Lagrangian of the electron-lattice interaction where the electron coordinate is $\vec{r}$ and m is its effective mass. The longitudinal optical phonon generalized coordinates are $q_{v_{\vec{\kappa}j}}$ for a mode of frequency $\omega_{v_{\vec{\kappa}j}}$. The amplitude of the electron-lattice coupling is $\gamma_{v_{\vec{\kappa}j}}(\tau)$.

18.3.1 Polaron Model Lagrangian

The physical motivation of our variational method is that a classical potential for the polaron problem might well be expected to be a good approximation when tight binding is considered. So, to imitate the polaron problem, a good model is one where instead of the electron being coupled to the lattice, it is coupled by some "**spring**" to another particle (fictitious particle) and the pair of particles are free to wander about the crystal (Figure 18.1).

For such a system, we select the model Lagrangian in the one oscillatory approximation where the interaction potential is replaced by a parabolic potential:

$$L_0 = -\frac{m\dot{r}^2}{2\hbar^2} - \frac{M\dot{R}^2}{2\hbar^2} - \frac{M\omega_f^2\left(\vec{R}-\vec{r}\right)^2}{2} \tag{18.16}$$

The first term describes the translational motion of the centre of mass; the second and third describe the oscillatory motion. It is an approximation of the full Lagrangian of the system, L. Here M and ω_f are, respectively, the mass of the fictitious particle and the frequency of the elastic coupling that will serve as variational parameters; $\vec{R}$ is the coordinate of the fictitious particle. The model system conserves the translational symmetry of the system and a more judicious choice of L_0 is to simulate a physical situation that may give a better upper bound for the energy.

18.3.2 Polaron Partition Function

The partition function of the model system can be obtained from the relation:

$$Z_0 = \int_{-\infty}^{+\infty} \delta\left(\vec{r}-\vec{r}'\right)\delta\left(\vec{R}-\vec{R}'\right)d\vec{r}d\vec{r}'d\vec{R}d\vec{R}' \int_{r'}^{r}\int_{R'}^{R} \exp\left\{S_0\right\} D\vec{r}D\vec{R} \tag{18.17}$$

Here, the action functional of the model system which is non-local in time:

$$S_0 = \int_0^{\beta} d\tau \left[-\frac{m}{2\hbar^2}\left(\frac{d\vec{r}}{d\tau}\right)^2 - \frac{M}{2\hbar^2}\left(\frac{d\vec{R}}{d\tau}\right)^2 - \frac{M\omega_f^2\left(\vec{R}-\vec{r}\right)^2}{2} \right] \tag{18.18}$$

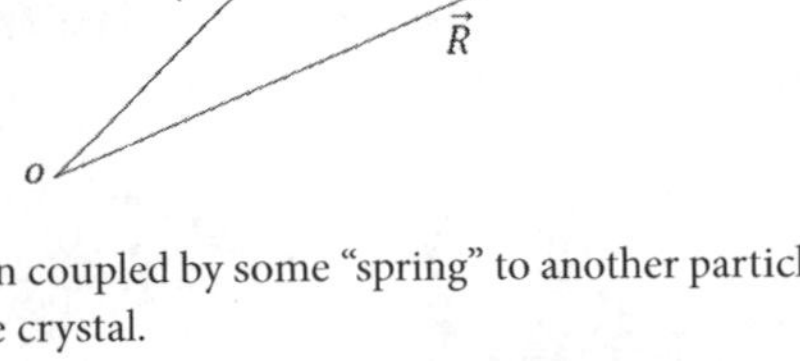

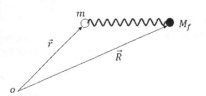

FIGURE 18.1 Showing the electron coupled by some "spring" to another particle (fictitious particle) and the pair of particles are free to wander about the crystal.

It is possible to evaluate the action functional of the model system in terms of only the path $\vec{r}$ by eliminating the fictitious particle subsystem variables $\vec{R}$. This may be done via the partition function of the model system:

$$Z_0 = \int \delta\left(\vec{r}-\vec{r}^{\,'}\right)\delta\left(\vec{R}-\vec{R}^{'}\right)d\vec{r}d\vec{r}^{\,'}d\vec{R}d\vec{R}^{'}\int_{r'}^{r}D\vec{r}\int_{R'}^{R}D\vec{R}\exp\left\{\int_0^\beta\left[-\frac{m\dot{\vec{r}}^2}{2\hbar^2}-\frac{M\dot{\vec{R}}^2}{2\hbar^2}-\frac{M\omega_f^2}{2}\left(\vec{r}^{\,2}+\vec{R}^2-2\vec{r}\cdot\vec{R}\right)\right]d\tau\right\} \quad (18.19)$$

After path integration, then:

$$Z_0 = Z_{\omega_f}\int\delta\left(\vec{r}-\vec{r}^{\,'}\right)d\vec{r}d\vec{r}^{\,'}\int_{r'}^{r}D\vec{r}\exp\left\{\int_0^\beta\left[-\frac{m\dot{\vec{r}}^2}{2\hbar^2}-\frac{M\dot{\vec{R}}^2}{2\hbar^2}-\frac{M\omega_f^2}{2}\vec{r}^{\,2}\right]d\tau-\Phi_{\omega_f}\left[\vec{r}\right]\right\} \quad (18.20)$$

Here, the partition function due to the fictitious lattice

$$Z_{\omega_f} = \left[2\sinh\left(\frac{\beta\hbar\omega_f}{2}\right)\right]^{-3} \quad (18.21)$$

and the **influence phase** of the interaction of the electron (hole) with the fictitious particle:

$$\Phi_{\omega_f}\left[\vec{r}\right] = \frac{\hbar M\omega_f^3}{4}\int_0^\beta\int_0^\beta d\sigma d\sigma'\vec{r}(\sigma)\vec{r}(\sigma')\mathrm{F}_{\omega_f}\left(\left|\sigma-\sigma'\right|\right) \quad (18.22)$$

where

$$\mathrm{F}_{\omega_f}\left(\left|\sigma-\sigma'\right|\right) = \frac{\cosh\hbar\omega_f\left(\left|\sigma-\sigma'\right|-\frac{\beta}{2}\right)}{\sinh\left(\frac{\hbar\omega_f\beta}{2}\right)} \quad (18.23)$$

The action functional of the interaction of the electron (hole) with the fictitious particle via its influence phase is:

$$\underline{S}_0 = \int_0^\beta\left[-\frac{m\dot{\vec{r}}^2}{2\hbar^2}-\frac{M\omega_f^2}{2}\vec{r}^{\,2}\right]d\tau+\Phi_{\omega_f}\left[\vec{r}\right]-3\ln\left[2\sinh\left(\frac{\beta\hbar\omega_f}{2}\right)\right] \quad (18.24)$$

The behavior of our system within the system-plus-environment model can be described by the density matrix as well as partition function formalism that we see later in this chapter. Since we are interested only in the dynamics of the relevant system then we trace out the environment degrees of freedom from the density matrix as well as partition function. In this manner we find the **influence phase** functional of the phonon subsystem on the electronic subsystem. From the Lagrangian L of the full system 18.15, we evaluate the partition function of the full system:

$$Z = \prod_{\nu\vec{\kappa}j}\int\delta\left(q_{\nu\vec{\kappa}j}-q'_{\nu\vec{\kappa}j}\right)\delta\left(\vec{r}-\vec{r}^{\,'}\right)dq_{\nu\vec{\kappa}j}dq'_{\nu\vec{\kappa}j}d\vec{r}d\vec{r}^{\,'}\int_{\vec{r}^{\,'}}^{\vec{r}}\int_{q'_{\nu\vec{\kappa}j}}^{q_{\nu\vec{\kappa}j}}\exp\left\{\underline{S}\right\}D\vec{r}Dq_{\nu\vec{\kappa}j} \quad (18.25)$$

This expression is the product of the partition functions over all possible oscillators:

$$Z=\prod_{\nu_{\vec{\kappa}j}} Z_L \int \delta\left(\vec{r}-\vec{r}'\right) d\vec{r} d\vec{r}' \int_{\vec{r}'}^{\vec{r}} D\vec{r} \exp\left\{\int_0^{\beta}\left(-\frac{m\dot{\vec{r}}^2}{2\hbar^2}\right) d\tau+\Phi_{\omega_{\nu_{\vec{\kappa}j}}}\left[\vec{r}\right]\right\} \tag{18.26}$$

Here, the lattice partition function is

$$Z_L=\left[2\sinh\left(\frac{\hbar\omega_{\nu_{\vec{\kappa}j}}\beta}{2}\right)\right]^{-3} \tag{18.27}$$

and the **electron-phonon influence phase:**

$$\Phi_{\omega_{\nu_{\vec{\kappa}j}}}\left[\vec{r}\right]=\frac{\hbar}{4\omega_{\nu_{\vec{\kappa}j}}}\int_0^{\beta}\int_0^{\beta} d\sigma d\sigma' \gamma_{\nu_{\vec{\kappa}j}}\left(\vec{r}(\sigma)\right)\gamma_{\nu_{\vec{\kappa}j}}\left(\vec{r}(\sigma')\right) \mathrm{F}_{\omega_{\nu_{\vec{\kappa}j}}}\left(\left|\sigma-\sigma'\right|\right) \tag{18.28}$$

where

$$\mathrm{F}_{\omega_{\nu_{\vec{\kappa}j}}}\left(\left|\sigma-\sigma'\right|\right)=\frac{\cosh\hbar\omega_{\nu_{\vec{\kappa}j}}\left(\left|\sigma-\sigma'\right|-\frac{\beta}{2}\right)}{\sinh\left(\frac{\hbar\omega_{\nu_{\vec{\kappa}j}}\beta}{2}\right)} \tag{18.29}$$

The expression

$$\Phi\left[\vec{r}\right]=\sum_{\nu_{\vec{\kappa}j}}\Phi_{\omega_{\nu_{\vec{\kappa}j}}}\left[\vec{r}\right] \tag{18.30}$$

is the **full functional of the electron-phonon interaction (influence phase)** with the double integral in 18.28 describing a non-local contribution where the system trajectory interacts with itself. This self-interaction is mediated by the environment as can be seen from the product $\gamma_{\nu_{\vec{\kappa}j}}\left(\vec{r}(\sigma)\right)\gamma_{\nu_{\vec{\kappa}j}}\left(\vec{r}(\sigma')\right)$. Hence, the coupling to the environment leads to an additional, non-local term to the action and so we may now rewrite the action of the full functional of the full system via its influence phase as follows:

$$\underline{S}\left[\vec{r}\right]=-\int_0^{\beta}\frac{m\dot{\vec{r}}^2}{2\hbar^2}d\tau+\sum_{\nu_{\vec{\kappa}j}}\Phi_{\omega_{\nu_{\vec{\kappa}j}}}\left[\vec{r}\right] \tag{18.31}$$

We observe that the Feynman approach for the polaron has an advantage because the phonon coordinates are adequately eliminated and, as a consequence, the polaron problem is reduced to an effective one-particle problem with a retarded interaction as described by the action functional in 18.31. This action functional 18.31, describes a driven harmonic oscillator.

The full partition function in 18.26 could not be possibly evaluated with ease via path integral as it is non-Gaussian. The Feynman solution to the polaron problem provides a technical mathematical tool on how to calculate this non-Gaussian path integral. This tool allows the building of a class of exactly solvable models corresponding to quadratic functionals on whose bases is built a variational technique that gives a good upper bound to the polaron ground state energy for all coupling regimes. For this purpose, Feynman introduced the notion of a model system with a fictitious particle that we examine later under this chapter.

18.4 Influence Phase via Feynman Functional Integral in The Density Matrix Representation

The interaction of the electron with lattice vibrations is a system coupled to a large number of environmental degrees of freedom. In most cases, the environment mimics a large heat bath characterized by a temperature T and so the environment state will therefore be described by an equilibrium density matrix. Sometimes, interest is also in this equilibrium density matrix of the system itself where such a state may be achieved after equilibration resulting from a weak coupling with a heat bath. To describe such thermal equilibrium states and dynamics of the system simultaneously, it is necessary to express the equilibrium density matrices via path integrals. Certainly, this is possible since one can write the equilibrium density operator in position representation.

We examine the motion of an electron in a polar crystal where we also consider its interaction with lattice vibrations. The interaction of the electron with its polarization in the phonon cloud forms the quasi-particle called the **polaron** defined earlier and may be characterized by its mass, energy, mobility, and so on. The interaction of the electron with lattice vibrations seems a complex problem and may involve two different effects: this renormalizes the energy spectrum as well as the mass of the electron. Different approaches may give different values of these renormalized quantities. In the polaron theory, it is necessary to have one unique approach and, in particular, the method of the density matrix in the language of Feynman functional integration is one of them [2,50]. The kinetic equation in principle may be helpful. The presentation of the problem for this case may not be sequential.

Interest may first be on the Schrödinger equation for which the problem may be approximately solvable. Further, we write absolutely the independent Boltzmann equation. If the first part of the problem is exactly solvable then we may be left with the scattering problem. Consequently, the problem is left to be solved by the density matrix that seems to be more appropriate than via the Boltzmann equation.

18.4.1 Expectation Value of a Physical Quantity

18.4.1.1 Density matrix

We examine now the electron-phonon system with the Hamiltonian

$$\hat{H}(\vec{r},q) = \hat{H}_e(\vec{r}) + \hat{H}_L(q) + \hat{H}_{e-L}(\vec{r},q) \tag{18.32}$$

dependent on the electronic $\vec{r}$ and phonon q coordinates, where $\hat{H}_e(\vec{r})$ is the electronic, $\hat{H}_L(q)$ -lattice and $\hat{H}_{e-L}(\vec{r},q)$-electron-lattice interaction Hamiltonians respectively. For brevity, we consider one electron with the wave function $\psi(\vec{r},q)$ corresponding to the Hamiltonian $\hat{H}(\vec{r},q)$. For a given operator, $\hat{F}(\vec{r},q)$ of the physical quantity, F the quantum mechanical-statistical expectation value can be found:

$$F \equiv \bar{F} = \int \psi^*(\vec{r},q)\hat{F}(\vec{r},q)\psi(\vec{r},q)d\vec{r}dq \tag{18.33}$$

or

$$\bar{F} = \int \hat{F}(\vec{r},q)\psi^*(\vec{r}',q')\psi(\vec{r},q)\delta(\vec{r}-\vec{r}')\delta(q-q')d\vec{r}d\vec{r}'dqdq' \tag{18.34}$$

Then the matrix density in the general case:

$$\rho(\vec{r},\vec{r}';q,q') \equiv \psi^*(\vec{r}',q')\psi(\vec{r},q) \tag{18.35}$$

So,

$$\overline{F} = \int \hat{F}(\vec{r},q)\rho(\vec{r},\vec{r}';q,q')\delta(\vec{r}-\vec{r}')\delta(q-q')d\vec{r}d\vec{r}'dqdq' = \mathrm{Tr}\left[\hat{F}\rho\right] \tag{18.36}$$

We see from here that the density matrix ρ gives the quantum mechanical expectation value of any physical quantity as defined in relations 14.7, 14.13, 14.28 and 18.36.

Suppose $\Psi(\vec{r},q)$ is one of the solutions of the Schrödinger equation then the **resulting density matrix ρ is that of a pure ensemble**, as seen above. With the knowledge of $\Psi(\vec{r},q)$ we may then find ρ that operates only on the state described by $\vec{r}$ and q.

The exact solution of the Schrödinger equation may be obtained with the help of the full Hamiltonian $\hat{H}(\vec{r},q)$ in 18.32. In most cases, the exact solution may not be feasible and a convenient approach should be applied with the help of the system's evolution via the functional-evolution-integral –operator $\int K$ to be defined later.

Let us consider the full Hamiltonian for the problem:

$$\hat{H}(\vec{r},q) = \hat{H}_e(\vec{r}) + \hat{H}_L(q) + \hat{H}_{e-L}(\vec{r},q) \tag{18.37}$$

Then the Schrödinger equation may be easily solvable at time moment $t = t_1$. For the Hamiltonian

$$\hat{H}_0 = \hat{H}_e(\vec{r}) + \hat{H}_L(q) \tag{18.38}$$

the solution corresponds to the wave function

$$\Psi_n(\vec{r},q) = \phi(\vec{r})\psi_n(q) \tag{18.39}$$

where the phonon wave function:

$$\psi_n(q) = \frac{1}{\sqrt{q_0 2^n n!\sqrt{\pi}}} H_n\left(\frac{q}{q_0}\right)\exp\left\{-\frac{q^2}{2q_0^2}\right\}, q_0 = \sqrt{\frac{\hbar}{m\omega}} \tag{18.40}$$

and $H_n(x)$ are the Hermite-polynomials with argument x.

We have seen earlier that the transition amplitude K governs how a wave function $\Psi(\vec{r},q)$ evolves with time moment t. If the wave function at an initial time moment t_1 is given as $\Psi(\vec{r}_1,q_1)$ and K is also provided, then we can obtain the wave function $\Psi(\vec{r}_2,q_2)$ at an arbitrary time moment t_2 via the following equation:

$$\Psi(\vec{r}_2,q_2) = \iint d\vec{r}_1 dq_1 \int_{\vec{r}_1}^{\vec{r}_2}\int_{q_1}^{q_2} D\vec{r}Dq \exp\left\{\frac{i}{\hbar}S[\vec{r},q]\right\}\Psi(\vec{r}_1,q_1) \tag{18.41}$$

where the Euclidean action corresponding to the full Hamiltonian 18.37:

$$S[\vec{r},q] = S_e[\vec{r}] + S_L[q] + S_{e-L}[\vec{r},q] \tag{18.42}$$

Here, $S_e[\vec{r}]$, $S_L[q]$ with $S_{e-L}[\vec{r},q]$ are, respectively, the electronic, lattice and electron-lattice action functional. Indeed, equation 18.41 is the integral equation of quantum mechanics equivalent to the time-dependent Schrödinger differential equation as seen earlier. Relation 18.41 may be rewritten via the functional-evolution-integral-operator $^{\int}\mathrm{K}$:

$$\Psi(\vec{r}_2,q_2) \equiv {}^{\int}\mathrm{K}\Psi(\vec{r}_1,q_1) = {}^{\int}\mathrm{K}_e\phi(\vec{r}_1)\,{}^{\int}\mathrm{K}_{L,e-L}\psi(\vec{r}_1,q_1) \tag{18.43}$$

We may now represent the **density matrix** ρ_i **of the pure ensemble** via the functional-evolution-integral-operator $^{\int}\mathrm{K}$:

$$\rho_i = \Psi_i'^{*}\Psi_i = {}^{\int}\mathrm{K}'^{*}\,{}^{\int}\mathrm{K}\Psi^{*}(\vec{r}'_1,q'_1)\Psi(\vec{r}_1,q_1) = {}^{\int}\mathrm{K}'^{*}\,{}^{\int}\mathrm{K}\rho_i(\vec{r}_1,q_{1};\vec{r}'_1,q'_1) \tag{18.44}$$

Here $\rho_i(\vec{r}_1,q_{1};\vec{r}'_1,q'_1)$ is the initial density matrix of the system.

We find now the density matrix of the mixed ensemble by considering, first, that the initial state has the probability W_i. Considering that the representation in 18.35 shows ρ to be Hermitian and that the probability W_i is coordinate-independent then the matrix density ρ can be diagonalized with the complete orthonormal set of eigenstates Ψ_i and the real eigenvalues W_i to give the **density matrix of the mixed ensemble**:

$$\rho = \sum_i \mathrm{W}_i\rho_i = \sum_i {}^{\int}\mathrm{K}'^{*}\,{}^{\int}\mathrm{K}W_i\rho_i(\vec{r}_1,q_{1};\vec{r}'_1,q'_1) = \rho(\vec{r}_2,q_{2};\vec{r}'_2,q'_2) \tag{18.45}$$

This is the **full density matrix of system and environment** that is tailored by two functional-evolution-integral-operators $^{\int}\mathrm{K}$ and $^{\int}\mathrm{K}'^{*}$. The environment is assumed to be in thermal equilibrium described by the density matrix:

$$\rho_L = {}^{\int}\mathrm{K}'^{*}_{L,e-L}\,{}^{\int}\mathrm{K}_{L,e-L}\mathrm{W}_{i'L}\rho_{i'L}(q_1,q'_1) \tag{18.46}$$

while the system in a nonequilibrium state can be described by:

$$\rho_e = {}^{\int}\mathrm{K}'^{*}_{e}\,{}^{\int}\mathrm{K}_e\mathrm{W}_{ie}\rho_{ie}(\vec{r}_1,\vec{r}'_1) \tag{18.47}$$

We assume the times of relaxation of the environment are faster than the time scales of the dynamics of the relevant system and so neglecting the initial correlations between system and environment, i.e. switching on the coupling after preparation of the initial state, then the initial density matrix ρ may be rewritten as the product of 18.46 and 18.47. Since we are only interested in the dynamics of the system degree of freedom, we trace out the environment and in our case by eliminating the lattice variables. We can now calculate the expectation value of the operator $\hat{\mathrm{F}}(\vec{r})$ of a physical quantity F via the full density matrix in 18.45:

$$\langle \mathrm{F}\rangle \equiv \bar{\mathrm{F}} = \mathrm{Tr}\left[\hat{\mathrm{F}}(\vec{r})\rho(\vec{r}_2,q_{2};\vec{r}'_2,q'_2)\right] = \mathrm{Tr}\left[\hat{\mathrm{F}}(\vec{r})\,{}^{\int}\mathrm{K}'^{*}\,{}^{\int}\mathrm{K}W_i\rho_i(\vec{r}_1,q_{1};\vec{r}'_1,q'_1)\right] \tag{18.48}$$

If the time t is sufficiently far removed in the past, then we assume only that the phonon subsystem is in thermal equilibrium at temperature, . The energy of the single electron and of the electron–phonon coupling is infinitesimal compared to the phonon energy. With this choice of the initial distribution, the phonon coordinates can be traced out from 18.48, and the entire expression is reduced to a path integral over the electron coordinates only.

So, tracing out the lattice variable q then

$$\bar{\mathrm{F}} = \mathrm{Tr}_e\left[\hat{\mathrm{F}}(\vec{r})\,{}^{\int}\mathrm{K}'^{*}_{e}\,{}^{\int}\mathrm{K}_e\mathrm{W}_{ie}\rho_{ie}(\vec{r}_1,\vec{r}'_1)\right]\mathrm{Tr}_L\left[{}^{\int}\mathrm{K}'^{*}_{L,e-L}\,{}^{\int}\mathrm{K}_{L,e-L}\mathrm{W}_{i'L}\rho_{i'L}(q_1,q'_1)\right] \tag{18.49}$$

Here, the tracing over the environmental degrees of freedom (partial tracing of the entire system) is obtained from

$$\mathrm{Tr}_L\left[\,{}^{\int}\mathrm{K}'^{*}_{L,e-L}\,{}^{\int}\mathrm{K}_{L,e-L}\mathrm{W}_{i'L}\rho_{i'L}\left(q_1,q'_1\right)\right]\equiv\int\delta\left(q_2-q'_2\right)dq_2dq'_2\,{}^{\int}\mathrm{K}'^{*}_{L,e-L}\,{}^{\int}\mathrm{K}_{L,e-L}\mathrm{W}_{i'L}\rho_{i'L}\left(q_1,q'_1\right) \quad (18.50)$$

After tracing out the environment this yields

$$\mathrm{Tr}_L\left[\,{}^{\int}\mathrm{K}'^{*}_{L,e-L}\,{}^{\int}\mathrm{K}_{L,e-L}\mathrm{W}_{i'L}\rho_{i'L}\left(q_1,q'_1\right)\right]=\exp\left\{\frac{i}{\hbar}\Phi\left[\vec{r},\vec{r}'\right]\right\} \quad (18.51)$$

where, $\Phi\left[\vec{r},\vec{r}'\right]$ is the functional of the influence phase of the electron–lattice interaction or the functional depending on the system path. The important point is that this functional has all information about the environment required to determine the system dynamics, i.e., it describes the influence of the environment on the system. So,

$$\bar{\mathrm{F}}=\mathrm{Tr}_e\,\hat{\mathrm{F}}\left(\vec{r}\right)\,{}^{\int}\mathrm{K}'^{*}_{e}\,{}^{\int}\mathrm{K}_e\mathrm{W}_{ie}\rho_{ie}\left(\vec{r}_1,\vec{r}'_1\right)\exp\left\{\frac{i}{\hbar}\Phi\left[\vec{r},\vec{r}'\right]\right\} \quad (18.52)$$

The quantity

$${}^{\int}\mathrm{K}'^{*}_{e}\,{}^{\int}\mathrm{K}_e\mathrm{W}_{ie}\rho_{ie}\left(\vec{r}_1,\vec{r}'_1\right)\exp\left\{\frac{i}{\hbar}\Phi\left[\vec{r},\vec{r}'\right]\right\} \quad (18.53)$$

is the reduced density matrix that describes the pure electronic system:

$$\int d\vec{r}'_1\int D\vec{r}'\exp\left\{-\frac{i}{\hbar}\mathrm{S}_e\left[\vec{r}'\right]\right\}\int d\vec{r}_1\int_{\vec{r}_1}^{\vec{r}_2}D\vec{r}\exp\left\{\frac{i}{\hbar}\mathrm{S}_e\left[\vec{r}\right]\right\}\mathrm{W}_{ie}\phi'^{*}_i\left(\vec{r}'\right)\phi_i\left(\vec{r}\right)\exp\left\{\frac{i}{\hbar}\Phi\left[\vec{r},\vec{r}'\right]\right\}\equiv\rho_e\left[\vec{r}_2,\vec{r}'_2\right] \quad (18.54)$$

or

$$\iint d\vec{r}_1d\vec{r}'_1\rho_{ie}\left(\vec{r}_1,\vec{r}'_1\right)\iint D\vec{r}D\vec{r}'\exp\left\{\frac{i}{\hbar}\left(\mathrm{S}_e\left[\vec{r}\right]-\mathrm{S}_e\left[\vec{r}'\right]+\Phi\left[\vec{r},\vec{r}'\right]\right)\right\}\equiv\rho_e\left[\vec{r}_2,\vec{r}'_2\right] \quad (18.55)$$

This reduced density matrix, $\rho_e\left[\vec{r}_2,\vec{r}'_2\right]$, gives the time evolution of the system under the influence of an environment described by the influence phase $\Phi\left[\vec{r},\vec{r}'\right]$. The advantage of functional integration is that, if we need to evaluate the expectation value $\bar{\mathrm{F}}$ dependent on the electronic variables, then we have to eliminate beforehand the phonon variables. This permits us to have, in addition, the influence phase $\Phi\left[\vec{r},\vec{r}'\right]$ of the electron–phonon system:

$$\exp\left\{\frac{i}{\hbar}\Phi\left[\vec{r},\vec{r}'\right]\right\}=\mathrm{Tr}_L\,{}^{\int}\mathrm{K}'^{*}_{L,e-L}\,{}^{\int}\mathrm{K}_{L,e-L}\mathrm{W}_{i'L}\rho_{i'L}\left(q_1,q'_1\right) \quad (18.56)$$

or

$$\exp\left\{\frac{i}{\hbar}\Phi\left[\vec{r},\vec{r}'\right]\right\}=\iint\delta\left(q_2-q'_2\right)dq_2dq'_2\iint Dq'Dqdq_1dq'_1\exp\left\{\frac{i}{\hbar}\mathrm{S}\left[\vec{r},\vec{r}',q,q'\right]\right\}\mathrm{W}_{i'L}\psi^{*}_{i'}\left(q_1\right)\psi_{i'}\left(q_1\right) \quad (18.57)$$

The following action functional

$$\mathrm{S}\left[\vec{r},\vec{r}',q,q'\right]=S_L\left[q\right]-S_L\left[q'\right]+S_{e-L}\left[\vec{r},q\right]-S_{e-L}\left[\vec{r}',q'\right] \quad (18.58)$$

is considered linear in the electron–phonon interaction and quadratic in the lattice vibration coordinates:

$$S_L\left[q'\right]=\frac{1}{2}\sum_{\nu\vec{\kappa}j}\int\left(\dot{q}^2_{\nu\vec{\kappa}j}-\omega^2_{\nu\vec{\kappa}j}q^2_{\nu\vec{\kappa}j}\right)dt$$

$$S_{e-L}\left[\vec{r},q\right]=\sum_{\nu\vec{\kappa}j}\int\gamma_{\nu\vec{\kappa}j}\left(t\right)q_{\nu\vec{\kappa}j}\left(t\right)dt,S_e\left[\vec{r}\right]=\frac{m}{2}\int\dot{\vec{r}}^{\,2}\left(t\right)dt \tag{18.59}$$

and the amplitude of the electron–phonon coupling interaction $\gamma_{\nu\vec{\kappa}j}\left(t\right)$ is to be defined later.

We assume that at time moment $t=t_1$, the phonon is found in an equilibrium state. So, the occupation probability of $n_{\nu\kappa j}$ bath oscillators not coupled among each other, may be decomposed into factors corresponding to the individual bath oscillators:

$$W_{iL}=\prod_{n_{\nu\vec{\kappa}j}}W_{n_{\nu\vec{\kappa}j}} \tag{18.60}$$

Here, the selection of quantum numbers:

$$\left\{i\right\}\equiv\left\{n_{\nu_1},\ldots,n_{\nu_n}\right\} \tag{18.61}$$

and

$$W_{n_{\nu\vec{\kappa}j}}=\frac{\exp\left\{-\beta\left(n_{\nu\vec{\kappa}j}+\frac{1}{2}\right)\hbar\omega_{\nu\vec{\kappa}j}\right\}}{Z_{\nu\vec{\kappa}j}} \tag{18.62}$$

The partition function of a single bath oscillator (phonon) is found to be:

$$Z_{\nu\vec{\kappa}j}=\sum_{n_{\nu\vec{\kappa}j}}\exp\left\{-\beta\left(n_{\nu\vec{\kappa}j}+\frac{1}{2}\right)\hbar\omega_{\nu\vec{\kappa}j}\right\}=\exp\left\{-\frac{\beta\hbar\omega_{\nu\vec{\kappa}j}}{2}\right\}\frac{1}{1-\exp\left\{-\beta\hbar\omega_{\nu\vec{\kappa}j}\right\}} \tag{18.63}$$

From here, it follows that

$$W_{n_{\nu\vec{\kappa}j}}=\frac{1}{1-\exp\left\{-\beta\hbar\omega_{\nu\vec{\kappa}j}\right\}}\exp\left\{-\beta\hbar\omega_{\nu\vec{\kappa}j}n_{\nu\vec{\kappa}j}\right\} \tag{18.64}$$

In relations 18.39 and 18.44 the phonon wave functions:

$$\psi_{i'}\left(q_1\right)=\prod_{\nu\vec{\kappa}j}\psi_{n_{\nu\vec{\kappa}j}}\left(q_1\right) \tag{18.65}$$

From here we have the completeness relation

$$\delta\left(q'_{\nu_2}-q_{\nu_2}\right)=\sum_{n_{\nu\vec{\kappa}j}}\psi^*_{n_{\nu'\vec{\kappa}j}}\left(q'_{\nu_2}\right)\psi_{n_{\nu'\vec{\kappa}j}}\left(q_{\nu_2}\right) \tag{18.66}$$

The electron–phonon influence phase is computed

$$\exp\left\{\frac{i}{\hbar}\Phi\left[\vec{r},\vec{r}'\right]\right\}=\prod_{\nu_{\vec{\kappa}j}}\exp\left\{\frac{i}{\hbar}\Phi_{\nu_{\vec{\kappa}j}}\left[\vec{r},\vec{r}'\right]\right\} \tag{18.67}$$

where

$$\exp\left\{\frac{i}{\hbar}\Phi_{\nu_{\vec{\kappa}j}}\left[\vec{r},\vec{r}'\right]\right\}=\sum_{n_{\nu_{\vec{\kappa}j}}}\mathrm{W}_{n_{\nu_{\vec{\kappa}j}}}\iint\psi^{*}_{n_{\nu'_{\vec{\kappa}j}}}\left(q'_{\nu_2}\right)\exp\left\{\frac{i}{\hbar}\mathrm{S}\left[\vec{r},\vec{r}',q,q'\right]\right\}\psi_{n_{\nu'_{\vec{\kappa}j}}}\left(q_{\nu_2}\right)dq'_{\nu_2}dq_{\nu_2}Dq'_{\nu_{\vec{\kappa}j}} \tag{18.68}$$

The transition amplitude in the matrix representation $\mathrm{G}_{n_{\nu'_{\vec{\kappa}j}}n_{\nu_{\vec{\kappa}j}}}$ to go from the initial state with $n_{\nu'_{\vec{\kappa}j}}$ to the final state with $n_{\nu'_{\vec{\kappa}j}}$ is computed from the relation:

$$\mathrm{G}_{n_{\nu'_{\vec{\kappa}j}}n_{\nu'_{\vec{\kappa}j}}}=\iint\psi^{*}_{n_{\nu'_{\vec{\kappa}j}}}\left(q'_{\nu_2}\right)\exp\left\{\frac{i}{\hbar}\mathrm{S}\left[\vec{r},\vec{r}',q,q'\right]\right\}\psi_{n_{\nu'_{\vec{\kappa}j}}}\left(q_{\nu_2}\right)dq'_{\nu_2}dq_{\nu_2}Dq'_{\nu_{\vec{\kappa}j}} \tag{18.69}$$

This represent all matrix elements between the energy eigenstates representing the full solution of the problem of a forced harmonic oscillator.

From 18.58 then

$$\exp\left\{\frac{i}{\hbar}\Phi_{\nu_{\vec{\kappa}j}}\left[\vec{r},\vec{r}'\right]\right\}=\left(1-\exp\left\{-\beta\hbar\omega_{\nu_{\vec{\kappa}j}}\right\}\right)\sum_{n_{\nu_{\vec{\kappa}j}}n_{\nu'_{\vec{\kappa}j}}}\exp\left\{-\beta\hbar\omega_{\nu_{\vec{\kappa}j}}n_{\nu_{\vec{\kappa}j}}\right\}\mathrm{G}_{n_{\nu'_{\vec{\kappa}j}}n_{\nu_{\vec{\kappa}j}}} \tag{18.70}$$

or

$$\exp\left\{\frac{i}{\hbar}\Phi_{\nu_{\vec{\kappa}j}}\left[\vec{r},\vec{r}'\right]\right\}=\left(1-\exp\left\{-\beta\hbar\omega_{\nu_{\vec{\kappa}j}}\right\}\right)\mathrm{G}_{00}\mathrm{G}'^{*}_{00}\exp\left\{-b_{\nu_{\vec{\kappa}j}}b^{*}_{\nu_{\vec{\kappa}j}}\right\}$$

$$\times\sum_{n,m}^{n,m}\sum_{r=0}^{\infty}\sum_{p=n,m}\frac{p!\left(ib_{\nu_{\vec{\kappa}j}}\right)^{p-m}\left(ib^{*}_{\nu_{\vec{\kappa}j}}\right)^{p-n}}{(p-n)!(p-m)!}\frac{\left(-ib'_{\nu_{\vec{\kappa}j}}\right)^{m-r}\left(-ib'^{*}_{\nu_{\vec{\kappa}j}}\right)^{n-r}}{r!(m-r)!(n-r)!}\exp\left\{-\beta\hbar\omega_{\nu_{\vec{\kappa}j}}n_{\nu_{\vec{\kappa}j}}\right\} \tag{18.71}$$

where

$$b_{\nu_{\vec{\kappa}j}}=\frac{1}{\sqrt{2m\hbar\omega_{\nu_{\vec{\kappa}j}}}}\int\gamma_{\nu_{\vec{\kappa}j}}(t)\exp\left\{i\omega_{\nu_{\vec{\kappa}j}}t\right\}dt \tag{18.72}$$

We do the change of summation parameters:

$$m-r\equiv l';n-r\equiv l;p-n\equiv l-l'';p-m\equiv l-l' \tag{18.73}$$

So,

$$\exp\left\{\frac{i}{\hbar}\Phi_{\nu_{\vec{\kappa}j}}\left[\vec{r},\vec{r}'\right]\right\}=\left(1-\exp\left\{-\beta\hbar\omega_{\nu_{\vec{\kappa}j}}\right\}\right)\mathrm{G}_{00}\mathrm{G}'^{*}_{00}\exp\left\{b_{\nu_{\vec{\kappa}j}}b^{*}_{\nu_{\vec{\kappa}j}}\right\}\times$$

$$\sum_{r=0}^{\infty}\sum_{l=0}^{\infty}\sum_{l',l''=0}^{l}\frac{(l+r)!\left(ib_{\nu_{\vec{\kappa}j}}\right)^{l-l'}}{(l-l'')!}\frac{\left(ib^{*}_{\nu_{\vec{\kappa}j}}\right)^{l-l''}}{(l-l')!}\frac{\left(-ib'_{\nu_{\vec{\kappa}j}}\right)^{l'}\left(-ib'^{*}_{\nu_{\vec{\kappa}j}}\right)^{l''}}{r!l'!l''!}\exp\left\{-\beta\hbar\omega_{\nu_{\vec{\kappa}j}}\left(l''+r\right)\right\}\frac{l!}{l!}\frac{l!}{l!} \tag{18.74}$$

Here, the amplitude G_{00} is that for which with an external perturbation, the oscillator does not change its state from the ground state:

$$G_{00} = \exp\left\{-\frac{1}{2m\hbar\omega_{\nu_{\vec{\kappa}j}}}\int_0^T\int_0^\tau \gamma_{\nu_{\vec{\kappa}j}}(t)\gamma_{\nu_{\vec{\kappa}j}}(s)\exp\left\{-i\omega_{\nu_{\vec{\kappa}j}}|t-s|\right\}d\tau d\sigma\right\} \tag{18.75}$$

It is easily seen that:

$$\left(-ib'^{*}_{\nu_{\kappa j}}\right)^{l''}\exp\left\{-\beta\hbar\omega_{\nu_{\vec{\kappa}j}}l''\right\} = \left(-ib'^{*}_{\nu_{\vec{\kappa}j}}\exp\left\{-\beta\hbar\omega_{\nu_{\vec{\kappa}j}}\right\}\right)^{l''} \tag{18.76}$$

then

$$\exp\left\{\frac{i}{\hbar}\Phi_{\nu_{\vec{\kappa}j}}\left[\vec{r},\vec{r}'\right]\right\} = \left(1-\exp\left\{-\beta\hbar\omega_{\nu_{\vec{\kappa}j}}\right\}\right)\times G_{00}G'^{*}_{00}\exp\left\{b_{\nu_{\vec{\kappa}j}}b^{*}_{\nu_{\vec{\kappa}j}}\right\}$$
$$\times\sum_{r=0}^{\infty}\sum_{l=0}^{\infty}\frac{(l+r)!(i)^l}{r!l!l!}\left(b_{\nu_{\vec{\kappa}j}}-b'_{\nu_{\vec{\kappa}j}}\right)^l(i)^l\left(b^{*}_{\nu_{\vec{\kappa}j}}-b'^{*}_{\nu_{\vec{\kappa}j}}\exp\left\{-\beta\hbar\omega_{\nu_{\vec{\kappa}j}}\right\}\right)^l\exp\left\{-\beta\hbar\omega_{\nu_{\vec{\kappa}j}}r\right\} \tag{18.77}$$

or

$$\exp\left\{\frac{i}{\hbar}\Phi_{\nu_{\vec{\kappa}j}}\left[\vec{r},\vec{r}'\right]\right\} = \left(1-\exp\left\{-\beta\hbar\omega_{\nu_{\vec{\kappa}j}}\right\}\right)\times G_{00}G'^{*}_{00}\exp\left\{b_{\nu_{\vec{\kappa}j}}b^{*}_{\nu_{\vec{\kappa}j}}\right\}$$
$$\times\sum_{l=0}^{\infty}\frac{1}{l!}\sum_{r=0}^{\infty}\frac{(l+r)!}{r!l!}\left[\left(ib_{\nu_{\vec{\kappa}j}}-ib'_{\nu_{\vec{\kappa}j}}\right)\left(ib^{*}_{\nu_{\vec{\kappa}j}}-ib'^{*}_{\nu_{\vec{\kappa}j}}\exp\left\{-\beta\hbar\omega_{\nu_{\vec{\kappa}j}}\right\}\right)\right]^l\exp\left\{-\beta\hbar\omega_{\nu_{\vec{\kappa}j}}r\right\} \tag{18.78}$$

or

$$\exp\left\{\frac{i}{\hbar}\Phi_{\nu_{\vec{\kappa}j}}\left[\vec{r},\vec{r}'\right]\right\} = \left(1-\exp\left\{-\beta\hbar\omega_{\nu_{\vec{\kappa}j}}\right\}\right)\times G_{00}G'^{*}_{00}\exp\left\{b_{\nu_{\vec{\kappa}j}}b^{*}_{\nu_{\vec{\kappa}j}}\right\}$$
$$\times\sum_{l=0}^{\infty}\frac{1}{l!}\left[\left(ib_{\nu_{\vec{\kappa}j}}-ib'_{\nu_{\vec{\kappa}j}}\right)\left(ib^{*}_{\nu_{\vec{\kappa}j}}-ib'^{*}_{\nu_{\vec{\kappa}j}}\exp\left\{-\beta\hbar\omega_{\nu_{\vec{\kappa}j}}\right\}\right)\right]^l\sum_{r=0}^{\infty}\frac{(1+r)!}{r!l!}\exp\left\{-\beta\hbar\omega_{\nu_{\vec{\kappa}j}}r\right\} \tag{18.79}$$

But,

$$\sum_{r=0}^{\infty}\frac{(l+r)!}{r!l!}\exp\left\{-\beta\hbar\omega_{\nu_{\vec{\kappa}j}}r\right\} = \frac{1}{\left(1-\exp\left\{-\beta\hbar\omega_{\nu_{\vec{\kappa}j}}\right\}\right)^{l+1}} \tag{18.80}$$

so,

$$\exp\left\{\frac{i}{\hbar}\Phi_{\nu_{\vec{\kappa}j}}\left[\vec{r},\vec{r}'\right]\right\} = G_{00}G'^{*}_{00}\exp\left\{b_{\nu_{\vec{\kappa}j}}b^{*}_{\nu_{\vec{\kappa}j}}\right\}\exp\left\{-\frac{\left(b_{\nu_{\vec{\kappa}j}}-b'_{\nu_{\vec{\kappa}j}}\right)\left(b^{*}_{\nu_{\vec{\kappa}j}}-b'^{*}_{\nu_{\vec{\kappa}j}}\exp\left\{-\beta\hbar\omega_{\nu_{\vec{\kappa}j}}\right\}\right)}{1-\exp\left\{-\beta\hbar\omega_{\nu_{\vec{\kappa}j}}\right\}}\right\} \tag{18.81}$$

We transform the influence phase $\Phi_{v_{\vec{\kappa}_j}}\left[\vec{r},\vec{r}'\right]$ of the electron–phonon interaction via trigonometric functions:

$$\Phi\left[\vec{r},\vec{r}'\right]=\frac{1}{2}\sum_{\vec{\kappa}_j}\iint_{t_1}^{t_2} dtds\left(\gamma_{v_{\vec{\kappa}_j}}(t)-\gamma'_{v_{\vec{\kappa}_j}}(t)\right)$$
$$\times\left[\left(\gamma_{v_{\vec{\kappa}_j}}(s)+\gamma'_{v_{\vec{\kappa}_j}}(s)\right)\mathrm{I}\left(\omega_{v_{\vec{\kappa}_j}},t-s\right)+\left(\gamma_{v_{\vec{\kappa}_j}}(s)-\gamma'_{v_{\vec{\kappa}_j}}(s)\right)i\mathrm{A}\left(\omega_{v_{\vec{\kappa}_j}},t-s\right)\right] \tag{18.82}$$

Here, the amplitude of the electron–phonon coupling interaction, $\gamma_{v_{\vec{\kappa}_j}}(t)$:

$$\gamma_{v_{\vec{\kappa}_j}}(t)=\left|\mathrm{V}_{v_{\vec{\kappa}_j}}\right|\chi_{v_{\vec{\kappa}_j}}(t) \tag{18.83}$$

where

$$\chi_{v_{\vec{\kappa}_j}}(t)=\sqrt{\frac{2}{V}}\begin{cases}\sin\left(\vec{\kappa}_j,\vec{r}\right), & \kappa_x\geq 0\\ \cos\left(\vec{\kappa}_j,\vec{r}\right), & \kappa_x<0\end{cases} \tag{18.84}$$

with the two real waves constituting a complete set when the values of $\vec{\kappa}_j$ are restricted to run only over a half-space. This implies a space in which, if a vector $\vec{\kappa}_j$ occurs, then $-\vec{\kappa}_j$ does not. The functions in $\mathrm{I}\left(\omega_{\vec{\kappa}_j},t-s\right)$ and $\mathrm{A}\left(\omega_{\vec{\kappa}_j},t-s\right)$ in 18.82 are expressed as

$$\mathrm{I}\left(\omega_{\vec{\kappa}_j},t-s\right)=\frac{1}{\omega_{\vec{\kappa}_j}}\theta(t-s)\sin\omega_{\vec{\kappa}_j}(t-s) \tag{18.85}$$

with

$$\mathrm{A}\left(\omega_{\vec{\kappa}_j},t-s\right)=\frac{1}{\omega_{\vec{\kappa}_j}}\mathrm{F}_{\omega_{\vec{\kappa}_j}}(0)\cos\omega_{\vec{\kappa}_j}(t-s),\mathrm{F}_{\omega_{\vec{\kappa}_j}}(0)=\frac{\cosh\hbar\omega_{\vec{\kappa}_j}\left(\frac{\beta}{2}\right)}{\sinh\left(\frac{\hbar\omega_{\vec{\kappa}_j}\beta}{2}\right)} \tag{18.86}$$

and the **Heaviside step function**:

$$\theta(t-s)=\begin{cases}1, & t>s\\ 0, & t<s\end{cases} \tag{18.87}$$

entails compelling the system at a starting point to be evolving towards the future. In the above, V is the volume of the system. Also,

$$\left(\gamma_{v_{\vec{\kappa}_j}}(s)+\gamma'_{v_{\vec{\kappa}_j}}(s)\right)\mathrm{I}\left(\omega_{v_{\vec{\kappa}_j}},t-s\right)+\left(\gamma_{v_{\vec{\kappa}_j}}(s)-\gamma'_{v_{\vec{\kappa}_j}}(s)\right)i\mathrm{A}\left(\omega_{v_{\vec{\kappa}_j}},t-s\right)$$
$$\equiv\phi^*_{v_{\vec{\kappa}_j}}\left(\omega_{v_{\vec{\kappa}_j}},t-s\right)\gamma_{v_{\vec{\kappa}_j}}(s)+\phi_{v_{\vec{\kappa}_j}}\left(\omega_{v_{\vec{\kappa}_j}},t-s\right)\gamma'_{v_{\vec{\kappa}_j}}(s) \tag{18.88}$$

where

$$\left(\mathrm{I}\left(\omega_{\nu_{\vec{\kappa}_j}},t-s\right)+i\mathrm{A}\left(\omega_{\nu_{\vec{\kappa}_j}},t-s\right)\right)\equiv\phi^*_{\nu_{\vec{\kappa}_j}}\left(\omega_{\nu_{\vec{\kappa}_j}},t-s\right) \tag{18.89}$$

$$\left(\mathrm{I}\left(\omega_{\nu_{\vec{\kappa}_j}},t-s\right)-i\mathrm{A}\left(\omega_{\nu_{\vec{\kappa}_j}},t-s\right)\right)\equiv\phi_{\nu_{\vec{\kappa}_j}}\left(\omega_{\nu_{\vec{\kappa}_j}},t-s\right) \tag{18.90}$$

Then

$$\begin{aligned}&\left(\gamma_{\nu_{\vec{\kappa}_j}}(t)-\gamma'_{\nu_{\vec{\kappa}_j}}(t)\right)\left(\phi^*_{\nu_{\vec{\kappa}_j}}\left(\omega_{\nu_{\vec{\kappa}_j}},t-s\right)\gamma_{\nu_{\vec{\kappa}_j}}(s)+\phi_{\nu_{\vec{\kappa}_j}}\left(\omega_{\nu_{\vec{\kappa}_j}},t-s\right)\gamma'_{\nu_{\vec{\kappa}_j}}(s)\right)\\&=\gamma_{\vec{\kappa}_j}\left(\omega_{\vec{\kappa}_j},\vec{r},\vec{r}'\right)\phi_{\nu_{\vec{\kappa}_j}}\left(\omega_{\nu_{\vec{\kappa}_j}},t-s\right)+\tilde{\gamma}_{\vec{\kappa}_j}\left(\omega_{\vec{\kappa}_j},\vec{r},\vec{r}'\right)\phi^*_{\nu_{\vec{\kappa}_j}}\left(\omega_{\nu_{\vec{\kappa}_j}},t-s\right)\end{aligned} \tag{18.91}$$

Here,

$$\gamma_{\vec{\kappa}_j}\left(\omega_{\vec{\kappa}_j},\vec{r},\vec{r}'\right)=\gamma_{\nu_{\vec{\kappa}_j}}(t)\gamma'_{\nu_{\vec{\kappa}_j}}(s)-\gamma'_{\nu_{\vec{\kappa}_j}}(t)\gamma'_{\nu_{\vec{\kappa}_j}}(s) \tag{18.92}$$

$$\tilde{\gamma}_{\vec{\kappa}_j}\left(\omega_{\vec{\kappa}_j},\vec{r},\vec{r}'\right)=\gamma_{\nu_{\vec{\kappa}_j}}(t)\gamma_{\nu_{\vec{\kappa}_j}}(s)-\gamma'_{\nu_{\vec{\kappa}_j}}(t)\gamma_{\nu_{\vec{\kappa}_j}}(s) \tag{18.93}$$

From here, then

$$\gamma_{\nu_{\vec{\kappa}_j}}(t)\gamma_{\nu_{\vec{\kappa}_j}}(s)=\left|V_{\nu_{\vec{\kappa}_j}}\right|^2\frac{2}{V}\begin{cases}\sin\left(\vec{\kappa}_j\vec{r}(t)\right)\sin\left(\vec{\kappa}_j\vec{r}(t)\right), & \kappa_x\geq 0\\ \cos\left(\vec{\kappa}_j\vec{r}(t)\right)\cos\left(\vec{\kappa}_j\vec{r}(t)\right), & \kappa_x<0\end{cases} \tag{18.94}$$

This is an even function with respect to κ and we can do the sum in 18.82 over an entire sphere ●. Considering,

$$\mathrm{M}_1=\{\cdots\}\sin\left(\vec{\kappa}_j\vec{r}(t)\right)\sin\left(\vec{\kappa}_j\vec{r}(t)\right),\mathrm{M}_2=\{\cdots\}\cos\left(\vec{\kappa}_j\vec{r}(t)\right)\cos\left(\vec{\kappa}_j\vec{r}(t)\right) \tag{18.95}$$

then

$$\sum_{\text{●}}\mathrm{M}=\sum_{\text{◗}}\mathrm{M}_1+\sum_{\text{◖}}\mathrm{M}_2=\sum_{\text{◗}}\mathrm{M}_1+\sum_{\overrightarrow{\kappa_j}=-\overrightarrow{\kappa_j}\text{◖}}\mathrm{M}_2=\sum_{\text{◗}}\mathrm{M}_1+\sum_{\text{◗}}\mathrm{M}_2=\sum_{\text{◗}}\left(\mathrm{M}_1+\mathrm{M}_2\right) \tag{18.96}$$

Consequently,

$$\sum_{\text{◗}}\left(\mathrm{M}_1+\mathrm{M}_2\right)=\frac{1}{2}\sum_{\text{●}}\left(\mathrm{M}_1+\mathrm{M}_2\right) \tag{18.97}$$

and so,

$$\sum_{\text{◗}}\{\cdots\}\left(\sin\left(\vec{\kappa}_j\vec{r}(t)\right)\sin\left(\vec{\kappa}_j\vec{r}(t)\right)+\cos\left(\vec{\kappa}_j\vec{r}(t)\right)\cos\left(\vec{\kappa}_j\vec{r}(t)\right)\right)=\frac{1}{2}\sum_{\text{●}}\{\cdots\}\cos\left(\vec{\kappa}_j,\vec{r}(t)-\vec{r}(s)\right) \tag{18.98}$$

The influence phase of the electron–phonon interaction in 18.82 may then be written

$$\Phi\left[\vec{r},\vec{r}'\right]=\frac{1}{2}\sum_{\vec{\kappa}_j}\int\int_{t_1}^{t_2}dtds\left[\gamma_{\vec{\kappa}_j}\left(\omega_{\vec{\kappa}_j},\vec{r},\vec{r}'\right)\phi_{\nu_{\vec{\kappa}_j}}\left(\omega_{\nu_{\vec{\kappa}_j}},t-s\right)+\tilde{\gamma}_{\vec{\kappa}_j}\left(\omega_{\vec{\kappa}_j},\vec{r},\vec{r}'\right)\phi^*_{\nu_{\vec{\kappa}_j}}\left(\omega_{\nu_{\vec{\kappa}_j}},t-s\right)\right] \tag{18.99}$$

or

$$\Phi\left[\vec{r},\vec{r}'\right]=\frac{1}{2}\sum_j\frac{Vd\vec{\kappa}}{(2\pi)^3}\left|V_{v_{\vec{\kappa}j}}\right|^2\frac{2}{V}\frac{1}{2}\sum_{\vec{\kappa}j}\int\int_{t_1}^{t_2}dtds$$
$$\times\left[\mathrm{F}_{\vec{\kappa}_j}\left(\omega_{\vec{\kappa}_j},\vec{r},\vec{r}'\right)\phi_{v_{\kappa_j}}\left(\omega_{v_{\kappa j}},t-s\right)+\tilde{\mathrm{F}}_{\vec{\kappa}_j}\left(\omega_{\vec{\kappa}_j},\vec{r},\vec{r}'\right)\phi^*_{v_{\kappa j}}\left(\omega_{v_{\kappa j}},t-s\right)\right] \tag{18.100}$$

where,

$$\mathrm{F}_{\vec{\kappa}_j}\left(\omega_{\vec{\kappa}_j},\vec{r},\vec{r}'\right)=\cos\left(\vec{\kappa}_j,\vec{r}(t)-\vec{r}(s)\right)-\cos\left(\vec{\kappa}_j,\vec{r}'(t)-\vec{r}(s)\right) \tag{18.101}$$

$$\tilde{\mathrm{F}}_{\vec{\kappa}_j}\left(\omega_{\vec{\kappa}_j},\vec{r},\vec{r}'\right)=\cos\left(\vec{\kappa}_j,\vec{r}(t)-\vec{r}'(s)\right)-\cos\left(\vec{\kappa}_j,\vec{r}'(t)-\vec{r}'(s)\right) \tag{18.102}$$

So, the full functional of the electron-phonon interaction **influence phase** is:

$$\Phi\left[\vec{r},\vec{r}'\right]=\frac{1}{2}\sum_j\frac{d\vec{\kappa}}{(2\pi)^3}\left|V_{v_{\vec{\kappa}j}}\right|^2\sum_{\vec{\kappa}j}\int\int_{t_1}^{t_2}dtds$$
$$\times\left[\mathrm{F}_{\vec{\kappa}_j}\left(\omega_{\vec{\kappa}_j},\vec{r},\vec{r}'\right)\phi_{v_{\vec{\kappa}j}}\left(\omega_{v_{\vec{\kappa}j}},t-s\right)+\tilde{F}_{\vec{\kappa}_j}\left(\omega_{\vec{\kappa}_j},\vec{r},\vec{r}'\right)\phi^*_{v_{\vec{\kappa}j}}\left(\omega_{v_{\vec{\kappa}j}},t-s\right)\right] \tag{18.103}$$

18.5 Full System Polaron Partition Function in a 3D Structure

The full system polaron Lagrangian is rewritten as in 18.15:

$$L=-\frac{m\dot{r}^2}{2\hbar^2}-\frac{1}{2}\sum_{v_{\vec{\kappa}j}}\left(\frac{\dot{q}^2_{v_{\vec{\kappa}j}}}{\hbar^2}+\omega^2_{v_{\vec{\kappa}j}}q^2_{v_{\vec{\kappa}j}}\right)+\sum_{v_{\vec{\kappa}j}}\gamma_{v_{\vec{\kappa}j}}(\tau)q_{v_{\vec{\kappa}j}}(\tau) \tag{18.104}$$

Here, the electron coordinate is $\vec{ }$ and m is its effective mass. The longitudinal optical phonon generalized coordinates are $q_{v_{\vec{\kappa}j}}$ for a mode of frequency $\omega_{v_{\vec{\kappa}j}}$ numbered by the oscillator number $v_{\vec{\kappa}_j}$ where $\vec{\kappa}$ is the wave number and j the polarization while the amplitude of the electron–lattice coupling is $\gamma_{v_{\vec{\kappa}j}}(\tau)$:

$$\gamma_{v_{\vec{\kappa}j}}(t)=\left|V_{v_{\vec{\kappa}j}}\right|\chi_{v_{\vec{\kappa}j}}(t) \tag{18.105}$$

where

$$\chi_{v_{\vec{\kappa}j}}(t)=\sqrt{\frac{2}{V}}\begin{cases}\sin\left(\vec{\kappa}_j,\vec{r}\right), & \kappa_x\geq 0\\ \cos\left(\vec{\kappa}_j,\vec{r}\right), & \kappa_x<0\end{cases} \tag{18.106}$$

The partition function for the full system is rewritten as in 18.26:

$$Z=\prod_{v_{\vec{\kappa}j}}Z_L\int\delta\left(\vec{r}-\vec{r}'\right)d\vec{r}d\vec{r}'\int_{\vec{r}'}^{\vec{r}}D\vec{r}\exp\left\{\int_0^\beta\left(-\frac{m\dot{\vec{r}}^2}{2\hbar^2}\right)d\tau+\Phi_{\omega_{v_{\vec{\kappa}j}}}\left[\vec{r}\right]\right\} \tag{18.107}$$

Here, the lattice partition function is rewritten:

$$Z_L = \left[2\sinh\left(\frac{\hbar\omega_{\nu_{\vec{\kappa}j}}\beta}{2} \right) \right]^{-3} \tag{18.108}$$

and the **electron–phonon influence phase** $\Phi_{\omega_{\nu_{\vec{\kappa}j}}}\left[\vec{r}\right]$ is defined in 18.103. Relation 18.107 permits us to rewrite the polaron action functional for the full system:

$$\underline{S}\left[\vec{r}\right] = -\int_0^{\beta} \frac{m\dot{\vec{r}}^2}{2\hbar^2} d\tau + \sum_{\nu_{\vec{\kappa}j}} \Phi_{\omega_{\nu_{\vec{\kappa}j}}}\left[\vec{r}\right] \tag{18.109}$$

18.6 Model System Polaron Partition Function in a 3D Structure

We select the one oscillatory Lagrangian to imitate the Lagrangian in 18.15:

$$L_0 = -\frac{m\dot{\vec{r}}^2}{2\hbar^2} - \frac{M\dot{\vec{R}}^2}{2\hbar^2} - \frac{M\omega_f^2\left(\vec{R}-\vec{r}\right)^2}{2} \tag{18.110}$$

That permits the writing of the action functional of the model system as:

$$\underline{S}_0 = \int_0^{\beta} d\tau \left[-\frac{m}{2\hbar^2}\left(\frac{d\vec{r}}{d\tau}\right)^2 - \frac{M}{2\hbar^2}\left(\frac{d\vec{R}}{d\tau}\right)^2 - \frac{M\omega_f^2\left(\vec{R}-\vec{r}\right)^2}{2} \right] \tag{18.111}$$

This will permit us to find the following quantities:

$$E_0 = -T\ln Z_0 \tag{18.112}$$

where,

$$Z_0 = \int_{-\infty}^{+\infty} \delta\left(\vec{r}-\vec{r}'\right)\delta\left(\vec{R}-\vec{R}'\right) d\vec{r}d\vec{r}'d\vec{R}d\vec{R}' \int_{r'}^{r}\int_{R'}^{R} \exp\left\{\underline{S}_0\right\} D\vec{r}D\vec{R} \tag{18.113}$$

This can be conveniently solved by considering the motion of the center of mass and also the relative motion:

$$\vec{\rho}_0 = \frac{\vec{r}m + \vec{R}M}{n+M}, \quad \vec{\rho} = \vec{r} - \vec{R} \tag{18.114}$$

from where

$$\vec{r} = \vec{\rho}_0 + \frac{M}{m+M}\vec{\rho}, \quad \vec{R} = \vec{\rho}_0 - \frac{m}{m+M}\vec{\rho} \tag{18.115}$$

The quantities $\vec{\rho}_0$ and $\vec{\rho}$ are, respectively, the coordinates of the center of mass and of the relative motion. Let

$$\mu = \frac{Mm}{m+M}, \quad v^2 = u^2\omega_f^2, \quad u^2 = \frac{m+M}{m} \tag{18.116}$$

Here μ is a reduced mass and v is a scaled frequency. So, from 18.111 to 18.116 then

$$\underline{S}_0 = -\frac{M+m}{2\hbar^2}\int_0^{\beta}\dot{\rho}_0^2 d\sigma - \frac{\mu}{2\hbar^2}\int_0^{\beta}\dot{\rho}^2 d\sigma - \frac{\mu v^2}{2}\int_0^{\beta}\rho^2 d\sigma \tag{18.117}$$

This transforms the model Lagrangian to quadratic form with one term describing the translational motion of the center of mass and the other two an oscillatory motion.

From 18.115 then the path integral:

$$Z_0 = \int_{-\infty}^{+\infty}\delta\left(\vec{\rho}_0 - \vec{\rho}_0'\right)\delta\left(\vec{\rho} - \vec{\rho}'\right)d\vec{\rho}_0 d\vec{\rho}_0' d\vec{\rho}d\vec{\rho}'\int_{\vec{\rho}_0'}^{\vec{\rho}_0}\int_{\vec{\rho}'}^{\vec{\rho}}\exp\{\underline{S}_0\}D\vec{\rho}_0 D\vec{\rho} \tag{18.118}$$

or

$$Z_0 = Vu^3 l_e^{-3}\left(\hbar\beta\right)\left[2\sinh\left(\frac{\beta\hbar v}{2}\right)\right]^{-3} \tag{18.119}$$

Considering the procedure from 18.113 to 18.119 the action functional $\underline{S}_0\left[\vec{r}\right]$ of the fictitious particle is:

$$\underline{S}_0\left[\vec{r}\right] = \int_0^{\beta}\left[-\frac{m\dot{\vec{r}}^2}{2\hbar^2} - \frac{M\omega_f^2}{2}\vec{r}^2\right]d\tau + \Phi_{\omega_f}\left[\vec{r}\right] - 3\ln\left[2\sinh\left(\frac{\beta\hbar\omega_f}{2}\right)\right] \tag{18.120}$$

18.7 Feynman Inequality and Generating Functional

We have so far calculated the full action functional $\underline{S}$ that is not easily calculated due to retardation effects and this effect may be resolved by first introducing an optimal variational action $\underline{S}_0$ that will minimize the following equation for the polaron energy and consequently the effective mass:

$$\ln Z \geq \ln Z_0 + \left\langle \underline{S} - \underline{S}_0 \right\rangle_{\underline{S}_0} \tag{18.121}$$

Here,

$$\ln Z_0 = \ln\left(Vu^3 l_e^{-3}\left(\hbar\beta\right)\left[2\sinh\left(\frac{\beta\hbar v}{2}\right)\right]^{-3}\right) \tag{18.122}$$

and

$$\left\langle \underline{S} - \underline{S}_0 \right\rangle_{\underline{S}_0} = \Phi\left[\vec{r}\right] - \Phi_{\omega_f}\left[\vec{r}\right]_{\underline{S}_0} + \frac{M\omega_f^2}{2}\int_0^{\beta}\vec{r}^2\left(\tau\right)_{\underline{S}_0} d\tau + 3\ln\left[2\sinh\left(\frac{\beta\hbar\omega_f}{2}\right)\right] - 3\ln\left[2\sinh\left(\frac{\hbar\omega_{v_{\vec{\kappa}_j}}\beta}{2}\right)\right] \tag{18.123}$$

To do this, it is necessary to evaluate the following average quantities:

$$\left\langle \vec{r}^2\left(\tau\right)\right\rangle_{\underline{S}_0}, \quad \left\langle \vec{r}\left(\tau\right)\vec{r}\left(\sigma\right)\right\rangle_{\underline{S}_0}, \quad \left\langle \cos\left(\vec{\kappa}\cdot\left(\vec{r}\left(\tau\right) - \vec{r}\left(\sigma\right)\right)\right)\right\rangle_{\underline{S}_0} \tag{18.124}$$

This can be obtained more straightforwardly with the generating function where source terms ξ and η add to simplify the algebra:

$$\Psi_{\vec{\kappa}}(\xi,\eta)=\left\langle \exp\left\{i\vec{\kappa}\left(\xi\vec{r}(\tau)-\eta\vec{r}(\sigma)\right)\right\}_{\underline{S}}\right\rangle_0=\frac{\mathrm{Tr}\left\{\int_{R'}^{R}\int_{r'}^{r}D\vec{r}D\vec{R}\exp\left\{i\vec{\kappa}\left(\xi\vec{r}(\tau)-\eta\vec{r}(\sigma)\right)+\underline{S}_0\right\}\right\}}{\mathrm{Tr}\left\{\int_{R'}^{R}\int_{r'}^{r}D\vec{r}D\vec{R}\exp\left\{\underline{S}_0\right\}\right\}} \tag{18.125}$$

Here, in its evaluation we consider the variables of the full system $\vec{r}$ and that of the model system $\vec{R}$. The exponential function in 18.125 is not a functional. However, it may be expressed via a functional having a general form:

$$\exp\left\{i\vec{\kappa}\left(\int_0^\beta dt\left[\xi\vec{r}(t)\delta(t-\tau)-\eta\vec{r}(t)\delta(t-\sigma)\right]\right)\right\}\equiv\exp\left\{\int_0^\beta\gamma(t)\vec{r}(t)dt\right\} \tag{18.126}$$

Here,

$$\gamma(t)\equiv i\vec{\kappa}\left[\xi\vec{r}(t)\delta(t-\tau)-\eta\vec{r}(t)\delta(t-\sigma)\right] \tag{18.127}$$

We average 18.126 by the action $\underline{S}_0$, i.e.,

$$\left\langle\exp\left\{\int_0^\beta\gamma(t)\vec{r}(t)dt\right\}\right\rangle_{\underline{S}_0}=\frac{\mathrm{Tr}\left\{\int_{R'}^{R}\int_{r'}^{r}D\vec{r}D\vec{R}\exp\left\{\int_0^\beta\gamma(t)\vec{r}(t)dt+\underline{S}_0\right\}\right\}}{\mathrm{Tr}\left\{\int_{R'}^{R}\int_{r'}^{r}D\vec{r}D\vec{R}\exp\left\{\underline{S}_0\right\}\right\}}\equiv\Psi_{\vec{\kappa}}(\xi,\eta) \tag{18.128}$$

This expression of the generating functional is conveniently evaluated by expanding the variables in quadratic form and may be done by making a change of variables considering the motion of the center of mass and relative motion as above. The generating functional is then obtained by integrating out the electronic as well the fictitious particle variables.

The expression of the generating functional can conveniently calculate the following by the functional derivatives with respect to ξ and η brought out of the functional integral over the electronic as well the fictitious particle variables:

$$\left\langle\vec{r}(\tau)\vec{r}(\sigma)\right\rangle_{\underline{S}_0}=\frac{1}{\kappa^2}\frac{\partial\Psi}{\partial\xi\partial\eta}\bigg|_{\xi=\eta=0},\left\langle\cos\left(\vec{\kappa}\left(\vec{r}(\tau)-\vec{r}(\sigma)\right)\right)\right\rangle_{\underline{S}_0}=\frac{1}{2}\left[\Psi_{\vec{\kappa}}(1,1)+\Psi_{\vec{\kappa}}(-1,-1)\right] \tag{18.129}$$

We apply this to our system described by the action functional 18.18 considering the change of variables 18.115 and permitted us to have two driven harmonic oscillators with frequencies $\nu_0 = 0$, $\nu_1 = \nu$. The generating function can then be written

$$\Psi_{\vec{\kappa}}(\xi,\eta)=\prod_{i=0,1}\exp\left\{\frac{\hbar}{4m_i\nu_i}\int_0^\beta\int_0^\beta dtds\gamma(t)\gamma(s)\mathrm{F}_{\nu_i}\left(|t-s|\right)\right\} \tag{18.130}$$

where

$$\gamma_0(t)=i\vec{\kappa}\left[\xi\delta(t-\tau)-\eta\delta(t-\sigma)\right]\quad,\quad\gamma_1(t)=\frac{M}{m+M}\gamma_0(t) \tag{18.131}$$

$$\nu_0=0,\quad\nu_1=\nu,\quad m_0=M+m,\quad m_1=\mu=\frac{mM}{m+M} \tag{18.132}$$

From here, then 18.130 becomes:

$$\Psi_{\vec{\kappa}}(\xi,\eta)=\exp\left\{-\frac{\hbar\kappa^2}{4m_0v_0}\left[\left(\xi^2+\eta^2\right)F_{v_0}(0)-2\xi\eta F_{v_0}\left(|\tau-\sigma|\right)\right]\right.$$
$$\left.-\frac{\hbar\kappa^2}{4m_1v_1}\left[\left(\xi^2+\eta^2\right)F_{v_1}(0)-2\xi\eta F_{v_1}\left(|\tau-\sigma|\right)\right]\right\} \quad (18.133)$$

Considering the Taylor series expansion of the following functions:

$$F_{v_0}(0)\cong\frac{2}{\beta\hbar v_0}+\frac{\beta\hbar v_0}{6}+\cdots,\quad F_{v_0}\left(|\tau-\sigma|\right)\cong\frac{2}{\beta\hbar v_0}+\frac{\hbar v_0}{\beta}\left(|\tau-\sigma|-\frac{\beta}{2}\right)^2-\frac{1}{6}\frac{\beta\hbar v_0}{2}+\cdots \quad (18.134)$$

then the generating functional

$$\Psi_{\vec{\kappa}}(\xi,\eta)=\exp\left\{\begin{array}{l}-\frac{\hbar\kappa^2a_2}{4mv}\left[\left(\xi^2+\eta^2\right)F_v(0)-2\xi\eta F_v\left(|\tau-\sigma|\right)\right]-\\ -\frac{\hbar^2\kappa^2a_1\beta}{4m}\left[\frac{1}{6}\left(\xi^2+\eta^2\right)-\frac{2\xi\eta}{\beta^2}\left(|\tau-\sigma|-\frac{\beta}{2}\right)^2+\frac{1}{6}\xi\eta\right]\end{array}\right\} \quad (18.135)$$

$$a_1=\frac{1}{u^2},\quad a_2=1-\frac{1}{u^2},\quad u^2=\frac{M+m}{m} \quad (18.136)$$

from where

$$\left\langle\vec{r}(\tau)\vec{r}(\sigma)\right\rangle_{S_0}=\frac{3\hbar a_2}{2mv}F_v\left(|\tau-\sigma|\right)-\frac{3\hbar^2a_1\beta}{4m}\left[-\frac{2}{\beta^2}\left(|\tau-\sigma|-\frac{\beta}{2}\right)^2+\frac{1}{6}\right] \quad (18.137)$$

$$\left\langle\vec{r}^2(\tau)\right\rangle_{S_0}=\frac{3\hbar a_2}{2mv}F_v(0)-\frac{3\hbar^2a_1\beta}{4m}\left[-\frac{2}{\beta^2}\frac{\beta^2}{4}+\frac{1}{6}\right] \quad (18.138)$$

The above formulae are observed to be functions of the variable $|\tau-\sigma|$ which is a property retarded functions. The retarded function shows the interaction with the past. The significance of the interaction with the past shows the perturbation to be caused by the moving electron that takes "**time**" to propagate in the crystal medium.

18.8 Polaron Characteristics in a 3D Structure

The polaron characteristics are conveniently obtained for low temperatures, $\beta\to\infty$ via $\ln\left(\frac{Z_F}{Z_L}\right)$, where the coefficient of β is the polaron ground state energy E_0 and the term independent of β in the argument of the logarithmic function, the polaron dimensionless effective mass $\frac{m^*}{m}$. This type of relation as mentioned earlier is useful in relation with the experimentally determined polaron effective mass to compute the electron band-mass. So, from

$$\lim_{\beta\to\infty}3\ln\frac{\sinh\frac{\hbar\omega_f\beta}{2}}{\sinh\frac{\hbar v\beta}{2}}=\lim_{\beta\to\infty}3\ln\frac{\exp\left\{\frac{\hbar\omega_f\beta}{2}\right\}}{\exp\left\{\frac{\hbar v\beta}{2}\right\}}=\frac{3\hbar\beta}{2}\left(\omega_f-v\right)=\frac{3}{2}\hbar v\beta\left(1-\frac{1}{u}\right) \quad (18.139)$$

then

$$\ln\left(\frac{Z_F}{Z_L}\right) = \ln\left(Vu^3 l_e^{-3}\left(\hbar\beta\right)\right) + \frac{3}{2}\hbar\nu\beta\left(1-\frac{1}{u}\right) + \frac{3\hbar\nu\beta}{4}\frac{u^2-1}{u^2}\coth\frac{\hbar\nu\beta}{2}$$
$$-\frac{3\left(u^2-1\right)}{2u^2} + \sum_{\vec{\kappa}_j}\frac{\hbar}{4V\omega_{\vec{\kappa}_j}}\left|V_{\nu\vec{\kappa}_j}\right|^2\int_0^\beta\int_0^\beta F\left(\left|\tau-\sigma\right|-\frac{\beta}{2}\right)d\sigma d\tau \quad (18.140)$$

Here, $V_{\nu\vec{\kappa}_j}$ are the Fourier components of the electron–phonon coupling amplitude to be defined later and

$$F\left(\left|\tau-\sigma\right|-\frac{\beta}{2}\right) = F_{\omega_{\vec{\kappa}_j}}\left(\left|\sigma-\sigma'\right|\right)\exp\left\{-\kappa_j^2 A\right\} \quad (18.141)$$

with

$$F_{\omega_{\vec{\kappa}_j}}\left(\left|\sigma-\sigma'\right|\right) = \frac{\cosh\hbar\omega_{\vec{\kappa}_j}\left(\left|\tau-\sigma\right|-\frac{\beta}{2}\right)}{\sinh\left(\frac{\hbar\omega_{\vec{\kappa}_j}\beta}{2}\right)} \quad (18.142)$$

$$A = \frac{\hbar a_2}{2m\nu}\left(F_\nu(0) - F_\nu\left(\left|\tau-\sigma\right|\right)\right) + \frac{\hbar^2 a_1}{2m\beta}\left(\frac{1}{6} - \frac{1}{\beta^2}\left(\left|\tau-\sigma\right|-\frac{\beta}{2}\right)^2\right) \quad (18.143)$$

We observe the resultant functions to have the common argument $\left(\left|\tau-\sigma\right|-\frac{\beta}{2}\right)$ and so we generalize the functions to be $F\left(\left|\tau-\sigma\right|-\frac{\beta}{2}\right)$ to aid in the simplification of the evaluation of the twofold integral in the problem. Setting the functions with argument $\left(\left|\tau-\sigma\right|-\frac{\beta}{2}\right)$ as integrands then:

$$f(\beta) = \int_0^\beta\int_0^\beta F\left(\left|\tau-\sigma\right|-\frac{\beta}{2}\right)d\sigma d\tau \quad (18.144)$$

Doing change of variables

$$\tau-\sigma \equiv \rho, \quad 0 \le \sigma \le \tau \quad (18.145)$$

then

$$f(\beta) = 2\int_0^\beta\int_0^\rho F\left(\left|\rho-\frac{\beta}{2}\right|\right)d\tau d\rho = 2\int_0^\beta F\left(\left|\rho-\frac{\beta}{2}\right|\right)d\rho\int_\rho^\beta d\tau \quad (18.146)$$

or

$$f(\beta) = 2\int_0^\beta F\left(\left|\rho-\frac{\beta}{2}\right|\right)(\beta-\rho)d\rho \quad (18.147)$$

Do another change of variable $\rho = \beta x$, where $0 \le x \le 1$. If $x = 1$ then $\rho = \beta$. So,

$$f(\beta) = 2\beta^2 \int_0^1 \mathrm{F}\left(\beta\left|x - \frac{1}{2}\right|\right)(1-x)\,dx \tag{18.148}$$

Setting,

$$x - \frac{1}{2} = y, 1 - x = \frac{1}{2} - y \tag{18.149}$$

then,

$$f(\beta) = 2\beta^2 \int_{-\frac{1}{2}}^{\frac{1}{2}} \mathrm{F}(\beta|y|)\left(\frac{1}{2} - y\right)dy \tag{18.150}$$

Integrating an odd function over symmetric limits yields zero and so

$$f(\beta) = \beta^2 \int_{-\frac{1}{2}}^{\frac{1}{2}} \mathrm{F}(\beta|y|)\,dy = 2\beta^2 \int_0^{\frac{1}{2}} \mathrm{F}(\beta|y|)\,dy = \beta^2 \int_0^1 \mathrm{F}\left(\frac{\beta y}{2}\right)dy \tag{18.151}$$

Therefore, the **twofold integral law that mimics Stokes' theorem**:

$$\int_0^\beta \int_0^\beta \mathrm{F}\left(|\tau - \sigma| - \frac{\beta}{2}\right)d\sigma\, d\tau = \beta^2 \int_0^1 \mathrm{F}\left(\frac{\beta y}{2}\right)dy \tag{18.152}$$

Here,

$$\mathrm{F}\left(\frac{\beta y}{2}\right) = \frac{\cosh\dfrac{\beta y \hbar\omega_{\vec{\kappa}_j}}{2}}{\sinh\left(\dfrac{\hbar\omega_{\vec{\kappa}_j}\beta}{2}\right)} \exp\left\{-\kappa_j^2 \mathrm{A}\right\} \tag{18.153}$$

and

$$\mathrm{A} = \frac{\hbar a_2}{2mv}\left(\coth\frac{\hbar v\beta}{2} - \frac{\cosh\dfrac{\beta y \hbar v}{2}}{\sinh\dfrac{\hbar v\beta}{2}}\right) + \frac{\hbar^2 \beta a_1}{8m}\left(1 - y^2\right) \tag{18.154}$$

We now examine the polaron parameters for low temperatures $\beta \to \infty$ ($T \to 0$):

$$\lim_{\beta\to\infty} \coth\frac{\hbar v\beta}{2} \to 1, \lim_{\beta\to\infty} \frac{\cosh\dfrac{\beta y \hbar\omega_{\vec{\kappa}_j}}{2}}{\sinh\dfrac{\hbar\omega_{\vec{\kappa}_j}\beta}{2}} \to \exp\left\{-\frac{\hbar\omega_{\vec{\kappa}_j}\beta}{2}(1-y)\right\} \tag{18.155}$$

It follows that

$$\mathrm{F}\left(\frac{\beta y}{2}\right)=\exp\left\{-\frac{\hbar\omega_{\vec{\kappa}_j}\beta}{2}(1-y)\right\}\exp\left\{-\kappa_j^2\mathrm{A}\right\} \tag{18.156}$$

Here,

$$\mathrm{A}=\frac{\hbar a_2}{2mv}\left(1-\exp\left\{-\frac{\hbar v\beta}{2}(1-y)\right\}\right)+\frac{\hbar^2\beta a_1}{8m}\left(1-y^2\right) \tag{18.157}$$

In an ionic crystal, the conduction electron interacts strongly with polar optical vibrations. We limit ourselves to the case when the electron interactions only with one branch of the polar vibrations in a cubic crystal. We show that the most important role in the interaction of the electron with the polar vibrations is played by long wavelength vibrations having a small dispersion. For this reason, we set $\omega_{\vec{\kappa}_j}$ to be equal to the limiting dispersionless optical phonon mode with frequency ω_0.

We do now the following change of variables in the above relations:

$$\frac{\beta}{2}(1-y)\equiv-\tau,d\tau=-\frac{\beta}{2}dy,0\leq\tau<\infty \tag{18.158}$$

This change of variable permits equation 18.157 to take the form:

$$f(\beta)=2\beta\int_0^\infty d\tau\exp\{-\hbar\omega_0\tau\}\exp\left\{-\kappa^2\mathrm{A}\right\} \tag{18.159}$$

Here,

$$\mathrm{A}\equiv\mathrm{A}_p-\frac{\hbar^2a_1\tau^2}{2m\beta} \tag{18.160}$$

$$\mathrm{A}_p=\frac{\hbar a_2}{2mv}\left(1-\exp\{-\hbar v\tau\}\right)+\frac{\hbar^2a_1\tau}{2m} \tag{18.161}$$

For $\beta\rightarrow\infty$, then

$$\exp\left\{\frac{\hbar^2a_1\tau^2\kappa^2}{2m\beta}\right\}\cong1+\frac{\hbar^2a_1\tau^2\kappa^2}{2m\beta} \tag{18.162}$$

So,

$$f(\beta)=2\beta\int_0^\infty d\tau\exp\{-\hbar\omega_0\tau\}\exp\left\{-\kappa^2\mathrm{A}_p\right\}\left(1+\frac{\hbar^2a_1\tau^2\kappa^2}{2m\beta}\right) \tag{18.163}$$

From here and considering equation 18.140 and the fact that

$$\lim_{\beta\rightarrow\infty}\coth\frac{\hbar v\beta}{2}=1 \tag{18.164}$$

then

$$\ln\left(\frac{Z_F}{Z_L}\right)=\ln\left(Vu^3 l_e^{-3}(\hbar\beta)\right)+\frac{3}{2}\hbar\nu\beta\left(1-\frac{1}{u}\right)+\frac{3\hbar\nu\beta}{4}\frac{u^2-1}{u^2}-\frac{3\left(u^2-1\right)}{2u^2}$$
$$+\frac{V}{(2\pi)^3}\sum_j\frac{\hbar}{4V\omega_{\vec{\kappa}_j}}\int d\vec{\kappa}_j\left|V_{\nu_{\vec{\kappa}_j}}\right|^2 2\beta\int_0^\infty d\tau\exp\{-\hbar\omega_0\tau\}\exp\{-\kappa^2 A_p\}$$
$$+\frac{V}{(2\pi)^3}\sum_j\frac{\hbar}{4V\omega_{\vec{\kappa}_j}}\int d\vec{\kappa}_j\left|V_{\nu_{\vec{\kappa}_j}}\right|^2 2\beta\int_0^\infty d\tau\frac{\hbar^2 a_1\tau^2\kappa^2}{2m\beta}\exp\{-\hbar\omega_0\tau\}\exp\{-\kappa^2 A_p\} \quad (18.165)$$

Comparing with equation 18.6 then the polaron variational energy is obtained:

$$E=\frac{3}{4}\hbar\nu\left(1-\frac{1}{u}\right)^2-\frac{V}{(2\pi)^3}\sum_j\frac{\hbar}{4V\omega_{\vec{\kappa}_j}}\int d\vec{\kappa}_j\left|V_{\nu_{\vec{\kappa}_j}}\right|^2\int_0^\infty d\tau\exp\{-\hbar\omega_0\tau\}\exp\{-\kappa^2 A_p\} \quad (18.166)$$

and the polaron effective mass:

$$\frac{m^*}{m}=\exp\left\{-\frac{u^2-1}{u^2}+\frac{V}{3(2\pi)^3}\sum_j\frac{\hbar}{4V\omega_{\vec{\kappa}_j}}\int d\vec{\kappa}_j\left|V_{\nu_{\vec{\kappa}_j}}\right|^2\int_0^\infty d\tau\tau^2\exp\{-\hbar\omega_0\tau\}\exp\{-\kappa^2 A_p\}\right\} \quad (18.167)$$

Consider

$$\left|V_{\vec{\kappa}}\right|^2=\frac{V_0^2}{\kappa^2}\ ,V_0^2=\alpha_F\pi\left(\frac{R_p^5\omega_0^5\hbar^3}{m}\right)^{\frac{1}{2}},\quad \alpha_F=\frac{e^2}{2R_p\hbar\omega_0}\left(\frac{1}{\varepsilon_\infty}-\frac{1}{\varepsilon_0}\right) \quad (18.168)$$

and

$$d\vec{\kappa}_j=4\pi\kappa_j^2 d\kappa_j \quad (18.169)$$

where,

$$\alpha_F=\frac{e^2}{2R_p\hbar\omega_0}\left(\frac{1}{\varepsilon_\infty}-\frac{1}{\varepsilon_0}\right) \quad (18.170)$$

is a dimensionless **Fröhlich electron–phonon coupling constant** and

$$R_p=\left(\frac{\hbar}{2m\omega_0}\right)^{\frac{1}{2}} \quad (18.171)$$

the polaron radius; ε_∞ and ε_0 are, respectively, the electronic and the static dielectric constant of the polar crystal; ω_0, is the dispersionless longitudinal (LO) optical phonon frequency. For a crystal such as sodium chloride, $\alpha_F = 5$. The Feynman method is for the entire range of coupling strength and applying the Feynman variational principle results in an upper bound for the polaron self-energy at all α_F. Actual crystals are observed to have α_F-values typically ranging from $\alpha_F = 0.02$ (InSb) to $\alpha_F \approx 3$ to 4 (alkali halides) [38].

In writing 18.168 we assume:

1. **spatial extension of the polaron is large compared to the lattice parameters,**

2. **continuum approximation,**
3. **spin and relativistic effects are neglected,**
4. **band-electron has parabolic dispersion,**
5. **longitudinal optical phonons to be long-wavelength phonons with constant frequency, ω_0.**

Over the years the original polaron concept has been generalized to include polarization fields other than the LO-phonon field. The continuum approximation is not appropriate for some materials when the polarization is confined to a region of the order of a unit cell. In such a case, the so-called small polaron is a more adequate quasi-particle. Usually, Fröhlich-polarons or large polarons are understood for quasi particles consisting of the electron (or hole) and the polarization due to the LO-phonons. For large polarons, the large-polaron wave functions and corresponding lattice distortions spread over many lattice sites. However, the term Landau- or Pekar- or Landau-Fröhlich-polaron would be a more appropriate appellation. It is worth noting that in a perfect lattice, the self-trapping is never complete. In addition, for a finite phonon frequency, ion polarization can follow polaron motion if motion is sufficiently slow. So, a large polaron with a low kinetic energy propagates through a crystal lattice as a free electron with enhanced effective mass.

18.8.1 Polaron Asymptotic Characteristics

In the present example we wish to compute the energy and effective mass of the polaron (electron dressed with polarization and moving in the phonon cloud). In this polaron example we will be interested in the region described by the almost constant part of the optical branch near zero wave vector. This is the branch for which positive ions move in a direction opposite to that of the neighboring ions. For convenience of computation of the polaron problem we consider dispersionless optical phonon modes with frequency ω_0 with the polaron variational energy being:

$$E=\frac{3}{4}\hbar v\left(1-\frac{1}{u}\right)^2-\frac{\pi^{\frac{3}{2}}V_0^2}{(2\pi)^3\omega_0}\int_0^\infty d\tau\frac{\exp\{-\hbar\omega_0\tau\}}{\sqrt{A_p(\tau)}} \tag{18.172}$$

Here,

$$A_p(\tau)=\frac{\hbar a_2}{2mv}\left(1-\exp\{-\hbar v\tau\}\right)+\frac{\hbar^2 a_1\tau}{2m} \tag{18.173}$$

For strong coupling polarons the fictitious particle is more massive than the electron (hole),

$$M>>m \tag{18.174}$$

So, the strong coupling polaron energy is

$$a_1\to 0, a_2=1-a_1\to 1 \tag{18.175}$$

$$A_p(\tau)=\frac{\hbar}{2mv}\left(1-\exp\{-\hbar v\tau\}\right)\to\frac{\hbar}{2mv} \tag{18.176}$$

and

$$E=\frac{3}{4}\hbar v-\frac{\pi^{\frac{3}{2}}V_0^2}{(2\pi)^3\omega_0^2}\left(\frac{2m}{\hbar}\right)^{\frac{1}{2}}v^{\frac{1}{2}} \tag{18.177}$$

So, from

$$\frac{dE}{dv}=0 \tag{18.178}$$

then for the 3D case [1,42], the polaron strong coupling energy has the form

$$E=-\frac{1}{3\pi}\alpha_F^2\hbar\omega_0 \cong -0.1061\hbar\omega_0\alpha_F^2 \tag{18.179}$$

18.9 Polaron Characteristics in a Quasi-1D Quantum Wire

We examine the polaron problem now in a cylindrical geometry and, in particular, a quasi-1D quantum wire where the motion of the electron, interacting with lattice vibrations is subjected to a transversal parabolic confinement potential. That favors the reduction of the lower bound of the electron-phonon coupling constant's α_F, threshold value in the quasi-1D quantum wire to within weak or intermediate-coupling observed in real crystals. The physical properties of the polaron are different from those of the band electron and depend on this electron–lattice interaction strength α_F with the quantities of interest being the ground-state energy and the effective mass. Some other quasi-particle characteristics of the polaron such as the mobility, the impedance and so on will be examined under a different chapter. Though a great number of studies are devoted to the ground state energy and the effective mass of the polaron, the problem of their derivation simultaneously by one method is very important. This is because, if a better upper bound for the polaron ground state energy is obtained by this method, then likely we get a better effective mass directly from the same method. We investigate this for the quasi 1D quantum wire with transversal parabolic confinement potential.

18.9.1 Hamiltonian of the Electron in a Quasi 1D Quantum Wire

The motion of the electron, interacting with lattice vibrations and subjected to a parabolic potential, is described by the Hamiltonian [2]:

$$\hat{\mathrm{H}}=\frac{\hat{p}_\perp^2+\hat{p}_\parallel^2}{2m}+\frac{m\Omega^2\rho^2}{2}+\sum_{\vec{\kappa}}\hbar\omega_{\vec{\kappa}}\hat{b}_{\vec{\kappa}}^\dagger\hat{b}_{\vec{\kappa}}+\frac{\hbar}{\sqrt{V}}\sum_{\vec{\kappa}}\left(C_{\vec{\kappa}}\hat{b}_{\vec{\kappa}}\exp\{i\vec{\kappa}\vec{r}\}+h.c.\right) \tag{18.180}$$

The parameter, Ω, characterizes the parabolic confinement potential and selected for technological reasons and aid as well for path integration to be performed exactly. For the parabolic confinement potential, surface phonon modes are smoothly distributed in the bulk of the crystal and so we consider only bulk phonons. So, the parabolic confinement potential eliminates surface phonon modes. The position coordinate operator $\vec{r}$, $\rho=\left(x^2+y^2\right)^{\frac{1}{2}}$ of the electron with effective mass m; $\hat{b}_{\vec{\kappa}}^\dagger$ $(\hat{b}_{\vec{\kappa}})$ are the creation (and annihilation) operators for longitudinal optical phonons of wave vector $\vec{\kappa}$ and phonon frequency mode $\omega_{\vec{\kappa}}$. The quantity

$$C_{\vec{\kappa}}=\left(\frac{4R_p\pi\alpha_F\omega_0}{\left(\kappa_\perp^2+\kappa_\parallel^2\right)}\right)^{\frac{1}{2}} \tag{18.181}$$

are Fourier components of the electron–phonon coupling amplitude.

The quantities, $\hbar\kappa_\perp$ and $\hbar\kappa_\parallel$, are, respectively, the components of the momentum of the electron in the transversal and longitudinal directions of the wire. In our problem, the electronic state is described by the variational wave function for the transversal motion of the electron in the strong confinement limit:

$$\Psi_0 = A\exp\left\{-\frac{\lambda\rho^2}{2}\right\} \tag{18.182}$$

Here λ is a variational parameter and A, a normalization constant found from the condition of normalization of the wave function:

$$A = \left(\frac{\lambda}{\pi}\right)^{\frac{1}{2}} \tag{18.183}$$

Averaging the Hamiltonian 18.180 by the wave function 18.182 we have

$$\tilde{\text{H}} = \frac{\hat{p}_\parallel^2}{2m} + \sum_{\vec{\kappa}} \hbar\omega_{\vec{\kappa}} \hat{b}_{\vec{\kappa}}^\dagger \hat{b}_{\vec{\kappa}} + \frac{\hbar}{\sqrt{V}} \sum_{\vec{\kappa}} \left(C_{\vec{\kappa}} \exp\{i\vec{\kappa}_\perp\vec{\varrho}\} \hat{b}_{\vec{\kappa}} \exp\{i\kappa_\parallel z\} + h.c. \right) \tag{18.184}$$

Considering that

$$\exp\{i\vec{\kappa}_\perp\vec{\varrho}\} = \exp\left\{-\frac{\kappa_\perp^2}{4\lambda}\right\} \tag{18.185}$$

then 18.184 becomes

$$\tilde{\text{H}} = \frac{\hat{p}_\parallel^2}{2m} + \sum_{\vec{\kappa}} \hbar\omega_{\vec{\kappa}} \hat{b}_{\vec{\kappa}}^\dagger \hat{b}_{\vec{\kappa}} + \frac{\hbar}{\sqrt{V}} \sum_{\vec{\kappa}} \left(\widetilde{C_{\vec{\kappa}}} \hat{b}_{\vec{\kappa}} \exp\{i\kappa_\parallel z\} + h.c. \right) \tag{18.186}$$

where the amplitude of the electron–phonon coupling is modified:

$$\widetilde{C_{\vec{\kappa}}} = C_{\vec{\kappa}} \exp\left\{-\frac{\kappa_\perp^2}{4\lambda}\right\} \tag{18.187}$$

18.9.1.1 Lagrangian of the Electron in a Quasi-1D Quantum Wire

We write the Lagrangian corresponding to the Hamiltonian in 18.184:

$$L(-i\hbar\tau) = -\frac{m\dot{z}^2}{2\hbar^2} - \frac{1}{2}\sum_{\vec{\kappa}} \left(\frac{\dot{q}_{\vec{\kappa}}^2}{\hbar^2} + \omega_{\vec{\kappa}}^2 q_{\vec{\kappa}}^2 \right) + \sum_{\vec{\kappa}} \gamma_{\vec{\kappa}}(\tau) q_{\vec{\kappa}}(\tau) \tag{18.188}$$

Here the electron coordinate is z, the longitudinal optical phonon generalized coordinates are $q_{\vec{\kappa}}$ for a frequency mode $\omega_{\vec{\kappa}}$ .The amplitude of the electron–phonon coupling interaction $\gamma_{\vec{\kappa}}(\tau)$ for the $\vec{\kappa}$ moving wave:

$$\gamma_{\vec{\kappa}}(\tau) = \frac{\hbar}{\sqrt{V}} \widetilde{C_{\vec{\kappa}}}(\tau) \exp\{i\kappa z\}, \kappa_\parallel \equiv \kappa \tag{18.189}$$

From the Feynman variational principle, the interaction of the electron with the polarized vibrations of the crystal is modelled by an elastic coupling of the electron and a hypothetical (fictitious) particle that attracts the electron to itself. The complex pair can now move in the direction of the wire axis. So, the effect of the polarized crystal lattice on the electron is approximated to an elastic attraction of the second particle. From these analyses, the model Lagrangian may be selected in the one oscillatory approximation:

$$L_0 = -\frac{m\dot{z}^2}{2\hbar^2} - \frac{M\dot{Z}_f^2}{2\hbar^2} - \frac{M\omega_f^2\left(Z_f - z\right)^2}{2} \tag{18.190}$$

Here M and ω_f are, respectively, the mass of a fictitious particle and the frequency of the elastic coupling serving as variational parameters; Z_f is the coordinate of the fictitious particle and $t = -i\hbar\tau$ with $\dot{z} = \frac{dz}{d\tau}$.

18.9.1.2 Partition function of the Electron in a Quasi-1D Quantum Wire

In order to find the polaron energy we proceed to find the following quantities:

$$E_0 = -T\ln Z_0 \tag{18.191}$$

where

$$Z_0 = \int_{-\infty}^{+\infty} \delta\left(z - z'\right)\delta\left(Z_f - Z'_f\right)dzdz'dZ_f dZ'_f \int_{z'}^{z}\int_{Z'_f}^{Z_f} \exp\left\{S_0\right\}DzDZ_f \tag{18.192}$$

The model Lagrangian may be conveniently represented in quadratic form by doing a change of variables:

$$\rho_1 = Z_f - z, \quad \rho_2 = \frac{zm + Z_f M}{m + M} \tag{18.193}$$

and

$$z = \rho_2 + \mu_1\rho_1, \quad Z_f = \rho_2 - \mu_2\rho_1 \tag{18.194}$$

with

$$\mu_1 = \frac{M}{m+M}, \quad \mu_2 = \frac{m}{m+M}, \mu = \frac{Mm}{m+M}, \quad v^2 = u^2\omega_f^2, \quad u^2 = \frac{m+M}{m} \tag{18.195}$$

The quantities ρ_1 and ρ_2 are, respectively, the coordinates of the relative motion and of the center of mass. The parameter μ is a reduced mass and v is a scaled frequency.

So,

$$S_0 = -\frac{M+m}{2\hbar^2}\int_0^\beta \dot{\rho}_2^2 d\sigma - \frac{\mu}{2\hbar^2}\int_0^\beta \dot{\rho}_1^2 d\sigma - \frac{\mu v^2}{2}\int_0^\beta \rho_1^2 d\sigma \tag{18.196}$$

The coordinate ρ_2 describes the free motion of the particle with mass $M + m$, while the coordinate ρ_1, the harmonic oscillator with frequency v. Hence the partition function in 18.192 now becomes

$$Z_0 = \int_{-\infty}^{+\infty} \prod_{i=1}^{2} d\rho_i d\rho'_i \delta\left(\rho_i - \rho'_i\right)\int_{\rho_i}^{\rho_i} \exp\left\{S_0\right\}D\rho_i \tag{18.197}$$

or

$$Z_0 = \frac{ul_z}{l_e(\hbar\beta)} \frac{1}{2\sinh\left(\frac{\beta\hbar\nu}{2}\right)} \quad (18.198)$$

Here, l_z is the length of the quantum wire and the **thermal de Broglie wavelength**:

$$l_e(\hbar\beta) \equiv \left(\frac{2\pi\hbar^2\beta}{m}\right)^{\frac{1}{2}} \quad (18.199)$$

From the Lagrangian of the full system

$$Z = \prod_{\vec{\kappa}} \int \delta\left(q_{\vec{\kappa}} - q'_{\vec{\kappa}}\right)\delta\left(z - z'\right) dq_{\vec{\kappa}} dq'_{\vec{\kappa}} dz dz' \int_{z'}^{z} \int_{q'_{\vec{\kappa}}}^{q_{\vec{\kappa}}} \exp\left\{\underline{S}\right\} Dz Dq_{\vec{\kappa}} \quad (18.200)$$

we have

$$Z = \prod_{\vec{\kappa}} Z_L \int \delta\left(z - z'\right) dz dz' \int_{z'}^{z} Dz \exp\left\{\int_0^\beta \left(-\frac{m\dot{z}^2}{2\hbar^2}\right) d\tau + \Phi_{\omega_{\vec{\kappa}}}\left[z\right]\right\} \quad (18.201)$$

where

$$Z_L = \left[2\sinh\left(\frac{\beta\hbar\omega_{\vec{\kappa}}}{2}\right)\right]^{-1} \quad (18.202)$$

is the lattice partition function and the functional of the electron–phonon interaction influence phase is:

$$\Phi_{\omega_{\vec{\kappa}}}\left[z\right] = \frac{\hbar}{4\omega_{\vec{\kappa}}} \int_0^\beta \int_0^\beta d\sigma d\sigma' \gamma\left(z(\sigma)\right) \gamma\left(z(\sigma')\right) \mathrm{F}_{\omega_{\vec{\kappa}}}\left(\left|\sigma - \sigma'\right|\right) \quad (18.203)$$

and

$$\mathrm{F}_{\omega_{\vec{\kappa}}}\left(\left|\sigma - \sigma'\right|\right) = \frac{\cosh\hbar\omega_{\vec{\kappa}}\left(\left|\sigma - \sigma'\right| - \frac{\beta}{2}\right)}{\sinh\left(\frac{\hbar\omega_{\vec{\kappa}}\beta}{2}\right)} \quad (18.204)$$

We may now write the full action functional of the full system:

$$\underline{S}\left[z\right] = -\int_0^\beta \frac{m\dot{z}^2}{2\hbar^2} d\tau + \Phi\left[z\right] \quad (18.205)$$

We may now write the influence phase of the full system:

$$\Phi\left[z\right] = \sum_{\vec{\kappa}} \Phi_{\omega_{\vec{\kappa}}}\left[z\right] \quad (18.206)$$

Consider

$$\gamma_{\vec{\kappa}}(\sigma')\gamma_{\vec{\kappa}}(\sigma) = \frac{\hbar^2}{V}\left|\widetilde{C_{\vec{\kappa}}}\right|^2 \exp\{i\kappa(z(\sigma) - z(\sigma'))\} \tag{18.207}$$

The influence phase of the electron–phonon interaction in 18.206 can then be rewritten

$$\Phi[z] = \frac{\hbar^2}{V}\sum_{\vec{\kappa}}\frac{\hbar}{8\omega_{\vec{\kappa}}}\left|\widetilde{C_{\vec{\kappa}}}\right|^2 \int_0^\beta\int_0^\beta d\sigma d\sigma' \exp\{i\kappa(z(\sigma) - z(\sigma'))\} \mathrm{F}_{\omega_{\vec{\kappa}}}(|\sigma - \sigma'|) \tag{18.208}$$

The action functional of the model system $\underset{\sim}{S}_0$ is obtained via the partition function of the model system:

$$Z_0 = Z_{\omega_f}\int\delta(z - z')\int dz dz' \int_{z'}^{z} Dz \exp\left\{\int_0^\beta\left[-\frac{m\dot{z}^2}{2\hbar^2} + \frac{M\omega_f^2}{2}z^2\right]d\tau + \Phi_{\omega_f}[z]\right\} \tag{18.209}$$

Here

$$Z_{\omega_f} = \left[2\sinh\left(\frac{\beta\hbar\omega_f}{2}\right)\right]^{-1} \tag{18.210}$$

and

$$\Phi_{\omega_f}[z] = \frac{\hbar M\omega_f^3}{4}\int_0^\beta\int_0^\beta d\tau d\sigma z(\sigma)z(\sigma')\mathrm{F}_{\omega_f}(|\tau - \sigma|) \tag{18.211}$$

So,

$$\underset{\sim}{S}_0 = \int_0^\beta\left[-\frac{m\dot{z}^2}{2\hbar^2} - \frac{M\omega_f^2}{2}z^2\right]d\tau + \Phi_{\omega_f}[z] - \ln\left[2\sinh\left(\frac{\beta\hbar\omega_f}{2}\right)\right] \tag{18.212}$$

and

$$\underset{\sim}{S} - \underset{\sim}{S}_0 = -\int_0^\beta \frac{M\omega_f^2}{2}z^2 d\tau + \sum_{\vec{\kappa}}\Phi_{\omega_{\vec{\kappa}}}[z] - \Phi_{\omega_f}[z] + \ln\left[2\sinh\left(\frac{\beta\hbar\omega_f}{2}\right)\right] - \ln\left[2\sinh\left(\frac{\beta\hbar\omega_{\vec{\kappa}}}{2}\right)\right] \tag{18.213}$$

18.10 Polaron Generating Function

Knowledge of the full action functional $\underset{\sim}{S}$ is not sufficient to calculate the polaron characteristics easily due to retardation effects. This is overcome by introducing an optimal variational action functional $\underset{\sim}{S}_0$ that will minimize the following equation for the polaron characteristics:

$$\ln Z \geq \ln Z_0 + \left\langle \underset{\sim}{S} - \underset{\sim}{S}_0 \right\rangle_{\underset{\sim}{S}_0} \tag{18.214}$$

To do this, it is necessary to evaluate

$$\left\langle z^2(\sigma)\right\rangle_{\underset{\sim}{S}_0}, \quad \left\langle z(\sigma')z(\sigma)\right\rangle_{\underset{\sim}{S}_0}, \quad \left\langle \exp\{i\kappa(z(\sigma) - z(\sigma'))\}\right\rangle_{\underset{\sim}{S}_0}, \kappa_\parallel \equiv \kappa \tag{18.215}$$

with the help of the generating function:

$$\Psi_{\vec{\kappa}}(\xi,\eta)=\left\langle \exp\left\{i\left(\kappa\left(\xi z(\tau)-\eta z(\sigma)\right)\right)\right\}\right\rangle_{S_0} \quad (18.216)$$

For example, we conveniently calculate the following by the functional derivatives with respect to ξ and η brought out of the functional integral over the electronic as well the fictitious particle variables:

$$\left\langle z(\sigma)z(\sigma')\right\rangle_{S_0}=\frac{1}{\kappa^2}\frac{\partial\Psi}{\partial\xi\partial\eta},\left\langle \exp\left\{i\kappa\left(z(\sigma)-z(\sigma')\right)\right\}\right\rangle_{S_0}=\Psi_{\kappa}(1,1) \quad (18.217)$$

We find the expression of the generating function by expanding the variables in quadratic form. This may be done by making a change of variables considering the motion of the center of mass and relative motion as seen earlier. This enables us to have two driven harmonic oscillators with frequencies $\nu_0 = 0$, $\nu_1 = \nu$. The generating function is conveniently obtained as follows:

$$\Psi_{\vec{\kappa}}(\xi,\eta)=\prod_{i=0,1}\exp\left\{\frac{\hbar}{4m_i\nu_i}\int_0^\beta\int_0^\beta dtds\gamma_i(t)\gamma_i(s)F_{\nu_i}\left(|t-s|\right)\right\} \quad (18.218)$$

where

$$\gamma_0(t)=i\kappa\left[\xi\delta(t-\sigma')-\eta\delta(t-\sigma)\right],\ \gamma_1(t)=\frac{M}{m+M}i\kappa\left[\xi\delta(t-\sigma')-\eta\delta(t-\sigma)\right] \quad (18.219)$$

$$\nu_0=0,\quad \nu_1=\nu,\quad m_0=M+m,\quad m_1=\frac{mM}{m+M} \quad (18.220)$$

If we consider

$$F_{\nu_0}(0)\cong\frac{2}{\beta\hbar\nu_0}+\frac{\beta\hbar\nu_0}{6}+\cdots,F_{\nu_0}\left(|\sigma-\sigma'|\right)\cong\frac{2}{\beta\hbar\nu_0}+\frac{\hbar\nu_0}{\beta}\left[|\sigma-\sigma'|-\frac{\beta}{2}\right]-\frac{1}{12}\beta\hbar\nu_0+\cdots \quad (18.221)$$

$$\Psi_{\kappa}(\xi,\eta)=\exp\left\{\begin{array}{l}-\frac{\hbar\kappa^2a_2}{4m\nu}\left[\left(\xi^2+\eta^2\right)F_\nu(0)-2\xi\eta F_\nu\left(|\sigma-\sigma'|\right)\right]-\\ -\frac{\hbar^2\kappa^2a_1\beta}{4m}\left[\frac{1}{6}\left(\xi^2+\eta^2\right)-\frac{2\xi\eta}{\beta^2}\left(|\sigma-\sigma'|-\frac{\beta}{2}\right)^2+\frac{1}{6}\xi\eta\right]\end{array}\right\} \quad (18.222)$$

$$a_1=\frac{1}{u^2},\quad a_2=1-\frac{1}{u^2},\quad u^2=\frac{M+m}{m} \quad (18.223)$$

18.11 Polaron Asymptotic Characteristics

The generating function in 18.222 is observed to be devoid of the electronic and fictitious particles variables but dependent on the field variables ξ and η. So, to find the polaron characteristics we calculate first the expressions

$$\left\langle z(\sigma')z(\sigma)\right\rangle_{S_0}=\frac{\hbar a_2}{2m\nu}F_\nu\left(|\sigma-\sigma'|\right)+\frac{\hbar^2a_1\beta}{2m}\left[\frac{1}{4}-\frac{2}{\beta^2}\left(|\sigma-\sigma'|-\frac{\beta}{2}\right)^2\right] \quad (18.224)$$

and

$$\left\langle \exp\left\{ i\kappa\left(z(\sigma)-z(\sigma')\right)\right\}\right\rangle_{S_0} = \exp\left\{-\kappa^2 A\right\} \tag{18.225}$$

where,

$$A = \frac{\hbar a_2}{2mv}\left[F_v(0) - F_v\left(|\sigma-\sigma'|\right)\right] + \frac{\hbar^2 a_1 \beta}{2m}\left[\frac{1}{4} - \frac{1}{\beta^2}\left(|\sigma-\sigma'| - \frac{\beta}{2}\right)^2\right] \tag{18.226}$$

We then substitute these expressions into 18.214 for the ground state polaron characteristics considering $\beta \to 0$ and we have

$$\ln\left(\frac{Z_F}{Z_L}\right) = \ln\left[\frac{ul_z}{l_e(\hbar\beta)}\right] - \frac{\hbar v \beta}{4}\left(1-\frac{1}{u}\right)^2 - \frac{u^2-1}{2u^2} + \beta\frac{\hbar^2}{V}\sum_{\vec{\kappa}}\frac{\left|\widetilde{C_{\vec{\kappa}}}\right|^2 \hbar}{8\omega_{\vec{\kappa}}}\int_0^\beta\int_0^\beta d\sigma' d\sigma F\left(|\sigma-\sigma'| - \frac{\beta}{2}\right) \tag{18.227}$$

where

$$F\left(|\sigma-\sigma'| - \frac{\beta}{2}\right) \equiv F_{\omega_{\vec{\kappa}}}\left(|\sigma-\sigma'|\right)\exp\left\{-\kappa^2 A\right\} \tag{18.228}$$

The double integral in 18.227 describes a non-local contribution where the system trajectory interacts with itself and this self-interaction is mediated by the environment (phonon cloud) as indicative from the factor $\left|\widetilde{C_{\vec{\kappa}}}\right|^2$. This shows the polaron problem mimics a one-particle problem where the interaction, non-local in time or "**retarded**," occurs between the electron and itself. The formulae depend on the quantity $|\sigma - \sigma'|$ and is an indication that the quantities (retarded functions) depend on the past. The significance of interaction with the past is that the perturbation due to the moving electron (hole) takes "**time**" to propagate in the crystal. From the expression of the full action functional, the potential which the electron feels at any "**time**" depends on its position at previous times. This implies the effect that the electron has on the crystal propagates at a finite velocity and can make itself felt on the electron at a later time.

We observe that the functions in the integrand 18.227 have a common argument $\left(|\sigma-\sigma'| - \frac{\beta}{2}\right)$ and for this reason we generalize the function in the integrand in 18.227 to be $G\left(|\sigma-\sigma'| - \frac{\beta}{2}\right)$ (retarded or advanced function) to aid in the simplification of the evaluation of the integral. So, we look at the following function that yields the value of the integral for a given $G\left(|\sigma-\sigma'| - \frac{\beta}{2}\right)$ function [1,2] (**retarded or advanced function twofold integral law that mimics Stokes' theorem**):

$$\int_0^\beta\int_0^\beta F\left(|\sigma-\sigma'| - \frac{\beta}{2}\right) d\sigma d\sigma' = \beta^2\int_0^1 F\left(\frac{\beta y}{2}\right) dy \tag{18.229}$$

Here,

$$F\left(\frac{\beta y}{2}\right) = \frac{\cosh\dfrac{\beta y \hbar\omega_{\vec{\kappa}_j}}{2}}{\sinh\dfrac{\hbar\omega_{\vec{\kappa}_j}\beta}{2}}\exp\left\{-\kappa_j^2 A\right\} \tag{18.230}$$

and

$$\mathrm{A} = \frac{\hbar a_2}{2mv}\left(\coth\frac{\hbar v\beta}{2} - \frac{\cosh\frac{\beta y\hbar v}{2}}{\sinh\frac{\hbar v\beta}{2}}\right) + \frac{\hbar^2\beta a_1}{8m}\left(1-y^2\right) \tag{18.231}$$

We examine the polaron parameters for low temperatures $\beta \to \infty$:

$$\lim_{\beta\to\infty}\coth\frac{\hbar v\beta}{2} \to 1, \lim_{\beta\to\infty}\frac{\cosh\frac{\beta y\hbar\omega_{\vec{\kappa}_j}}{2}}{\sinh\frac{\hbar\omega_{\vec{\kappa}_j}\beta}{2}} \to \exp\left\{-\frac{\hbar\omega_{\vec{\kappa}_j}\beta}{2}(1-y)\right\} \tag{18.232}$$

So,

$$\mathrm{F}\left(\frac{\beta y}{2}\right) = \exp\left\{-\frac{\hbar\omega_{\vec{\kappa}_j}\beta}{2}(1-y)\right\}\exp\left\{-\kappa_j^2\mathrm{A}\right\} \tag{18.233}$$

where,

$$\mathrm{A} = \frac{\hbar a_2}{2mv}\left(1-\exp\left\{-\frac{\hbar v\beta}{2}(1-y)\right\}\right) + \frac{\hbar^2\beta a_1}{8m}\left(1-y^2\right) \tag{18.234}$$

Since the most important role in the interaction of the electron with the polar vibrations is played by long wavelength vibrations having a small dispersion, then for this reason, we set $\omega_{\vec{\kappa}_j}$ to be equal to the limiting dispersionless optical phonon mode with frequency ω_0.

As seen earlier, we do the following change of variables:

$$\frac{\beta}{2}(1-y) \equiv -\tau, d\tau = -\frac{\beta}{2}dy, 0 \le \tau < \infty \tag{18.235}$$

and 18.229 takes the form:

$$f(\beta) = 2\beta\int_0^\infty d\tau \exp\{-\hbar\omega_0\tau\}\exp\left\{-\kappa^2\mathrm{A}\right\} \tag{18.236}$$

Here,

$$\mathrm{A} \equiv \mathrm{A}_p - \frac{\hbar^2 a_1\tau^2}{2m\beta}, \mathrm{A}_p = \frac{\hbar a_2}{2mv}\left(1-\exp\{-\hbar v\tau\}\right) + \frac{\hbar^2 a_1\tau}{2m} \tag{18.237}$$

For $\beta \to \infty$, then

$$\exp\left\{\frac{\hbar^2 a_1\tau^2\kappa^2}{2m\beta}\right\} \cong 1 + \frac{\hbar^2 a_1\tau^2\kappa^2}{2m\beta} \tag{18.238}$$

So,

$$f(\beta) = 2\beta\int_0^\infty d\tau \exp\{-\hbar\omega_0\tau\}\exp\left\{-\kappa^2\mathrm{A}_p\right\}\left(1+\frac{\hbar^2 a_1\tau^2\kappa^2}{2m\beta}\right) \tag{18.239}$$

Considering

$$\lim_{\beta\to\infty} \coth\frac{\hbar\nu\beta}{2} = 1 \tag{18.240}$$

and 18.227 then
for $\beta \to 0$, and from 18.214 we have

$$\ln\left(\frac{Z_F}{Z_L}\right) = \ln\left[\frac{ul_z}{l_e(\hbar\beta)}\right] - \beta\left[\frac{\hbar\nu}{4}\left(1-\frac{1}{u}\right)^2 - \sum_{\vec{\kappa}}\frac{\left|\widetilde{C}_{\vec{\kappa}}\right|^2\hbar^3}{4V\omega_0}\int_0^\infty d\tau\exp\{-\hbar\omega_0\}\exp\{-\kappa^2 A_p\}\right]$$

$$+\left[-\frac{u^2-1}{2u^2} + \sum_{\vec{\kappa}}\frac{\hbar^5\kappa^2\left|\widetilde{C}_{\vec{\kappa}}\right|^2 a_1}{8Vm\omega_0}\int_0^\infty d\tau\tau^2\exp\{-\hbar\omega_0\}\exp\{-\kappa^2 A_p\}\right] \tag{18.241}$$

From 18.6 and 18.241 as well as $\beta \to \infty$ ($T \to 0$), the polaron variational energy as well as the effective mass respectively may be obtained as follows [2]:

$$E_p = \frac{\hbar\nu}{4}\left(1-\frac{1}{u}\right)^2 - \sum_{\vec{\kappa}}\frac{\left|\widetilde{C}_{\vec{\kappa}}\right|^2\hbar^3}{4V\omega_0}\int_0^\infty d\tau\exp\{-\hbar\omega_0\}\exp\{-\kappa^2 A_p\} \tag{18.242}$$

$$\frac{m_p}{m} = \exp\left\{-\frac{u^2-1}{2u^2} + \sum_{\vec{\kappa}}\frac{\hbar^5\kappa^2\left|\widetilde{C}_{\vec{\kappa}}\right|^2 a_1}{8Vm\omega_0}\int_0^\infty d\tau\tau^2\exp\{-\hbar\omega_0\}\exp\{-\kappa^2 A_p\}\right\} \tag{18.243}$$

18.12 Strong Coupling Regime Polaron Characteristics

We evaluate the asymptotic expansions of Feynman's polaron characteristics in 18.242 and 18.243. For the strong coupling regime in our problem the transversal polaron energy $\varepsilon_\perp$ is greater than the energy $\varepsilon(\nu)$ of oscillatory motion:

$$a_1 \to 0, a_2 = 1 - a_1 \to 1, A = \frac{\hbar}{2m\nu}, 2A\lambda = \frac{\hbar\varepsilon_\perp}{2\varepsilon(\nu)} > 1, \varepsilon_\perp = \frac{\hbar^2\lambda}{2m} > \varepsilon(\nu) = \hbar\nu \tag{18.244}$$

From here, the polaron energy takes the form

$$E_p = \frac{\varepsilon(\nu)}{4} - \alpha_F\left(\frac{\hbar\omega_0\varepsilon(\nu)}{4\pi}\right)^{\frac{1}{2}}\ln\left(\frac{4\varepsilon_\perp}{\varepsilon(\nu)}\right) \tag{18.245}$$

Extremizing 18.245 over the energy of the oscillatory motion $\varepsilon(\nu)$ we have

$$\frac{dE_p}{d\varepsilon(\nu)} = 0 \tag{18.246}$$

We write the solution of equation 18.246 in the form [2]:

$$-y = q\ln y, y = \frac{e}{2}\left(\frac{\varepsilon(v)_0}{\varepsilon_{\perp 0}}\right)^{\frac{1}{2}}, q = \frac{e}{\pi^{\frac{1}{2}}}\alpha_F\left(\frac{\hbar\omega_0}{\varepsilon_{\perp 0}}\right)^{\frac{1}{2}} \tag{18.247}$$

Here, $\varepsilon(v)_0$ and $\varepsilon_{\perp 0}$ are the extremal parameters, respectively, of $\varepsilon(v)$ and $\varepsilon_\perp$ corresponding to the minimum energy at the ground state 18.245 for $\beta \to \infty$. The solution of 18.247 can be obtained via iteration method [2]:

$$y_1 = +q, y_2 = -q\ln(+q) \tag{18.248}$$

$$y_3 = -q\ln\left[-q\ln(+q)\right], \cdots, y_n = -q\ln\left[-q\ln\left[-q\ln\cdots\ln q\right]\right] \tag{18.249}$$

From here,

$$y = qf(q), f(q) \equiv -q\ln\left[-q\ln\left[-q\ln\cdots\ln q\right]\right] \tag{18.250}$$

and letting

$$\frac{\varepsilon_{\perp 0}}{\hbar\omega_0} = \varepsilon_\perp, \frac{\overline{E_p}}{\hbar\omega_0} = E_p \tag{18.251}$$

then the minimum dimensionless polaron energy [2]:

$$E_p = -\frac{\alpha_F^2}{\pi} f(q)\ln\left(\frac{e(\pi\varepsilon_\perp)^{\frac{1}{2}}}{\alpha_F f(q)}\right) \tag{18.252}$$

or

$$E_p = -\frac{\alpha_F^2}{\pi} f(q)\ln\left(\frac{e^2}{qf(q)}\right) \tag{18.253}$$

Since, $f(q) > 1$ and $\varepsilon_\perp > 1$, then the values of α_F are greater than those in the 3D case:

$$E_p = -\frac{1}{3\pi}\hbar\omega_0\alpha_F^2 \tag{18.254}$$

For the strong coupling regime, the polaron is characterized by Franck-Condon (F. C.) excited states that correspond to the electron excitations in the potential adapted to the ground state with the energy of the lowest F. C. state within the Produkt-Ansatz [51]:

$$E_p = -\frac{1}{9\pi}\hbar\omega_0\alpha_F^2 = -0.0354\hbar\omega_0\alpha_F^2 \tag{18.255}$$

When the lattice polarization is allowed to relax or adapt to the electronic distribution of the excited electron than itself then adapts its wave function to the new potential, this leads to a self-consistent final state. This results in the so-called relaxed excited state [42,52]:

$$E_p = -0.041\hbar\omega_0\alpha_F^2 \tag{18.256}$$

The dimensionless effective polaron mass is obtained by substituting in 18.243 the values of $\varepsilon(v)_0$ and $\varepsilon_{\perp 0}$ for which the energy 18.242 is minimum [2]:

$$\frac{m_p}{m} = \frac{4\alpha_F^4}{\pi^2}\ln^2\left(\frac{\pi\varepsilon_\perp}{\alpha_F^2}\right) \tag{18.257}$$

In the general case,

$$\frac{m_p}{m} = 200\times 10^{-4}\alpha_F^4 \tag{18.258}$$

Since,

$$\frac{4}{\pi^2}\ln^2\left(\frac{\pi\varepsilon_\perp}{\alpha_F^2}\right) \geq 1 \tag{18.259}$$

then we observe that the effective polaron mass in the quasi-1D quantum wire is twice more than that in the 3D case. The strong-coupling polaron mass within the Produkt-Ansatz is given as:

$$\frac{m_p}{m} = 1 + 0.0200\alpha_F^4 \tag{18.260}$$

In another strong coupling limit [1], Feynman found the energy to be:

$$E_p = -0.1061\hbar\omega_0\alpha_F^2 - 2.83\hbar\omega_0 \tag{18.261}$$

where the last term in the energy accounts for the weak coupling limit. The last term in 18.261 relates two effects:

- **change in the system's degree of freedom;**
- **3*N* electrons relate 3 degrees of freedom of the crystal and also relate an arbitrary polarization that dresses the electron leading to the creation of the phonon cloud.**

Some strong-coupling expansions for ground state energy and effective mass respectively are obtained [53]:

$$E_p = -0.108513\hbar\omega_0\alpha_F^2 - 2.836\hbar\omega_0 \tag{18.262}$$

$$\frac{m_p}{m} = 1 + 0.0227019\alpha_F^4 \tag{18.263}$$

In another weak coupling limit, [1] the electron (hole) is almost completely free and

$$E_p = -\alpha_F\hbar\omega_0 - 0.0123\alpha_F^2\hbar\omega_0 \tag{18.264}$$

Fröhlich [38] and Lee, Low and Pines (L.L.P) [54] found the weak coupling polaron energy and the polaron effective mass, respectively:

$$E_p = -\alpha_F \hbar\omega_0, \frac{m_p}{m} = \frac{1}{1 - \frac{\alpha_F}{6}} \tag{18.265}$$

These results are exactly correct to the first order in α_F for the weak coupling regime.

Fröhlich provided the first weak-coupling perturbation-theory results while Lee, Low and Pines used two elegant successive canonical transformation formulations for its derivation. The Lee, Low and Pines method that puts the Fröhlich result on a variational basis has often been called **intermediate-coupling theory** although its range of validity is in principle not larger than that of the weak-coupling approximation. In his article on polarons, Feynman [34] found the following higher-order weak-coupling expansions:

$$E_p = -\alpha_F \hbar\omega_0 - 0.0123\alpha_F^2 \hbar\omega_0 - 0.00064\alpha_F^3 \hbar\omega_0 - \cdots \tag{18.266}$$

$$\frac{m_p}{m} = 1 + \frac{\alpha_F}{6} + 0.025\alpha_F^2 + \cdots \tag{18.267}$$

The strong-coupling variational calculations of Landau and Pekar [35,37] obtained the analytic polaron energy and effective mass respectively:

$$E_p = -0.1\alpha_F^2 \hbar\omega_0, \frac{m_p}{m} = 1 + 0.02\alpha_F^2 \tag{18.268}$$

The analytic form of the polaron energy is observed to be asymptotically correct. The polaron effective mass expression is obtained considering an alternative definition of the effective mass equivalent to Fröhlich's definition only when the trial ground state wave function is an exact eigenfunction of the total wave vector which is not the case for the Landau-Pekar approximation trial wave function. Since then more exact coefficients in this expansion have been obtained in a lot of theoretical works. The Feynman polaron characteristics are in accordance with the LLP weak coupling regime and also with the Landau-Pekar strong coupling regime and possess a smooth transitional behavior for the intermediate coupling regime. For all values of the coupling constant, α_F, the Feynman polaron self-energy is an upper-bound approximation that is less than, or approximately equal to, all known polaron self-energy upper bounds. So, the Feynman approach is very effective when calculating the polaron ground-state energy for all coupling strengths.

We observe from our evaluation that instead of the electron being coupled to the lattice, it is coupled by some "**spring**" to another particle and the resultant pair of particles would be free to move in the crystal. The effect of the transversal electronic confinement potential can be deduced from Figures 18.2 and 18.3, respectively, for the polaron energy and the polaron effective mass. It is deduced from these figures that for confinement enhancement, the domain of weak coupling polaron tends to overlap with that of the strong coupling polaron. For confinement enhancement (reducing radius of quantum wire) as well as increasing electron–phonon coupling constant, α_F, the energy and the effective mass has a monotonic increase. For $R \to 0$, the domain of the weak coupling polaron regime completely disappears.

18.13 Bipolaron Characteristics in a Quasi-1D Quantum Wire

18.13.1 Introduction

We study the bipolaron coupling energy, $W = 2E_p - E_{bip}$ as well as the bipolaron energy E_{bip} tailored by the radius of the quasi-1D quantum wire as well as the Fröhlich electron–phonon coupling constant, α_F

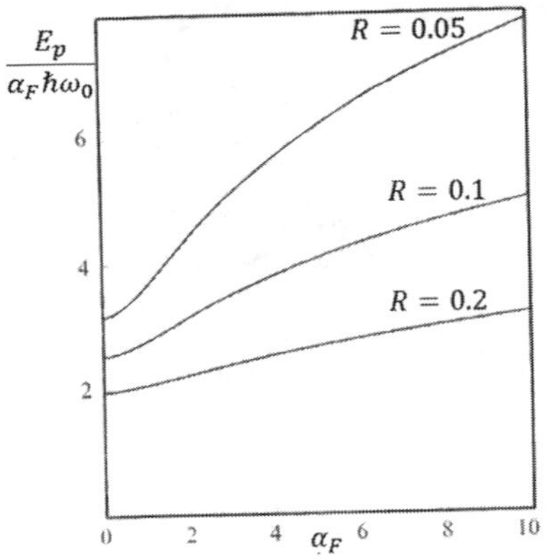

FIGURE 18.2 Depicts the dimensionless polaron ground state energy, $\frac{E_p}{\alpha_F \hbar\omega_0}$ versus the electron-phonon interaction coupling constant, $\boldsymbol{\alpha}_F$ for different radii of the quantum wire taken from reference [2].

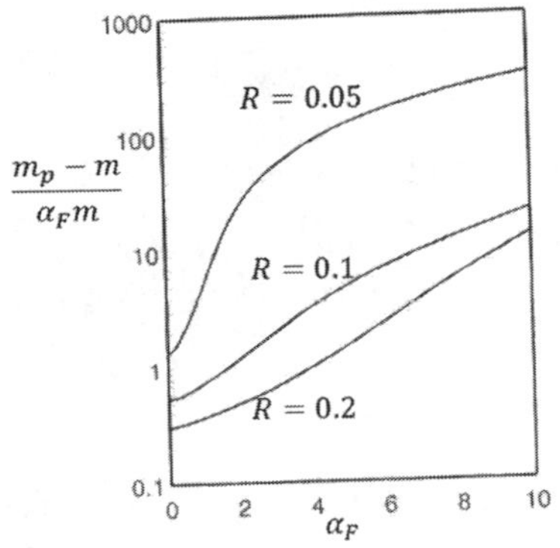

FIGURE 18.3 Depicts the dimensionless polaron effective mas, $\frac{m_p - m}{\alpha_F m}$ versus the electron-phonon interaction coupling constant, $\boldsymbol{\alpha}_F$ for different radii of the quantum wire taken from reference[2].

. The **bipolaron is a stable pair of two polarons [33,55]**. From literature, we know that when the dimension of the structure is reduced as

$$3D \to 2D \to 1D \to 0D \tag{18.269}$$

the polaronic effects are enhanced and the values of α_F are shifted to to within weak or intermediate-coupling range (realistic for strong ionic crystals such as alkali halides). However, the number of independent directions of charge transport is reduced for this case. For this reason, the quasi-1D quantum wire represents a great interest as a system with the maximal polaronic effect with the condition that charge transport at large distances is maintained. So, the quasi-1D quantum wire seems to be promising for the existences in it strong bipolarons and the superconductivity bipolaron effects.

We examine the quasi-1D quantum wire with the electronic confinement provided by the cylindrical-symmetrical parabolic confinement potential. The effective radius

$$R = \left(\frac{\hbar}{2m\Omega}\right)^{\frac{1}{2}} \tag{18.270}$$

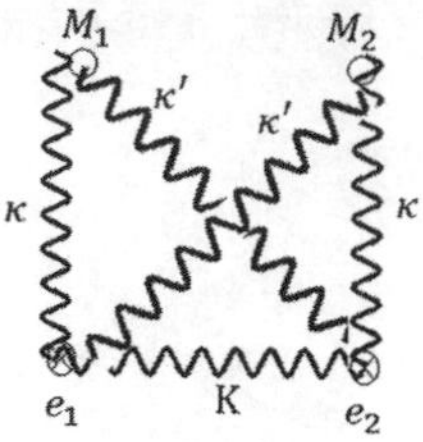

FIGURE 18.4 Showing the diagrammatic representation of the bipolaron with $\boldsymbol{e}_1$ and $\boldsymbol{e}_2$, being the first and second electron; $\boldsymbol{\kappa}$, $\boldsymbol{\kappa}'$, K are the elastic constants linked to the particles; $\boldsymbol{\kappa}$ and $\boldsymbol{\kappa}'$ attract the particles to each other and K is responsible for the electron repulsion.

may be observed to tailor the bipolaron effects as was for the case of the polaron. Here, Ω , is the frequency characterizing the parabolic confinement potential. We consider that the interaction between the electrons is described by *3D* Coulomb potential and the polar interactions by *3D* phonons. For the Coulomb repulsion between electrons in the bipolaron to disappear,

$$W = 0 \tag{18.271}$$

18.13.2 Bipolaron Diagrammatic Representation

In the bipolaron problem we assume as we did for the case of the polaron, the interaction of the two electrons with the lattice vibrations is modulated by two fictitious particles each with mass (Figure 18.4):

$$M = M_1 = M_2 \tag{18.272}$$

For brevity in Figure 18.4 we depict all particles on a plane, i.e., e_1 and e_2, the first and second electron, κ, κ', K are the elastic constants linked to the particles; κ and κ' attract the particles to each other and K is responsible for the electron repulsion. So, in our problem we have four variational parameters κ, κ', K and M.

18.13.3 Bipolaron Lagrangian

The model Lagrangian for the bipolaron problem is selected in the one oscillatory approximation:

$$L_0 = \sum_{i=1}^{2}\left(-\frac{m\dot{r}_i^2}{2\hbar^2} - \frac{M\dot{R}_i^2}{2\hbar^2} - \frac{m\Omega^2 r_i^2}{2} - \frac{\kappa\left(\vec{r}_i - \vec{R}_i\right)^2}{2} \right) - \sum_{i\neq k}^{2} \frac{\kappa'\left(\vec{r}_i - \vec{R}_k\right)^2}{2} + \frac{\mathrm{K}\left(\vec{r}_1 - \vec{r}_2\right)^2}{2} \tag{18.273}$$

A typical bipolaron Hamiltonian has two polarons and a Coulomb repulsion of electrons:

$$\hat{\mathrm{H}} = \frac{\hat{p}_1^2}{2m} + \frac{\hat{p}_2^2}{2m} + U\left(\left|\vec{r}_1 - \vec{r}_2\right|\right) + \frac{m\Omega^2\left(r_1^2 + r_2^2\right)}{2} + \sum_{\vec{\kappa}} \hbar\omega_{\vec{\kappa}} \hat{b}_{\vec{\kappa}}^{\dagger} \hat{b}_{\vec{\kappa}}$$
$$+ \frac{\hbar}{\sqrt{V}} \sum_{\vec{\kappa}} \left(\left[C_{\vec{\kappa}} \hat{b}_{\vec{\kappa}} \left(\exp\{i\vec{\kappa}\vec{r}_1\} + \exp\{i\vec{\kappa}\vec{r}_2\} \right) \right] + \text{h.c.} \right) \tag{18.274}$$

Here, the potential energy for the Coulomb repulsion is

$$U\left(\left|\vec{r}_1-\vec{r}_2\right|\right)=\frac{e^2}{\varepsilon_\infty\left|\vec{r}_1-\vec{r}_2\right|} \tag{18.275}$$

Considering

$$\alpha_F=\frac{e^2}{2\hbar\omega_0 R_p}\left(\frac{1}{\varepsilon_\infty}-\frac{1}{\varepsilon_0}\right) \tag{18.276}$$

from where

$$\frac{e^2}{\varepsilon_\infty}=\hbar\omega_0 R_p U_0, U_0=\frac{\sqrt{2}\alpha_F}{1-\eta}, \eta=\frac{\varepsilon_\infty}{\varepsilon_0} \tag{18.277}$$

then

$$U\left(\left|\vec{r}_1-\vec{r}_2\right|\right)=\frac{\hbar\omega_0 R_p U_0}{\left|\vec{r}_1-\vec{r}_2\right|}=\frac{4\pi\hbar\omega_0 R_p U_0}{V}\sum_{\vec{\kappa}}\exp\left\{i\vec{\kappa}\left(\vec{r}_1-\vec{r}_2\right)\right\} \tag{18.278}$$

The transition from many-body collapse to the existence of a thermodynamic limit for many polarons occurs exactly at [2,22]:

$$U_0=\sqrt{2}\alpha_F \tag{18.279}$$

Here, U_0 is the electronic Coulomb repulsion and α_F is the polaron coupling constant as earlier indicated. If U_0 is large enough, there is no multipolaron binding of any kind. Considering the known fact [2,22] that there is binding for some

$$U_0>\sqrt{2}\alpha_F \tag{18.280}$$

These conclusions are not obvious and their proof has been an open problem for some time. Usually [2,55], there is binding for the condition 18.280. It is apparent that

$$\varepsilon_0>\varepsilon_\infty \tag{18.281}$$

So, the investigation of bipolarons, η falls within the range

$$0\le\eta<1 \tag{18.282}$$

and, hence, the formation of the bipolaron is favored by smaller η.

We proceed as in the previous problem and consider the strong confinement regime where the motion in the transversal direction of the wire is fast and that in the longitudinal direction (axis of the wire) slow (adiabatic approximation). This permits us to average the Hamiltonian 18.274 by the wave function of the transversal motion:

$$\Psi=A\exp\left\{-\frac{\lambda}{2}\left(\rho_1^2+\rho_2^2\right)\right\} \tag{18.283}$$

So, the resultant Hamiltonian has the form

$$\tilde{\mathrm{H}}=\frac{\hat{p}_{\parallel 1}^2}{2m}+\frac{\hat{p}_{\parallel 2}^2}{2m}+\tilde{U}\left(\left|z_1-z_2\right|\right)+\sum_{\vec{\kappa}}\hbar\omega_{\vec{\kappa}}\hat{b}_{\vec{\kappa}}^{\dagger}\hat{b}_{\vec{\kappa}}+\frac{\hbar}{\sqrt{V}}\sum_{\vec{\kappa}}\left(\left[\tilde{C}_{\vec{\kappa}}\hat{b}_{\vec{\kappa}}\left(\exp\left\{i\kappa_\parallel z_1\right\}+\exp\left\{i\kappa_\parallel z_2\right\}\right)\right]+\mathrm{h.c.}\right) \tag{18.284}$$

Here,

$$C_{\vec{\kappa}}\left\langle \Psi \middle| \exp\left\{i\vec{\kappa}\vec{r}\right\} \middle| \Psi \right\rangle = \tilde{C}_{\vec{\kappa}} \tag{18.285}$$

$$\left\langle \Psi \middle| U\left(\left|\vec{r}_1 - \vec{r}_2\right|\right) \middle| \Psi \right\rangle = \tilde{U}\left(\left|z_1 - z_2\right|\right) \tag{18.286}$$

$$\tilde{U}\left(\left|z_1 - z_2\right|\right) = \sum_{\vec{\kappa}} \tilde{U}\left(\vec{\kappa}\right) \exp\left\{i\kappa_{\parallel}\left(z_1 - z_2\right)\right\} \tag{18.287}$$

where

$$\tilde{U}\left(\vec{\kappa}\right) = \frac{4\pi\omega_0 R_p U_0}{\left(\kappa_\perp^2 + \kappa_\parallel^2\right)} \exp\left\{-\frac{\kappa_\perp^2}{4\lambda}\right\} \tag{18.288}$$

The Hamiltonian 18.284 corresponds to the one oscillatory model Lagrangian:

$$L_0 = \sum_{i=1}^{2}\left(-\frac{m\dot{z}_i^2}{2\hbar^2} - \frac{M\dot{Z}_i^2}{2\hbar^2} - \frac{\kappa\left(z_i - Z_i\right)^2}{2}\right) - \sum_{i \neq k}^{2} \frac{\kappa'\left(z_i - Z_k\right)^2}{2} + \frac{\mathrm{K}\left(z_1 - z_2\right)^2}{2} \tag{18.289}$$

In order to perform Feynman path integration, over the electronic coordinates, we transform the model Lagrangian 18.289 to one with normal coordinates. This renders the model Lagrangian separable and constitutes one with independent oscillators that elegantly facilitates functional integration.

18.13.4 Bipolaron Equation of Motion

The model Lagrangian 18.289 of the system considers the variables of the electronic as well as those of the fictitious particle system that are coupled to each other. Path integration will be done with ease if we decouple these variables by diagonalizing the Lagrangian. To do this we find the eigenmodes of the electronic and fictitious particle system via the Lagrange equation of motion

$$\frac{d}{dt}\frac{\partial}{\partial \dot{q}} L_0 - \frac{\partial}{\partial q} L_0 = 0 \tag{18.290}$$

Considering 18.289, we have

$$m\ddot{z}_1 + \kappa\left(z_1 - Z_1\right) - \mathrm{K}\left(z_1 - z_2\right) + \kappa'\left(z_1 - Z_2\right) = 0 \tag{18.291}$$

$$m\ddot{z}_2 + \kappa\left(z_2 - Z_2\right) + \mathrm{K}\left(z_1 - z_2\right) + \kappa'\left(z_2 - Z_1\right) = 0 \tag{18.292}$$

$$M\ddot{Z}_1 - \kappa\left(z_1 - Z_1\right) - \kappa'\left(z_2 - Z_1\right) = 0 \tag{18.293}$$

$$M\ddot{Z}_2 - \kappa\left(z_2 - Z_2\right) - \kappa'\left(z_1 - Z_2\right) = 0 \tag{18.294}$$

Suppose

$$\ddot{z}_i = -s z_i \tag{18.295}$$

where s is the square of the frequency. We introduce new coordinates:

$$\xi_1 = \frac{mz_1 + mz_2}{m + m} = \frac{1}{2}\left(z_1 + z_2\right) \tag{18.296}$$

the center of mass of two electrons:

$$\xi_2 = z_1 - z_2 \tag{18.297}$$

the relative distance between the electrons:

$$\xi_{1f} = \frac{MZ_1 + MZ_2}{M + M} = \frac{1}{2}(Z_1 + Z_2) \tag{18.298}$$

the centre of mass of two fictitious particle and

$$\xi_{2f} = Z_1 - Z_2 \tag{18.299}$$

the relative distance between the fictitious particles.

This change of variables transforms the system of equations in 18.294 into two pairs of equations:

$$-ms\xi_1 + \kappa(\xi_1 - \xi_{1f}) + \kappa'(\xi_1 - \xi_{1f}) = 0 \tag{18.300}$$

$$-Ms\xi_{1f} - \kappa(\xi_2 - \xi_{2f}) + \kappa'(\xi_2 + \xi_{2f}) = 0 \tag{18.301}$$

and

$$-ms\xi_2 + \kappa(\xi_2 - \xi_{2f}) + \kappa'(\xi_2 + \xi_{2f}) - 2\mathrm{K}\xi_2 = 0 \tag{18.302}$$

$$-Ms\xi_{2f} - \kappa(\xi_2 - \xi_{2f}) + \kappa'(\xi_2 + \xi_{2f}) = 0 \tag{18.303}$$

Solving the system of equations in 18.303 yields

$$(P - ms)\xi_2 - p'\xi_{2f} = 0, -p'\xi_2 + (p - Ms)\xi_{2f} = 0 \tag{18.304}$$

This permits us to obtain the equations for the normal frequencies

$$[mMs - p(m + M)]s = 0, mMs^2 - (mp + MP)s + Pp - p'^2 = 0 \tag{18.305}$$

The first equation yields the solution

$$s_1 = 0, s_2 = \frac{m + M}{mM} p = \frac{p}{\mu} \tag{18.306}$$

while the second equation of 18.305 yields the solution:

$$s_{3,4} = \frac{PM + pm}{2Mm} \pm \frac{1}{2}\left[\left(\frac{PM + pm}{Mm}\right)^2 + 4\frac{Pp - p'^2}{Mm}\right]^{\frac{1}{2}} \tag{18.307}$$

where

$$P = p - 2\mathrm{K}, p' = \kappa - \kappa' \tag{18.308}$$

Summing the two equations in 18.302 we have

$$-s\left(m\xi_1 + M\xi_{1f}\right) = 0 \tag{18.309}$$

This confirms one free motion with

$$s_1 = 0 \tag{18.310}$$

The normal coordinate

$$x_1 = \frac{m\xi_1 + M\xi_{1f}}{m + M} \tag{18.311}$$

is the center of mass of all four particles moving freely. This gives the reason why the frequency of this motion is zero. The square of the second mode of the oscillator, s_2, does not have the Coulomb variable K . This is because it describes only the relative oscillations of the electronic and fictitious particle subsystems. The Coulomb variable K relates only to the internal and inter-electronic interaction as described by the square of the modes $s_{3,4}$. To conclude, we have one free motion with $s_1 = 0$ and three internal oscillations with square of frequencies, s_2 and $s_{3,4}$.

18.13.5 Transformation into Normal Coordinates

With the help of the frequency eigenmodes in 18.306 and 18.307 we can now diagonalize the Lagrangian in 18.289 via a unified technique that will be used throughout this book by a continuous coordinate transformation that involves spatial translations as well as rotations. In this technique, even if one selects a direction for the transformation, there is always the continuous coordinate and, in addition, this technique conserves all quantities. It is instructive to note that such a continuous coordinate transformation has no effect on the particle state since the particle state is an abstract object and makes no reference to a given coordinate system.

18.13.5.1 Diagonalization of the Lagrangian

One now applies the continuous coordinate transformation transforming the non-harmonic variables ξ_2 and ξ_{2f} via two harmonic coordinates dependent on the following frequency eigenmodes:

$$\omega_3 = \sqrt{s_3}, \omega_4 = \sqrt{s_4} \tag{18.312}$$

and

$$\xi_2 = \xi_2\left(\omega_3\right) + \xi_2\left(\omega_4\right), \xi_{2f} = \xi_{2f}\left(\omega_3\right) + \xi_{2f}\left(\omega_4\right) \tag{18.313}$$

This will help to diagonalize the model Lagrangian in 18.289. From this transformation, equation 18.304 can then be rewritten:

$$\left(P - ms_i\right)\xi_2\left(\omega_i\right) - p'\xi_{2f}\left(\omega_i\right) = 0, -p'\xi_2\left(\omega_i\right) + \left(p - Ms_i\right)\xi_{2f}\left(\omega_i\right) = 0 \tag{18.314}$$

We consider the continuous coordinate transformation that involves spatial translations as well as rotations for ξ_2 and ξ_{2f} in 18.318. For brevity, the transformation via normal variables x_i can be represented as follows:

$$\xi_2 = C_3x_3\left(\omega_3\right) - C_4F_2\left(\omega_4\right)x_4\left(\omega_4\right), \xi_{2f} = -C_3F_1\left(\omega_3\right)x_3\left(\omega_3\right) + C_4x_4\left(\omega_4\right) \tag{18.315}$$

The explicit expressions of the coefficients C_i and of the functions $F_1(\omega_3)$ and $F_2(\omega_4)$ can be obtained from the equation of motion 18.304 by substituting first $\xi_2(\omega_3)$ and $\xi_{2f}(\omega_3)$ into the second equation of 18.304:

$$F_1(\omega_3) = -\frac{p'}{p - Ms_3} \tag{18.316}$$

and, similarly, $\xi_{2f}(\omega_4)$ and $\xi_2(\omega_4)$ into the first equation of 18.304:

$$F_2(\omega_4) = -\frac{p'}{P - ms_4} \tag{18.317}$$

Considering the conservation of the kinetic energy in any of the representations then

$$\frac{m\dot{\xi}_2^2(\omega_3)}{4\hbar^2} + \frac{M\dot{\xi}_{2f}^2(\omega_3)}{4\hbar^2} = \frac{m\dot{x}_3^2(\omega_3)}{4\hbar^2} \tag{18.318}$$

$$\frac{m\dot{\xi}_2^2(\omega_4)}{4\hbar^2} + \frac{M\dot{\xi}_{2f}^2(\omega_4)}{4\hbar^2} = \frac{M\dot{x}_4^2(\omega_4)}{4\hbar^2} \tag{18.319}$$

and

$$C_3^2 = \frac{m}{m + MF_1^2(s_3)}, C_4^2 = \frac{M}{M + mF_2^2(s_4)}, F_2(\omega_4) = -\frac{M}{m}F_1(\omega_3) \tag{18.320}$$

The model Lagrangian now has the new form

$$L_0 = -\frac{M'\dot{x}_1^2}{\hbar^2} - \frac{\mu\dot{x}_2^2}{\hbar^2} - \mu s_2 x_2^2 - \frac{m\dot{x}_3^2}{4\hbar^2} - \frac{M\dot{x}_4^2}{4\hbar^2} - \frac{ms_3x_3^2}{4} - \frac{Ms_4x_4^2}{4} \tag{18.321}$$

where,

$$M' = m + M \tag{18.322}$$

It is obvious from the Lagrangian 18.321 that we have one free motion and three oscillatory motions. The form of the Lagrangian can be a proof of an extraordinary transformation transforming real to normal coordinates. This permits an elegant diagonalization of the Lagrangian and renders it perfectly quadratic. So, the Feynman path integral will be a Gaussian one and computable exactly. This transformation of real to normal coordinates applied to the polaron problem will be observed to be an elegant and effective way of computing problems of particles interacting with the environment.

18.13.6 Bipolaron Partition Function

As done previously for the polaron problem, the energy and effective mass of the bipolaron can be found from the relation:

$$E \le F_F = -\frac{1}{\beta}\ln Z_0 - \frac{1}{\beta}\langle S - S_0\rangle_{S_0} \tag{18.323}$$

Considering 18.321, the partition function Z_0 is obtained:

$$Z_0 = \int_{-\infty}^{+\infty} \prod_{i=1}^{4} dx_i dx'_i \delta\left(x_i - x'_i\right) \int_{x'_i}^{x_i} \exp\left\{\underline{S_0}\right\} Dx_i \tag{18.324}$$

So, the associated partition function has therefore the product of the contributions from one free motion and three oscillatory motions:

$$Z_0 = \frac{l_z}{\sqrt{2} l_e\left(\hbar\beta\right)} \prod_{i=2}^{4} \frac{1}{2\sinh\left(\frac{\beta\hbar\omega_i}{2}\right)} \tag{18.325}$$

where l_z is the length of the wire,

$$l_e\left(\hbar\beta\right) \equiv \left(\frac{2\pi\hbar^2\beta}{M'}\right)^{\frac{1}{2}} \tag{18.326}$$

is the **thermal de Broglie wavelength** and frequency describing the relative oscillations of the electronic and fictitious particle subsystems:

$$\omega_2 = \left(\frac{p}{\mu}\right)^{\frac{1}{2}} = u\omega_{2f} \tag{18.327}$$

We write the Lagrangian corresponding to the Hamiltonian 18.284:

$$L\left(-i\hbar\tau\right) = -\frac{m\dot{z}_1^2}{2\hbar^2} - \frac{m\dot{z}_2^2}{2\hbar^2} - \frac{1}{2}\sum_{\vec{\kappa}}\left(\frac{\dot{q}_{\vec{\kappa}}^2}{\hbar^2} + \omega_{\vec{\kappa}}^2 q_{\vec{\kappa}}^2\right) + \sum_{\vec{\kappa}, j=1,2} \gamma_{\vec{\kappa}}\left(z_j\right) q_{\vec{\kappa}}\left(\tau\right) - \tilde{U}\left(\left|z_1 - z_2\right|\right) \tag{18.328}$$

The amplitude of the electron–phonon coupling interaction $\gamma_{\vec{\kappa}}\left(\tau\right)$ for the $\vec{\kappa}$ moving wave is defined:

$$\gamma_{\vec{\kappa}}\left(\tau\right) = \frac{\hbar}{\sqrt{V}} \widetilde{C_{\vec{\kappa}}} \chi_{\vec{\kappa}}\left(\tau\right) \tag{18.329}$$

with

$$\chi_{\vec{\kappa}}\left(\tau\right) = \begin{cases} \sin\kappa_{\|} z, & \kappa_{\|} \geq 0 \\ \cos\kappa_{\|} z, & \kappa_{\|} < 0 \end{cases} \tag{18.330}$$

From equation 18.328, the action functional for the full system after eliminating the phonon variables is obtained similarly as for the polaron problem:

$$\underline{S}\left[z_1, z_2\right] = -\int_0^{\beta} \frac{m\dot{z}_1^2}{2\hbar^2} d\tau - \int_0^{\beta} \frac{m\dot{z}_2^2}{2\hbar^2} d\tau + \Phi\left[z_1, z_2\right] - \int_0^{\beta} \tilde{U}\left(\left|z_1 - z_2\right|\right) d\tau \tag{18.331}$$

Here the influence phase of the exact system is:

$$\Phi\left[z_1, z_2\right] = \sum_{\vec{\kappa}} \Phi_{\omega_{\vec{\kappa}}}\left[z_1, z_2\right] \tag{18.332}$$

with

$$\Phi\left[z_1,z_2\right]=\sum_{\vec{\kappa},j=1,2}\frac{\hbar}{8\omega_{\vec{\kappa}}}\int_0^{\beta}\int_0^{\beta}d\sigma d\sigma'\gamma_{\vec{\kappa}}\left(z_j\left(\sigma'\right)\right)\gamma_{\vec{\kappa}}\left(z_j\left(\sigma\right)\right)F_{\omega_{\vec{\kappa}}}\left(\left|\sigma-\sigma'\right|\right) \tag{18.333}$$

From 18.289, following the same procedure like for the polaron when finding the partition function then considering 18.324 we have:

$$Z_0=\int\prod_{i=1}^{2}dz_idz'_idZ_{if}dZ'_{if}\delta\left(z_i-z'_i\right)\delta\left(Z_{if}-Z'_{if}\right)\int_{z'_i}^{z_i}Dz_i\int_{Z'_{if}}^{Z_{if}}DZ_{if}\exp\left\{\underline{S}_0\right\} \tag{18.334}$$

or

$$Z_0=\int\prod_{i=1}^{2}\delta\left(z_i-z'_i\right)dz_idz'_i\int_{z'_i}^{z_i}Dz_i\exp\left\{\underline{S}_0\left[z_1,z_2\right]\right\} \tag{18.335}$$

Here, the model bipolaron action functional is obtained:

$$\begin{aligned}\underline{S}_0\left[z_1,z_2\right]=&-\frac{m}{2\hbar^2}\int_0^{\beta}\left(\dot{z}_1^2+\dot{z}_2^2\right)d\tau-\frac{p}{2}\int_0^{\beta}\left(z_1^2+z_2^2\right)d\tau+\frac{\mathrm{K}}{2}\int_0^{\beta}\left(z_1-z_2\right)^2d\tau\\&+\frac{\hbar}{4M\omega_{1f}}\int_0^{\beta}\int_0^{\beta}\left[\left(\kappa z_1\left(\tau\right)+\kappa' z_2\left(\tau\right)\right)\left(\kappa z_1\left(\sigma\right)+\kappa' z_2\left(\sigma\right)\right)\right]F_{\omega_{1f}}\left(\left|\tau-\sigma\right|\right)d\tau d\sigma\\&+\frac{\hbar}{4M\omega_{2f}}\int_0^{\beta}\int_0^{\beta}\left[\left(\kappa' z_1\left(\tau\right)+\kappa z_2\left(\tau\right)\right)\left(\kappa' z_1\left(\sigma\right)+\kappa z_2\left(\sigma\right)\right)\right]F_{\omega_{2f}}\left(\left|\tau-\sigma\right|\right)d\tau d\sigma\\&+\ln\left[2\sinh\left(\frac{\beta\hbar\omega_{1f}}{2}\right)\right]+\ln\left[2\sinh\left(\frac{\beta\hbar\omega_{2f}}{2}\right)\right]\end{aligned} \tag{18.336}$$

18.13.7 Bipolaron Generating Function

We find now $\left\langle \underline{S}-\underline{S}_0\right\rangle_{\underline{S}_0}$ with the help of the generating function:

$$\Psi_{\vec{\kappa}}\left(\xi_1,\eta_1,\xi_2,\eta_2\right)=\left\langle\exp\left\{i\kappa\left(\xi_1z_1\left(\tau\right)-\eta_1z_1\left(\sigma\right)+\xi_2z_2\left(\tau\right)-\eta_2z_2\left(\sigma\right)\right)\right\}\right\rangle_{\underline{S}_0} \tag{18.337}$$

From change of variables

$$z_1=x_1+\frac{M}{M'}x_2+\frac{1}{2}C_3x_3\left(\omega_3\right)-\frac{1}{2}C_4F_4\left(\omega_4\right)x_4\left(\omega_4\right) \tag{18.338}$$

$$z_2=x_1+\frac{M}{M'}x_2-\frac{1}{2}C_3F_3\left(\omega_3\right)x_3\left(\omega_3\right)+\frac{1}{2}C_4x_4\left(\omega_4\right) \tag{18.339}$$

and considering

$$f_1^+=\left(\xi_1+\xi_2\right)^2+\left(\eta_1+\eta_2\right)^2,\ f_2^+=\left(\xi_1+\xi_2\right)\left(\eta_1+\eta_2\right) \tag{18.340}$$

$$f_1^-=\left(\xi_1-\xi_2\right)^2+\left(\eta_1-\eta_2\right)^2,\ f_2^-=\left(\xi_1-\xi_2\right)\left(\eta_1-\eta_2\right) \tag{18.341}$$

then

$$\Psi_{\vec{\kappa}}(\xi_1,\eta_1,\xi_2,\eta_2)=\left\langle \exp\left\{\kappa^2(f_1+f_2+f_3+f_4)\right\}\right\rangle_{S_0} \tag{18.342}$$

where

$$f_1=-\frac{\hbar a_1\beta}{8m}\left[\frac{1}{6}f_1^+-f_2^+\left(\frac{2}{\beta^2}\left(|\tau-\sigma|-\frac{\beta}{2}\right)^2+\frac{1}{6}\right)\right] \tag{18.343}$$

$$f_2=-\frac{\hbar a_2}{8m\omega_2}\left[f_1^+\mathrm{F}_{\omega_2}(0)-2f_2^+\mathrm{F}_{\omega_2}(|\tau-\sigma|)\right] \tag{18.344}$$

$$f_3=-\frac{\hbar C_3^2}{8M\omega_3}\left[f_1^-\mathrm{F}_{\omega_3}(0)-2f_2^-\mathrm{F}_{\omega_3}(|\tau-\sigma|)\right] \tag{18.345}$$

$$f_4=-\frac{\hbar C_4^2F_4^2}{8M\omega_4}\left[f_1^-\mathrm{F}_{\omega_4}(0)-2f_2^-\mathrm{F}_{\omega_4}(|\tau-\sigma|)\right] \tag{18.346}$$

18.13.8 Bipolaron Asymptotic Characteristics

From 18.323, we now evaluate the quantities in $\left\langle \underline{S}-\underline{S}\right\rangle_{S_0}$ to permit the evaluation of the polaron characteristics. In order to do this, we first find the following expectation values:

$$\begin{aligned}\left\langle z_1(\tau)z_1(\sigma)\right\rangle_{S_0}&=\frac{\hbar a_1\beta}{4m}\left[\frac{2}{\beta^2}\left(|\tau-\sigma|-\frac{\beta}{2}\right)^2+\frac{1}{6}\right]+\frac{\hbar a_2}{4m\omega_2}F_{\omega_2}(|\tau-\sigma|)\\&+\frac{\hbar C_3^2}{4M\omega_3}\mathrm{F}_{\omega_3}(|\tau-\sigma|)+\frac{\hbar C_4^2F_4^2}{4M\omega_4}\mathrm{F}_{\omega_4}(|\tau-\sigma|)\end{aligned} \tag{18.347}$$

$$\left\langle z_1(\tau)z_2(\sigma)\right\rangle_{S_0}=\left\langle z_2(\tau)z_1(\sigma)\right\rangle_{S_0} \tag{18.348}$$

$$\left\langle z_1^2(\tau)\right\rangle_{S_0}\equiv\left\langle z_1(\tau)z_1(\sigma)\right\rangle_{S_0}\Big|_{\tau=\sigma} \tag{18.349}$$

$$\left\langle \exp\left\{i\kappa\left(z_1(\tau)-z_1(\sigma)\right)\right\}\right\rangle_{S_0}=\Psi_{\vec{\kappa}}(\xi_1=1,\eta_1=1,\xi_2=0,\eta_2=0) \tag{18.350}$$

$$\left\langle \exp\left\{i\kappa\left(z_2(\tau)-z_1(\sigma)\right)\right\}\right\rangle_{S_0}=\Psi_{\vec{\kappa}}(\xi_1=1,\eta_1=0,\xi_2=0,\eta_2=1) \tag{18.351}$$

$$\left\langle \exp\left\{i\kappa\left(z_2(\tau)-z_1(\sigma)\right)\right\}\right\rangle_{S_0}=\left\langle \exp\left\{i\kappa\left(z_1(\tau)-z_2(\sigma)\right)\right\}\right\rangle_{S_0} \tag{18.352}$$

All the formulae in the bipolaron have the same dependence on the quantity $|\tau-\sigma|$. This again confirms the fact that these quantities (retarded functions) depend on the past with the significance of interaction with the past being the perturbation due to the moving electrons (holes) that take **"time"** propagating in the crystal. The exact action functional shows again that the potential of the electrons (holes) feel at any **"time"** the dependence on its position at previous times. This confirms the effect of the electron (hole) on the crystal as it propagates at a finite velocity and can make itself felt by the crystal at a later time.

We again generalize the function in the integrand of the bipolaron to be $G\left(|\tau-\sigma|-\frac{\beta}{2}\right)$ to help in the simplification of the evaluation of the integral. So, we look again at the following function that yields the value of the integral for a given $G\left(|\tau-\sigma|-\frac{\beta}{2}\right)$ function [1,2]:

$$f(\beta)=\int_0^\beta\int_0^\beta G\left(|\tau-\sigma|-\frac{\beta}{2}\right)d\sigma d\tau \tag{18.353}$$

We do the change of variables

$$\tau-\sigma\equiv\rho \quad , \quad 0\le\sigma\le\tau \tag{18.354}$$

then

$$f(\beta)=2\int_0^\beta G\left(\rho-\frac{\beta}{2}\right)(\beta-\rho)d\rho \tag{18.355}$$

We consider again the change of variable $\rho=\beta y$, where $0\le y\le 1$. If $y=1$ then $\rho=\beta$. Again, if we also do the change of variable

$$\frac{\beta(1-y)}{2}\equiv-\tau,\quad d\tau=-\frac{\beta}{2}dy,\quad 0\le\tau\le\infty \tag{18.356}$$

then from 18.323 considering $\beta\to\infty$ ($T\to 0$), we evaluate the asymptotic expansions of Feynman bipolaron dimensionless energy and consider strong coupling for bipolarons where the fictitious particle is more massive than the electron (hole) [2]:

$$M>>m \tag{18.357}$$

For brevity, we exclude the Coulomb interaction by letting:

$$v_2=v_3 \tag{18.358}$$

and

$$E_{bip}=\frac{\varepsilon(v)}{2}-4\alpha_F\left(\frac{\varepsilon(v)}{\pi}\right)^{\frac{1}{2}}\ln\left(\frac{4\varepsilon_\perp}{\varepsilon(v)}\right)^{\frac{1}{2}} \tag{18.359}$$

where

$$v_2^2=\frac{p}{\mu},v_3=\sqrt{s_3},\frac{\hbar^2\lambda}{m}\frac{1}{R_p^2}=\varepsilon_\perp,\frac{\hbar v_2}{\hbar\omega_0}=\varepsilon(v),\frac{\varepsilon_{bip}}{\hbar\omega_0}=E_{bip} \tag{18.360}$$

Extremizing 18.359 over the energy of the oscillatory motion ε_2:

$$\frac{dE_{bip}}{d\varepsilon(v)} = 0 \tag{18.361}$$

then

$$-y = q\ln y, \frac{1}{y} = \left(\frac{4\varepsilon_\perp}{e^2\varepsilon(v)}\right)^{\frac{1}{2}}, q = 4\alpha_F\left(\frac{e^2}{\pi\varepsilon(v)}\right)^{\frac{1}{2}} \tag{18.362}$$

The solution of 18.362 can be obtained via the iteration method as done for the polaron problem:

$$y = qf(q), f(q) \equiv -q\ln\left[-q\ln\left[-q\ln\cdots\ln q\right]\right] \tag{18.363}$$

The bipolaron coupling energy without the Coulomb contribution:

$$E_{bip} = -\frac{8\alpha_F^2 f(q)}{\pi}\ln\left[\frac{e(\pi\varepsilon_\perp)^{\frac{1}{2}}}{2\alpha_F f(q)}\right] \tag{18.364}$$

We introduce the effective electron–phonon coupling constant

$$\tilde{\alpha}_F = 2\alpha_F \tag{18.365}$$

then, the bipolaron coupling energy without the Coulomb contribution:

$$E_{bip} = -\frac{2\tilde{\alpha}_F^2 f(q)}{\pi}\ln\left[\frac{e(\pi\varepsilon_\perp)^{\frac{1}{2}}}{\tilde{\alpha}_F f(q)}\right] \tag{18.366}$$

The polaron counterpart is

$$E_p = -\frac{\alpha_F^2}{\pi} f(q)\ln\left(\frac{e(\pi\varepsilon_\perp)^{\frac{1}{2}}}{\alpha_F f(q)}\right) \tag{18.367}$$

From these two results we see the bipolaron energy appears to be twice the polaron energy [2]. This is an indication that the induced polarization due to the two electrons in the bipolaron is proportional to $2\alpha_F$ while each of the electrons interacts with this polarization. This gives the reason why the interaction energy is described as $\tilde{\alpha}_F = 2\alpha_F$. So, the simple doubling of the polaron energy by $\alpha_F \to 2\alpha_F$ does not take place.

From Figures 18.5 and 18.6 we observe that the bipolaron coupling ground state energy has a monotonic decrease with an increase in the radius of the quantum wire. This implies the bipolaron coupling energy increases with confinement enhancement. In a similar fashion, the bipolaron effective mass has a monotonic decrease with an increased radius of the quantum wire and implies an increase of the bipolaron effective mass with confinement enhancement.

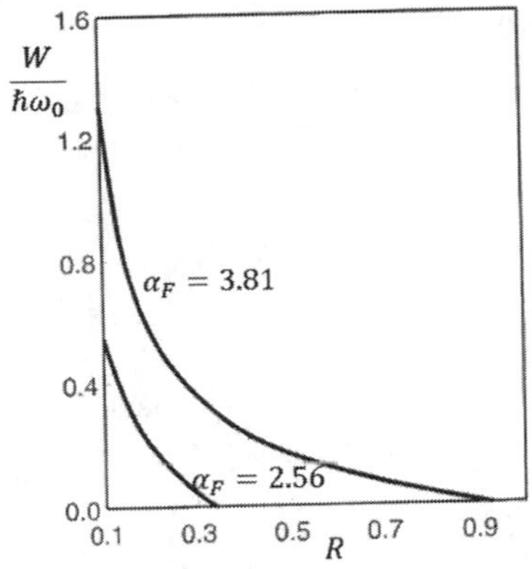

FIGURE 18.5 Depicts the dimensionless bipolaron coupling energy, $\frac{W}{\hbar\omega_0}$ versus the radius, ***R*** of the quantum wire for different electron-phonon interaction coupling constant, $\boldsymbol{\alpha}_F$ taken from reference [2].

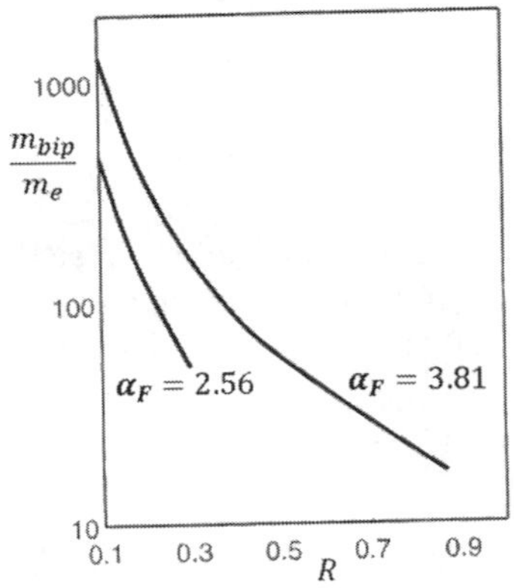

FIGURE 18.6 Depicts the dimensionless bipolaron effective mass, $\frac{m_{bip}}{m_e}$ versus the radius, ***R*** of the quantum wire for different electron-phonon interaction coupling constant, $\boldsymbol{\alpha}_F$ taken from reference [2].

18.14 Polaron Characteristics in a Quasi-0D Spherical Quantum Dot

18.14.1 Introduction

We continue to examine the effect of the electronic confinement on the polaron and bipolaron in a quasi-0D quantum dot. The problem of the quantum dot will be seen to have reduced variational parameters compared to the case of the quasi-1D quantum wire.

From the Feynman variation principle framework, we examine the polaron energy and effective mass in a quantum dot. For the moment we examine a quantum dot with spherical geometry. In this problem we example the polaron characteristics for low temperatures.

18.14.2 Polaron Lagrangian

The system of conducting electron interacts with the lattice vibrations in the field of a parabolic confinement potential where the system is described by the Lagrangian:

$$L = -\frac{m\dot{\vec{r}}^2}{2\hbar^2} - \frac{m\Omega^2\vec{r}^2}{2} - \frac{1}{2}\sum_{\vec{\kappa}}\left(\frac{\dot{q}_{\vec{\kappa}}^2}{\hbar^2} + \omega_{\vec{\kappa}}^2 q_{\vec{\kappa}}^2\right) + \sum_{\vec{\kappa}}\gamma_{\vec{\kappa}}(\tau)q_{\vec{\kappa}}(\tau) \tag{18.368}$$

Here, the electron coordinate is $\vec{r}$ and m is its effective mass. The model Lagrangian may be selected in the one oscillatory approximation as seen earlier:

$$L_0 = -\frac{m\dot{\vec{r}}^2}{2\hbar^2} - \frac{m\Omega^2\vec{r}^2}{2} - \frac{M\dot{\vec{R}}^2}{2\hbar^2} - \frac{M\omega_f^2\left(\vec{R}-\vec{r}\right)^2}{2} \tag{18.369}$$

where M and ω_f are, respectively, the mass of a fictitious particle and the frequency of the elastic coupling that will serve as variational parameters while $\vec{R}$ is the coordinate of the fictitious particle. The elastic constant describing the force of attraction between the electron and the fictitious particle may also be represented in the form:

$$\kappa = M\omega_f^2 \tag{18.370}$$

18.14.3 Normal Modes

The model Lagrangian 18.369 considers the electronic as well as the fictitious particle variables. Path integration will be easily done by moving from real to normal coordinates that will be feasible via frequency eigenmodes. This is possible via the equation of motion in 18.290 and considering 18.369:

$$m\ddot{\vec{r}} + m\Omega^2\vec{r} - \kappa\left(\vec{R}-\vec{r}\right) = 0, M\ddot{\vec{R}} + \kappa\left(\vec{R}-\vec{r}\right) = 0 \tag{18.371}$$

Suppose as done earlier for the problems in the quantum wire that

$$\ddot{\vec{r}} = -\omega^2\vec{r}, M\ddot{\vec{R}} = -\omega^2\vec{R} \tag{18.372}$$

Here ω is the frequency of the normal modes of the system. Substituting 18.372 into 18.371 then we have

$$-m\omega^2\vec{r} + m\Omega^2\vec{r} - \kappa\left(\vec{R}-\vec{r}\right) = 0, -M\omega^2\vec{R} + \kappa\left(\vec{R}-\vec{r}\right) = 0 \tag{18.373}$$

From the determinant of the system of equations 18.373 we have the following normal frequency eigenmodes:

$$\omega_{1,2}^2 = \frac{1}{2}\left(\frac{\kappa}{\mu} + {}'\;{}^{2}\right) \pm \frac{1}{2}\left[\left(\frac{\kappa}{\mu} + \Omega^2\right)^2 - \frac{4\kappa\Omega^2}{M}\right]^{\frac{1}{2}}, \frac{1}{\mu} = \frac{1}{m} + \frac{1}{M} \tag{18.374}$$

18.14.4 Lagrangian Diagonalization

With the help of the frequency eigenmodes in 18.374 we can now diagonalize the Lagrangian in 18.369. This will be done via the unified technique applied in 18.315 by the transformation of real to normal coordinates. As observed earlier, such a coordinate transformation has no effect on the particle state as the particle state is an abstract object and makes no reference to a given coordinate system.

18.14.4.1 Transformation to Normal Coordinates

As seen earlier, to perform Feynman path integration elegantly, we move to normal coordinates via normal modes in 18.374. For the normal mode, ω_1 we have the normal coordinates:

$$\vec{r}(\omega_1) = C_1(\omega_1)\vec{\xi}_1(\omega_1), \vec{R}(\omega_1) = C_1(\omega_1)F_1(\omega_1)\vec{\xi}_1(\omega_1) \tag{18.375}$$

Substituting this into 18.373 then we have

$$F_1(\omega_1) = \frac{\kappa}{-M\omega_1^2 + \kappa} \tag{18.376}$$

Considering the conservation of the kinetic energy:

$$m\dot{\vec{\xi}}_1^{\,2}(\omega_1) = m\dot{\vec{r}}^{\,2}(\omega_1) + M\dot{\vec{R}}^{\,2}(\omega_1) \tag{18.377}$$

Then from 18.375 and 18.376 we have

$$C_1^2(\omega_1) = \frac{m}{m + MF_1^2(\omega_1)} \tag{18.378}$$

We do the same change of variables to normal variables via the normal mode ω_2:

$$\vec{R}(\omega_2) = C_2(\omega_2)\vec{\xi}_2(\omega_2), \vec{r}(\omega_2) = C_2(\omega_2)F_2(\omega_2)\vec{\xi}_2(\omega_2) \tag{18.379}$$

From here considering the kinetic energy

$$M\dot{\vec{\xi}}_2^{\,2}(\omega_2) = m\dot{\vec{r}}^{\,2}(\omega_2) + M\dot{\vec{R}}^{\,2}(\omega_2) \tag{18.380}$$

then

$$C_2^2(\omega_2) = \frac{M}{M + mF_2^2(\omega_2)} \tag{18.381}$$

The coordinates $\vec{r}$ and $\vec{R}$ are not harmonic variables but, however, they represent the superposition of two harmonic oscillations with the frequencies ω_1 and ω_2:

$$\vec{r} = \vec{r}(\omega_1) + \vec{r}(\omega_2) = C_1(\omega_1)\vec{\xi}_1(\omega_1) + C_2(\omega_2)F_2(\omega_2)\vec{\xi}_2(\omega_2) \tag{18.382}$$

$$\vec{R} = \vec{R}(\omega_1) + \vec{R}(\omega_2) = C_1(\omega_1)F_1(\omega_1)\vec{\xi}_1(\omega_1) + C_2(\omega_2)\vec{\xi}_2(\omega_2) \tag{18.383}$$

So, from 18.375 to 18.383 the model Lagrangian in 18.369 takes the form

$$L_0 = \sum_{i=1}^{2}\left[-\frac{\dot{\vec{\xi}}_i^{\,2}}{2\hbar^2} - \frac{\omega_i^2\vec{\xi}_i^{\,2}}{2}\right] \tag{18.384}$$

The model Lagrangian has two oscillators with $\vec{\xi}_1$ and $\vec{\xi}_2$ being normal coordinates as earlier seen for the previous polaron problem in a quasi-1D quantum wire. The form of the Lagrangian is a proof of an extraordinary transformation transforming real to normal coordinates and permits an elegant diagonalization of the Lagrangian, rendering it perfectly quadratic. Feynman path integral will then be a Gaussian one and computable exactly.

18.14.5 Polaron Partition Function

To proceed with the polaron energy we find the following quantities:

$$E_0 = -T\ln Z_0 \tag{18.385}$$

where

$$Z_0 = \int_{-\infty}^{+\infty}\prod_{i=1}^{2} d\xi_i d\xi'_i\,\delta\left(\vec{\xi}_i - \vec{\xi}'_i\right)\int_{\xi'_i}^{\xi_i}\exp\{S_0\}D\xi_i \tag{18.386}$$

or

$$Z_0 = \prod_{i=1}^{2}\left(2\sinh\left(\frac{\beta\hbar\omega_i}{2}\right)\right)^{-3} \tag{18.387}$$

and

$$S_0 = \sum_{i=1}^{2}\int_0^{\beta}d\tau\left[-\frac{\dot{\vec{\xi}}_i^{\,2}}{2\hbar^2} - \frac{\omega_i^2\vec{\xi}_i^{\,2}}{2}\right] \tag{18.388}$$

As done previously for the polaron in a quasi-1D quantum wire, considering 18.369 then

$$Z_0 = \int\delta\left(\vec{r}-\vec{r}'\right)\delta\left(\vec{R}-\vec{R}'\right)d\vec{r}d\vec{r}'d\vec{R}d\vec{R}'\int_{\vec{r}'}^{\vec{r}}\int_{\vec{R}'}^{\vec{R}}\exp\{S_0\}D\vec{r}D\vec{R} \tag{18.389}$$

Eliminating the fictitious particle variable $\vec{R}$ then

$$Z_0 = Z_{\omega_f}\int\delta\left(\vec{r}-\vec{r}'\right)d\vec{r}d\vec{r}'\int_{\vec{r}'}^{\vec{r}}\exp\left\{\int_0^{\beta}\left[-\frac{m\dot{\vec{r}}^{2}}{2\hbar^2} - \frac{m\Omega^2\vec{r}^{2}}{2} + \frac{M\omega_f^2}{2}\vec{r}^{2}\right]d\tau + \Phi_{\omega_f}\left[\vec{r}\right]\right\}D\vec{r} \tag{18.390}$$

where

$$Z_{\omega_f} = \left[2\sinh\left(\frac{\beta\hbar\omega_f}{2}\right)\right]^{-3} \tag{18.391}$$

and

$$\Phi_{\omega_f}\left[\vec{r}\right] = \frac{\hbar M\omega_f^3}{4}\int_0^\beta\int_0^\beta d\tau d\sigma \vec{r}(\sigma)\vec{r}(\sigma')F_{\omega_f}\left(|\tau-\sigma|\right) \tag{18.392}$$

So,

$$\underline{S_0} = \int_0^\beta \left[-\frac{m\dot{\vec{r}}^2}{2\hbar^2} - \frac{m\Omega^2\vec{r}^2}{2} + \frac{M\omega_f^2}{2}\vec{r}^2\right]d\tau + \Phi_{\omega_f}\left[\vec{r}\right] - 3\ln\left[2\sinh\left(\frac{\beta\hbar\omega_f}{2}\right)\right] \tag{18.393}$$

Also, from 18.368 we have

$$Z = \prod_{\vec{\kappa}}\int\delta\left(q_{\vec{\kappa}} - q_{\vec{\kappa}}'\right)\delta\left(\vec{r} - \vec{r}'\right)dq_{\vec{\kappa}}dq_{\vec{\kappa}}'d\vec{r}d\vec{r}'\int_{\vec{r}'}^{\vec{r}}\int_{q_{\vec{\kappa}}'}^{q_{\vec{\kappa}}}\exp\left\{\underline{S}\right\}D\vec{r}Dq_{\vec{\kappa}} \tag{18.394}$$

or

$$Z = \prod_{\vec{\kappa}} Z_L\int\delta\left(\vec{r} - \vec{r}'\right)d\vec{r}d\vec{r}'\int_{\vec{r}'}^{\vec{r}}D\vec{r}\exp\left\{\int_0^\beta\left(-\frac{m\dot{\vec{r}}^2}{2\hbar^2} - \frac{m\Omega^2\vec{r}^2}{2}\right)d\tau + \Phi_{\omega_{\vec{\kappa}}}\left[\vec{r}\right]\right\} \tag{18.395}$$

where

$$Z_L = \left[2\sinh\left(\frac{\beta\hbar\omega_{\vec{\kappa}}}{2}\right)\right]^{-3} \quad (18.396$$

is the lattice partition function and the functional of the electron-phonon interaction influence phase:

$$\Phi_{\omega_{\vec{\kappa}}}\left[\vec{r}\right] = \frac{\hbar}{4\omega_{\vec{\kappa}}}\int_0^\beta\int_0^\beta d\sigma d\sigma'\gamma\left(\vec{r}(\sigma)\right)\gamma\left(\vec{r}(\sigma')\right)F_{\omega_{\vec{\kappa}}}\left(|\sigma-\sigma'|\right) \tag{18.397}$$

18.14.6 Generating Function

From the partition function obtained above, the polaron characteristics can be evaluated via the relation:

$$\ln Z \geq \ln Z_0 + \left\langle \underline{S} - \underline{S_0}\right\rangle_{S_0} \tag{18.398}$$

where,

$$\underline{S}-\underline{S_0} = -\int_0^{\beta} \frac{M\omega_f^2}{2}\vec{r}^2 d\tau + \sum_{\vec{\kappa}} \Phi_{\omega_{\vec{\kappa}}}\left[\vec{r}\right] - \Phi_{\omega_f}\left[\vec{r}\right] + 3\ln\left[2\sinh\left(\frac{\beta\hbar\omega_f}{2}\right)\right] \tag{18.399}$$

To evaluate this, it is necessary to evaluate

$$\left\langle \vec{r}^2(\tau)\right\rangle_{\underline{S_0}}, \quad \left\langle \vec{r}(\tau)\vec{r}(\sigma)\right\rangle_{\underline{S_0}}, \quad \left\langle \cos\left(\vec{\kappa},\vec{r}(\tau)-\vec{r}(\sigma)\right)\right\rangle_{\underline{S_0}} \tag{18.400}$$

We have to do this via the generating function:

$$\Psi_{\vec{\kappa}}(\xi,\eta) = \left\langle \exp\left\{ i\left(\vec{\kappa}\left(\xi\vec{r}(\tau)-\eta\vec{r}(\sigma)\right)\right)\right\}\right\rangle_{\underline{S_0}} \tag{18.401}$$

The generating function is obtained in the same manner as done for the polaron in a quasi-1D quantum wire via the model Lagrangian 18.368:

$$\Psi_{\vec{\kappa}}(\xi,\eta) = \exp\left\{ \begin{array}{l} -\dfrac{\hbar\kappa^2 C_1^2(\omega_1)}{4\omega_1}\left[\left(\xi^2+\eta^2\right)\mathrm{F}_{\omega_1}(0) - 2\xi\eta\mathrm{F}_{\omega_1}\left(|\tau-\sigma|\right)\right] - \\ -\dfrac{\hbar\kappa^2 C_2^2(\omega_2)F_2^2(\omega_2)}{4\omega_2}\left[\left(\xi^2+\eta^2\right)\mathrm{F}_{\omega_2}(0) - 2\xi\eta\mathrm{F}_{\omega_2}\left(|\tau-\sigma|\right)\right] \end{array}\right\} \tag{18.402}$$

We then find

$$\left\langle \vec{r}(\tau)\vec{r}(\sigma)\right\rangle_{\underline{S_0}} = \frac{3\hbar C_1^2(\omega_1)}{2\omega_1}\mathrm{F}_{\omega_1}\left(|\tau-\sigma|\right) - \frac{3\hbar C_2^2(\omega_2)F_2^2(\omega_2)}{4\omega_2}\mathrm{F}_{\omega_2}\left(|\tau-\sigma|\right) \tag{18.403}$$

$$\left\langle \cos\left(\vec{\kappa},\vec{r}(\tau)-\vec{r}(\sigma)\right)\right\rangle_{\underline{S_0}} = \exp\left\{ \begin{array}{l} -\dfrac{\hbar\kappa^2 C_1^2(\omega_1)}{2\omega_1}\left[\mathrm{F}_{\omega_1}(0) - \mathrm{F}_{\omega_1}\left(|\tau-\sigma|\right)\right] - \\ -\dfrac{\hbar\kappa^2 C_2^2(\omega_2)F_2^2(\omega_2)}{2\omega_2}\left[\mathrm{F}_{\omega_2}(0) - \mathrm{F}_{\omega_2}\left(|\tau-\sigma|\right)\right] \end{array}\right\} \tag{18.404}$$

All the formulae so far for the quasi-0D quantum dot are observed to depend on the quantity $|\tau-\sigma|$. This implies, the formulae express retarded functions that depend on the past histories of the particle. This signifies the interaction with the past where a perturbative motion of the electron (hole) takes "**time**" to propagate in the crystal lattice. Since the retarded functions have a common argument $\left(|\tau-\sigma|-\frac{\beta}{2}\right)$, so we again generalize them by the functions $G\left(|\tau-\sigma|-\frac{\beta}{2}\right)$ and then perform the simplifications of the double integral [1,2]. For $\beta\to\infty$ ($T\to 0$) and considering the equation

$$E \le \mathrm{F}_F = -\frac{1}{\beta}\ln Z_0 - \frac{1}{\beta}\left\langle \underline{S}-\underline{S_0}\right\rangle_{\underline{S_0}} \tag{18.405}$$

and

$$\ln Z = \ln\left[\frac{V}{(2\pi\hbar)^3}\left(\frac{2m\pi}{\beta}\right)^{\frac{3}{2}}\right] - \beta\mathrm{E}_0(v) + \frac{3}{2}\ln\left(\frac{m^*}{m}\right) \tag{18.406}$$

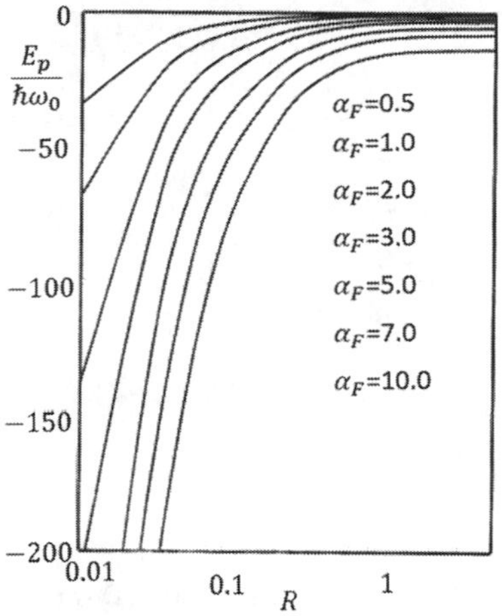

FIGURE 18.7 Depicts the dimensionless polaron ground state energy, $\frac{E_p}{\hbar\omega_0}$ versus the radius, **R** of the quantum dot for different electron-phonon interaction coupling constant, $\boldsymbol{\alpha}_F$ taken from reference [2].

the polaron variational energy can then be obtained as follows [2]:

$$E_p = \frac{3}{2}\hbar\left(\omega_1 + \omega_2 + \omega_f\right) + \frac{3\hbar m\Omega^2}{4}\left(\frac{C_1^2\left(\omega_1\right)}{\omega_1} + \frac{C_2^2\left(\omega_2\right)F_2^2\left(\omega_2\right)}{\omega_2}\right)$$

$$+\frac{3\hbar M\omega_f^2}{4}\left(\frac{C_1^2\left(\omega_1\right)}{\omega_1+\omega_f} + \frac{C_2^2\left(\omega_2\right)F_2^2\left(\omega_2\right)}{\omega_2+\omega_f}\right) + \frac{\hbar^2\omega_0^2 R_p \alpha_F}{4}\int_0^\infty d\tau \frac{\exp\left\{-\hbar\omega_0\tau\right\}}{\sqrt{A(\tau)}} \tag{18.407}$$

Here,

$$A(\tau) = \frac{\hbar C_1^2\left(\omega_1\right)}{2\omega_1}\left(1-\exp\left\{-\hbar\omega_2\tau\right\}\right) + \frac{\hbar C_2^2\left(\omega_2\right)F_2^2\left(\omega_2\right)}{2\omega_2}\left(1-\exp\left\{-\hbar\omega_2\tau\right\}\right) \tag{18.408}$$

The effective mass is conveniently obtained from the equation of motion [2]:

$$\frac{m_p}{m} = \frac{\left(\omega_1^2 - \Omega^2\right)\left(\Omega^2 - \omega_2^2\right)}{\omega_1^2\omega_2^2} \tag{18.409}$$

From Figure 18.7, we observe that the polaron energy increases with reducing radius of the quantum dot and also increasing electron-phonon coupling constant. From here we observe also increase polaron energy as a result of enhanced confinement.

18.15 Bipolaron Characteristics in a Quasi-0D Spherical Quantum Dot

18.15.1 Introduction

We examine the bipolaron characteristics in a spherical quantum dot with an all-sided parabolic electronic confinement potential. The confinement potential is selected for technological reasons and aid also for path integration to be performed exactly.

18.15.2 Model Lagrangian

The model Lagrangian for the bipolaron problem is selected in the same fashion as that of the quasi-1D quantum dot in the one oscillatory form:

$$L_0 = -\frac{m\dot{r}_1^2}{2\hbar^2} - \frac{m\dot{r}_2^2}{2\hbar^2} - \frac{M\dot{R}_1^2}{2\hbar^2} - \frac{M\dot{R}_2^2}{2\hbar^2} - \frac{\kappa\left(\vec{r}_1 - \vec{R}_1\right)^2}{2} - \frac{\kappa\left(\vec{r}_2 - \vec{R}_2\right)^2}{2}$$
$$+\frac{K\left(\vec{r}_1 - \vec{r}_2\right)^2}{2} - \frac{\kappa'\left(\vec{r}_1 - \vec{R}_2\right)^2}{2} - \frac{\kappa'\left(\vec{r}_2 - \vec{R}_1\right)^2}{2} - \frac{m\Omega^2\left(r_1^2 + r_2^2\right)}{2} \tag{18.410}$$

18.15.3 Model Lagrangian

18.15.3.1 Equation of Motion and Normal Modes

With the same procedure as for the bipolaron in the quasi-1D quantum wire, we find the equation of motion from

$$\frac{d}{dt}\frac{\partial}{\partial \dot{q}}L_0 - \frac{\partial}{\partial q}L_0 = 0 \tag{18.411}$$

and then set

$$\ddot{\vec{r}}_i = -\omega^2\vec{r}_i, \ddot{\vec{R}}_i = -\omega^2\vec{R}_i \tag{18.412}$$

where, ω is the frequency of the normal modes. W then introduce new coordinates with

$$\vec{\eta}_1 = \frac{\vec{r}_1 + \vec{r}_2}{2} \tag{18.413}$$

the centre of mass of the two electrons:

$$\vec{r} = \vec{r}_1 - \vec{r}_2 \tag{18.414}$$

the relative distance between the electrons:

$$\vec{\eta}_2 = \frac{\vec{R}_1 + \vec{R}_2}{2} \tag{18.415}$$

the centre of mass of two fictitious particle and

$$\vec{R} = \vec{R}_1 - \vec{R}_2 \tag{18.416}$$

the relative distance between the fictitious particles.

This change of variables transforms the resultant system of equations:

$$\left(P + m\Omega^2 - m\omega^2\right)\vec{\eta}_1 - p'\vec{\eta}_2 = 0, -p'\vec{\eta}_1 + \left(p - M\omega^2\right)\vec{\eta}_2 = 0 \tag{18.417}$$

$$\left(P + m\Omega^2 - m\omega^2\right)\vec{r} - p'\vec{R} = 0, -p'\vec{r} + \left(p - M\omega^2\right)\vec{R} = 0 \tag{18.418}$$

This permits us to have the equations for the normal modes with frequencies:

$$\omega_{1,2}^2 = \frac{1}{2}\left(\frac{p}{\mu}+\Omega^2\right) \pm \frac{1}{2}\left[\left(\frac{p}{\mu}+\Omega^2\right)^2 - \frac{4p\Omega^2}{M}\right]^{\frac{1}{2}}, \frac{1}{\mu} = \frac{1}{m}+\frac{1}{M} \tag{18.419}$$

$$\omega_{3,4}^2 = \frac{1}{2}\left(\Omega^2 + \frac{PM+pm}{2Mm}\right) \pm \frac{1}{2}\left[\left(\Omega^2 + \frac{PM+pm}{2Mm}\right)^2 + 4\frac{\left(m\Omega^2+P\right)p - p'^2}{Mm}\right]^{\frac{1}{2}} \tag{18.420}$$

$$\omega_1^2\omega_2^2 = \omega_f^2\Omega^2 \tag{18.421}$$

where

$$P = p - 2\mathrm{K}, p' = \kappa - \kappa', p = \kappa + \kappa' \tag{18.422}$$

The **effective bipolaron mass** may be conveniently obtained from 18.419:

$$\frac{M}{m} = \frac{\left(\omega_1^2 - \Omega^2\right)\left(\Omega^2 - \omega_2^2\right)}{\omega_1^2\omega_2^2} \tag{18.423}$$

18.15.4 Diagonalization of the Lagrangian

The variables $\vec{\eta}_1$ and $\vec{\eta}_2$ are the superposition of normal modes with frequencies ω_1 and ω_2:

$$\vec{\eta}_1(\omega_1) = C_1(\omega_1)\vec{x}_1(\omega_1), \vec{\eta}_2(\omega_1) = C_1(\omega_1)F_1(\omega_1)\vec{x}_1(\omega_1) \tag{18.424}$$

$$\vec{\eta}_1(\omega_2) = C_2(\omega_2)F_2(\omega_2)\vec{x}_2(\omega_2), \vec{\eta}_2(\omega_2) = C_2(\omega_2)\vec{x}_2(\omega_2) \tag{18.425}$$

Substituting these equations into 18.417 then

$$\mathrm{F}_1(\omega_1) = \frac{p}{-M\omega_1^2 + p}, \mathrm{F}_2(\omega_2) = \frac{p}{-m\omega_2^2 + m'^{\,2} + p} \tag{18.426}$$

Similarly, considering 18.420 we represent $\vec{r}$ and $\vec{R}$ as a superposition of normal modes and the result substituted into 18.418 then we have

$$\mathrm{F}_3(\omega_3) = \frac{p'}{-M\omega_3^2 + p}, \mathrm{F}_4(\omega_4) = \frac{p'}{-m\omega_4^2 + m\Omega^2 + P} \tag{18.427}$$

Consider,

$$\vec{\xi}_1 = 2\vec{\eta}_1, \vec{\xi}_2 = 2\vec{\eta}_2 \tag{18.428}$$

then from the conservation of kinetic energy

$$m\dot{\vec{x}}_1^{\,2}(\omega_1) = m\dot{\vec{\xi}}_1^{\,2}(\omega_1) + M\dot{\vec{\xi}}_2^{\,2}(\omega_1) \tag{18.429}$$

we have

$$C_1^2(\omega_1) = \frac{m}{m + MF_1^2(\omega_1)} \tag{18.430}$$

Similarly,

$$m\dot{\vec{x}}_2^{\,2}(\omega_1) = m\dot{\vec{\xi}}_1^{\,2}(\omega_2) + M\dot{\vec{\xi}}_2^{\,2}(\omega_2) \tag{18.431}$$

and

$$C_2^2(\omega_1) = \frac{M}{M + mF_2^2(\omega_2)} \tag{18.432}$$

In a similar fashion for

$$m\dot{\vec{x}}_3^{\,2}(\omega_3) = m\dot{\vec{r}}^{\,2}(\omega_3) + M\dot{\vec{R}}^{\,2}(\omega_3) \tag{18.433}$$

and

$$m\dot{\vec{x}}_4^{\,2}(\omega_4) = m\dot{\vec{r}}^{\,2}(\omega_4) + M\dot{\vec{R}}^{\,2}(\omega_4) \tag{18.434}$$

We then have

$$C_3^2(\omega_3) = \frac{m}{m + MF_3^2(\omega_3)} \tag{18.435}$$

$$C_4^2(\omega_4) = \frac{M}{M + mF_4^2(\omega_4)} \tag{18.436}$$

Substituting all the normal coordinates into the model Lagrangian 18.273 and considering

$$\omega_1^2 = C_1^2(\omega_1)\left[m\Omega^2 + P + pF_1^2(\omega_1) - 2p'F_1(\omega_1)\right] \tag{18.437}$$

$$\omega_2^2 = C_2^2(\omega_2)\left[\left(m\Omega^2 + P\right)F_2^2(\omega_2) + p - 2p'F_2(\omega_2)\right] \tag{18.438}$$

$$\omega_3^2 = C_3^2(\omega_3)\left[m\Omega^2 + P + pF_3^2(\omega_3) - 2p'F_3(\omega_3)\right] \tag{18.439}$$

$$\omega_4^2 = C_4^2(\omega_4)\left[\left(m\Omega^2 + P\right)F_4^2(\omega_4) + p - 2p'F_4(\omega_4)\right] \tag{18.440}$$

then

$$L_0 = \sum_{i=1}^{4}\left[-\frac{\dot{x}_i^2}{4\hbar^2} - \frac{\omega_i^2 x_i^2}{4}\right] \tag{18.441}$$

18.15.5 Partition Function

To find the bipolaron energy we start with

$$E_0 = -T \ln Z_0 \tag{18.442}$$

where

$$Z_0 = \int_{-\infty}^{+\infty} \prod_{i=1}^{4} d\xi_i dx_i \delta\left(x_i - x'_i\right) \int_{x'_i}^{x_i} \exp\left\{\underline{S}_0\right\} Dx_i D\xi_i \tag{18.443}$$

The partition function has the product form:

$$Z_0 = \prod_{i=1}^{4} \left(2\sinh\left(\frac{\beta\hbar\omega_i}{2}\right)\right)^{-3} \tag{18.444}$$

From the model Lagrangian 18.410, we follow the same procedure as for the bipolaron in a quasi-1D quantum wire to find the partition function:

$$Z_0 = \int \prod_{i=1}^{2} d\vec{r}_i d\vec{r}'_i dZ_{if} dZ'_{if} \delta\left(\vec{r}_i - \vec{r}'_i\right) \delta\left(\vec{R}_i - \vec{R}'_i\right) \int_{\vec{r}'_i}^{\vec{r}_i} D\vec{r}_i \int_{\vec{R}'_i}^{\vec{R}_i} D\vec{R}_i \exp\left\{\underline{S}_0\right\} \tag{18.445}$$

or

$$Z_0 = \int \prod_{i=1}^{2} \delta\left(\vec{r}_i - \vec{r}'_i\right) d\vec{r}_i d\vec{r}'_i \int_{\vec{r}'_i}^{\vec{r}_i} D\vec{r}_i \exp\left\{\underline{S}_0\left[\vec{r}_1, \vec{r}_2\right]\right\} \tag{18.446}$$

Here,

$$\begin{aligned} \underline{S}_0\left[\vec{r}_1,\vec{r}_2\right] = & -\frac{m}{2\hbar^2}\int_0^{\beta}\left(\dot{\vec{r}}_1^{\,2} + \dot{\vec{r}}_2^{\,2}\right)d\tau - \frac{m\Omega^2 + p}{2}\int_0^{\beta}\left(r_1^2 + r_2^2\right)d\tau + \frac{\mathrm{K}}{2}\int_0^{\beta}\left(\vec{r}_1 - \vec{r}_2\right)^2 d\tau \\ & + \frac{\hbar}{4M\omega_{1f}}\int_0^{\beta}\int_0^{\beta}\left[\left(\kappa\vec{r}_1(\tau) + \kappa'\vec{r}_2(\tau)\right)\left(\kappa\vec{r}_1(\sigma) + \kappa'\vec{r}_2(\sigma)\right)\right]\mathrm{F}_{\omega_f}\left(\left|\tau - \sigma\right|\right)d\tau d\sigma \\ & + \frac{\hbar}{4M\omega_{2f}}\int_0^{\beta}\int_0^{\beta}\left[\left(\kappa'\vec{r}_1(\tau) + \kappa\vec{r}_2(\tau)\right)\left(\kappa'\vec{r}_1(\sigma) + \kappa\vec{r}_2(\sigma)\right)\right]\mathrm{F}_{\omega_f}\left(\left|\tau - \sigma\right|\right)d\tau d\sigma \\ & + 6\ln\left[2\sinh\left(\frac{\beta\hbar\omega_f}{2}\right)\right] \end{aligned} \tag{18.447}$$

We write the Lagrangian of the electron-lattice interaction:

$$L(-i\hbar\tau) = -\frac{m\dot{\vec{r}}_1^{\,2}}{2\hbar^2} - \frac{m\dot{\vec{r}}_2^{\,2}}{2\hbar^2} - \frac{m\Omega^2\left(\vec{r}_1^{\,2} + \vec{r}_2^{\,2}\right)}{2} - \frac{1}{2}\sum_{\vec{\kappa}}\left(\frac{\dot{q}_{\vec{\kappa}}^2}{\hbar^2} + \omega_{\vec{\kappa}}^2 q_{\vec{\kappa}}^2\right) + \sum_{\vec{\kappa}, j=1,2} \gamma_{\vec{\kappa}}\left(\vec{r}_j\right) q_{\vec{\kappa}}(\tau) - U\left(\left|\vec{r}_1 - \vec{r}_2\right|\right) \tag{18.448}$$

The amplitude of the electron-phonon coupling interaction, $\gamma_{\vec{\kappa}}(\tau)$ for the $\vec{\kappa}$ moving wave is defined as:

$$\gamma_{\vec{\kappa}}(\tau) = \frac{\hbar}{\sqrt{V}} C_{\vec{\kappa}} \chi_{\vec{\kappa}}(\tau) \tag{18.449}$$

with

$$C_{\vec{\kappa}} = \left(\frac{4R_p\pi\alpha_F\omega_0}{\kappa^2}\right)^{\frac{1}{2}}, \chi_{\vec{\kappa}}(\tau) = \begin{cases} \sin\vec{\kappa}\vec{r}, & \kappa \geq 0 \\ \cos\vec{\kappa}\vec{r}, & \kappa < 0 \end{cases} \quad (18.450)$$

18.15.6 Full System Influence Phase

After eliminating the phonon variables from the action functional for the full system we obtain:

$$\underline{S}[\vec{r}_1,\vec{r}_2] = -\frac{m}{2\hbar^2}\int_0^\beta \left(\dot{\vec{r}}_1^{\,2} + \dot{\vec{r}}_2^{\,2}\right)d\tau + \frac{m\Omega^2}{2}\int_0^\beta \left(\vec{r}_1^{\,2} + \vec{r}_2^{\,2}\right)d\tau + \Phi[\vec{r}_1,\vec{r}_2] + \int_0^\beta U(|\vec{r}_1 - \vec{r}_2|)d\tau \quad (18.451)$$

Here the influence phase of the full system is:

$$\Phi[\vec{r}_1,\vec{r}_2] = \sum_{\vec{\kappa}} \Phi_{\omega_{\vec{\kappa}}}[\vec{r}_1,\vec{r}_2] \quad (18.452)$$

with

$$\Phi[\vec{r}_1,\vec{r}_2] = \sum_{\vec{\kappa},j=1,2} \frac{\hbar}{8\omega_{\vec{\kappa}}}\int_0^\beta\int_0^\beta d\sigma d\sigma' \gamma_{\vec{\kappa}}(\vec{r}_j(\sigma'))\gamma_{\vec{\kappa}}(\vec{r}_j(\sigma))F_{\omega_{\vec{\kappa}}}(|\sigma - \sigma'|) \quad (18.453)$$

18.16 Bipolaron Energy

18.16.1 Generating Function

In a similar manner as done previously, we find now $\left\langle \underline{S} - \underline{S}_0 \right\rangle_{\underline{S}_0}$ where it is necessary to evaluate

$$\left\langle\left(\vec{r}_1^{\,2} + \vec{r}_2^{\,2}\right)\right\rangle_{\underline{S}_0}, \left\langle\left(\vec{r}_1(\tau)\vec{r}_2(\sigma) + \vec{r}_2(\tau)\vec{r}_1(\sigma)\right)\right\rangle_{\underline{S}_0}, \left\langle\left(\vec{r}_1(\tau)\vec{r}_1(\sigma) + \vec{r}_2(\tau)\vec{r}_2(\sigma)\right)\right\rangle_{\underline{S}_0} \eta \quad (18.454)$$

$$\left\langle\cos\left(\vec{\kappa},\vec{r}_1(\tau) - \vec{r}_2(\sigma)\right)\right\rangle_{\underline{S}_0}, \left\langle\cos\left(\vec{\kappa},\vec{r}_1(\tau) - \vec{r}_1(\sigma)\right)\right\rangle_{\underline{S}_0}, \left\langle\cos\left(\vec{\kappa},\vec{r}_2(\tau) - \vec{r}_2(\sigma)\right)\right\rangle_{\underline{S}_0} \quad (18.455)$$

with the help of the generating function:

$$\Psi_{\vec{\kappa}}(\xi_1,\eta_1,\xi_2,\eta_2) = \left\langle \exp\left\{i\vec{\kappa}\left(\xi_1\vec{r}_1(\tau) - \eta_1\vec{r}_1(\sigma) + \xi_2\vec{r}_2(\tau) - \eta_2\vec{r}_2(\sigma)\right)\right\}\right\rangle_{\underline{S}_0} \quad (18.456)$$

or

$$\Psi_{\vec{\kappa}}(\xi_1,\eta_1,\xi_2,\eta_2) = \left\langle \exp\left\{\kappa^2\left(f_1 + f_2 + f_3 + f_4\right)\right\}\right\rangle_{\underline{S}_0} \quad (18.457)$$

where

$$f_1 = -\frac{\hbar C_1^2(\omega_1)}{8m\omega_1}\left[f_1^{+}\mathrm{F}_{\omega_1}(0) - 2f_2^{+}\mathrm{F}_{\omega_1}(|\tau - \sigma|)\right] \quad (18.458)$$

$$f_2 = -\frac{\hbar C_2^2(\omega_2)F_2^2(\omega_2)}{8m\omega_2}\left[f_1^+\mathrm{F}_{\omega_2}(0) - 2f_2^+\mathrm{F}_{\omega_2}(|\tau-\sigma|)\right] \tag{18.459}$$

$$f_3 = -\frac{\hbar C_3^2(\omega_3)}{8M\omega_3}\left[f_1^-\mathrm{F}_{\omega_3}(0) - 2f_2^-\mathrm{F}_{\omega_3}(|\tau-\sigma|)\right] \tag{18.460}$$

$$f_4 = -\frac{\hbar C_4^2(\omega_4)F_4^2(\omega_4)}{8M\omega_4}\left[f_1^-\mathrm{F}_{\omega_4}(0) - 2f_2^-\mathrm{F}_{\omega_4}(|\tau-\sigma|)\right] \tag{18.461}$$

$$f_1^+ = (\xi_1+\xi_2)^2 + (\eta_1+\eta_2)^2,\ f_2^+ = (\xi_1+\xi_2)(\eta_1+\eta_2) \tag{18.462}$$

$$f_1^- = (\xi_1-\xi_2)^2 + (\eta_1-\eta_2)^2,\ f_2^- = (\xi_1-\xi_2)(\eta_1-\eta_2) \tag{18.463}$$

18.16.2 Bipolaron Characteristics

From here, we now evaluate the quantities in $\langle \underline{S} - \underline{S_0}\rangle_{S_0}$:

$$\langle \vec{r}_1(\tau)\vec{r}_1(\sigma)\rangle_{S_0} = \frac{3\hbar C_1^2}{4\omega_1}\mathrm{F}_{\omega_1}(|\tau-\sigma|) + \frac{3\hbar C_2^2F_2^2}{4\omega_2}\mathrm{F}_{\omega_2}(|\tau-\sigma|) + \frac{3\hbar C_3^2}{4\omega_3}\mathrm{F}_{\omega_3}(|\tau-\sigma|) + \frac{3\hbar C_4^2F_4^2}{4\omega_4}\mathrm{F}_{\omega_4}(|\tau-\sigma|) \tag{18.464}$$

$$\langle \vec{r}_1(\tau)\vec{r}_2(\sigma)\rangle_{S_0} = \frac{3\hbar C_1^2}{4\omega_1}\mathrm{F}_{\omega_1}(|\tau-\sigma|) + \frac{3\hbar C_2^2F_2^2}{4\omega_2}\mathrm{F}_{\omega_2}(|\tau-\sigma|) - \frac{3\hbar C_3^2}{4\omega_3}\mathrm{F}_{\omega_3}(|\tau-\sigma|) - \frac{3\hbar C_4^2F_4^2}{4\omega_4}\mathrm{F}_{\omega_4}(|\tau-\sigma|) \tag{18.465}$$

$$\langle \vec{r}_1(\tau)\vec{r}_2(\tau)\rangle_{S_0} = \frac{3\hbar C_1^2}{4\omega_1}F_{\omega_1}(|0|) + \frac{3\hbar C_2^2F_2^2}{4\omega_2}F_{\omega_2}(|0|) - \frac{3\hbar C_3^2}{4\omega_3}F_{\omega_3}(|0|) - \frac{3\hbar C_4^2F_4^2}{4\omega_4}F_{\omega_4}(|0|) \tag{18.466}$$

$$\langle(\vec{r}_1(\tau) - \vec{r}_2(\tau))\rangle_{S_0} = \frac{3\hbar C_1^2}{4\omega_1}\mathrm{F}_{\omega_1}(|0|) + \frac{3\hbar C_4^2F_4^2}{4\omega_4}\mathrm{F}_{\omega_4}(|0|) \tag{18.467}$$

$$\langle \vec{r}_1(\tau)\vec{r}_2(\sigma)\rangle_{S_0} = \langle \vec{r}_2(\tau)\vec{r}_1(\sigma)\rangle_{S_0} \tag{18.468}$$

$$\langle \vec{r}_1^{\,2}(\tau)\rangle_{S_0} \equiv \langle \vec{r}_1(\tau)\vec{r}_1(\sigma)\rangle_{S_0}\Big|_{\tau=\sigma} \tag{18.469}$$

$$\left\langle \exp\{i\kappa(z_1(\tau) - z_1(\sigma))\}\right\rangle_{S_0} = \Psi_{\vec{\kappa}}(\xi_1 = 1, \eta_1 = 1, \xi_2 = 0, \eta_2 = 0) \tag{18.470}$$

$$\left\langle \exp\{i\kappa(z_2(\tau) - z_1(\sigma))\}\right\rangle_{S_0} = \Psi_{\vec{\kappa}}(\xi_1 = 1, \eta_1 = 0, \xi_2 = 0, \eta_2 = 1) \tag{18.471}$$

$$\left\langle \exp\{i\kappa(z_2(\tau) - z_1(\sigma))\}\right\rangle_{S_0} = \left\langle \exp\{i\kappa(z_1(\tau) - z_2(\sigma))\}\right\rangle \tag{18.472}$$

Again as seen previously, all the formulae in the bipolaron has the dependence on the quantity $|\tau-\sigma|$ confirming the fact that the quantities (retarded functions) depend on the past with the significance of interaction with the past being the perturbation due to the moving electrons (holes) that take "**time**" propagating in the crystal lattice.

We again generalize the function in the integrand of the bipolaron to be $G\left(|\tau-\sigma| - \frac{\beta}{2}\right)$ to aid in the simplification of the evaluation of the integral [1,2]:

$$f(\beta) = \int_0^\beta\int_0^\beta G\left(|\tau-\sigma| - \frac{\beta}{2}\right)d\sigma d\tau \tag{18.473}$$

We do the change of variables

$$\tau - \sigma \equiv \rho \quad , \quad 0 \le \sigma \le \tau \tag{18.474}$$

then

$$f(\beta) = 2\int_0^{\beta} G\left(\rho - \frac{\beta}{2}\right)(\beta - \rho)\,d\rho \tag{18.475}$$

We consider again the change of variable $\rho = \beta y$, where $0 \le y \le 1$. If $y = 1$ then $\rho = \beta$. Again, if we do the change of variable

$$\frac{\beta(1-y)}{2} \equiv -\tau, \quad d\tau = -\frac{\beta}{2}dy, \quad 0 \le \tau \le \infty \tag{18.476}$$

then from $\beta \to \infty$ ($T \to 0$), we evaluate the asymptotic expression of Feynman bipolaron dimensionless energy [1,2]:

$$\begin{aligned} E_{bip} = {} & \frac{3}{2}\hbar\sum_{i=1}^{4}\omega_i + 3\hbar\omega_f - \frac{3\hbar p}{4}\left(\frac{C_1^2(\omega_1)}{\omega_1+\omega_f} + \frac{C_2^2(\omega_2)F_2^2(\omega_2)}{\omega_2+\omega_f}\right) \\ & - \frac{3\hbar P}{4}\left(\frac{C_3^2(\omega_3)}{\omega_3} + \frac{C_4^2(\omega_4)F_4^2(\omega_4)}{\omega_4}\right) + \frac{3\hbar p'^2\omega_f}{4p}\left(\frac{C_3^2(\omega_3)}{\omega_3(\omega_3+\omega_f)} + \frac{C_4^2(\omega_4)F_4^2(\omega_4)}{\omega_4(\omega_4+\omega_f)}\right) \\ & + \frac{2^{\frac{1}{2}}\hbar^{\frac{1}{2}}\omega_0 R_p \alpha_F}{(1-\eta)\left(\frac{C_3^2(\omega_3)}{\omega_3} + \frac{C_4^2(\omega_4)F_4^2(\omega_4)}{\omega_4}\right)^{\frac{1}{2}}} + \frac{\hbar^2\omega_0^2 R_p \alpha_F}{\pi^{\frac{1}{2}}}\int_0^{\infty} d\tau \frac{\exp\{-\hbar\omega_0\tau\}}{\sqrt{A(\tau)}} \end{aligned} \tag{18.477}$$

Here,

$$A(\tau) = \sum_{i=1}^{4} d_i^2\left(1 - \exp\{-\hbar\omega_i\tau\}\right) \tag{18.478}$$

where,

$$d_1^2 = \frac{\omega_1(\Omega^2 - \omega_2^2)}{m'^{\,2}(\omega_1^2 - \omega_2^2)}, d_2^2 = \frac{\omega_f^2(\omega_1^2 - \Omega^2)}{m\omega_2(\omega_1^4 - \omega_f^2\Omega^2)}, d_3^2 = \frac{\omega_3^2 - \omega_f^2}{m\omega_3(\omega_3^2 - \omega_4^2)}, d_4^2 = \frac{\omega_f^2 - \omega_4^2}{m\omega_1(\omega_3^2 - \omega_4^2)} \tag{18.479}$$

The effective mass is conveniently obtained from the equation of motion:

$$\frac{m_p}{m} = \frac{(\omega_1^2 - \Omega^2)(\Omega^2 - \omega_2^2)}{\omega_1^2\omega_2^2} \tag{18.480}$$

For the transition to the polaron problem for the quasi-0D spherical quantum dot we set:

$$K = 0, \kappa' = 0, P = p, p' = \kappa, p = \kappa \tag{18.481}$$

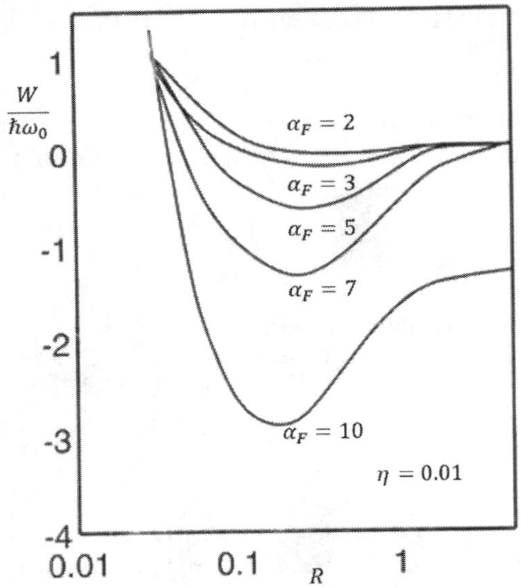

FIGURE 18.8 Depicts the dimensionless bipolaron coupling energy, $\frac{W}{\hbar\omega_0}$ versus the radius, $\boldsymbol{R}$ of the quantum dot for different electron-phonon interaction coupling constant, $\boldsymbol{\alpha}_F$ and fixed $\boldsymbol{\eta} = 0.01$ taken from reference [2].

We observe after substituting this into the bipolaron energy that

$$\omega_1 = \omega_3, \omega_2 = \omega_4 \tag{18.482}$$

$$F_1 = F_3, F_2 = F_4 \tag{18.483}$$

$$C_1^2 = C_3^2, C_2^2 = C_4^2 \tag{18.484}$$

From here

$$E_{bip} = 2E_p \tag{18.485}$$

This shows the bipolaron to constitute a stable pair of two polarons.

In [2], from the analytical results of the bipolaron coupling energy as well as Figure 18.8, it is observed from the curve of the bipolaron coupling energy versus the radius of the quantum dot, that for different coupling constants, α_F there exist a non-monotonic dependence. The bipolaron coupling energy faster decreases through a minimum at the neighborhood of the confinement radius $R \approx 0$, and then starts increasing. This effect may be explained by the fact that, when electrons are forcefully brought closer together, there arises enhanced polarization as well as increased Coulomb repulsion. For small R, when the bipolaron radius becomes greater than the radius of the quantum dot, this obviously leads to bringing the electrons very near to each other and obviously mutual repulsion is enhanced and, consequently, there is an increase of the bipolaron energy. This phenomenon is observed in references [2,56].

18.17 Polaron Characteristics in a Cylindrical Quantum Dot

We apply path integration in the same fashion as done for the geometries above but now for a cylindrical quantum dot. We consider the motion of the electron in the z-axis direction to be bounded by an infinite high rectangular potential well while bounded on the oxy-plane by a parabolic potential (Figure 18.9).

18.17.1 System Hamiltonian

The Hamiltonian of the system is written in the form:

$$\hat{\mathrm{H}} = \frac{\hat{p}_{\perp}^2}{2m_{\perp}} + \frac{\hat{p}_{\parallel}^2}{2m_{\parallel}} + V(z) + \frac{m_{|}\Omega^2\vec{\rho}^2}{2} + \sum_{\vec{\kappa}} \hbar\omega_{\vec{\kappa}}\hat{b}_{\vec{\kappa}}\hat{b}_{\vec{\kappa}} + \frac{\hbar}{\sqrt{V}}\sum_{\vec{\kappa}}\left(C_{\vec{\kappa}}\hat{b}_{\vec{\kappa}}\exp\{i\vec{\kappa}\vec{r}\} + \text{h.c.}\right) \quad (18.486)$$

Here, $V(z)$ is the confinement potential (infinite high rectangular potential well) in the direction of the oz-axis. The fourth summand of the Hamiltonian is the transversal parabolic confinement potential.

The state of the electron is described by the variational wave function that has a large spread compared to the ground state wave function:

$$\Psi_0 = \left(\frac{\lambda}{\pi}\right)^{\frac{1}{2}} A\exp\left\{-\frac{\lambda z^2}{2}\right\} \quad (18.487)$$

Averaging the Hamiltonian 18.486 by the wave function 18.487 we have

$$\tilde{\mathrm{H}} = \frac{\hat{p}_{\perp}^2}{2m_{\perp}} + \tilde{V}(z) + \frac{m_{\perp}\Omega^2\vec{\rho}^2}{2} + \sum_{\vec{\kappa}} \hbar\omega_{\vec{\kappa}}\hat{b}_{\vec{\kappa}}\hat{b}_{\vec{\kappa}} + \frac{\hbar}{\sqrt{V}}\sum_{\vec{\kappa}}\left(\widetilde{C_{\vec{\kappa}}}\hat{b}_{\vec{\kappa}}\exp\{i\vec{\kappa}_{\perp}\vec{\rho}\} + \text{h.c.}\right) \quad (18.488)$$

Here,

$$\widetilde{C_{\vec{\kappa}}} = C_{\vec{\kappa}}\exp\left\{-\frac{\kappa_{\parallel}^2}{4\lambda}\right\} \quad (18.489)$$

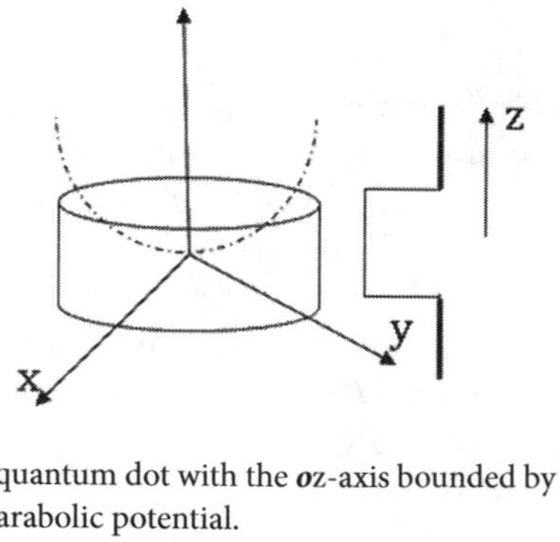

FIGURE 18.9 Depicts a cylindrical quantum dot with the **oz**-axis bounded by an infinite high rectangular potential while the oxy-plane is bounded by a parabolic potential.

From the Feynman variational principle, the interaction of the electron with the polarized vibrations of the crystal is modelled by an elastic coupling of the electron and a fictitious particle that attracts the electron to itself. So, the effect of the polarized crystal lattice on the electron is approximated to an elastic attraction of the second particle. From these analyses, the model Lagrangian may be selected in the one oscillatory approximation:

$$L_0 = -\frac{m_\perp \dot{\vec{\rho}}^{\,2}}{2\hbar^2} - \frac{m_\perp \Omega^2 \vec{\rho}^{\,2}}{2} - \frac{M\dot{\vec{\mathrm{R}}}^2}{2\hbar^2} - \frac{M\omega_f^2\left(\vec{\mathrm{R}}-\vec{\rho}\right)^2}{2} \tag{18.490}$$

Here M and ω_f are, respectively, the mass of the fictitious particle and the frequency of the elastic coupling serving as variational parameters; $\vec{\mathrm{R}}$ is the coordinate of the fictitious particle.

18.17.2 Transformation to Normal Coordinates

18.17.2.1 Lagrangian Diagonalization

For normal modes we substitute the following harmonic coordinates:

$$\ddot{\vec{\rho}} = -\omega^2\vec{\rho}, \ddot{\vec{\mathrm{R}}} = -\omega^2\vec{\mathrm{R}} \tag{18.491}$$

into the equation of motion

$$m_\perp \ddot{} + m_\perp\Omega_\perp^2\vec{\rho} - \kappa_\perp\left(\vec{\mathrm{R}}-\vec{\rho}\right) = 0, M\ddot{\vec{\mathrm{R}}} + \kappa_\perp\left(\vec{\mathrm{R}}-\vec{\rho}\right) = 0 \tag{18.492}$$

and solving for the frequency eigenmodes we have

$$\omega_{1,2}^2 = \frac{1}{2}\left(\frac{\kappa_\perp}{\mu_\perp} + \Omega_\perp^2\right) \pm \frac{1}{2}\left[\left(\frac{\kappa_\perp}{\mu_\perp} + \Omega_\perp^2\right)^2 - \frac{4\kappa_\perp\Omega_\perp^2}{M}\right]^{\frac{1}{2}}, \frac{1}{\mu_\perp} = \frac{1}{m_\perp} + \frac{1}{M} \tag{18.493}$$

with

$$\omega_1^2\omega_2^2 = \omega_f^2\Omega_\perp^2 \tag{18.494}$$

The **effective polaron mass** is conveniently obtained from the frequency eigenmodes in 18.493:

$$\frac{M}{m_\perp} = \frac{\left(\omega_1^2 - \Omega_\perp^2\right)\left(\Omega_\perp^2 - \omega_2^2\right)}{\omega_1^2\omega_2^2} \tag{18.495}$$

The frequency eigenmodes in 18.493 permit us to move to normal mode coordinates as previously done:

$$\vec{\rho}(\tau) = \vec{\rho}_1(\omega_1) + \vec{\rho}_2(\omega_2) = C_1(\omega_1)\vec{\xi}_1(\omega_1) + C_2(\omega_2)F_2(\omega_2)\vec{\xi}_2(\omega_2) \tag{18.496}$$

$$\vec{\mathrm{R}}(\tau) = \vec{\mathrm{R}}_1(\omega_1) + \vec{\mathrm{R}}_2(\omega_2) = C_1(\omega_1)F_1(\omega_1)\vec{\xi}_1(\omega_1) + C_2(\omega_2)\vec{\xi}_2(\omega_2) \tag{18.497}$$

Inserting these equations for the normal coordinates into the equation of motion 18.492 and also considering the conservation of kinetic energy in any of the representations then

$$F_1(\omega_1) = \frac{\kappa_\perp}{-M\omega_1^2 + \kappa_\perp}, C_1^2(\omega_1) = \frac{1}{m_\perp + MF_1^2(\omega_1)} \tag{18.498}$$

$$F_2(\omega_2) = \frac{\kappa_\perp}{-m_\perp\omega_2^2 + \kappa_\perp + m_\perp\Omega_\perp^2}, C_2^2(\omega_2) = \frac{1}{M + m_\perp F_2^2(\omega_2)} \tag{18.499}$$

$$C_2^2(\omega_2) - \frac{m_\perp}{M} C_1^2(\omega_1), F_2(\omega_2) = -\frac{M}{m_\perp} F_1(\omega_1) \tag{18.500}$$

Substituting 18.497 and 18.496 into the model Lagrangian 18.490 is then diagonalized:

$$L_0 = -\frac{\dot{\xi}_1^2}{2\hbar^2} - \frac{\omega_1^2\xi_1^2}{2} - \frac{\dot{\xi}_2^2}{2\hbar^2} - \frac{\omega_2^2\xi_2^2}{2} \tag{18.501}$$

18.17.3 Polaron Energy/Partition Function

We follow the same procedure for the evaluation of the energy and effective mass of the polaron via the relation:

$$E \le F_F = -\frac{1}{\beta}\ln Z_0 - \frac{1}{\beta}\langle \underline{S} - \underline{S_0}\rangle_{\underline{S_0}} \tag{18.502}$$

The partition function Z_0:

$$Z_0 \equiv Z_{0\perp}, \underline{S_0} \equiv \underline{S_{0\perp}} \tag{18.503}$$

and

$$Z_0 = \int d\xi_1 d\xi'_1 d\xi_2 d\xi'_2 \delta(\xi_1 - \xi'_1)\delta(\xi_2 - \xi'_2)\int_{\xi'_1}^{\xi_1}\int_{\xi'_2}^{\xi_2} \exp\{\underline{S_0}\} D\xi_1 D\xi_2 \tag{18.504}$$

or

$$Z_{0\perp} = \prod_{i=1}^{2}\left(2\sinh\left(\frac{\beta\hbar\omega_i}{2}\right)\right)^{-2} \tag{18.505}$$

From the model Lagrangian 18.490 then

$$Z_{0\perp} = \int d\vec{\rho}\, d\vec{\rho}'\, d\vec{R}\, d\vec{R}'\,\delta(\vec{\rho} - \vec{\rho}')\delta(\vec{R} - \vec{R}')\int_{\vec{\rho}'}^{\vec{\rho}}\int_{\vec{R}'}^{\vec{R}} \exp\{\underline{S_0}\} D\vec{\rho} D\vec{R} \tag{18.506}$$

from where,

$$Z_{0\perp} = \int d\vec{\rho}\, d\vec{\rho}'\,\delta(\vec{\rho} - \vec{\rho}')\int_{\vec{\rho}'}^{\vec{\rho}} \exp\{\underline{S_{0\perp}}[\vec{\rho}]\} D\vec{\rho} \tag{18.507}$$

Here,

$$\underline{S}_{0\perp}[\vec{\rho}] = \int_0^\beta \left(-\frac{m_\perp \dot{\vec{\rho}}^2}{2\hbar^2} - \frac{m_\perp \Omega_\perp^2 + \kappa_\perp}{2}\vec{\rho}^2 \right) d\tau + \Phi_{\omega_f}[\vec{\rho}] - 2\ln\left[2\sinh\left(\frac{\beta\hbar\omega_f}{2}\right)\right] \tag{18.508}$$

where,

$$\Phi_{\omega_f}[\vec{\rho}] = \frac{\hbar\kappa_\perp}{4M\omega_f}\int_0^\beta\int_0^\beta \vec{\rho}(\tau)\vec{\rho}(\sigma)F_{\omega_f}(|\tau-\sigma|)d\tau d\sigma \tag{18.509}$$

From 18.488 we have

$$Z = \prod_{\vec{\kappa}} \int \delta(q_{\vec{\kappa}} - q'_{\vec{\kappa}})\delta(\vec{\rho} - \vec{\rho}')dq_{\vec{\kappa}}dq'_{\vec{\kappa}}d\vec{\rho}d\vec{\rho}' \int_{\vec{\rho}'}^{\vec{\rho}}\int_{q_{\vec{\kappa}}}^{q_{\vec{\kappa}}} \exp\{\underline{S}\}D\vec{\rho}Dq_{\vec{\kappa}} \tag{18.510}$$

or

$$Z = \prod_{\vec{\kappa}} Z_L\delta(\vec{\rho} - \vec{\rho}')d\vec{\rho}d\vec{\rho}' \int \exp\{\underline{S}[\rho] + \Phi_{\omega_{\vec{\kappa}}}[\rho]\}D \tag{18.511}$$

where lattice partition function is

$$Z_L = \left[2\sinh\left(\frac{\beta\hbar\omega_{\vec{\kappa}}}{2}\right)\right]^{-2} \tag{18.512}$$

and the functional of the electron-phonon interaction influence phase is

$$\Phi[\vec{\rho}] = \sum_{\vec{\kappa}} \frac{\hbar^2}{8V\omega_{\vec{\kappa}}}|C_{\vec{\kappa}}|^2 \exp\left\{-\frac{\kappa_\parallel^2}{2\lambda}\right\}\int_0^\beta\int_0^\beta d\sigma d\sigma' \exp\{i\vec{\kappa}_\perp(\vec{\rho}(\tau) - \vec{\rho}(\sigma))\}F_{\omega_{\vec{\kappa}}}(|\sigma - \sigma'|) \tag{18.513}$$

The action functional is:

$$\underline{S}[\vec{\rho}] = \int_0^\beta \left(-\frac{m_\perp \dot{\vec{\rho}}^2}{2\hbar^2} - \frac{m_\perp \Omega_\perp^2 \vec{\rho}^2}{2} \right) d\tau + \Phi[\vec{\rho}] \tag{18.514}$$

18.17.4 Polaron Generating Function

We find now $\langle \underline{S} - \underline{S}_0 \rangle_{\underline{S}_0}$ with the help of the generating function:

$$\Psi_{\vec{\kappa}}(\xi,\eta) = \left\langle \exp\{i\vec{\kappa}_\perp(\xi\vec{\rho}(\tau) - \eta\vec{\rho}(\sigma))\}\right\rangle_{\underline{S}_0} \equiv \frac{\mathrm{Tr}\int \exp\{i\vec{\kappa}_\perp(\xi\vec{\rho}(\tau) - \eta\vec{\rho}(\sigma))\}\exp\{\underline{S}_0\}D\vec{\rho}D\vec{R}}{\mathrm{Tr}\int \exp\{\underline{S}_0\}D\vec{\rho}D\vec{R}} \tag{18.515}$$

From equation 18.496 then

$$\xi\vec{\rho}(\tau) - \eta\vec{\rho}(\sigma) = C_1(\omega_1)\int\gamma(t)\vec{\xi}_1(t)dt + C_2(\omega_2)F_2(\omega_2)\int\gamma(t)\vec{\xi}_2(t)dt \tag{18.516}$$

where

$$\gamma(t) \equiv \xi\delta(\tau - t) - \eta\delta(\tau - \sigma) \tag{18.517}$$

then

$$\Psi_{\vec{\kappa}}(\xi,\eta) \equiv \frac{\mathrm{Tr}\int \prod_{i=1}^{2} \Psi_i(\vec{\xi}_i) D\vec{\xi}_i}{\mathrm{Tr}\int \exp\{\underline{S}_0\} D^{\prime\prime} D\vec{R}} \tag{18.518}$$

where,

$$\Psi_1(\vec{\xi}_1) \equiv \exp\left\{i\vec{\kappa}_\perp C_1 \int \gamma(t)\vec{\xi}_1(t)dt\right\} \exp\left\{-\int_0^\beta \left(\frac{\dot{\vec{\xi}}_1^{\,2}}{2\hbar^2} + \frac{\omega_1^2 \vec{\xi}_1^{\,2}}{2}\right) d\tau\right\} \tag{18.519}$$

$$\Psi_2(\vec{\xi}_2) \equiv \exp\left\{i\vec{\kappa}_\perp C_2 F_2 \int \gamma(t)\vec{\xi}_2(t)dt\right\} \exp\left\{-\int_0^\beta \left(\frac{\dot{\vec{\xi}}_2^{\,2}}{2\hbar^2} + \frac{\omega_1^2 \vec{\xi}_2^{\,2}}{2}\right) d\tau\right\} \tag{18.520}$$

So,

$$\Psi_{\kappa_\perp}(\xi,\eta) = \exp\{f_{1\perp} + f_{2\perp}\} \tag{18.521}$$

where

$$f_{1\perp} = -\frac{\hbar C_1^2 \kappa_\perp^2}{4\omega_1}\left[\left(\xi^2 + \eta^2\right)\mathrm{F}_{\omega_1}(0) - 2\xi\eta \mathrm{F}_{\omega_1}\left(|\tau - \sigma|\right)\right] \tag{18.522}$$

$$f_{2\perp} = -\frac{\hbar C_2^2 F_2^2}{4\omega_2}\left[\left(\xi^2 + \eta^2\right)\mathrm{F}_{\omega_2}(0) - 2\xi\eta \mathrm{F}_{\omega_2}\left(|\tau - \sigma|\right)\right] \tag{18.523}$$

18.17.5 Polaron Energy

We now calculate all quantities in $\left\langle \underline{S} - \underline{S}_0 \right\rangle_{\underline{S}_0}$ with the generating function and, in particular,

$$\left\langle \Phi[\vec{\rho}] \right\rangle_{\underline{S}_0} = \sum_{\vec{\kappa}} \frac{\hbar^2}{8V\omega_{\vec{\kappa}}} |C_{\vec{\kappa}}|^2 \exp\left\{-\frac{\kappa_\parallel^2}{2\lambda}\right\} \int_0^\beta \int_0^\beta d\sigma d\sigma \Psi_{\kappa_\perp}(\xi = 1, \eta = 1) \mathrm{F}_{\omega_{\vec{\kappa}}}\left(|\sigma - \sigma'|\right) \tag{18.524}$$

We observe again all the formulae in the polaron have the same dependence on the quantity $|\tau - \sigma|$, confirming the fact that the quantities (retarded functions) depend on the past with the significance of interaction with the past being the perturbation due to the moving electrons (holes) that take **"time"** propagating in the crystal lattice. We again generalize the functions to be $G\left(|\tau - \sigma| - \frac{\beta}{2}\right)$ that aid in the evaluation of the twofold integral [1,2]:

$$f(\beta)=\int_0^\beta\int_0^\beta G\left(|\tau-\sigma|-\frac{\beta}{2}\right)d\sigma d\tau \tag{18.525}$$

that after the change of variables

$$\tau-\sigma\equiv\rho,\quad 0\le\sigma\le\tau \tag{18.526}$$

then

$$f(\beta)=2\int_0^\beta G\left(\rho-\frac{\beta}{2}\right)(\beta-\rho)d\rho \tag{18.527}$$

and again, considering the change of variable $\rho=\beta y$, where $0\le y\le 1$. If $y=1$ then $\rho=\beta$. Subsequently, we also do a the change of variable

$$\frac{\beta(1-y)}{2}\equiv-\tau,\quad d\tau=-\frac{\beta}{2}dy,\quad 0\le\tau\le\infty \tag{18.528}$$

This renders all our integrals convergent and, consequently,

$$\langle\Phi[\rho]\rangle_{S_0}=\frac{\beta(\hbar\omega_0)^2\alpha_F\lambda_0}{\pi^{\frac{1}{2}}}\int_0^\infty d\tau\Phi(\mathrm{A}_\perp(\tau))\exp\{-\hbar\omega_0\tau\} \tag{18.529}$$

$$\Phi(\mathrm{A}_\perp(\tau))=\frac{1}{\left|1-2\lambda_0^2\mathrm{A}_\perp(\tau)\right|^{\frac{1}{2}}}\begin{cases}\sinh^{-1}\left(\frac{1}{2\lambda_0^2\mathrm{A}_\perp(\tau)}-1\right)^{\frac{1}{2}}, & \frac{1}{2\lambda_0^2\mathrm{A}_\perp(\tau)}>1\\ \sin^{-1}\left(1-\frac{1}{2\lambda_0^2\mathrm{A}_\perp(\tau)}\right)^{\frac{1}{2}}, & \frac{1}{2\lambda_0^2\mathrm{A}_\perp(\tau)}\le 1\end{cases} \tag{18.530}$$

where

$$\mathrm{A}_\perp(\tau)=\frac{1}{2}\sum_{i=1}^{2}\frac{d_{i,\perp}^2}{\omega_{i,\perp}}\left(1-\exp\{-\hbar\omega_{i,\perp}\tau\}\right),\lambda_0=\left(\frac{\hbar}{\mathrm{m}_\perp\omega_0}\right)^{\frac{1}{2}}\lambda \tag{18.531}$$

and

$$d_{1,\perp}^2=\frac{\omega_{1,\perp}^2-\omega_{f,\perp}^2}{\omega_{1,\perp}^2-\omega_{2,\perp}^2},d_{2\perp}^2=\frac{\omega_{f,\perp}^2-\omega_{2,\perp}^2}{\omega_{1,\perp}^2-\omega_{2,\perp}^2} \tag{18.532}$$

The polaron energy:

$$E_p=\frac{\hbar\omega_0\lambda_0^2}{4}+\sum_{i=1}^{2}\hbar\omega_{i,\perp}\left(1-\frac{\omega_{i,\perp}^2-\Omega_\perp^2}{\omega_{i,\perp}^2}d_{i,\perp}^2\right)-\frac{\langle\Phi[\rho]\rangle_{S_0}}{\beta} \tag{18.533}$$

18.18 Bipolaron Characteristics in a Cylindrical Quantum Dot

We consider the motion of the electron in the z -axis direction to be bounded by an infinite high rectangular potential well and bounded on the oxy–plane by a transversal parabolic potential.

18.18.1 System Hamiltonian

The Hamiltonian of the system is written in the form:

$$\hat{\mathrm{H}} = \sum_{i-1}^{2}\left(\frac{\hat{p}_{i\perp}^{2}}{2m_{\perp}} + \frac{\hat{p}_{i\|}^{2}}{2m_{\|}} + \frac{m_{i\perp}\Omega_{i\perp}^{2}\vec{r}_{i}^{\,2}}{2} + V(z_i)\right) + U(|\vec{r}_1 - \vec{r}_2|)$$
$$+\sum_{\vec{\kappa}}\hbar\omega_{\vec{\kappa}}\hat{b}_{\vec{\kappa}}\hat{b}_{\vec{\kappa}} + \frac{\hbar}{\sqrt{V}}\sum_{\vec{\kappa}}\left(C_{\vec{\kappa}}\hat{b}_{\vec{\kappa}}\exp\{i\vec{\kappa}\vec{r}\} + \text{h.c.}\right) \tag{18.534}$$

Here, $V(z)$ is the confinement potential (infinite high rectangular potential well) in the direction of the oz-axis. The state of the electron is described by the variational wave function that has a large spread compared to the ground state wave function:

$$\Psi_0 = \left(\frac{\lambda}{\pi}\right)^{\frac{1}{2}}\exp\left\{-\frac{\lambda z_1^2}{2} - \frac{\lambda z_2^2}{2}\right\} \tag{18.535}$$

Averaging the Lagrangian 18.534 by the wave function 18.535 we have

$$\tilde{\mathrm{H}} = \sum_{i=1}^{2}\left(\frac{\hat{p}_{i\perp}^{2}}{2m_{\perp}} + \frac{m_{i\perp}\Omega_{i\perp}^{2}\vec{\rho}_i^{\,2}}{2} + \tilde{V}(z_i)\right) + \tilde{U}(|\vec{r}_1 - \vec{r}_2|) + \sum_{\vec{\kappa}}\hbar\omega_{\vec{\kappa}}\hat{b}_{\vec{\kappa}}\hat{b}_{\vec{\kappa}} + \frac{\hbar}{\sqrt{V}}\sum_{\vec{\kappa}}\left(\widetilde{C_{\vec{\kappa}}}\hat{b}_{\vec{\kappa}}\exp\{i\vec{\kappa}_{\perp}\vec{\rho}\} + h.c.\right) \tag{18.536}$$

Here,

$$\widetilde{C_{\vec{\kappa}}} = \frac{\hbar}{\sqrt{V}}C_{\vec{\kappa}}\exp\left\{-\frac{\kappa_{\|}^{2}}{4\lambda}\right\} \tag{18.537}$$

The model Lagrangian of the transversal motion:

$$L_{\perp 0} = \sum_{i=1}^{2}\left(-\frac{m_{i\perp}\dot{\vec{\rho}}_i^{\,2}}{2\hbar^2} - \frac{M_{i\perp}\dot{\vec{R}}_i^{\,2}}{2\hbar^2} - \frac{m_{i\perp}\Omega_{i\perp}^{2}\vec{\rho}_i^{\,2}}{2} - \frac{\kappa_{i\perp}(\vec{R}_i - \vec{\rho}_i)^2}{2}\right) - \sum_{i\neq k}^{2}\frac{\kappa'_{i\perp}(\vec{R}_k - \vec{\rho}_i)^2}{2} + \frac{\mathrm{K}_{\perp}(\vec{\rho}_1 - \vec{\rho}_2)^2}{2} \tag{18.538}$$

18.18.1.1 Model System Action Functional

From the model Lagrangian 18.538 we have the model action functional $\underline{S_0}[\vec{\rho}_1, \vec{\rho}_2]$:

$$Z_{0\perp} = \int d\vec{\rho}_1\vec{\rho}'_1 d\vec{\rho}_2\vec{\rho}'_2 d\vec{R}_1\vec{R}'_1 d\vec{R}_2\vec{R}'_2\delta(\vec{\rho}_1 - \vec{\rho}'_1)\delta(\vec{\rho}_2 - \vec{\rho}'_2)$$
$$\times\delta(\vec{R}_1 - \vec{R}'_1)\delta(\vec{R}_2 - \vec{R}'_2)\int_{\vec{\rho}'}^{\vec{\rho}}\int_{\vec{R}'}^{\vec{R}}\exp\{\underline{S_0}\}D\vec{\rho}_1 D\vec{\rho}_2 D\vec{R}_1 D\vec{R}_2 \tag{18.539}$$

Following the procedure of path integration seen earlier then

$$S_0\left[\vec{\rho}_1,\vec{\rho}_2\right]=-\frac{m}{2\hbar^2}\int_0^{\beta}\sum_{i=1}^{2}\dot{\vec{\rho}}_i^{\,2}\,d\tau-m\Omega^2\int_0^{\beta}\sum_{i=1}^{2}\vec{\rho}_i^{\,2}d\tau-\frac{p}{2}\int_0^{\beta}\sum_{i=1}^{2}\vec{\rho}_i^{\,2}d\tau$$
$$+\frac{K_{\perp}}{2}\int_0^{\beta}\left(\vec{\rho}_1-\vec{\rho}_2\right)^2d\tau+\frac{\hbar}{4M\omega_f}\int_0^{\beta}\int_0^{\beta}\Big[\left(\kappa_{i\perp}^2+\kappa_{i\perp}'^2\right)\vec{\rho}_1(\sigma)\vec{\rho}_1(\tau)$$
$$+2\kappa_{i\perp}\kappa'_{i\perp}\vec{\rho}_1(\sigma)\vec{\rho}_2(\tau)\Big]F_{\omega_f}\left(|\tau-\sigma|\right)d\tau d\sigma-42\ln\left[2\sinh\left(\frac{\beta\hbar\omega_f}{2}\right)\right] \tag{18.540}$$

18.18.1.2 Equation of Motion / Normal Modes

The equations of motion for the transversal motion are

$$m_{\perp}\ddot{\vec{\rho}}_1+m_{\perp}\Omega_{\perp}^2\vec{\rho}_1-\kappa_{\perp}\left(\vec{R}_1-\vec{\rho}_1\right)-\kappa_{\perp}'\left(\vec{R}_2-\vec{\rho}_1\right)-K_{\perp}\left(\vec{\rho}_1-\vec{\rho}_2\right)=0 \tag{18.541}$$

$$m_{\perp}\ddot{\vec{\rho}}_2+m_{\perp}\Omega_{\perp}^2\vec{\rho}_2-\kappa_{\perp}\left(\vec{R}_2-\vec{\rho}_2\right)-\kappa_{\perp}'\left(\vec{R}_1-\vec{\rho}_2\right)+K_{\perp}\left(\vec{\rho}_1-\vec{\rho}_2\right)=0 \tag{18.542}$$

$$M_{\perp}\ddot{\vec{R}}_1+\kappa_{\perp}\left(\vec{R}_1-\vec{\rho}_1\right)+\kappa_{\perp}'\left(\vec{R}_1-\vec{\rho}_2\right)=0 \tag{18.543}$$

$$M_{\perp}\ddot{\vec{R}}_2+\kappa_{\perp}\left(\vec{R}_2-\vec{\rho}_2\right)+\kappa_{\perp}'\left(\vec{R}_2-\vec{\rho}_1\right)=0 \tag{18.544}$$

For normal modes we substitute the following

$$\ddot{\vec{\rho}}=-\omega^2\vec{\rho},\ddot{\vec{R}}=-\omega^2\vec{R} \tag{18.545}$$

into the equation of motion and letting,

$$\vec{\rho}_1+\vec{\rho}_2=\vec{\xi}_1,\vec{R}_1+\vec{R}_2=\vec{\xi}_2,\vec{\rho}_1-\vec{\rho}_2=\vec{\rho},\vec{R}_1-\vec{R}_2=\vec{R} \tag{18.546}$$

then

$$\left(-m\omega^2+m\Omega^2+p\right)\vec{\xi}_1-p\vec{\xi}_2=0,-p\vec{\xi}_1+\left(M\omega^2+m\Omega^2+p\right)\vec{\xi}_2=0 \tag{18.547}$$

$$\left(-m\omega^2+m\Omega^2+P\right)\vec{\rho}-p'\vec{R}=0,-p'\vec{\rho}+\left(-M\omega^2+p\right)\vec{R}=0 \tag{18.548}$$

$$\Omega_{\perp}^2=\Omega^2,p=\kappa_{\perp}+\kappa'_{\perp},p=\kappa_{\perp}-\kappa_{\perp}',P=p-2K_{\perp} \tag{18.549}$$

Solving for the eigenmodes we have

$$\omega_{1,2}^2=\frac{1}{2}\left(\frac{\kappa_{\perp}}{\mu}+\Omega^2\right)\pm\frac{1}{2}\left[\left(\frac{\kappa_{\perp}}{\mu}+\Omega^2\right)^2-\frac{4\kappa_{\perp}\Omega^2}{M}\right]^{\frac{1}{2}},\frac{1}{\mu}=\frac{1}{m}+\frac{1}{M} \tag{18.550}$$

$$\omega_{3,4}^2=\frac{1}{2}\left(\Omega^2+\frac{PM+pm}{2Mm}\right)\pm\frac{1}{2}\left[\left(\Omega^2+\frac{PM+pm}{2Mm}\right)^2+4\frac{\left(m\Omega^2+P\right)p-p'^2}{Mm}\right]^{\frac{1}{2}} \tag{18.551}$$

with

$$\omega_1^2\omega_2^2 = \omega_f^2\Omega^2 \tag{18.552}$$

The **bipolaron effective mass** is conveniently obtained from the eigenmode equations:

$$\frac{M}{m} = \frac{\left(\omega_1^2 - \Omega^2\right)\left(\Omega^2 - \omega_2^2\right)}{\omega_1^2\omega_2^2} \tag{18.553}$$

18.18.1.3 Lagrangian Diagonalization

To diagonalize the model Lagrangian, we move to normal coordinates:

$$\vec{\xi}_1 = C_1(\omega_1)\vec{x}_1(\omega_1) + C_2(\omega_2)F_2(\omega_2)\vec{x}_2(\omega_2) \tag{18.554}$$

$$\vec{\xi}_2 = C_1(\omega_1)F_1(\omega_1)\vec{x}_1(\omega_1) + C_2(\omega_2)\vec{x}_2(\omega_2) \tag{18.555}$$

$$\vec{\rho} = C_3\vec{x}_3(\omega_3) + C_4F_2(\omega_4)\vec{x}_4(\omega_4) \tag{18.556}$$

$$\vec{R} = C_3F_3(\omega_3)\vec{x}_3(\omega_3) + C_4\vec{x}_4(\omega_4) \tag{18.557}$$

Substituting these equations into the equation of motion and also considering the conservation of the kinetic energy in any representation then

$$F_1(\omega_1) = \frac{v^2}{-\omega_1^2 + v^2}, F_2(\omega_2) = \frac{p}{m\left(\Omega^2 + v - \omega_2^2\right)} = \frac{M}{m}F_1(\omega_1), v = \frac{p}{M} \tag{18.558}$$

$$C_2^2(\omega_1) = \frac{1}{M + mF_2^2(\omega_2)}, C_1^2(\omega_1) = \frac{M}{m}C_2^2(\omega_1) \tag{18.559}$$

$$F_3(\omega_3) = \frac{p'}{-M\omega_3^2 + p}, F_4(\omega_2) = \frac{p'}{m\omega_4^2 + \Omega^2 + P} \tag{18.560}$$

$$C_3^2(\omega_1) = \frac{1}{m + MF_3^2(\omega_3)}, C_4^2(\omega_4) = \frac{1}{M + mF_4^2(\omega_4)} \tag{18.561}$$

and the model diagonalized Lagrangian in normal coordinates:

$$L_0 = \sum_{i=1}^{4}\left[-\frac{\dot{\vec{x}}_i^2}{4\hbar^2} - \frac{\omega_i^2\vec{x}_i^2}{4}\right] \tag{18.562}$$

It shows that in the motion of the bipolaron we have four oscillators indicating four internal motions.

18.18.1.4 Bipolaron Partition Function

The bipolaron partition function can be obtained from the equation:

$$Z_0 = \prod_{i=1}^{4}\int d\vec{x}_i d\vec{x}'_i \delta\left(\vec{x}_i - \vec{x}'_i\right)\int_{\vec{x}'_i}^{\vec{x}_i} \exp\{S_0\} D\vec{x}_i \tag{18.563}$$

or

$$Z_0 = \prod_{i=1}^{4}\left(2\sinh\left(\frac{\beta\hbar\omega_i}{2}\right)\right)^{-2} \tag{18.564}$$

The Coulomb interaction can be written in the form:

$$U\left(\left|\vec{r}_1-\vec{r}_2\right|\right)=\frac{(2)^{\frac{1}{2}}\hbar\omega_0 4\pi\alpha_F}{V(1-\eta)}\sum_{\vec{\kappa}}\frac{1}{\kappa^2}\exp\left\{-\frac{\kappa_{\|}^2}{2\lambda}\right\}\exp\left\{i\vec{\kappa}_{\perp}\left(\vec{\rho}_1(\tau)-\vec{\rho}_2(\tau)\right)\right\} \tag{18.565}$$

18.18.1.5 Bipolaron Generating Function

We find now $\left\langle \underline{S}-\underline{S_0}\right\rangle_{S_0}$ via the generating function:

$$\Psi_{\vec{\kappa}}\left(\xi_1,\eta_1,\xi_2,\eta_2\right)=\left\langle\exp\left\{i\vec{\kappa}_{\perp}\left(\xi_1\vec{\rho}_1(\tau)-\eta_1\vec{\rho}_1(\sigma)+\xi_2\vec{\rho}_2(\tau)-\eta_2\vec{\rho}_2(\sigma)\right)\right\}\right\rangle_{S_0} \tag{18.566}$$

Letting,

$$f_1^+=\left(\xi_1+\xi_2\right)^2+\left(\eta_1+\eta_2\right)^2,\ f_2^+=\left(\xi_1+\xi_2\right)\left(\eta_1+\eta_2\right) \tag{18.567}$$

$$f_1^-=\left(\xi_1-\xi_2\right)^2+\left(\eta_1-\eta_2\right)^2,\ f_2^-=\left(\xi_1-\xi_2\right)\left(\eta_1-\eta_2\right) \tag{18.568}$$

then

$$\Psi_{\vec{\kappa}}\left(\xi_1,\eta_1,\xi_2,\eta_2\right)=\exp\left\langle\left\{\kappa_{\perp}^2\left(f_1+f_2+f_3+f_4\right)\right\}\right\rangle_{S_0} \tag{18.569}$$

where

$$f_1=-\frac{\hbar C_1^2}{16\omega_1 m_1}\left[f_1^+\mathrm{F}_{\omega_1}(0)-2f_2^+\mathrm{F}_{\omega_1}\left(\left|\tau-\sigma\right|\right)\right] \tag{18.570}$$

$$f_2=-\frac{\hbar C_2^2F_2^2}{16\omega_2 m_2}\left[f_1^+\mathrm{F}_{\omega_2}(0)-2f_2^+\mathrm{F}_{\omega_2}\left(\left|\tau-\sigma\right|\right)\right] \tag{18.571}$$

$$f_3=-\frac{\hbar C_3^2}{16\omega_3 m_3}\left[f_1^-\mathrm{F}_{\omega_3}(0)-2f_2^-\mathrm{F}_{\omega_3}\left(\left|\tau-\sigma\right|\right)\right] \tag{18.572}$$

$$f_4=-\frac{\hbar C_4^2F_4^2}{16\omega_4 m_4}\left[f_1^-\mathrm{F}_{\omega_4}(0)-2f_2^-\mathrm{F}_{\omega_4}\left(\left|\tau-\sigma\right|\right)\right] \tag{18.573}$$

$$m_1=m_2=m_3=m_4=\frac{1}{2} \tag{18.574}$$

18.18.1.6 Bipolaron Energy

We again observe all the formulae in the bipolaron problem have the same dependence on the quantity $|\tau-\sigma|$ and confirm the retarded functions' dependence on the past with the significance of interaction with the past being the perturbation due to the moving electrons (holes) that take "**time**" propagating in the crystal lattice. We again generalize the functions to be $\left(|\tau-\sigma|-\frac{\beta}{2}\right)$. This generalization is appropriate to establish the twofold integral theorem [1] and consequently the bipolaron energy as:

$$E_{bip}=\sum_{j=1}^{4}\omega_j\left(1-\frac{\omega_j^2-\Omega^2}{2\omega_j^2}\mathrm{d}_j^2\right)-\frac{2\hbar\omega_0\alpha_F\lambda_0}{(1-\eta)\pi^{\frac{1}{2}}}\Phi\left(\mathrm{A}_{12}(0)\right)$$
$$+\frac{2\hbar\omega_0\alpha_F\lambda_0}{\pi^{\frac{1}{2}}}\int_0^{\infty}d\tau\exp\{-\tau\}\left[\Phi\left(\mathrm{A}_{11}(\tau)\right)+\Phi\left(\mathrm{A}_{12}(\tau)\right)\right] \quad (18.575)$$

where,

$$\Phi\left(\mathrm{A}_{12}(0)\right)=\frac{1}{\left|\mathrm{A}_{12}(0)-1\right|^{\frac{1}{2}}}\begin{cases}\ln\left[\dfrac{1+\left(1-\mathrm{A}_{12}(0)\right)^{\frac{1}{2}}}{\left(\mathrm{A}_{12}(0)\right)^{\frac{1}{2}}}\right], & \mathrm{A}_{12}(0)<1\\ \sin^{-1}\left(\dfrac{\mathrm{A}_{12}(0)-1}{\mathrm{A}_{12}(0)}\right)^{\frac{1}{2}}, & \mathrm{A}_{12}(0)\geq 1\end{cases} \quad (18.576)$$

$$\mathrm{A}_{12}(0)=4\lambda_0^2\left(\frac{d_3^2}{\omega_3}+\frac{d_4^2}{\omega_4}\right) \quad (18.577)$$

$$\Phi\left(\mathrm{A}_{12}(\tau)\right)=\frac{1}{\left|\mathrm{A}_{12}(\tau)-1\right|^{\frac{1}{2}}}\begin{cases}\sinh^{-1}\left(\dfrac{1}{\mathrm{A}_{12}(\tau)}-1\right)^{\frac{1}{2}}, & \dfrac{1}{\mathrm{A}_{12}(\tau)}>1\\ \sin^{-1}\left(1-\dfrac{1}{\mathrm{A}_{12}(\tau)}\right)^{\frac{1}{2}}, & \dfrac{1}{\mathrm{A}_{12}(\tau)}\leq 1\end{cases} \quad (18.578)$$

$$\Phi\left(\mathrm{A}_{11}(\tau)\right)=\frac{1}{\left|\mathrm{A}_{11}(\tau)-1\right|^{\frac{1}{2}}}\begin{cases}\sinh^{-1}\left(\dfrac{1}{\mathrm{A}_{11}(\tau)}-1\right)^{\frac{1}{2}}, & \dfrac{1}{\mathrm{A}_{11}(\tau)}>1\\ \sin^{-1}\left(1-\dfrac{1}{\mathrm{A}_{11}(\tau)}\right)^{\frac{1}{2}}, & \dfrac{1}{\mathrm{A}_{11}(\tau)}\leq 1\end{cases} \quad (18.579)$$

and

$$\mathrm{A}_{11}(\tau)=\lambda_0^2\sum_{i=1}^{4}\frac{d_i^2}{\omega_i}\left(1-\exp\{-\omega_i\tau\}\right) \quad (18.580)$$

$$\mathrm{A}_{12}(\tau)=\lambda_0^2\sum_{i=1}^{2}\frac{d_i^2}{\omega_i}\left(1-\exp\{-\omega_i\tau\}\right)+\lambda_0^2\sum_{i=3}^{4}\frac{d_i^2}{\omega_i}\left(1+\exp\{-\omega_i\tau\}\right) \quad (18.581)$$

with

$$d_1^2=\frac{\omega_1^2-\omega_f^2}{\omega_1^2-\omega_2^2},d_2^2=\frac{\omega_f^2-\omega_2^2}{\omega_1^2-\omega_2^2},d_3^2=\frac{\omega_3^2-\omega_f^2}{\omega_3^2-\omega_4^2},d_4^2=\frac{\omega_f^2-\omega_4^2}{\omega_3^2-\omega_4^2} \quad (18.582)$$

For $\lambda_0 \gg 1$ then

$$\left|A_{12}(0)-1\right|^{\frac{1}{2}} \cong \left|A_{12}(0)\right|^{\frac{1}{2}} \tag{18.583}$$

$$\sin^{-1}\left(\frac{A_{12}(0)-1}{A_{12}(0)}\right)^{\frac{1}{2}} = \frac{\pi}{2} \tag{18.584}$$

$$U\left(\left|\vec{r}_1-\vec{r}_2\right|\right)_{S_0} = \frac{2\hbar\omega_0\alpha_F\lambda_0}{(1-\eta)\pi^{\frac{1}{2}}}\Phi\left(A_{12}(0)\right) = \frac{2\hbar\omega_0\alpha_F\lambda_0}{(1-\eta)\pi^{\frac{1}{2}}\left|A_{12}(0)\right|^{\frac{1}{2}}}\frac{\pi}{2} \tag{18.585}$$

18.19 Polaron Characteristics in a Quasi-0D Cylindrical Quantum Dot with Asymmetrical Parabolic Potential

We apply path integration in the same fashion as done for other geometries seen earlier. We consider the motion of the electron in the direction of the *oz*-axis to be bounded by a longitudinal parabolic confinement potential while bounded on the oxy-plane by a transversal parabolic potential (Figure 18.10).

The exact Hamiltonian for the full system is written:

$$\hat{H} = \frac{\hat{p}_\perp^2}{2m_\perp} + \frac{\hat{p}_\parallel^2}{2m_\parallel} + \frac{m_\perp\Omega_\perp^2\vec{\rho}^2}{2} + \frac{m_\parallel\Omega_\parallel^2 z^2}{2} + \sum_{\vec{\kappa}}\hbar\omega_{\vec{\kappa}}\hat{b}_{\vec{\kappa}}^{+}\hat{b}_{\vec{\kappa}} + \frac{\hbar}{\sqrt{V}}\sum_{\vec{\kappa}}\left(C_{\vec{\kappa}}\hat{b}_{\vec{\kappa}}\exp\{i\vec{\kappa}\vec{r}\} + h.c.\right) \tag{18.586}$$

The model Lagrangian of the system is written as:

$$L_0 = -\frac{m_\perp\dot{\vec{\rho}}^2}{2\hbar^2} - \frac{m_\parallel\dot{z}^2}{2\hbar^2} - \frac{M_\parallel\dot{Z}^2}{2\hbar^2} - \frac{M_\perp\dot{\vec{R}}^2}{2\hbar^2} - \frac{m_\perp\Omega_\perp^2\vec{\rho}^2}{2} - \frac{m_\parallel\Omega_\parallel^2 z^2}{2} - \frac{\kappa_\perp\left(\vec{R}-\vec{\rho}\right)^2}{2} - \frac{\kappa_\parallel\left(Z-z\right)^2}{2} \tag{18.587}$$

For normal modes we substitute the following

$$\ddot{\vec{\rho}} = -\omega^2\vec{\rho}, \ddot{z} = -\omega^2 z, \ddot{Z} = -\omega^2 Z, \ddot{\vec{R}} = -\omega^2\vec{R} \tag{18.588}$$

into the equation of motion

$$m_\perp\ddot{\vec{\rho}} + m_\perp\Omega_\perp^2\vec{\rho} - \kappa_\perp\left(\vec{R}-\vec{\rho}\right) = 0, M_\perp\ddot{\vec{R}} + \kappa_\perp\left(\vec{R}-\vec{\rho}\right) = 0 \tag{18.589}$$

$$m_\parallel\ddot{z} + m_\parallel\Omega_\parallel^2 z - \kappa_\parallel\left(Z-z\right) = 0, M_\parallel\ddot{Z} + \kappa_\parallel\left(Z-z\right) = 0 \tag{18.590}$$

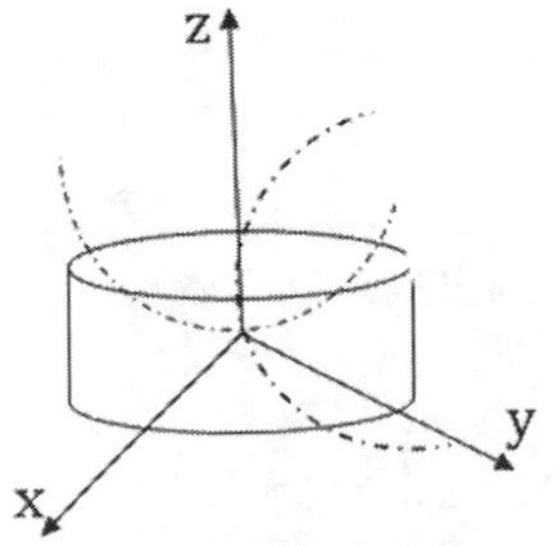

FIGURE 18.10 Depicts an asymmetric cylindrical quantum dot where the oz–axis is bounded by a longitudinal parabolic confinement potential and bounded on the oxy–plane by a transversal parabolic potential.

and solving for the frequency eigenmodes we have

$$\omega_{i1,2}^2 = \frac{1}{2}\left(\frac{\kappa_i}{\mu_i} + \Omega_i^2\right) \pm \frac{1}{2}\left[\left(\frac{\kappa_i}{\mu_i} + \Omega_i^2\right)^2 - \frac{4\kappa_i\Omega_i^2}{M_i}\right]^{\frac{1}{2}}, \frac{1}{\mu_i} = \frac{1}{m_i} + \frac{1}{M_i}, i = \perp, \| \tag{18.591}$$

with

$$\omega_{i1}^2\omega_{i2}^2 = \omega_{if}^2\Omega_i^2 \tag{18.592}$$

The **effective polaron mass** is conveniently obtained from 18.591:

$$\frac{M}{m} = \frac{1}{2}\sum_{i=\perp,\|}\frac{\left(\omega_{i1}^2 - \Omega_i^2\right)\left(\Omega_i^2 - \omega_{i2}^2\right)}{\omega_{i1}^2\omega_{i2}^2} \tag{18.593}$$

We observe from here that there are two transversal oscillatory and two longitudinal oscillatory motions. These frequency eigenmodes may permit us to move to normal mode coordinates as previously done. We do for one variable and apply similarly to all other variables:

$$Z(\tau) = Z_1(\omega_1) + Z_2(\omega_2) = C_1(\omega_1)\xi_1(\omega_1) + C_2(\omega_2)F_2(\omega_2)\xi_2(\omega_2) \tag{18.594}$$

$$z(\tau) = z_1(\omega_1) + z_2(\omega_2) = C_1(\omega_1)F_1(\omega_1)\xi_1(\omega_1) + C_2(\omega_2)\xi_2(\omega_2) \tag{18.595}$$

Substituting these equations for the normal coordinates into the equation of motion 18.590 and also considering the conservation of kinetic energy in any of the representations then

$$F_{i1}(\omega_{i1}) = \frac{\kappa_i}{-M_i\omega_{i1}^2 + \kappa_i}, C_{i1}^2(\omega_{i1}) = \frac{1}{m_i + M_iF_{i1}^2(\omega_{i1})} \tag{18.596}$$

$$F_{i2}(\omega_{i2}) = \frac{\kappa_i}{-m_i\omega_{i2}^2 + \kappa_i + m_i\Omega_i^2}, C_{i2}^2(\omega_{i2}) = \frac{1}{M_i + m_iF_{i2}^2(\omega_{i2})} \tag{18.597}$$

$$C_{i2}^2(\omega_{i2}) = \frac{m_i}{M_i}C_{i1}^2(\omega_{i1}), F_{i2}(\omega_{i2}) = -\frac{M_i}{m_i}F_{i1}(\omega_{i1}) \tag{18.598}$$

Considering 18.594 to 18.598 in the model Lagrangian 18.587 then

$$L_0 = \sum_{i=\perp,\|}\left[-\frac{\dot{\xi}_{i1}^2}{2\hbar^2} - \frac{\omega_{i1}^2\xi_{i1}^2}{2} - \frac{\dot{\xi}_{i2}^2}{2\hbar^2} - \frac{\omega_{i2}^2\xi_{i2}^2}{2}\right] \tag{18.599}$$

18.20 Polaron Energy

We follow the same procedure as previously seen above for the evaluation of the energy and effective mass of the polaron via the relation:

$$E \le F_F = -\frac{1}{\beta}\ln Z_0 - \frac{1}{\beta}\left\langle \underline{S} - \underline{S}_0 \right\rangle_{S_0} \tag{18.600}$$

For this problem, the partition function Z_0:

$$Z_0 = Z_{0\perp} Z_{0\parallel} \tag{18.601}$$

$$\underline{S}_0 = \underline{S}_{0\perp} + \underline{S}_{0\parallel} \tag{18.602}$$

and

$$Z_{0\parallel} = \prod_{i=1}^{2} \left(2\sinh\left(\frac{\beta\hbar\omega_{i\parallel}}{2} \right) \right)^{-1}, Z_{0\perp} = \prod_{i=1}^{2} \left(2\sinh\left(\frac{\beta\hbar\omega_{i\perp}}{2} \right) \right)^{-2} \tag{18.603}$$

Considering the model Lagrangian 18.587 then

$$Z_{0\parallel} = \int_{-\infty}^{+\infty} \prod_{i=1}^{2} dz_i dz'_i dZ_i dZ'_i \delta\left(z_i - z'_i\right) \delta\left(Z_i - Z'_i\right) \int_{z'_i}^{z_i} \int_{Z'_i}^{Z_i} \exp\left\{\underline{S}_0\right\} Dz_i DZ_i \tag{18.604}$$

from where,

$$Z_{0\parallel} = \int_{-\infty}^{+\infty} \prod_{i=1}^{2} dz_i dz'_i \delta\left(z_i - z'_i\right) \int_{z'_i}^{z_i} \exp\left\{\underline{S}_{0\parallel}\left[z\right]\right\} Dz_i \tag{18.605}$$

Here,

$$\underline{S}_{0\parallel}\left[z\right] = \int_0^{\beta} \left(-\frac{m_\parallel \dot{z}^2}{2\hbar^2} - \frac{m_\parallel \Omega_\parallel^2 + \kappa_\parallel}{2} z^2 \right) d\tau + \Phi_{\omega_{\parallel f}}\left[z\right] - \ln\left[2\sinh\left(\frac{\beta\hbar\omega_{\parallel f}}{2} \right) \right] \tag{18.606}$$

and

$$\Phi_{\omega_{\parallel f}}\left[z\right] = \frac{\hbar\kappa_\parallel}{4M_\parallel \omega_{\parallel f}} \int_0^{\beta} \int_0^{\beta} z(\tau) z(\sigma) F_{\omega_{\parallel f}}\left(\left|\tau - \sigma\right|\right) d\tau d\sigma \tag{18.607}$$

From 18.586 we have

$$Z = \prod_{\vec{\kappa}} \int \delta\left(q_{\vec{\kappa}} - q'_{\vec{\kappa}}\right) \delta\left(z - z'\right) \delta\left(\vec{\rho} - \vec{\rho}'\right) dq_{\vec{\kappa}} dq'_{\vec{\kappa}} dz dz' d\vec{\rho} d\vec{\rho}' \int_{z'}^{z} \int_{\vec{\rho}'}^{\vec{\rho}} \int_{q'_{\vec{\kappa}}}^{q_{\vec{\kappa}}} \exp\left\{\underline{S}\right\} Dz D\vec{\rho} Dq_{\vec{\kappa}} \tag{18.608}$$

or

$$Z = \prod_{\vec{\kappa}} Z_L \int \delta\left(z - z'\right) \delta\left(\vec{\rho} - \vec{\rho}'\right) dz dz' d\vec{\rho} d\vec{\rho}' \int_{z'}^{z} \int_{\vec{\rho}'}^{\vec{\rho}} \exp\left\{\underline{S}\left[z, \rho\right] + \Phi_{\omega_{\vec{\kappa}}}\left[z, \rho\right]\right\} Dz D\vec{\rho} \tag{18.609}$$

where lattice partition function is

$$Z_L = \left[2\sinh\left(\frac{\beta\hbar\omega_{\vec{\kappa}}}{2} \right) \right]^{-3} \tag{18.610}$$

and the functional of the electron-phonon interaction influence phase is

$$\Phi[z,\rho]=\sum_{\vec{\kappa}}\frac{\hbar}{8\omega_{\vec{\kappa}}}|C_{\vec{\kappa}}|^2\int_0^\beta\int_0^\beta d\sigma d\sigma'\exp\{i\kappa_\parallel(z(\tau)-z(\sigma))\}\exp\{i\vec{\kappa}_\perp(\vec{\rho}(\tau)-\vec{\rho}(\sigma))\}F_{\omega_{\vec{\kappa}}}(|\sigma-\sigma'|) \quad (18.611)$$

The action functional is:

$$\underline{S}[z,\rho]=\int_0^\beta\left(-\frac{m_\parallel\dot{z}^2}{2\hbar^2}-\frac{m_\parallel\Omega_\parallel^2+\kappa_\parallel}{2}z^2-\frac{m_\perp\dot{\vec{\rho}}^2}{2\hbar^2}-\frac{m_\perp\Omega_\perp^2\vec{\rho}^2+\kappa_\perp}{2}\right)d\tau+\Phi[z,\rho] \quad (18.612)$$

We find now $\langle\underline{S}-\underline{S}_0\rangle_{\underline{S}_0}$ with the help of the generating function:

$$\Psi_{\vec{\kappa}}(\xi,\eta)=\exp\left\langle\{i\kappa_\parallel(\xi z(\tau)-\eta z(\sigma))\}\right\rangle_{\underline{S}_0}\exp\left\langle\{i\vec{\kappa}_\perp(\xi^{\rightarrow}(\tau)-\eta^{\rightarrow}(\sigma))\}\right\rangle_{\underline{S}_0} \quad (18.613)$$

Here, for example,

$$\Psi_{\kappa_\parallel}(\xi,\eta)=\exp\left\langle\{i\kappa_\parallel(\xi z(\tau)-\eta z(\sigma))\}\right\rangle_{\underline{S}_0}=\exp\{f_{1\parallel}+f_{2\parallel}\} \quad (18.614)$$

where

$$f_{1\parallel}=-\frac{\hbar C_1^2\kappa_\parallel^2}{4\omega_{1\parallel}}\left[(\xi^2+\eta^2)F_{\omega_{1\parallel}}(0)-2\xi\eta F_{\omega_{1\parallel}}(|\tau-\sigma|)\right] \quad (18.615)$$

$$f_{2\parallel}=-\frac{\hbar C_2^2F_2^2}{4\omega_{2\parallel}}\left[(\xi^2+\eta^2)F_{\omega_{2\parallel}}(0)-2\xi\eta F_{\omega_{2\parallel}}(|\tau-\sigma|)\right] \quad (18.616)$$

The above polaron formulae have the dependence on the quantity $|\tau-\sigma|$, which is a property of retarded functions. We again generalize the functions with $G\left(|\tau-\sigma|-\frac{\beta}{2}\right)$ [1,2]:

$$f(\beta)=\int_0^\beta\int_0^\beta G\left(|\tau-\sigma|-\frac{\beta}{2}\right)d\sigma d\tau \quad (18.617)$$

that after the change of variables

$$\tau-\sigma\equiv\rho,\quad 0\le\sigma\le\tau \quad (18.618)$$

then

$$f(\beta)=2\int_0^\beta G\left(\rho-\frac{\beta}{2}\right)(\beta-\rho)d\rho \quad (18.619)$$

We do again the change of variable $\rho=\beta y$, where $0\le y\le 1$. If $y=1$ then $\rho=\beta$. We do another subsequent change of variable

$$\frac{\beta(1-y)}{2}\equiv-\tau,\quad d\tau=-\frac{\beta}{2}dy,\quad 0\le\tau\le\infty \quad (18.620)$$

This renders all our integrals convergent and consequently,

$$\left\langle \Phi[z,\rho] \right\rangle_{S_0} = \sum_{\vec{\kappa}} \frac{\hbar}{8\omega_{\vec{\kappa}}} |C_{\vec{\kappa}}|^2 \int_0^\beta \int_0^\beta d\sigma d\sigma' \left\langle \exp\left\{ i\kappa_{\|}\left(\xi z(\tau) - \eta z(\sigma)\right)\right\}\right\rangle_{S_0} \times \left\langle \exp\left\{ i\vec{\kappa}_{\perp}\left(\vec{\rho}(\tau) - \vec{\rho}(\sigma)\right)\right\}\right\rangle_{S_0} \mathrm{F}_{\omega_{\vec{\kappa}}}\left(|\sigma - \sigma'|\right) \quad (18.621)$$

or

$$\Phi[z,\rho]_{S_0} = \frac{\hbar^2 \pi^{\frac{1}{2}} C_0^2 \beta}{(2\pi)^2} \int_0^\infty d\tau \Phi\left(\mathrm{A}_{\perp}(\tau), \mathrm{A}_{\|}(\tau)\right) \exp\{-\hbar\omega_0\tau\} \quad (18.622)$$

$$\Phi\left(\mathrm{A}_{\perp}(\tau), \mathrm{A}_{\|}(\tau)\right) = \frac{1}{\left|\mathrm{A}_{\|}(\tau) - \mathrm{A}_{\perp}(\tau)\right|^{\frac{1}{2}}} \begin{cases} \sinh^{-1}\left(\dfrac{\mathrm{A}_{\|}(\tau)}{\mathrm{A}_{\perp}(\tau)} - 1\right)^{\frac{1}{2}}, & \dfrac{\mathrm{A}_{\|}(\tau)}{\mathrm{A}_{\perp}(\tau)} > 1 \\ \sin^{-1}\left(1 - \dfrac{\mathrm{A}_{\|}(\tau)}{\mathrm{A}_{\perp}(\tau)}\right)^{\frac{1}{2}}, & \dfrac{\mathrm{A}_{\|}(\tau)}{\mathrm{A}_{\perp}(\tau)} \le 1 \end{cases} \quad (18.623)$$

where

$$\mathrm{A}_{\perp,\|}(\tau) = \frac{1}{2} \sum_{i=1}^{2} \frac{d_{i,\perp,\|}^2}{\omega_{i,\perp,\|}} \left(1 - \exp\{-\hbar\omega_{i,\perp,\|}\tau\}\right) \quad (18.624)$$

and

$$d_{1,\perp,\|}^2 = \frac{\omega_{1,\perp,\|}^2 - \omega_{f,\perp,\|}^2}{\omega_{1,\perp,\|}^2 - \omega_{2,\perp,\|}^2}, \quad d_2^2 = \frac{\omega_{f,\perp,\|}^2 - \omega_{2,\perp,\|}^2}{\omega_{1,\perp,\|}^2 - \omega_{2,\perp,\|}^2} \quad (18.625)$$

For the symmetric case when

$$\mathrm{A}_{\perp}(\tau) = \mathrm{A}_{\|}(\tau) = \mathrm{A}(\tau) \quad (18.626)$$

Then we have the result, as seen earlier, for the spherical quantum dot:

$$\Phi[z,\rho]_{S_0} = \frac{\hbar^2 \pi^{\frac{1}{2}} C_0^2 \beta}{(2\pi)^2} \int_0^\infty d\tau \frac{\exp\{-\hbar\omega_0\tau\}}{\mathrm{A}^{\frac{1}{2}}(\tau)} \quad (18.627)$$

where the dimensionless variational polaron energy is obtained as:

$$E_p = \Omega_{\perp} + \frac{\Omega_{\|}}{2} + \frac{(\Omega_{\perp} - \omega_{1\perp})^2 (\Omega_{\perp} - \omega_{2\perp})^2}{2\Omega_{\perp}^2 (\omega_{1\perp} + \omega_{2\perp})} + \frac{(\Omega_{\|} - \omega_{1\|})^2 (\Omega_{\|} - \omega_{2\|})^2}{4\Omega_{\|}^2 (\omega_{1\|} + \omega_{2\|})} - \frac{\alpha_F}{\pi^{\frac{1}{2}}} \int_0^\infty d\tau \frac{\exp\{-\tau\}}{\mathrm{A}^{\frac{1}{2}}(\tau)} \quad (18.628)$$

18.21 Bipolaron Characteristics in a Quasi-0D Cylindrical Quantum Dot with Asymmetrical Parabolic Potential

The exact Hamiltonian for the system is written:

$$\hat{H} = \sum_{i=1}^{2} \left(\frac{\hat{p}_{i\perp}^2}{2m_\perp} + \frac{\hat{p}_{i\parallel}^2}{2m_\parallel} + \frac{m_{i\perp}\Omega_{i\perp}^2 \vec{\rho}_i^2}{2} + \frac{m_{i\parallel}\Omega_{i\parallel}^2 z_i^2}{2} \right) + U\left(\left|\vec{r}_1 - \vec{r}_2\right|\right)$$
$$+ \sum_{\vec{\kappa}} \hbar\omega_{\vec{\kappa}} \hat{b}_{\vec{\kappa}}^{+} \hat{b}_{\vec{\kappa}} + \frac{\hbar}{\sqrt{V}} \sum_{\vec{\kappa}} \left(C_{\vec{\kappa}} \hat{b}_{\vec{\kappa}} \exp\left\{ i\vec{\kappa}\vec{r} \right\} + h.c. \right) \tag{18.629}$$

The model Lagrangian of the system is written:

$$L_0 = L_{10} + L_{20} \tag{18.630}$$

where,

$$L_{10} = \sum_{i=1}^{2} \left(-\frac{m_{i\perp}\dot{\vec{\rho}}_i^2}{2\hbar^2} - \frac{m_{i\parallel}\dot{z}_i^2}{2\hbar^2} - \frac{M_{i\parallel}\dot{Z}_i^2}{2\hbar^2} - \frac{M_{i\perp}\dot{\vec{R}}_i^2}{2\hbar^2} - \frac{m_{i\perp}\Omega_{i\perp}^2\vec{\rho}_i^2}{2} \right.$$
$$\left. - \frac{m_{i\parallel}\Omega_{i\parallel}^2 z_i^2}{2} - \frac{\kappa_{i\perp}\left(\vec{R}_i - \vec{\rho}_i\right)^2}{2} - \frac{\kappa_{i\parallel}\left(Z_i - z_i\right)^2}{2} \right) \tag{18.631}$$

$$L_{20} = -\sum_{i\neq k}^{2} \left(\frac{\kappa'_{i\perp}\left(\vec{R}_k - \vec{\rho}_i\right)^2}{2} + \frac{\kappa'_{i\parallel}\left(Z_k - z_i\right)^2}{2} \right) + \frac{\mathrm{K}_\perp\left(\vec{\rho}_1 - \vec{\rho}_2\right)^2}{2} + \frac{\mathrm{K}_\parallel\left(Z_1 - z_2\right)^2}{2} \tag{18.632}$$

The equations of motion for the transversal motion are

$$m_\perp \ddot{\vec{\rho}}_1 + m_\perp\Omega_\perp^2\vec{\rho}_1 - \kappa_\perp\left(\vec{R}_1 - \vec{\rho}_1\right) - \kappa'_\perp\left(\vec{R}_2 - \vec{\rho}_1\right) - \mathrm{K}_\perp\left(\vec{\rho}_1 - \vec{\rho}_2\right) = 0 \tag{18.633}$$

$$m_\perp \ddot{\vec{\rho}}_2 + m_\perp\Omega_\perp^2\vec{\rho}_2 - \kappa_\perp\left(\vec{R}_2 - \vec{\rho}_2\right) - \kappa'_\perp\left(\vec{R}_1 - \vec{\rho}_2\right) + \mathrm{K}_\perp\left(\vec{\rho}_1 - \vec{\rho}_2\right) = 0 \tag{18.634}$$

$$M_\perp \ddot{\vec{R}}_1 + \kappa_\perp\left(\vec{R}_1 - \vec{\rho}_1\right) + \kappa'_\perp\left(\vec{R}_1 - \vec{\rho}_2\right) = 0 \tag{18.635}$$

$$M_\perp \ddot{\vec{R}}_2 + \kappa_\perp\left(\vec{R}_2 - \vec{\rho}_2\right) + \kappa'_\perp\left(\vec{R}_2 - \vec{\rho}_1\right) = 0 \tag{18.636}$$

and for the longitudinal motion:

$$m_\parallel \ddot{z}_1 + m_\parallel\Omega_\parallel^2 z_1 - \kappa_\parallel\left(Z_1 - z_1\right) - \kappa'_\parallel\left(Z_2 - z_1\right) - \mathrm{K}_\parallel\left(z_1 - z_2\right) = 0 \tag{18.637}$$

$$m_\parallel \ddot{z}_2 + m_\parallel\Omega_\parallel^2 z_2 - \kappa_\parallel\left(Z_2 - z_2\right) - \kappa'_\parallel\left(Z_1 - z_2\right) + \mathrm{K}_\parallel\left(z_1 - z_2\right) = 0 \tag{18.638}$$

$$M_\parallel \ddot{Z}_1 + \kappa_\parallel\left(Z_1 - z_1\right) + \kappa'_\parallel\left(Z_1 - z_2\right) = 0 \tag{18.639}$$

$$M_\parallel \ddot{Z}_2 + \kappa_\parallel\left(Z_2 - z_2\right) + \kappa'_\parallel\left(Z_2 - z_1\right) = 0 \tag{18.640}$$

For normal modes we substitute the following

$$\ddot{\vec{\rho}} = -\omega^2\vec{\rho}, \ddot{z} = -\omega^2 z, \ddot{Z} = -\omega^2 Z, \ddot{\vec{R}} = -\omega^2\vec{R} \tag{18.641}$$

into the equation of motion and letting,

$$z_1 + z_2 = \xi_{\|1}, Z_1 + Z_2 = \xi_{\|2}, z_1 - z_2 = z, Z_1 - Z_2 = Z \tag{18.642}$$

then

$$\left(-m_{\|}\omega_{\|}^2 + m_{\|}\Omega_{\|}^2 + p_{\|}\right)\xi_{\|1} - p_{\|}\xi_{\|2} = 0, -p_{\|}\xi_{\|1} + \left(-M_{\|}\omega_{\|}^2 + m_{\|}\Omega_{\|}^2 + p_{\|}\right)\xi_{\|2} = 0 \tag{18.643}$$

$$\left(-m_{\|}\omega_{\|}^2 + m_{\|}\Omega_{\|}^2 + P_{\|}\right)z - p_{\|}'Z = 0, -p_{\|}'z + \left(-M_{\|}\omega_{\|}^2 + p_{\|}\right)Z = 0 \tag{18.644}$$

Solving for the frequency eigenmodes we have

$$\omega_{i1,2}^2 = \frac{1}{2}\left(\frac{\kappa_i}{\mu_i} + \Omega_i^2\right) \pm \frac{1}{2}\left[\left(\frac{\kappa_i}{\mu_i} + \Omega_i^2\right)^2 - \frac{4\kappa_i\Omega_i^2}{M_i}\right]^{\frac{1}{2}}, \frac{1}{\mu_i} = \frac{1}{m_i} + \frac{1}{M_i}, i = \perp, \| \tag{18.645}$$

$$\omega_{i3,4}^2 = \frac{1}{2}\left(\Omega_i^2 + \frac{P_iM_i + p_im_i}{2M_im_i}\right) \pm \frac{1}{2}\left[\left(\Omega_i^2 + \frac{P_iM_i + p_im_i}{2M_im_i}\right)^2 + 4\frac{\left(m_i\Omega_i^2 + P_i\right)p_i - p_i'^2}{M_im_i}\right]^{\frac{1}{2}} \tag{18.646}$$

with

$$\omega_{i1}^2\omega_{i2}^2 = \omega_{if}^2\Omega_i^2 \tag{18.647}$$

The **bipolaron effective mass** is conveniently obtained from the frequency eigenmode equations:

$$\frac{M}{m} = \frac{1}{2}\sum_{i=\perp,\|}\frac{\left(\omega_{i1}^2 - \Omega_i^2\right)\left(\Omega_i^2 - \omega_{i2}^2\right)}{\omega_{i1}^2\omega_{i2}^2} \tag{18.648}$$

To diagonalize the model Lagrangian, we move to normal coordinates:

$$\xi_1 = C_1(\omega_1)x_1(\omega_1) + C_2(\omega_2)F_2(\omega_2)x_2(\omega_2) \tag{18.649}$$

$$\xi_2 = C_1(\omega_1)F_1(\omega_1)x_1(\omega_1) + C_2(\omega_2)x_2(\omega_2) \tag{18.650}$$

$$z = C_3x_3(\omega_3) + C_4F_2(\omega_4)x_4(\omega_4)$$

$$Z = C_3F_3(\omega_3)x_3(\omega_3) + C_4x_4(\omega_4) \tag{18.651}$$

Substituting these equations into the equation of motion and also considering the conservation of the kinetic energy in any representation then

$$F_{1i}(\omega_{1i}) = \frac{v_i^2}{-\omega_{1i}^2 + v_i^2}, F_{2i}(\omega_{2i}) = \frac{p_i}{m_i\left(\Omega_i^2 + v_i - \omega_{2i}^2\right)} = \frac{M_i}{m_i}F_{1i}(\omega_{1i}), v_i = \frac{p_i}{M_i} \tag{18.652}$$

$$C_{2i}^2(\omega_1) = \frac{1}{M_i + m_iF_{2i}^2(\omega_{2i})}, C_{1i}^2(\omega_{1i}) = \frac{M_i}{m_i}C_{2i}^2(\omega_{1i}) \tag{18.653}$$

$$F_{3i}(\omega_{3i}) = \frac{p'_i}{-M_i\omega_{3i}^2 + p_i}, F_{4i}(\omega_{2i}) = \frac{p'_i}{m_i\omega_{4i}^2 + \Omega_i^2 + P_i} \tag{18.654}$$

$$C_{3i}^2(\omega_1) = \frac{1}{m_i + M_i F_{3i}^2(\omega_{3i})}, C_{4i}^2(\omega_{4i}) = \frac{1}{M_i + m_i F_{4i}^2(\omega_{4i})} \tag{18.655}$$

The model Lagrangian in normal coordinates is obtained:

$$L_0 = \sum_{i=1}^{4}\left[-\frac{\dot{x}_i^2}{4\hbar^2} - \frac{\omega_i^2 x_i^2}{4}\right] \tag{18.656}$$

This shows that in the motion of the bipolaron we have four oscillators showing four internal motions. The Coulomb interaction can be written in the form:

$$U(|\vec{r}_1 - \vec{r}_2|) = \frac{(2)^{\frac{1}{2}}\hbar\omega_0 4\pi\alpha_F}{V(1-\eta)}\sum_{\vec{\kappa}}\frac{1}{\kappa^2}\exp\{i\kappa_\parallel(z_1(\tau) - z_2(\tau))\}\exp\{i\vec{\kappa}_\perp(\vec{\rho}_1(\tau) - \vec{\rho}_2(\tau))\} \tag{18.657}$$

We find now $\langle \underline{S} - \underline{S_0}\rangle_{S_0}$ via the generating function:

$$\Psi_{\vec{\kappa}}(\xi_1,\eta_1,\xi_2,\eta_2) = \left\langle \exp\{i\kappa_\parallel(\xi_1 z_1(\tau) - \eta_1 z_1(\sigma) + \xi_2 z_2(\tau) - \eta_2 z_2(\sigma))\}\right\rangle_{S_0} \tag{18.658}$$

Considering,

$$f_1^+ = (\xi_1 + \xi_2)^2 + (\eta_1 + \eta_2)^2, f_2^+ = (\xi_1 + \xi_2)(\eta_1 + \eta_2) \tag{18.659}$$

$$f_1^- = (\xi_1 - \xi_2)^2 + (\eta_1 - \eta_2)^2, f_2^- = (\xi_1 - \xi_2)(\eta_1 - \eta_2) \tag{18.660}$$

then

$$\Psi_{\vec{\kappa}}(\xi_1,\eta_1,\xi_2,\eta_2) = \exp\{\kappa_\parallel^2(f_1 + f_2 + f_3 + f_4)\}_{S_0} \tag{18.661}$$

where

$$f_1 = -\frac{\hbar C_{1i}^2}{16\omega_{1i}m_i}\left[f_1^+ \mathrm{F}_{\omega_{1i}}(0) - 2 f_2^+ \mathrm{F}_{\omega_{1i}}(|\tau - \sigma|)\right] \tag{18.662}$$

$$f_2 = -\frac{\hbar C_{2i}^2 F_{2i}^2}{16\omega_{2i}m_i}\left[f_1^+ \mathrm{F}_{\omega_{2i}}(0) - 2 f_2^+ \mathrm{F}_{\omega_{2i}}(|\tau - \sigma|)\right] \tag{18.663}$$

$$f_3 = -\frac{\hbar C_{3i}^2}{16\omega_{3i}m_i}\left[f_1^- \mathrm{F}_{\omega_{3i}}(0) - 2 f_2^- \mathrm{F}_{\omega_{3i}}(|\tau - \sigma|)\right] \tag{18.664}$$

$$f_4 = -\frac{\hbar C_{4i}^2 F_{4i}^2}{16\omega_{4i}m_i}\left[f_1^- \mathrm{F}_{\omega_{4i}}(0) - 2 f_2^- \mathrm{F}_{\omega_{4i}}(|\tau - \sigma|)\right] \tag{18.665}$$

$$m_i = \frac{1}{2} \tag{18.666}$$

The resultant formulae for the bipolaron are observed to have the same dependence on the quantity $|\tau-\sigma|$. This is the property of retarded functions as earlier indicated.. We generalize the resultant bipolaron functions $G\left(|\tau-\sigma|-\frac{\beta}{2}\right)$ which helps in the evaluation of the twofold integral [1,2] . This procedure renders all our integrals convergent and, consequently, the variational bipolaron energy:

$$E_{bip}=\sum_{j=1}^{4}\omega_{j\perp}\left(1-\frac{\omega_{j\perp}^2-\Omega_{\perp}^2}{2\omega_{j\perp}^2}\mathrm{d}_{j\perp}^2\right)+\frac{1}{2}\sum_{j=1}^{4}\omega_{j\parallel}\left(1-\frac{\omega_{j\parallel}^2-\Omega_{\parallel}^2}{2\omega_{j\parallel}^2}\mathrm{d}_{j\parallel}^2\right)-\frac{2\alpha_F}{V(1-\eta)\pi^{\frac{1}{2}}}\Phi$$

$$\left(\mathrm{A}_{12\perp}(0),\mathrm{A}_{12\parallel}(0)\right)+\frac{2\alpha_F}{\pi^{\frac{1}{2}}}\int_0^\infty d\tau\exp\{-\tau\}\left[\Phi\left(\mathrm{A}_{11\perp}(\tau),\mathrm{A}_{11\parallel}(\tau)\right)+\Phi\left(\mathrm{A}_{12\perp}(\tau),\mathrm{A}_{12\parallel}(\tau)\right)\right] \quad (18.667)$$

where,

$$\Phi\left(\mathrm{A}_{12\perp}(\tau),\mathrm{A}_{12\parallel}(\tau)\right)=\frac{1}{\left|\mathrm{A}_{12\perp}(\tau)-\mathrm{A}_{12\parallel}(\tau)\right|^{\frac{1}{2}}}\begin{cases}\sinh^{-1}\left(\frac{\mathrm{A}_{12\parallel}(\tau)}{\mathrm{A}_{12\perp}(\tau)}-1\right)^{\frac{1}{2}}, & \frac{\mathrm{A}_{12\parallel}(\tau)}{\mathrm{A}_{12\perp}(\tau)}>1\\ \sin^{-1}\left(1-\frac{\mathrm{A}_{12\parallel}(\tau)}{\mathrm{A}_{12\perp}(\tau)}\right)^{\frac{1}{2}}, & \frac{\mathrm{A}_{12\parallel}(\tau)}{\mathrm{A}_{12\perp}(\tau)}\le 1\end{cases} \quad (18.668)$$

$$\Phi\left(\mathrm{A}_{11\perp}(\tau),\mathrm{A}_{11\parallel}(\tau)\right)=\frac{1}{\left|\mathrm{A}_{11\perp}(\tau)-\mathrm{A}_{11\parallel}(\tau)\right|^{\frac{1}{2}}}\begin{cases}\sinh^{-1}\left(\frac{\mathrm{A}_{11\parallel}(\tau)}{\mathrm{A}_{11\perp}(\tau)}-1\right)^{\frac{1}{2}}, & \frac{\mathrm{A}_{11\parallel}(\tau)}{\mathrm{A}_{11\perp}(\tau)}>1\\ \sin^{-1}\left(1-\frac{\mathrm{A}_{11\parallel}(\tau)}{\mathrm{A}_{11\perp}(\tau)}\right)^{\frac{1}{2}}, & \frac{\mathrm{A}_{11\parallel}(\tau)}{\mathrm{A}_{11\perp}(\tau)}\le 1\end{cases} \quad (18.669)$$

and

$$\mathrm{A}_{11\perp,\parallel}(\tau)=\frac{1}{2}\sum_{i=1}^{4}\frac{d_{i,\perp,\parallel}^2}{\omega_{i,\perp,\parallel}}\left(1-\exp\{-\omega_{i,\perp,\parallel}\tau\}\right) \quad (18.670)$$

$$\mathrm{A}_{12\perp,\parallel}(\tau)=\frac{1}{2}\sum_{i=1}^{2}\frac{d_{i,\perp,\parallel}^2}{\omega_{i,\perp,\parallel}}\left(1-\exp\{-\omega_{i,\perp,\parallel}\tau\}\right)+\frac{1}{2}\sum_{i=3}^{4}\frac{d_{i,\perp,\parallel}^2}{\omega_{i,\perp,\parallel}}\left(1+\exp\{-\omega_{i,\perp,\parallel}\tau\}\right) \quad (18.671)$$

with

$$d_{1,\perp,\parallel}^2=\frac{\omega_{1,\perp,\parallel}^2-\omega_{f,\perp,\parallel}^2}{\omega_{1,\perp,\parallel}^2-\omega_{2,\perp,\parallel}^2},\ d_{2,\perp,\parallel}^2=\frac{\omega_{f,\perp,\parallel}^2-\omega_{2,\perp,\parallel}^2}{\omega_{1,\perp,\parallel}^2-\omega_{2,\perp,\parallel}^2} \quad (18.672)$$

$$d_{3,\perp,\parallel}^2=\frac{\omega_{3,\perp,\parallel}^2-\omega_{f,\perp,\parallel}^2}{\omega_{3,\perp,\parallel}^2-\omega_{4,\perp,\parallel}^2},\ d_{4,\perp,\parallel}^2=\frac{\omega_{f,\perp,\parallel}^2-\omega_{4,\perp,\parallel}^2}{\omega_{3,\perp,\parallel}^2-\omega_{4,\perp,\parallel}^2} \quad (18.673)$$

Theoretical studies show that the electronic confinement is observed to imitate the magnetic field subjected to the system [22,57]. In this case, similar effects of the cyclotron frequency [9] and confinement frequencies on the electronic and polaron as well as the bipolaron states are expected.

18.22 Polaron in a Magnetic Field

We have so far examined a lot of problems on the harmonic oscillator. We examine another problem when a charged particle, for example say an electron, is subjected to a homogenous magnetic field in a polar medium and interacts with lattice vibrations. This will be calculated using the same Feynman variational technique where, in this case, Feynman employed a special approach for the evaluation of the polaron in a magnetic field by introducing a transition amplitude permitting the action to be written in quadratic form over the electronic coordinates:

$$K\left(q,-i\hbar t;0,0\right)=\int_0^q Dq\left(t\right)\exp\left\{\frac{i}{\hbar}S\left[q\right]\right\} \tag{18.674}$$

We express now the transition amplitude K via the temperature dependence by using the analytic prolongation:

$$t=-i\hbar\tau, 0\le\tau\le\beta \tag{18.675}$$

So, for the action functional

$$S\left[q\right]=\int_0^T\left(\frac{m\dot{q}^2}{2}-\frac{m\omega^2q^2}{2}\right)dt \tag{18.676}$$

or

$$S\left[q\right]=-\frac{i\hbar}{2}\int_0^\beta\left(\frac{m\dot{q}^2}{\hbar^2}+m\omega^2q^2\right)d\tau \tag{18.677}$$

we employ a special approach via

$$S'\left[q\right]=-\frac{1}{2}\int_0^\beta\left(\frac{m\dot{q}^2}{\hbar^2}+m\omega^2q^2\right)d\tau \tag{18.678}$$

We consider first the kinetic energy term:

$$\mathcal{K}_{\text{kin}}=\int_0^\beta \dot{q}^2d\tau=\int_0^\beta\int_0^\beta \dot{q}\left(\tau\right)\dot{q}\left(\sigma\right)\delta\left(\tau-\sigma\right)d\tau d\sigma \tag{18.679}$$

and consider for the Dirac delta function the property

$$\delta\left(\tau-0,\beta\right)=0 \tag{18.680}$$

Apply twice integration by parts in 18.679 then

$$\mathcal{K}_{\text{kin}}=\int_0^\beta \dot{q}\left(\tau\right)d\tau\left[q\left(\sigma\right)\delta\left(\tau-\sigma\right)\Big|_0^\beta-\int_0^\beta q\left(\sigma\right)\frac{\partial\delta\left(\tau-\sigma\right)}{\partial\sigma}d\tau\right] \tag{18.681}$$

or

$$\mathcal{K}_{kin} = \int_0^\beta \int_0^\beta q(\tau)q(\sigma)\frac{\partial^2 \delta(\tau-\sigma)}{\partial\sigma\partial\tau} d\tau d\sigma = -\int_0^\beta \int_0^\beta q(\tau)q(\sigma)\frac{\partial^2 \delta(\tau-\sigma)}{\partial\tau^2} d\tau d\sigma \tag{18.682}$$

So,

$$S'[q] = -\frac{1}{2}\int_0^\beta \int_0^\beta \left[-\frac{1}{\hbar^2}\frac{\partial^2}{\partial\tau^2} + \omega^2\right] q(\tau)q(\sigma)\delta(\tau-\sigma) d\tau d\sigma \tag{18.683}$$

or

$$S'[q] = -\frac{\omega^2}{2}\int_0^\beta \int_0^\beta \left[1 - \frac{1}{\hbar^2\omega^2}\frac{\partial^2}{\partial\tau^2}\right] \delta(\tau-\sigma) q(\tau)q(\sigma) d\tau d\sigma \tag{18.684}$$

or

$$S'[q] = -\frac{1}{2}\int_0^\beta \int_0^\beta \hat{F}(\tau,\sigma) q(\tau)q(\sigma) d\tau d\sigma \tag{18.685}$$

Here, the differential operator $\hat{F}(\tau,\sigma)$:

$$\hat{F}(\tau,\sigma) = \omega^2\left[1 - \frac{1}{\hbar^2\omega^2}\frac{\partial^2}{\partial\tau^2}\right] \tag{18.686}$$

We have just seen that the functional is independent of the velocity but dependent only on the coordinates. For the excursion of the particle from the new path we select a fluctuating path

$$\delta q(\tau) \equiv y(\tau) \tag{18.687}$$

about the classical one:

$$q(\tau) = \bar{q}(\tau) + y(\tau) \tag{18.688}$$

We write now the actional functional considering the introduction of a perturbation $\gamma(\tau)$:

$$S[q] = -\frac{1}{2}\int_0^\beta \int_0^\beta \hat{F}(\tau,\sigma) q(\tau)q(\sigma) d\tau d\sigma + \int_0^\beta \gamma(\tau) q(\tau) d\tau \tag{18.689}$$

This should be the action functional of a driven harmonic oscillator. We find the equation of motion:

$$\delta S[q] = -\int_0^\beta \int_0^\beta \hat{F}(\tau,\sigma) q(\tau)\delta q(\sigma) d\tau d\sigma + \int_0^\beta \gamma(\sigma)\delta q(\sigma) d\sigma = 0 \tag{18.690}$$

or

$$-\int_0^\beta \hat{F}(\tau,\sigma) q(\tau)\delta q(\sigma) d\tau d\sigma + \gamma(\sigma) = 0 \tag{18.691}$$

From this equation of motion, we find first the homogenous solution. From definition of the inverse operator $\hat{F}^{-1}(\sigma,\tau)$:

$$\int_0^\beta \hat{F}(\tau,\sigma)\hat{F}^{-1}(\sigma,\tau')d\sigma = \delta(\tau-\tau') \quad (18.692)$$

then

$$\bar{q}(\tau) = \int_0^\beta \hat{F}^{-1}(\sigma,\tau')\gamma(\sigma)d\sigma \quad (18.693)$$

This gives us the solution of some partial solution for the classical path.

We write now the action in terms of the fluctuation $y(\tau)$ to the classical path $\bar{q}(\tau)$:

$$S[q] = -\frac{1}{2}\int_0^\beta\int_0^\beta \hat{F}(\tau,\sigma)q(\tau)q(\sigma)d\tau d\sigma + \int_0^\beta \gamma(\tau)q(\tau)d\tau \quad (18.694)$$

or

$$S[q] = -\frac{1}{2}\int_0^\beta\int_0^\beta \hat{F}(\tau,\sigma)\bar{q}(\tau)\bar{q}(\sigma)d\tau d\sigma - \int_0^\beta\int_0^\beta \hat{F}(\tau,\sigma)\bar{q}(\tau)y(\sigma)d\tau d\sigma$$
$$-\frac{1}{2}\int_0^\beta\int_0^\beta \hat{F}(\tau,\sigma)y(\tau)y(\sigma)d\tau d\sigma + \int_0^\beta \gamma(\tau)\bar{q}(\tau)d\tau + \int_0^\beta \gamma(\tau)y(\tau)d\tau \quad (18.695)$$

The fluctuation y may be found from the equation of motion after varying the action functional $S[q]$. Considering the problem of the oscillators in the previous chapters as well as the quasi-classical approximation then the transition amplitude is:

$$K = \Phi(\beta)\exp\left\{-\frac{1}{2}\int_0^\beta\int_0^\beta \hat{F}(\tau,\sigma)\bar{q}(\tau)\bar{q}(\sigma)d\tau d\sigma + \int_0^\beta \gamma(\tau)q(\tau)d\tau\right\} \quad (18.696)$$

Here, we introduce the following function:

$$\Phi(\beta) = \int_0^0 Dy(\tau)\exp\left\{-\frac{1}{2}\int_0^\beta\int_0^\beta \hat{F}(\tau,\sigma)y(\tau)y(\sigma)d\tau d\sigma\right\} \quad (18.697)$$

Considering 18.674, then

$$K = \Phi(\beta)\exp\left\{\frac{1}{2}\int_0^\beta\int_0^\beta \hat{F}^{-1}(\tau,\sigma)\gamma(\tau)\gamma(\sigma)d\tau d\sigma\right\} \quad (18.698)$$

where now,

$$\Phi(\beta) = \int_0^0 Dy(\tau)\exp\left\{-\frac{1}{2}\int_0^\beta\int_0^\beta \hat{F}(\tau,\sigma)y(\tau)y(\sigma)d\tau d\sigma\right\} \quad (18.699)$$

This first factor is due to the classical solution while the second is due to the fluctuation of the classical path. Expression 18.698 permit to find the partition function of a driven harmonic oscillator.

Let us find the properties of the operator $\hat{F}(\tau,\sigma)$:

$$\hat{F}(|\tau-\sigma|)=\sum_n \lambda_n \Phi_n(\tau)\Phi_n(\sigma) \tag{18.700}$$

Here, $\Phi_n(\tau)$ is the eigenfunction of the operator $\hat{F}(\tau,\sigma)$ and λ_n is the eigenvalue. From the above we have

$$\int_0^\beta \hat{F}(\tau,\sigma)\Phi_n(\tau)d\tau=\lambda_n\Phi_n(\sigma) \tag{18.701}$$

If this relation is satisfied then $\Phi_n(\tau)$ is obviously the eigenfunction and λ_n eigenvalue of the operator $\hat{F}(\tau,\sigma)$.

Considering $\hat{F}(\tau,\sigma)$ to be a symmetrical operator we examine the following properties:

$$\hat{F}^{-1}(\tau,\sigma)=\sum_{n'}\frac{1}{\lambda_{n'}}\Phi_{n'}(\tau)\Phi_{n'}(\sigma) \tag{18.702}$$

So,

$$\int_0^\beta \hat{F}(\tau,\sigma)\hat{F}^{-1}(\sigma,\tau')d\sigma=\int_0^\beta\sum_n \lambda_n\Phi_n(\tau)\Phi_n(\sigma)\sum_{n'}\frac{1}{\lambda_{n'}}\Phi_{n'}(\tau')\Phi_{n'}(\sigma)d\sigma \tag{18.703}$$

or

$$\int_0^\beta \hat{F}(\tau,\sigma)\hat{F}^{-1}(\sigma,\tau')d\sigma=\sum_{nn'}\frac{\lambda_n}{\lambda_{n'}}\Phi_n(\tau)\Phi_{n'}(\tau')\int_0^\beta\Phi_n(\sigma)\Phi_{n'}(\sigma)d\sigma \tag{18.704}$$

or

$$\int_0^\beta \hat{F}(\tau,\sigma)\hat{F}^{-1}(\sigma,\tau')d\sigma=\sum_{nn'}\frac{\lambda_n}{\lambda_{n'}}\Phi_n(\tau)\Phi_{n'}(\tau')\delta_{nn'} \tag{18.705}$$

This follows that the eigenfunctions $\Phi_n(\tau)$ and $\Phi_{n'}(\tau')$ satisfy, respectively, the orthogonality and completeness relations:

$$\int_0^\beta\Phi_n(\sigma)\Phi_{n'}(\sigma)d\sigma=\delta_{nn'},\int_0^\beta \hat{F}(\tau,\sigma)\hat{F}^{-1}(\sigma,\tau')d\sigma=\sum_n\Phi_n(\tau)\Phi_n(\tau')=\delta(\tau-\tau') \tag{18.706}$$

We will treat with operators of the form:

$$\hat{F}(\tau,\sigma)\equiv\hat{F}(|\tau-\sigma|)=\frac{1}{2}\sum_n{}_{-n}\cos(\Omega_n|\tau-\sigma|),\ \Omega_n=\frac{2\pi n}{\beta} \tag{18.707}$$

$$\delta(\tau-\sigma)=\frac{1}{\beta}\sum_n\cos(\Omega_n|\tau-\sigma|) \tag{18.708}$$

In order to find the inverse operator $\hat{F}^{-1}(\sigma,\tau)$, it is important to find α_n and α_n^{-1}:

$$\hat{F}^{-1}(\sigma,\tau) = \frac{1}{2}\sum_n \alpha_n^{-1}\cos\left(\Omega_n|\tau-\sigma|\right) \tag{18.709}$$

Multiplying 18.709 and 18.707 and then taking the integral of the resultant expression we have

$$\int_0^\beta \hat{F}(\tau,\sigma)\hat{F}^{-1}(\sigma,\tau')d\sigma = \frac{1}{\beta}\sum_n \cos\left(\Omega_n|\tau-\sigma|\right) \tag{18.710}$$

Comparing the coefficients of the left-hand and right-hand sides then we have:

$$\alpha_n^{-1} = \frac{4}{\beta^2}\frac{1}{-_n} \tag{18.711}$$

Considering the operator $\hat{F}(\tau,\sigma)$ in 18.709 and the definition of the Dirac delta function in 18.708 then

$$\hat{F}(\tau,\sigma) = \omega^2\left(1 - \frac{1}{\hbar^2\omega^2}\frac{\partial^2}{\partial\tau^2}\right)\cdot(\tau-\sigma) \tag{18.712}$$

or

$$\hat{F}(\tau,\sigma) = \omega^2\left(1 - \frac{1}{\hbar^2\omega^2}\frac{\partial^2}{\partial\tau^2}\right)\frac{1}{\beta}\sum_n \cos\left(\Omega_n|\tau-\sigma|\right) \tag{18.713}$$

or

$$\hat{F}(\tau,\sigma) = \frac{\omega^2}{\beta}\left(1 + \frac{\Omega_n^2}{\hbar^2\omega^2}\right)\sum_n \cos\left(\Omega_n|\tau-\sigma|\right) = \frac{1}{2}\sum_n \alpha_n\cos\left(\Omega_n|\tau-\sigma|\right) \tag{18.714}$$

So,

$$\alpha_n = \frac{2\omega^2}{\beta}\left(1 + \frac{\Omega_n^2}{\hbar^2\omega^2}\right) \tag{18.715}$$

$$\alpha_n^{-1} = \frac{4}{\beta^2}\frac{1}{\alpha_n} = \frac{2}{\beta\omega^2}\frac{1}{1+\frac{\Omega_n^2}{\hbar^2\omega^2}} \tag{18.716}$$

Thus, we find

$$\hat{F}^{-1}(\sigma,\tau) = \frac{1}{2}\frac{2}{\beta\omega^2}\sum_n \frac{1}{1+\frac{\Omega_n^2}{\hbar^2\omega^2}}\cos\left(\Omega_n\frac{\hbar}{\hbar}|\tau-\sigma|\right) \tag{18.717}$$

From,

$$\sum_{\kappa=-\infty}^{\infty}\frac{\cos\kappa\xi}{\kappa^2+\alpha^2} = \frac{\pi}{\xi}\frac{\cosh\left[\alpha(\pi-\xi)\right]}{\sinh\alpha\pi} \tag{18.718}$$

then

$$\hat{F}^{-1}(\sigma,\tau)=\frac{1}{2}\frac{2}{\beta\omega^2}\sum_n\frac{1}{1+\frac{\Omega_n^2}{\hbar^2\omega^2}}\cos\left(\Omega_n\frac{\hbar}{\hbar}|\tau-\sigma|\right)=\frac{\hbar}{2\omega}F_\omega\left(|\tau-\sigma|\right) \tag{18.719}$$

where the function $F_\omega(|\tau-\sigma|)$ as well as the influence phase $\Phi_\omega\left[\vec{r}\right]$ for the full system has been found this time via the operator method:

$$F_\omega\left(|\tau-\sigma|\right)=\frac{\cosh\hbar\omega\left(|\tau-\sigma|-\frac{\beta}{2}\right)}{\sinh\left(\frac{\hbar\omega\beta}{2}\right)} \tag{18.720}$$

$$\Phi_\omega\left[q\right]=\frac{\hbar}{4\omega}\int_0^\beta\int_0^\beta d\sigma d\tau\,\gamma_\omega(\sigma)\gamma_\omega(\tau)F_\omega\left(|\tau-\sigma|\right) \tag{18.721}$$

The transition amplitude for this problem is thus:

$$K=\exp\left\{\Phi_\omega\left[q\right]\right\}\int_0^0 Dy(\tau)\exp\left\{-\frac{1}{2}\int_0^\beta\int_0^\beta\hat{F}(\tau,\sigma)y(\tau)y(\sigma)d\tau d\sigma\right\} \tag{18.722}$$

We now apply the above procedure for the case of an electron interacting with lattice vibrations in a polar medium and subjected to a magnetic field which is described by the following Lagrangian:

$$L(-i\hbar\tau)=-\frac{m}{2\hbar^2}\left(\frac{dr}{d\tau}\right)^2+\frac{ie}{\hbar c}\left(\frac{dr_i}{d\tau},\vec{A}_i\right)-\frac{1}{2}\sum_{\vec{\kappa}j}\left(\frac{\dot{q}_{\vec{\kappa}j}^2}{\hbar^2}+\omega_{\vec{\kappa}j}^2q_{\vec{\kappa}j}^2\right)+\sum_{\vec{\kappa}j}\gamma_{\vec{\kappa}j}(\tau)q_{\vec{\kappa}j}(\tau) \tag{18.723}$$

Here the vector potential $\vec{A}$ is related to the magnetic field strength $\vec{H}$ expected to point in the oz-axis direction:

$$\vec{H}=curl\vec{A} \tag{18.724}$$

The Feynman approach can also be applicable for the case of a particle in a magnetic field where the full partition function:

$$Z=\prod_{\vec{\kappa}j}\int\delta\left(q_{\vec{\kappa}j}-q'_{\vec{\kappa}j}\right)\delta\left(\vec{r}-\vec{r}'\right)dq_{\vec{\kappa}j}dq'_{\vec{\kappa}j}d\vec{r}d\vec{r}'\int_{\vec{r}'}^{\vec{r}}\int_{q'_{\vec{\kappa}j}}^{q_{\vec{\kappa}j}}\exp\left\{\underline{S}\right\}D\vec{r}Dq_{\vec{\kappa}j} \tag{18.725}$$

This is the product of the partition functions over all possible oscillators:

$$Z=\prod_{\vec{\kappa}j}Z_L\int\delta\left(\vec{r}-\vec{r}'\right)d\vec{r}d\vec{r}'\int_{\vec{r}'}^{\vec{r}}D\vec{r}\exp\left\{-\int_0^\beta\frac{m\dot{\vec{r}}^2}{2\hbar^2}d\tau+\frac{ieH}{\hbar c}\int_0^\beta\frac{dy(\tau)}{d\tau}x(\tau)d\tau+\Phi\left[\vec{r}\right]\right\} \tag{18.726}$$

Here is the lattice partition function:

$$Z_L=\left[2\sinh\left(\frac{\hbar\omega_{\vec{\kappa}j}\beta}{2}\right)\right]^{-3} \tag{18.727}$$

and the **electron-phonon influence phase:**

$$\Phi_{\omega_{\vec{\kappa}j}}\left[\vec{r}\right]=\frac{\hbar}{4\omega_{\vec{\kappa}j}}\int_0^\beta\int_0^\beta d\sigma d\tau\gamma_{\nu_{\vec{\kappa}j}}\left(\vec{r}\left(\sigma\right)\right)\gamma_{\nu_{\vec{\kappa}j}}\left(\vec{r}\left(\tau\right)\right)\mathrm{F}_{\omega_{\nu_{\vec{\kappa}j}}}\left(\left|\tau-\sigma\right|\right) \tag{18.728}$$

and

$$\Phi\left[\vec{r}\right]=\sum_{\vec{\kappa}j}\Phi_{\omega_{\vec{\kappa}j}}\left[\vec{r}\right] \tag{18.729}$$

is the **full functional of the electron-phonon interaction (influence phase)**. So, the coupling to the environment leads to an additional, non-local term to the action where the original bonafide many-body problem is now reduced to a one-body problem and we rewrite the full action of the full system via its influence phase $\Phi\left[\vec{r}\right]$:

$$\underline{S}\left[\vec{r}\right]=-\int_0^\beta\frac{m\dot{\vec{r}}^2}{2\hbar^2}d\tau+\frac{ieH}{\hbar c}\int_0^\beta\frac{dy}{d\tau}xd\tau+\Phi\left[\vec{r}\right] \tag{18.730}$$

The action functional 18.730 describes a driven harmonic oscillator in a magnetic field. In the full action functional 18.730 the first term is the contribution of the kinetic energy of the electron, the second term that due to the magnetic interaction, and the last, the full functional of the electron-phonon interaction described by $\Phi\left[\vec{r}\right]$ and is observed to depend on the function $\mathrm{F}_\omega(|\tau-\sigma|)$ with the time difference $|\tau-\sigma|$ indicative of a retarded function depending on the past thereby signifying interaction with the past where the perturbative motion of the electron takes "**time**" to propagate in the crystal lattice. This function $\mathrm{F}_\omega(|\tau-\sigma|)$ suggests the self-interaction to be stronger in the near past than it is in the distant past.

As mentioned earlier, a classical potential for the polaron problem might well be expected to be a good approximation with tight binding and so, to imitate the polaron problem, a good model is one where instead of the electron being coupled to the lattice, it is coupled by some "**spring**" to another particle (fictitious particle) and the pair of particles are free to wander about the crystal where the model Lagrangian may be selected in the one oscillatory approximation:

$$L_0=-\frac{m\dot{r}^2}{2\hbar^2}+\frac{ieH}{\hbar c}\frac{dy}{d\tau}x-\frac{M\dot{R}^2}{2\hbar^2}-\frac{M\omega_f^2\left(\vec{R}-\vec{r}\right)^2}{2} \tag{18.731}$$

The first term describes the translational motion; the second term due to the magnetic field acting on the particle, the third and fourth describe the oscillatory motion. It is an approximation of the full Lagrangian of the system, L. Here M and ω_f are, respectively, the mass of a fictitious particle and the frequency of the elastic coupling that will serve as variational parameters; $\vec{R}$ is the coordinate of the fictitious particle. The model system conserves the translational symmetry of the system. The judicious choice of L_0 is to simulate a physical situation that may give a better upper bound for the energy, as mentioned earlier.

The partition function of the model system can be written:

$$Z_0=\int_{-\infty}^{+\infty}\delta\left(\vec{r}-\vec{r}'\right)\delta\left(\vec{R}-\vec{R}'\right)d\vec{r}d\vec{r}'d\vec{R}d\vec{R}'\int_{r'}^{r}\int_{R'}^{R}\exp\left\{\underline{S}_0\right\}D\vec{r}D\vec{R} \tag{18.732}$$

We find the action functional of the model system in terms of only the path $\vec{r}$ by eliminating the fictitious particle subsystem variables $\vec{R}$ in 18.732:

$$Z_0 = Z_{\omega_f} \int \delta\left(\vec{r}-\vec{r}'\right) d\vec{r} d\vec{r}' \int_{r'}^{r} D\vec{r} \exp\left\{ -\int_0^\beta \frac{m\dot{\vec{r}}^2}{2\hbar^2} d\tau + \frac{ieH}{\hbar c}\int_0^\beta \frac{dy}{d\tau} x d\tau - \Phi_{\omega_f}\left[\vec{r}\right] \right\} \tag{18.733}$$

The partition function due to the fictitious lattice is

$$Z_{\omega_f} = \left[2\sinh\left(\frac{\beta\hbar\omega_f}{2} \right) \right]^{-3} \tag{18.734}$$

and the **influence phase** of the interaction of the electron with the fictitious particle:

$$\Phi_{\omega_f}\left[\vec{r}\right] = \frac{\hbar M\omega_f^3}{4} \int_0^\beta \int_0^\beta d\sigma d\tau \vec{r}(\sigma)\vec{r}(\tau) \mathrm{F}_{\omega_f}\left(\left|\tau-\sigma\right|\right) \tag{18.735}$$

where

$$\mathrm{F}_{\omega_f}\left(\left|\tau-\sigma\right|\right) = \frac{\cosh\hbar\omega_f\left(\left|\tau-\sigma\right| - \frac{\beta}{2} \right)}{\sinh\left(\frac{\hbar\omega_f\beta}{2} \right)} \tag{18.736}$$

The action functional of the interaction of the electron (hole) with the fictitious particle via its influence phase:

$$\underline{S_0} = \int_0^\beta \frac{m\dot{\vec{r}}^2}{2\hbar^2} d\tau + \frac{ieH}{\hbar c}\int_0^\beta \frac{dy}{d\tau} x d\tau + \Phi_{\omega_f}\left[\vec{r}\right] - 3\ln\left[2\sinh\left(\frac{\beta\hbar\omega_f}{2} \right) \right] \tag{18.737}$$

From

$$\frac{Z}{Z_{\omega_f}} = Z_0 \left\langle \exp\left\{ \underline{S} - \underline{S_0} \right\} \right\rangle_{\underline{S_0}} \tag{18.738}$$

then

$$\ln\frac{Z}{Z_{\omega_f}} \equiv \ln\tilde{Z} = \ln Z_0 + \ln\left\langle \exp\left\{ \underline{S} - \underline{S_0} \right\} \right\rangle_{\underline{S_0}} \tag{18.739}$$

Apply the Feynman inequality

$$\left\langle \exp\left\{ \underline{S} - \underline{S_0} \right\} \right\rangle_{\underline{S_0}} \geq \exp\left\{ \left\langle \underline{S} - \underline{S_0} \right\rangle_{\underline{S_0}} \right\} \tag{18.740}$$

then from

$$F = -T\ln Z \tag{18.741}$$

we have

$$F = -T \ln Z_0 - T \ln \left\langle \exp\left\{\underline{S} - \underline{S}_0\right\}\right\rangle_{\underline{S}_0} \tag{18.742}$$

with

$$\ln \left\langle \exp\left\{\underline{S} - \underline{S}_0\right\}\right\rangle_{\underline{S}_0} \geq \ln \exp\left\{\left\langle \underline{S} - \underline{S}_0\right\rangle_{\underline{S}_0}\right\} = \left\langle \underline{S} - \underline{S}_0\right\rangle_{\underline{S}_0} \tag{18.743}$$

and

$$\mathrm{F} \leq -T \ln Z_0 - \left\langle T\underline{S} - \underline{S}_0\right\rangle_{\underline{S}_0} \equiv \mathrm{F}_\mathrm{F} \tag{18.744}$$

In the limit of zero temperature $T \to 0$, the free energy becomes the ground state energy $F = E$:

$$E \leq \mathrm{F}_F = -T \ln Z_0 - \left\langle T\underline{S} - \underline{S}_0\right\rangle_{\underline{S}_0} \tag{18.745}$$

or

$$E \leq \mathrm{F}_F = -\frac{1}{\beta} \ln Z_0 - \frac{1}{\beta}\left\langle \underline{S} - \underline{S}_0\right\rangle_{\underline{S}_0} \tag{18.746}$$

We evaluate Z_0 via the model action functional that considers the magnetic field:

$$\underline{S}_0\left[\vec{r},\vec{R},H\right] = \underline{S}_0^*\left[\vec{r},\vec{R}\right] + \frac{ieH}{\hbar c}\int_0^\beta \frac{dy}{d\tau} x d\tau \tag{18.747}$$

The second summand in 18.747 considers the interaction of the particle with the magnetic field and the first summand, the action functional that considers the interaction of the particle with the fictitious particle subsystem of oscillators:

$$\underline{S}_0^* = \int_0^\beta d\tau \left[-\frac{m}{2\hbar^2}\left(\frac{d\vec{r}}{d\tau}\right)^2 - \frac{M}{2\hbar^2}\left(\frac{d\vec{R}}{d\tau}\right)^2 - \frac{M\omega_f^2\left(\vec{R}-\vec{r}\right)^2}{2}\right] \tag{18.748}$$

So, after path integration, then:

$$Z_0 = Z_{\omega_f} \int \delta\left(\vec{r} - \vec{r}'\right) d\vec{r} d\vec{r}' \int_{r'}^{r} D\vec{r} \exp\left\{\underline{S}_0\left[\vec{r},H\right]\right\} \tag{18.749}$$

where,

$$\underline{S}_0\left[\vec{r},H\right] = -\frac{m}{2\hbar^2}\int_0^\beta \left(\frac{d\vec{r}}{d\tau}\right)^2 d\tau - \frac{M\omega_f^2}{2}\int_0^\beta \vec{r}^{\,2} d\tau + \frac{ieH}{\hbar c}\int_0^\beta \dot{y} x d\tau + \Phi_{\omega_f}\left[\vec{r}\right] \tag{18.750}$$

and the **influence phase** $\Phi_{\omega_f}\left[\vec{r}\right]$ of the interaction of the electron with the fictitious particle is defined in 18.735. The first three terms in 18.750 can be rewritten via the following kernel:

$$-\frac{m}{2\hbar^2}\int_0^\beta \int_0^\beta \hat{\mathrm{F}}(\tau,\sigma) x(\tau) x(\sigma) d\tau d\sigma \tag{18.751}$$

Here,

$$\hat{\mathrm{F}}\left(\left|\tau-\sigma\right|\right)=-\delta\left(\tau-\sigma\right)\frac{\partial^2}{\partial\tau^2}+\hbar^2\omega_f^2\frac{M}{m}\delta\left(\tau-\sigma\right)-\frac{\hbar^3\omega_f^3}{2m}F_{\omega_f}\left(\left|\tau-\sigma\right|\right) \tag{18.752}$$

Then from here and 18.749 we have

$$Z_0=Z_{\omega_f}Tr\int_{r'}^{r}D\vec{r}\exp\left\{-\frac{m}{2\hbar^2}\int_0^\beta\int_0^\beta\hat{\mathrm{F}}\left(\left|\tau-\sigma\right|\right)\vec{r}\left(\tau\right)\vec{r}\left(\sigma\right)d\tau d\sigma+\frac{ieH}{\hbar c}\int_0^\beta\dot{y}xd\tau\right\} \tag{18.753}$$

or

$$Z_0=Z_{\omega_f}Tr\int Dy\Phi_{\hat{F}}\left[y\right]\int Dx\Phi_{\hat{F}}\left[x\right]\int Dz\Phi_{\hat{F}}\left[z\right] \tag{18.754}$$

where,

$$\Phi_{\hat{F}}\left[y\right]=\exp\left\{-\frac{m}{2\hbar^2}\int_0^\beta\int_0^\beta\hat{\mathrm{F}}\left(\left|\tau-\sigma\right|\right)y\left(\tau\right)y\left(\sigma\right)d\tau d\sigma\right\} \tag{18.755}$$

$$\Phi_{\hat{\mathrm{F}}}\left[x\right]=\exp\left\{-\frac{m}{2\hbar^2}\int_0^\beta\int_0^\beta\hat{\mathrm{F}}\left(\left|\tau-\sigma\right|\right)x\left(\tau\right)x\left(\sigma\right)d\tau d\sigma-\frac{ieH}{\hbar c}\int_0^\beta\dot{y}xd\tau\right\} \tag{18.756}$$

$$\Phi_{\hat{\mathrm{F}}}\left[z\right]=\exp\left\{-\frac{m}{2\hbar^2}\int_0^\beta\int_0^\beta\hat{\mathrm{F}}\left(\left|\tau-\sigma\right|\right)z\left(\tau\right)z\left(\sigma\right)d\tau d\sigma\right\} \tag{18.757}$$

In the absence of the magnetic field term, then we have the normal polaron integral:

$$Z_0=Z_{\omega_f}\mathrm{Tr}\int Dz\Phi_{\hat{\mathrm{F}}}\left[z\right] \tag{18.758}$$

This integral can be sequentially transformed to

$$Z_0=\int DR_zdz\exp\left\{\underline{S}_0\left[R_z,z\right]\right\} \tag{18.759}$$

Here,

$$\underline{S}_0\left[R_z,z\right]=\int_0^\beta d\tau\left[-\frac{m}{2\hbar^2}\left(\frac{dz}{d\tau}\right)^2-\frac{M}{2\hbar^2}\left(\frac{dR_z}{d\tau}\right)^2-\frac{M\omega_f^2\left(R_z-z\right)^2}{2}\right] \tag{18.760}$$

To find the magneto-polaron characteristics we move to normal coordinates that permit the finding of the magneto-polaron ground state energy:

$$E_0=-T\ln Z_0\left(0\right) \tag{18.761}$$

where

$$Z_0\left(0\right)=\int_{-\infty}^{+\infty}\delta\left(z-z'\right)\delta\left(R_z-R'_z\right)dzdz'dR_zdR'_z\int_{z'}^{z}\int_{R'_z}^{R_z}\exp\left\{\underline{S}_0\right\}DzDR_z \tag{18.762}$$

The model Lagrangian may be conveniently represented in quadratic form by doing a change of variables:

$$\rho_1 = R_z - z, \quad \rho_2 = \frac{zm + R_z M}{n + M} \tag{18.763}$$

and

$$z = \rho_2 + \mu_1 \rho_1, \quad R_z = \rho_2 - \mu_2 \rho_1 \tag{18.764}$$

with

$$\mu_1 = \frac{M}{m+M}, \quad \mu_2 = \frac{m}{m+M}, \mu = \frac{Mm}{m+M}, \quad v^2 = u^2 \omega_f^2, \quad u^2 = \frac{m+M}{m} \tag{18.765}$$

The quantities ρ_1 and ρ_2 are, respectively, the coordinates of the relative motion and of the center of mass. The parameter μ is a reduced mass and v is a scaled frequency.

So,

$$\underline{S}_0 = -\frac{M+m}{2\hbar^2} \int_0^\beta \dot{\rho}_2^2 d\sigma - \frac{\mu}{2\hbar^2} \int_0^\beta \dot{\rho}_1^2 d\sigma - \frac{\mu v^2}{2} \int_0^\beta \rho_1^2 d\sigma \tag{18.766}$$

The coordinate ρ_2 describes the free motion of the particle with mass $M + m$, while the coordinate ρ_1, the harmonic oscillator with frequency v. Hence the partition function in 18.762 now becomes

$$Z_0(0) = \int_{-\infty}^{+\infty} \prod_{i=1}^{2} \mathrm{d}\rho_i d\rho'_i \delta(\rho_i - \rho'_i) \int_{\rho'_i}^{\rho_i} \exp\{\underline{S}_0\} D\rho_i \tag{18.767}$$

or

$$Z_0(0) = u \frac{l_z}{l_e(\hbar\beta)} \frac{1}{2\sinh\left(\frac{\beta\hbar v}{2}\right)} \equiv u \frac{l_z}{l_e(\hbar\beta)} Z_v \tag{18.768}$$

So, from equation 18.754 then

$$Z_0(H) = u \frac{l_z}{l_e(\hbar\beta)} Z_v Z_{\omega_f} \mathrm{Tr} \int Dy \Phi_{\hat{\mathrm{F}}}[y] \int Dx \Phi_{\hat{\mathrm{F}}}[x] \tag{18.769}$$

Here, the quantity l_z is the length in the oz-direction, $l_e(\hbar\beta)$ is the thermal de Broglie wavelength. The role of the driving force is represented by

$$\gamma(t) = \frac{ieH}{\hbar c} \dot{y} \tag{18.770}$$

From this denotation relation 18.769 can be rewritten:

$$Z_0(H) = \left(\frac{u l_z}{l_e(\hbar\beta)}\right)^2 Z_v Z_{\omega_f} \int Dy \exp\left\{-\frac{m}{2\hbar^2} \int_0^\beta \int_0^\beta \left[\hat{\mathrm{F}}(|\tau - \sigma|) y(\tau) y(\sigma) + \hat{\mathrm{F}}^{-1}(|\tau - \sigma|) \gamma(\tau) \gamma(\sigma)\right] d\tau d\sigma\right\} \tag{18.771}$$

Here, the additional factor l_z is due to integration over

$$x = \int_0^\beta \hat{F}^{-1}(|\tau - \sigma|)\gamma(\tau)d\tau \tag{18.772}$$

Introducing the operator

$$\hat{A}(|\tau - \sigma|) = \hat{F}(|\tau - \sigma|) + \hbar^2\omega^2 \frac{\partial^2}{\partial\tau\partial\sigma}\hat{F}^{-1}(|\tau - \sigma|) \tag{18.773}$$

then

$$Z_0(H) = \left(\frac{ul_z}{l_e(\hbar\beta)}\right)^2 Z_v Z_{\omega_f} \exp\left\{-\frac{m}{2\hbar^2}\int_0^\beta\int_0^\beta \hat{A}(|\tau - \sigma|)y(\tau)y(\sigma)d\tau d\sigma\right\} \tag{18.774}$$

We observe that the integral over y has a quadratic form and may be evaluated easily. The most convenient approach may be by introducing normal variables that will permit the integral to be transformed into the sum of independent quadratic parts. There is the possibility of writing the Lagrangian corresponding to 18.774 in the form:

$$\tilde{L} = -\frac{m}{2\hbar^2}\left(\frac{dy}{d\tau}\right)^2 - \frac{M}{2\hbar^2}\left(\frac{dy}{d\tau}\right)^2 - \frac{m_1}{2\hbar^2}\left(\frac{dy_1}{d\tau}\right)^2 - \frac{m_2}{2\hbar^2}\left(\frac{dy_2}{d\tau}\right)^2$$
$$\frac{M\omega_f^2(y-y)^2}{2} - \frac{m_1\varpi^2(y_1-y)^2}{2} - \frac{m_2 v^2(y_2-y)^2}{2} \tag{18.775}$$

Here,

$$m_1 = \left(\frac{\omega_c}{\varpi}\right)^2 u_1 m, m_2 = \left(\frac{\omega_c}{v}\right)^2 u_2 m, u_1 = \frac{1}{u^2}, u_2 = 1 - \frac{1}{u^2} \tag{18.776}$$

Taking the form of the Lagrangian in 18.775 and doing path integration over, , y_1 and y_2 then the result imitates the initial path integral over y:

$$\tilde{Z}_0(H) = Z_v Z_{\omega_f} Z_\varpi \exp\left\{-\frac{m}{2\hbar^2}\int_0^\beta\int_0^\beta \hat{A}(|\tau - \sigma|)y(\tau)y(\sigma)d\tau d\sigma\right\} \tag{18.777}$$

The expression for $\tilde{Z}_0$ is rather convenient than Z_0 due to the fact that if, from the very beginning, we took 18.775 before path integration over , y_1 and y_2 then we could have obtained the normal terms.

Let us select

$$y(t) = y_0 \exp\{ivt\} \tag{18.778}$$

and substitute into the equation of motion:

$$\frac{d}{dt}\frac{\partial}{\partial \dot{y}}\tilde{L}_0 - \frac{\partial}{\partial y}\tilde{L}_0 = 0 \tag{18.779}$$

then

$$m\ddot{y} + M\ddot{y} + m_1\ddot{y}_1 + m_2\ddot{y}_2 = 0, \ddot{y} \mp \omega_f^2 \left(y - y\right) = 0, \ddot{y}_1 \mp \varpi^2 \left(y_1 - y\right) = 0, \ddot{y}_2 \mp v^2 \left(y_2 - y\right) \tag{18.780}$$

Substituting 18.778 into 18.780 then the resulting system has non-trivial solutions when

$$\begin{vmatrix} m & M & m_1 & m_2 \\ \omega_f^2 & -\upsilon^2 + \omega_f^2 & 0 & 0 \\ \varpi^2 & 0-\upsilon^2 + \varpi^2 & 0 & \\ -v^2 & 0\,0 & v^2 - \upsilon^2 & \end{vmatrix} = 0 \tag{18.781}$$

From here,

$$\upsilon^2 \omega_c^2 \left(v^2 - \omega_f^2\right)\left(\upsilon^2 - u_2 \varpi^2 - \omega_f^2\right) = \upsilon^2 \left(v^2 - \upsilon^2\right)^2 \left(v^2 - \varpi^2\right) \tag{18.782}$$

then

$$\upsilon = 0 \tag{18.783}$$

and

$$\upsilon_2 = \varpi + \frac{u_1}{2\varpi} \omega_c^2 \frac{\varpi^2 - \omega_f^2}{\varpi^2 - v^2} \tag{18.784}$$

This is when υ_2 is at the neighborhood of ϖ. Also, for υ_3 at the neighborhood of v we have:

$$\upsilon_{3,4} = v \pm \left(\frac{u_2}{2} \omega_c + \frac{u_2}{u} \frac{\omega_c^2}{v} \left(\frac{u_2}{4} + u_1 \right) \right) \tag{18.785}$$

So, the four components of the frequency can permit us to move to normal coordinates. If we transform y via normal coordinates and substitute into 18.775, then we have four independent terms. In this case, the integral is easily calculated and for the partition function we have:

$$Z_0\left(H\right) = Z_0\left(0\right) \exp\left\{ -\left(\beta\hbar\omega_c\right)^2 \left[\frac{u_2}{2\beta\hbar v} \left(\frac{u_2}{4} + u_1 \right) F_v\left(0\right) - 2Z_v \left(\frac{u_2}{4} \right)^2 + \frac{2}{\varpi} \left(\frac{u_1}{4} \right)^2 \right] \right\} \tag{18.786}$$

19

Multiphoton Absorption by Polarons in a Spherical Quantum Dot

19.1 Theory of Multiphoton Absorption by Polarons

The discovery of quantum generators and fast development of quantum electronics have stimulated theoretical development and investigation of the interaction of electrons with matter as well as the electromagnetic field [2,58–60]. This is the basis of non-linear optics and spectroscopy. One of the fundamental problems in this domain is the investigation of multiphoton processes that may yield absorption or the generation of some quanta of the electromagnetic field in some elementary interaction.

The given problem is related to the theory of absorption of laser radiation by charge carriers in matter. The theory gives insights to the development of laser technology of new non-linear materials, the investigation of laser heating of plasma as well as the technology of fabrication of new mesoscopic structures. At low temperatures and in the weak-coupling regime, the optical absorption of a Fröhlich polaron is due to the elementary polaron scattering process with the absorption of incoming photon and emission of a phonon.

In the works, [2,61] is examined non-linear absorption by free charge carriers due to the laser field considering the scattering probability of electrons on ionized impurities. In [62] on the basis of the classical examination of the non-linear effects is studied the kinetic equation for the electron in a strong homogenous electric field. In the work [63] is obtained the quantum mechanical correction to the scattering cross-section due to two photon processes. The investigation of the asymptotic behavior of the high-frequency conductivity (or scattering cross-section) in these works show that, for exceedingly high intensities, I is described by the relation $CI^{-\frac{3}{2}}$. In the work [63], for the first time the multiphoton absorption of light by charge carriers in semiconductors is examined for a deformable scattering. The probability of the multiphoton absorption is computed by summing the perturbation series considering the interaction of charge carriers with light field. The coefficient of multiphoton light absorption can be obtained via the solution of the Schrödinger equation for the electron in a strong electromagnetic field in the first order approximation of the perturbation theory in the low-temperature limit

$$\hbar\Omega >> T \tag{19.1}$$

Here, Ω is the laser field frequency and T is the absolute temperature in the units of energy.

The new aspect of the multiphoton absorption problem involves the interaction of the strong electromagnetic radiation with nanocrystals and with longitudinal optical phonons. The dimensional quantization of the electronic energy spectrum in mesoscopic structures presents new particularities in the absorption process compared to ones in unbounded media. In quantum dots, the electronic spectrum completely quantizes due to the laser frequency dependence in the absorption coefficient with the absorption spectrum having stripes. The heights and intensities of these stripes depends on the laser intensity, dimension of the nanocrystal and the dominant absorption mechanism. So, we must develop a formula for the absorption coefficient of the multiphoton absorption in spherical quantum dots defining the peculiarity of the absorption spectrum as a function of their dimensions and interaction mechanism of the electron with the third body. This is done using the method of the density matrix in the representation of Feynman path integrals for the case of an electron. This will result in the multiphoton non-linear absorption light coefficient in the second order of interaction energy with polar optical phonons. This coefficient will be found to describe any electron interaction mechanism with phonons.

19.2 Basic Approximations

The electromagnetic field is examined in the dipole approximation, i.e., without considering space dispersion. This approximation is true when the free path length l of the electron for the space change of the electromagnetic field is small. In mesoscopic structures, the free path length is of the order of the dimension of the structure with

$$\lambda \gg l \tag{19.2}$$

or

$$\frac{v_T}{v_{eff}}\frac{v}{c} \ll 1 \tag{19.3}$$

Here, the speed v is the upper bound of the thermal velocity

$$v_T \approx \left(\frac{T}{m}\right)^{\frac{1}{2}} \tag{19.4}$$

and the velocity achieved by the electron in the field

$$v_T \approx \frac{eE}{m\Omega_0} \tag{19.5}$$

compared to the velocity due to dimensional quantization. Since, for the optical frequency, the velocity v_T, is much greater than the effective scattering velocity v_{eff}, the criterion of the dipole approximation is achieved for non-relativistic velocities:

$$\frac{v}{c} \ll 1 \tag{19.6}$$

For this reason, we are going to use only the non-relativistic Lagrangian.

The interaction of the electron with the third body leads to a possible absorption considered in the first order of the perturbation theory. In this case, when the role of the third body is played by the quantization of the field, such an approximation is true for small interactions. For the Coulomb scattering, the Born approximation is achieved by the criterion:

$$\frac{Ze^2}{\hbar v} \ll 1 \tag{19.7}$$

Such a criterion is also satisfied for multiphoton processes. For the case of strong dimensional quantization of the energy of interaction of the electron with charged impurities sufficiently less than the energy of dimensional quantization and the criterion of applicability of the perturbation theory is easier than for massive media.

The lifetime of the laser impulse τ is assumed to be sufficiently short lived since it is possible to neglect the influence of collective effects in plasma on the absorption and also the heating of the electrons and quasi-particles playing the role of the third body;

$$\tau \lll \tau_T \tag{19.8}$$

Here, τ_T, is the temperature relaxation time.

Finally, the screening Coulomb potential is not significant since the Debye screening length is of the order of the temperature domain and the concentration is much more than the characteristic distance, $\frac{v}{v_T}$, and the de Broglie wave of the electron and also the dimension of the nanostructure.

19.3 Absorption Coefficient

We examine via the Feynman path integral in the representation of the density matrix a system of charge carriers (electrons) in a strong electromagnetic field $\varepsilon_0 \cos \Omega t$ and absorbing the energy of this radiation due to the interaction with phonons or ions in plasma at arbitrary temperatures T. The absorption coefficient K is evaluated from the equation

$$d\mathrm{I} = -\mathrm{KI}dz \tag{19.9}$$

where I, is the average over the period $\frac{2\pi}{\Omega}$ of the value of the flux density of the electromagnetic energy; $d\mathrm{I}$ is the flux density decrease within the interval dz in the direction of the distribution of the wave. In the medium with dielectric permittivity $\varepsilon = n_0^2$, the flux density is expressed as:

$$\mathrm{I} = \frac{cn_0\varepsilon_0'^2}{8\pi} \tag{19.10}$$

Here n_0 is the refractive index of the medium and the speed of light is c and $\vec{\varepsilon}_0$ the amplitude of the field in the dipole approximation.

Consider N to be the electron concentration then

$$d\mathrm{I} = -\overline{\left\langle \frac{d\mathrm{E}}{d\tau} \right\rangle} N dz \tag{19.11}$$

with $\frac{d\mathrm{E}}{d\tau}$ being the energy absorbed by the electron per unit time and $\langle \cdots \rangle$ denotes the statistical average of the ensemble of electrons and also the quasi-particle forming the third body; and the bar over the statistical average, $\langle \overline{\cdots} \rangle$, implies the chronological average over the period of the electromagnetic wave. The statistical average will be calculated further via the Feynman path integration in the representation of the density matrix.

Consider the electronic potential energy to be time-dependent:

$$V = -e\vec{r}\vec{\varepsilon}_0 \cos \Omega\tau \tag{19.12}$$

Here, $\vec{r}$ is the electron coordinate and then from

$$\left\langle \frac{d\mathrm{E}}{d\tau} \right\rangle = \left\langle \frac{\partial \mathrm{V}}{\partial \tau} \right\rangle = -\Omega e \langle \vec{r} \rangle \vec{\varepsilon}_0 \sin \Omega\tau \tag{19.13}$$

we have the absorption coefficient [2,60,64]:

$$K = -\frac{N\Omega e}{I}\overline{\vec{\varepsilon}_0 \vec{r}(\tau)\sin\Omega\tau} \quad (19.14)$$

So, in this problem we first find the mean value of the electron coordinate $\vec{r}(\tau)$. We consider the Hamiltonian of the electron interacting with the lattice vibrations in the field of the laser field to be:

$$\hat{H}(\vec{r}) = \hat{H}_e[\vec{r}] + \hat{H}_L[q] + \hat{H}_{e-L}[\vec{r},q] \quad (19.15)$$

Here, q is the lattice variable. The solution of the Schrödinger equation for the system at time moment $t = t_1$ can be represented in the form:

$$\Psi(\vec{r}_1,q_1) = \phi(\vec{r}_1)\psi(q_1) \quad (19.16)$$

The wave function at time moment $t = t_2$ can be written via path integration:

$$\Psi(\vec{r}_2,q_2) = \iint d\vec{r}_1 dq_1 \int_{\vec{r}_1}^{\vec{r}_2}\int_{q_1}^{q_2} D\vec{r}Dq \exp\left\{\frac{i}{\hbar}S[\vec{r},q]\right\}\Psi(\vec{r}_1,q_1) \quad (19.17)$$

or

$$\Psi(\vec{r}_2,q_2) \equiv {}^{\int}K\Psi(\vec{r}_1,q_1) = {}^{\int}K_e\phi(\vec{r}_1)\,{}^{\int}K_{L,e-L}\psi(\vec{r}_1,q_1) \quad (19.18)$$

Here,

$$S[\vec{r},q] = S_e[\vec{r}] + S_L[q] + S_{e-L}[\vec{r},q] \quad (19.19)$$

where,

$$S_e[\vec{r}] = \int_0^\beta \left(\frac{m\dot{\vec{r}}^2}{2} - e\vec{\varepsilon}_0\vec{r}\cos\Omega t - \frac{m\Omega^2\vec{r}^2}{2}\right)dt \quad (19.20)$$

$$S_L[q] = \frac{1}{2}\sum_{\vec{\kappa}}\int_0^\beta \left(\dot{q}_{\vec{\kappa}}^2 - \omega_{\vec{\kappa}}^2 q_{\vec{\kappa}}^2\right)dt \quad (19.21)$$

with

$$S_{e-L}[\vec{r},q] = \sum_{\vec{\kappa}}\int_0^\beta \gamma_{\vec{\kappa}}(t)q_{\vec{\kappa}}(t)dt \quad (19.22)$$

are, respectively, the electronic, lattice and electron-lattice action functionals and $\gamma_{\vec{\kappa}}(t)$ is the amplitude of the electron-lattice coupling interaction; m is the electron bare band mass.

The density matrix ρ_i of the **pure ensemble** is:

$$\rho_i = \Psi_i'^{*}\Psi_i = {}^{\int}K'^{*}\,{}^{\int}K\Psi^{*}(\vec{r}'_1,q'_1)\Psi(\vec{r}_1,q_1) = {}^{\int}K'^{*}\,{}^{\int}K\rho_i(\vec{r}_1,q_1;\vec{r}'_1,q'_1) \quad (19.23)$$

and that of the **mixed ensemble (full density matrix)**:

$$\rho = \sum_i W_i \rho_i = \sum_i {}^{\int}K'^{*}\,{}^{\int}K W_i \rho_i \left(\vec{r}_1, q_1; \vec{r}'_1, q'_1\right) = \rho\left(\vec{r}_2, q_2; \vec{r}'_2, q'_2\right) \tag{19.24}$$

where, the initial state has the probability W_i.

We can now calculate the expectation value $\left\langle \vec{r}(\tau) \right\rangle$ of the coordinate $\vec{r}(\tau)$:

$$\left\langle \vec{r}(\tau) \right\rangle \equiv \mathrm{Tr}\left[\vec{r}(\tau)\,{}^{\int}K'^{*}\,{}^{\int}K \rho\left(\vec{r}_1, q_1; \vec{r}'_1, q'_1\right)\right] \tag{19.25}$$

Here Tr denotes the trace. We eliminate lattice variables via path integration over q:

$$\left\langle \vec{r}(\tau) \right\rangle = \mathrm{Tr}_e\left[\vec{r}(\tau)\,{}^{\int}K'^{*}_{e}\,{}^{\int}K_e W_{ie} \rho_{ie}\left(\vec{r}_1, \vec{r}'_1\right)\right]\mathrm{Tr}_L\left[{}^{\int}K'^{*}_{L,e-L}\,{}^{\int}K_{L,e-L} W_{i'L} \rho_{i'L}\left(q_1, q'_1\right)\right] \tag{19.26}$$

Here Tr_e and Tr_L denote, respectively, the trace over the electronic and lattice variables and path integration over the lattice variable q:

$$\mathrm{Tr}_L\left[{}^{\int}K'^{*}_{L,e-L}\,{}^{\int}K_{L,e-L} W_{i'L} \rho_{i'L}\left(q_1, q'_1\right)\right] \equiv \int \delta\left(q_2 - q'_2\right) dq_2 dq'_2\,{}^{\int}K'^{*}_{L,e-L}\,{}^{\int}K_{L,e-L} W_{i'L} \rho_{i'L}\left(q_1, q'_1\right) \tag{19.27}$$

or

$$\mathrm{Tr}_L\left[{}^{\int}K'^{*}_{L,e-L}\,{}^{\int}K_{L,e-L} W_{i'L} \rho_{i'L}\left(q_1, q'_1\right)\right] = \exp\left\{\frac{i}{\hbar}\Phi\left[\vec{r}, \vec{r}'\right]\right\} \tag{19.28}$$

Here, $\Phi\left[\vec{r}, \vec{r}'\right]$ is the functional of the influence phase of the electron–lattice interaction. So,

$$\left\langle \vec{r}(\tau) \right\rangle = \mathrm{Tr}_e\left[\vec{r}(\tau)\,{}^{\int}K'^{*}_{e}\,{}^{\int}K_e W_{ie} \rho_{ie}\left(\vec{r}_1, \vec{r}'_1\right) \exp\left\{\frac{i}{\hbar}\Phi\left[\vec{r}, \vec{r}'\right]\right\}\right] \equiv \mathrm{Tr}_e\left[\vec{r}(\tau)\rho_e\left(\vec{r}, \vec{r}'\right)\right] \tag{19.29}$$

Here, the pure electronic density matrix is:

$$\rho_e\left(\vec{r}, \vec{r}'\right) = {}^{\int}K'^{*}_{e}\,{}^{\int}K_e W_{ie} \rho_{ie}\left(\vec{r}_1, \vec{r}'_1\right) \exp\left\{\frac{i}{\hbar}\Phi\left[\vec{r}, \vec{r}'\right]\right\} \tag{19.30}$$

The reduced density matrix is:

$$\rho_e\left(\vec{r}_2, \vec{r}'_2\right) = \int d\vec{r}'_1 d\vec{r}_1 \int_{\vec{r}_1}^{\vec{r}_2} D\vec{r} \int_{\vec{r}_1}^{\vec{r}'_2} D\vec{r}' \exp\left\{\frac{i}{\hbar}\left(S_e\left[\vec{r}\right] - S_e\left[\vec{r}'\right] + \Phi\left[\vec{r}, \vec{r}'\right]\right)\right\} W_{ie} \phi_i'^{*}\left(\vec{r}'\right)\phi_i\left(\vec{r}\right) \tag{19.31}$$

or

$$\rho_e\left(\vec{r}_2, \vec{r}'_2\right) = \int d\vec{r}_1 d\vec{r}'_1 \rho_{ie}\left(\vec{r}_1, \vec{r}'_1\right) \int D\vec{r} D\vec{r}' \exp\left\{\frac{i}{\hbar}\left(S_e\left[\vec{r}\right] - S_e\left[\vec{r}'\right] + \Phi\left[\vec{r}, \vec{r}'\right]\right)\right\} \tag{19.32}$$

Here, the initial density matrix is $\rho_{ie}\left(\vec{r}_1, \vec{r}'_1\right)$ and the influence phase, $\Phi\left[\vec{r}, \vec{r}'\right]$ is found to be:

$$\Phi\left[\vec{r}, \vec{r}'\right] = \frac{1}{2}\sum_{\vec{\kappa}} \iint_{t_1}^{t_2} dt ds \left(\gamma_{\vec{\kappa}_j}(t) - \gamma'_{\vec{\kappa}_j}(t)\right)\left[\phi^{*}_{\vec{\kappa}_j}\left(\omega_{\vec{\kappa}_j}, t - s\right)\gamma_{\vec{\kappa}_j}(s) + \phi_{\vec{\kappa}_j}\left(\omega_{\vec{\kappa}_j}, t - s\right)\gamma'_{\vec{\kappa}_j}(s)\right] \tag{19.33}$$

where

$$\phi^*_{\vec{\kappa}_j}\left(\omega_{\vec{\kappa}_j},t-s\right)\equiv\left(\mathrm{I}\left(\omega_{\vec{\kappa}_j},t-s\right)+i\mathrm{A}\left(\omega_{\vec{\kappa}_j},t-s\right)\right) \tag{19.34}$$

$$\phi_{\vec{\kappa}_j}\left(\omega_{\vec{\kappa}_j},t-s\right)\equiv\left(\mathrm{I}\left(\omega_{\vec{\kappa}_j},t-s\right)-i\mathrm{A}\left(\omega_{\vec{\kappa}_j},t-s\right)\right) \tag{19.35}$$

Here the amplitude of the electron–phonon coupling interaction $\gamma_{\vec{\kappa}_j}(t)$ is:

$$\gamma_{\vec{\kappa}_j}(t)=\left|V_{\vec{\kappa}_j}\right|\chi_{\vec{\kappa}_j}(t) \tag{19.36}$$

where

$$\chi_{\vec{\kappa}_j}(t)=\sqrt{\frac{2}{V}}\begin{cases}\sin\left(\vec{\kappa}_j,\vec{r}\right), & \kappa_x\geq 0\\ \cos\left(\vec{\kappa}_j,\vec{r}\right), & \kappa_x<0\end{cases} \tag{19.37}$$

and

$$\mathrm{I}\left(\omega_{\vec{\kappa}_j},t-s\right)=\frac{1}{\omega_{\vec{\kappa}_j}}\theta(t-s)\sin\omega_{\vec{\kappa}_j}(t-s) \tag{19.38}$$

with

$$\mathrm{A}\left(\omega_{\vec{\kappa}_j},t-s\right)=\frac{1}{\omega_{\vec{\kappa}_j}}\mathrm{F}_{\omega_{\vec{\kappa}_j}}(0)\cos\omega_{\vec{\kappa}_j}(t-s) \tag{19.39}$$

and the **Heaviside step function**:

$$\theta(t-s)=\begin{cases}1, & t>s\\ 0, & t<s\end{cases} \tag{19.40}$$

entails compelling the system at a starting point to be evolving towards the future. In the above, V is the volume of the system.

Considering,

$$\gamma_{\vec{\kappa}_j}(t)\gamma_{\vec{\kappa}_j}(s)=\left|V_{\vec{\kappa}_j}\right|^2\frac{2}{V}\begin{cases}\sin\left(\vec{\kappa}_j\vec{r}(t)\right)\sin\left(\vec{\kappa}_j\vec{r}(s)\right), & \kappa_x\geq 0\,◗\\ \cos\left(\vec{\kappa}_j\vec{r}(t)\right)\cos\left(\vec{\kappa}_j\vec{r}(s)\right), & \kappa_x<0\,◗\end{cases} \tag{19.41}$$

is an even function with respect to κ, then we do the sum in 19.33 over an entire sphere, ●. Letting,

$$\mathrm{M}_1=\{\cdots\}\sin\left(\vec{\kappa}_j\vec{r}(t)\right)\sin\left(\vec{\kappa}_j\vec{r}(t)\right),\mathrm{M}_2=\{\cdots\}\cos\left(\vec{\kappa}_j\vec{r}(t)\right)\cos\left(\vec{\kappa}_j\vec{r}(t)\right) \tag{19.42}$$

then from chapter 18 we have

$$\sum_{◗}\left(\mathrm{M}_1+\mathrm{M}_2\right)=\frac{1}{2}\sum_{●}\left(\mathrm{M}_1+\mathrm{M}_2\right) \tag{19.43}$$

and so,

$$\sum\left(\sin\left(\vec{\kappa}_j\vec{r}(t)\right)\sin\left(\vec{\kappa}_j\vec{r}(t)\right)+\cos\left(\vec{\kappa}_j\vec{r}(t)\right)\cos\left(\vec{\kappa}_j\vec{r}(t)\right)\right)=\frac{1}{2}\sum\cos\left(\vec{\kappa}_j,\vec{r}(t)-\vec{r}(s)\right) \quad (19.44)$$

The exact functional **influence phase** of the electron–phonon interaction in 19.33 can be rewritten:

$$\Phi\left[\vec{r},\vec{r}^{'}\right]=\frac{1}{2}\sum_j\frac{Vd\vec{\kappa}}{(2\pi)^3}\left|V_{\vec{\kappa}_j}\right|^2\frac{2}{V}\frac{1}{2}\sum_{\vec{\kappa}_j}\iint_{t_1}^{t_2}dtds\left[\mathrm{F}_{\vec{\kappa}_j}\left(\omega_{\vec{\kappa}_j},\vec{r},\vec{r}^{'}\right)\phi_{\vec{\kappa}_j}\left(\omega_{\vec{\kappa}_j},t-s\right)+\mathrm{F}_{\vec{\kappa}_j}\left(\omega_{\vec{\kappa}_j},\vec{r},\vec{r}^{'}\right)\phi^*_{\vec{\kappa}_j}\left(\omega_{\vec{\kappa}_j},t-s\right)\right] \quad (19.45)$$

where,

$$\mathrm{F}_{\vec{\kappa}_j}\left(\omega_{\vec{\kappa}_j},\vec{r},\vec{r}^{'}\right)=\cos\left(\vec{\kappa}_j,\vec{r}(t)-\vec{r}(s)\right)-\cos\left(\vec{\kappa}_j,\vec{r}'(t)-\vec{r}(s)\right) \quad (19.46)$$

$$\tilde{\mathrm{F}}_{\vec{\kappa}_j}\left(\omega_{\vec{\kappa}_j},\vec{r},\vec{r}^{'}\right)=\cos\left(\vec{\kappa}_j,\vec{r}(t)-\vec{r}'(s)\right)-\cos\left(\vec{\kappa}_j,\vec{r}'(t)-\vec{r}'(s)\right) \quad (19.47)$$

To evaluate $\left\langle\vec{r}(\tau)\right\rangle$ together with the average quantities related to it, the following generating function is useful while applying similar procedures as done earlier:

$$g\left(\vec{\epsilon},\vec{\epsilon}^{'},\tau\right)=\mathrm{Tr}_e\int D\vec{r}D\vec{r}'\exp\left\{\frac{i}{\hbar}\left(\tilde{\mathrm{S}}_e\left[\vec{r}\right]-\tilde{\mathrm{S}}_e\left[\vec{r}^{'}\right]+\Phi\left[\vec{r},\vec{r}'\right]\right)\right\}\rho_e\left(\vec{r},\vec{r}'\right) \quad (19.48)$$

where,

$$\tilde{\mathrm{S}}_e\left[\vec{r}\right]=\int_\sigma^\tau\left[L_e\left(\vec{r}\right)-e\vec{\varepsilon}\vec{r}(t)\delta(t-\tau)\right]dt \quad (19.49)$$

From where,

$$-e\vec{\varepsilon}\left\langle\vec{r}(\tau)\right\rangle=\frac{\hbar}{i}\left(\vec{\varepsilon},\frac{\partial}{\partial\vec{\epsilon}}\right)g\left(\vec{\epsilon},\vec{\epsilon}^{'},\tau\right)\Big|_{\vec{\epsilon}=\vec{\epsilon}^{'}=0} \quad (19.50)$$

In the above the parameter ϵ is the source field.

We Taylor series expand 19.48 in a series of $\Phi\left[\vec{r},\vec{r}'\right]$ and use the first two terms of this series. From the generating function

$$\Psi\left(\xi,\xi',\eta,\eta'\right)=\mathrm{Tr}_e\int D\vec{r}D\vec{r}'\exp\left\{\frac{i}{\hbar}\left(\tilde{\mathrm{S}}_e\left[\vec{r}\right]-\tilde{\mathrm{S}}_e\left[\vec{r}'\right]\right)\right\}\exp\left\{i\left(\hbar\vec{\kappa},\xi\vec{r}(t)-\eta\vec{r}(s)-\xi'\vec{r}'(t)-\eta'\vec{r}'(s)\right)\right\}\rho_e\left(\vec{r},\vec{r}'\right) \quad (19.51)$$

we evaluate four terms in the integrand of 19.45 via the following generating functionals:

$$\Psi(1,1,0,0)=\Psi^{00},\Psi(1,0,0,1)=\Psi^{01},\Psi(0,-1,-1,0)=\Psi^{10},\Psi(0,0,-1,1)=\Psi^{11} \quad (19.52)$$

From here, then

$$e\vec{\varepsilon}_0\vec{r}(\tau)=\frac{1}{2}\sum_{\vec{\kappa},j}\frac{d\vec{\kappa}}{(2\pi)^3}\left|V_{\vec{\kappa}_j}\right|^2\int_{-\infty}^{\infty}\int_{-\infty}^{\infty}dtds\left(\vec{\varepsilon}_0,\frac{\partial}{\partial\vec{\epsilon}}\right)\sum_{j_1,j_2=0}^{1}(-1)^{j_1}\left(\mathrm{I}\left(\omega_{\vec{\kappa}_j},t-s\right)+(-1)^{j_2}i\mathrm{A}\left(\omega_{\vec{\kappa}_j},t-s\right)\right)\Psi^{j_1j_2}\Big|_{\vec{\epsilon}=\vec{\epsilon}^{'}=0} \quad (19.53)$$

This formula also depends on the quantity $|\tau-\sigma|$ which is a property of retarded functions that relate the past.

Absorption of Laser Radiation Enhanced by Polar Optical Phonons

From here and considering 19.14, the light absorption coefficient that considers the absorption or emission of a phonon is found to be:

$$K = N\Omega\sum_j \int \frac{d\vec{\kappa}}{(2\pi)^3}\left|V_{\vec{\kappa}_j}\right|^2 F_{\omega_{\vec{\kappa}_j}}(\vec{\kappa}_j,u)\sum_s 2sJ_s^2(\Delta_1\vec{\kappa}\cos\theta)\sin(su\Omega)du \equiv \sum_s K_s \tag{19.54}$$

Here,

$$F_{\omega_{\vec{\kappa}_j}}(\vec{\kappa}_j,t) = \cos\omega_{\vec{\kappa}_j}t\exp\left\{-\Delta_2\vec{\kappa}_j^2\right\} \tag{19.55}$$

$$\Delta_2 = \frac{\hbar}{2m\Omega_0}\frac{\cosh\frac{\beta\hbar\Omega_0}{2}-\cos\Omega_0 t}{\sinh\frac{\beta\hbar\Omega_0}{2}}, \Delta_1 = \frac{e\vec{\varepsilon}_0}{m}\frac{1}{\Omega^2-\Omega_0^2}, \quad \vec{\varepsilon}_0\vec{\kappa} = \varepsilon_0\kappa\cos\theta \tag{19.56}$$

and K_s is the coefficient of the *s*-photon absorption that describes the scattering of electrons by ions or all possible interactions with phonons at arbitrary temperatures:

$$K_s = 2N\Omega s\sinh\frac{\beta\hbar s\Omega}{2}\sum_j\int\frac{d\vec{\kappa}}{(2\pi)^3}\frac{\left|V_{\vec{\kappa}_j}\right|^2}{\omega_{\vec{\kappa}_j}\sinh\frac{\beta\hbar\omega_{\vec{\kappa}_j}}{2}}J_s^2(\Delta_1\vec{\kappa}\cos\theta)\Phi_{\omega_{\vec{\kappa}_j}}(\vec{\kappa}_j) \tag{19.57}$$

where,

$$\Phi_{\omega_{\vec{\kappa}_j}}(\vec{\kappa}_j) = \int_0^\infty F_{\omega_{\vec{\kappa}_j}}(\vec{\kappa}_j,t)\cos s\Omega t dt \tag{19.58}$$

Evaluating the integrals and the sum in 19.57, permit the rewriting of the coefficient of the *s*-photon absorption K_s:

$$K_s = K_s^1\sum_{n=0}^{\infty}\frac{\left(s+\frac{1}{2}+k\right)\cdots\left(s+\frac{1}{2}+k+\cdots+n\right)}{n!}\frac{1}{2^n\cosh^n\frac{\beta\hbar\Omega_0}{2}}$$
$$\times\sum_{n'=0}^{n}C_{n,n'}\left[\delta(s\Omega+\omega_0-m\Omega_0)+\delta(s\Omega+\omega_0+m\Omega_0)+\delta(s\Omega-\omega_0-m\Omega_0)+\delta(s\Omega-\omega_0+m\Omega_0)\right] \tag{19.59}$$

$$K_s^1 = \frac{N}{4\pi}V_0^2\left(\frac{\hbar}{2m\Omega_0}\right)^{-\frac{1}{2}}\frac{\Omega\sinh\frac{\beta\hbar s\Omega}{2}}{\omega_0\sinh\frac{\beta\hbar\omega_0}{2}}\frac{\Gamma\left(s+\frac{1}{2}\right)}{2^{s-1}\Gamma^2(s+1)}F_{sn}(\beta\hbar\Omega_0,\epsilon) \tag{19.60}$$

$$F_{sn}(\beta\hbar\Omega_0,\epsilon) \equiv \sum_k(-1)^k\frac{\left(s+\frac{1}{2}\right)_k^2}{(s+1)_k^2(2s+1)_k}\left(2\epsilon^k\frac{1}{2s+2k+1}\right)\left(\frac{1}{F_{\Omega_0}(0)}\right)^{s+\frac{1}{2}+k} \tag{19.61}$$

$$\left|V_{\vec{\kappa}_j}\right|^2 \equiv \frac{V_0^2}{\kappa^2}, \quad V_0^2 = \frac{2^{\frac{3}{2}} \pi\alpha\hbar^{\frac{5}{2}} \omega_0^{\frac{3}{2}}}{m^{\frac{1}{2}}} \tag{19.62}$$

$$C_{n,n'} = \frac{n!}{n'!(n-n')!}, \quad \epsilon = \epsilon_0 \frac{\frac{\Omega_0}{\Omega}}{\left(1-\frac{\Omega_0^2}{\Omega^2}\right)^2}, \quad \epsilon_0 = \frac{e^2\varepsilon_0^2}{m\hbar\Omega^3} \tag{19.63}$$

$$(x)_n = x(x+1)\cdots(x+n-1) = \frac{\Gamma(x+1)}{\Gamma(x)} \tag{19.64}$$

where the Gamma function

$$\Gamma(x) = \int_{-\infty}^{\infty} t^{x-1}\exp\{-t\}dt \tag{19.65}$$

$$\Gamma(x+1) = x\Gamma(x), \Gamma(x)\Gamma(1-x) = \frac{\pi}{\sin \pi x}, \Gamma\left(\frac{1}{2}\right) = \sqrt{\pi} \tag{19.66}$$

$$\Gamma(1) = 1, \Gamma(n+1) = n\Gamma(n) = n(n-1)\Gamma(n-1) = \cdots = n(n-1)\cdots 2\cdot 1\Gamma(1) = n! \tag{19.67}$$

$$\Gamma(x)\Gamma\left(x+\frac{1}{m}\right)\cdots\Gamma\left(x+\frac{m-1}{m}\right) = (2\pi)^{\frac{m-1}{2}} m^{\frac{1-2xm}{2}} \Gamma(mx) \tag{19.68}$$

$$m = 2, 2^{2x-1}\Gamma(x)\Gamma\left(x+\frac{1}{2}\right) = \sqrt{\pi}\,\Gamma(2x) \tag{19.69}$$

In the above, ϵ_0 has the sense of the ratio of the energy absorbed by the electron from the laser radiation with frequency Ω for the period $\frac{2\pi}{\Omega}$. The value $\epsilon \geq 1$ corresponds to a strong laser field. For $\Omega = \Omega_0$, there is dimensional resonance for which the electron continuously absorbs energy from the laser field due to synchronization of its oscillation with that of the field. During the interaction with the lattice vibrations it achieves the phonon frequency. This resonance corresponds to a strong absorption when the delta functions achieve the value zero.

Letting

$$\mathrm{K}_s = \sum_m \mathrm{K}_{s,m} \tag{19.70}$$

where,

$$\mathrm{K}_{s,m} = \mathrm{K}_s^0 \epsilon^s \delta\left(s\Omega \pm m\Omega_0 \pm \omega_0\right) \sum_{n'=0}^{\infty}\sum_{n'=0}^{n} C_{n,n'} \mathrm{F}_{sn}\left(\beta\hbar\Omega_0, \epsilon\right) \tag{19.71}$$

For the summation over n and n'we set the constraint

$$n - 2n' = m = \text{const} \tag{19.72}$$

Here, m defines the number of oscillator levels on which the electron makes a transition when it absorbs or emits s photons. For the formula of m we find that

$$n' = \frac{n-m}{2} \tag{19.73}$$

This constraint permits us to write

$$\mathrm{K}_{s,m} = \mathrm{K}_s^0 \epsilon^s \delta\left(s\Omega \pm m\Omega_0 \pm \omega_0\right) \sum_{n=0}^{\infty} \mathrm{C}_{n,\frac{n-m}{2}} \mathrm{F}_{sn}\left(\beta\hbar\Omega_0, \epsilon\right), \quad -n \le m \le n \tag{19.74}$$

which has various temperature dependences and from 19.70:

$$\frac{\Omega \sinh \frac{\beta\hbar s\Omega}{2}}{\omega_0 \sinh \frac{\beta\hbar\omega_0}{2}} \frac{1}{\cosh^n \frac{\beta\hbar\Omega_0}{2}} \tag{19.75}$$

for low temperatures we have

$$\frac{\Omega}{\omega_0} \exp\left\{\frac{\beta\hbar}{2}\left(s\Omega - \omega_0 - n\Omega_0\right)\right\} \tag{19.76}$$

The resonance frequency Ω is obtained when the argument of the delta function in 19.74 achieves the value zero:

$$\Omega = \frac{m\Omega_0 \pm \omega_0}{s} \tag{19.77}$$

This formula defines the resonance position on the Ω-axis.

For the general resonance of the type 19.77, at the intermediate state there is absorption of s photons by an electron that has not yet achieved the energy level but it is only due to the photon–phonon mechanism that the electron achieves the energy level. From the interaction mechanism, the main role is played by dimensional resonance when the electron continuously absorbs energy from the field as a result of synchronizing of its oscillation with the laser field where

$$\Omega = \Omega_0 = \frac{\pm m\Omega_0 \pm \omega_0}{s} \tag{19.78}$$

This is described by the delta functions in the expression of the absorption coefficient. The absorption is enhanced for the condition

$$\Omega = \Omega_0 = \frac{\pm\omega_0}{s \mp m} = \frac{\pm\omega_0}{\pm l}, \quad l = 1, 2, \cdots \tag{19.79}$$

This equation describes the dimensional resonance that is possible when the frequency characterizing the laser field is a multiple of the phonon frequency:

$$\Omega_0 = \frac{\omega_0}{l} \tag{19.80}$$

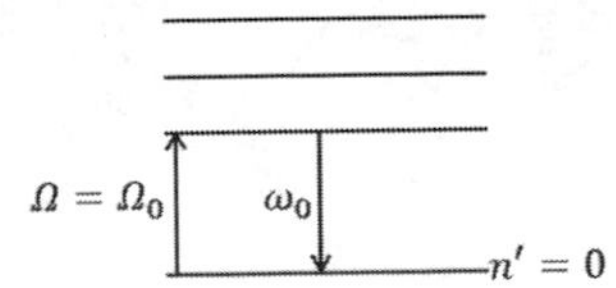

FIGURE 19.1 Describes the zero term of the sum over $\boldsymbol{n}$ in K_1 for $s = 1$, $m = 0$, $n = 0$

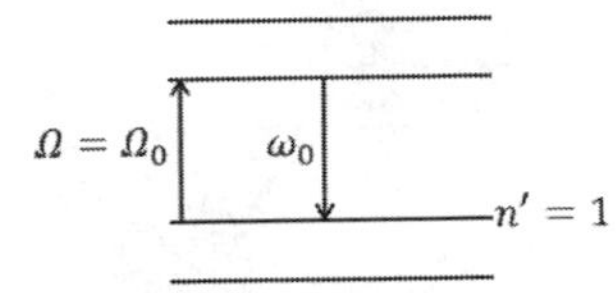

FIGURE 19.2 Describes the zero term of the sum over n in K_1 for $s = 1, m = 0, n = 2, n' = \dfrac{n-m}{2}$

We examine some particular cases of 19.80:

$$\Omega_0 = \omega_0, \quad \Omega = \Omega_0 = \omega_0, \quad l = 1 \tag{19.81}$$

Here we have a triple resonance. From 19.79 we examine other conditions (Figures 19.1 and 19.2):

$$s \mp m = \pm 1 \tag{19.82}$$

The transition in Figure 19.2 describes the zero term of the sum over n in K_1. This implies the intensity of this transition is proportional to ϵ (this is the lowest order). The temperature dependence (for low temperatures) considering formula 19.76 is absent. The transition in formula 19.76 is also proportional to ϵ but has temperature dependence $\exp\{-\beta\hbar\Omega_0\}$. This is due to the fact that it starts from the initial excited state $n' = 1$ and the other terms with $m = 0$ are proportional to ϵ^2. The temperature dependence in this case is $\exp\left\{-\dfrac{\beta\hbar\Omega_0}{2}\right\}$ (Figures 19.3 and 19.4).

We examine the case,

$$s = 1, \quad m = 2 \tag{19.83}$$

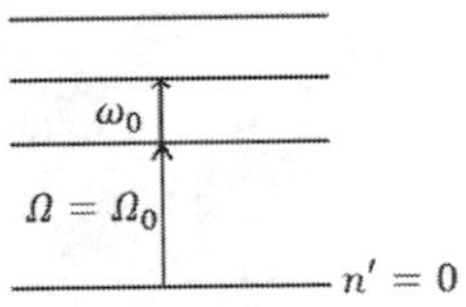

FIGURE 19.3 Describes the zero term of the sum over n in K_1 for $s = 1$, $m = 2$, $n = 2$, $n' = 0$

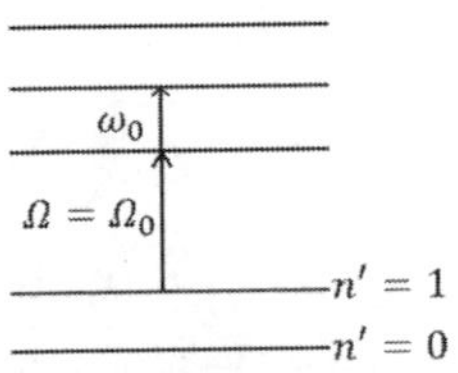

FIGURE 19.4 Describes the zero term of the sum over $\boldsymbol{n}$ in $\boldsymbol{K}_1$ for $s = 1$, $m = 2$, $n = 4$, $n' = 1$

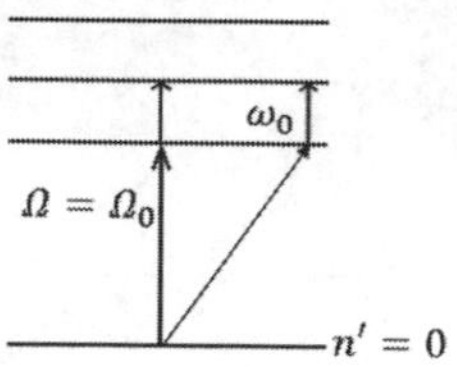

FIGURE 19.5 Describes the zero term of the sum over n in K_1 for $s = 1$, $m = 1$, $n = 1$

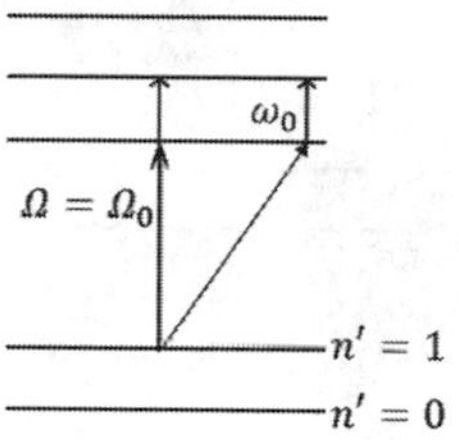

FIGURE 19.6 Describes the zero term of the sum over n in K_1 for $s = 1$, $m = 1$, $n = 3$

This type of transition is proportional to ϵ and the temperature dependence has the form

$$\exp\left\{-\beta\hbar\left(\omega_0 + \frac{n-m}{2}\Omega_0\right)\right\} = \exp\left\{-\frac{\beta\hbar\omega_0}{2}(2+n-m)\right\} \tag{19.84}$$

We also examine the following transition:

All the transitions of the type in Figures 19.5 and 19.6 are described by the lowest order proportional to ϵ and ϵ^2 while the temperature dependence is $\exp\left\{-\frac{\beta\hbar\omega_0}{2}(2+n-m)\right\}$. Other transitions may be written in a similar manner (Figures 19.7 and 19.8).

We calculate the peculiarity of the contribution of the absorption when

$$\Omega_0 = \frac{\omega_0}{2} \tag{19.85}$$

and

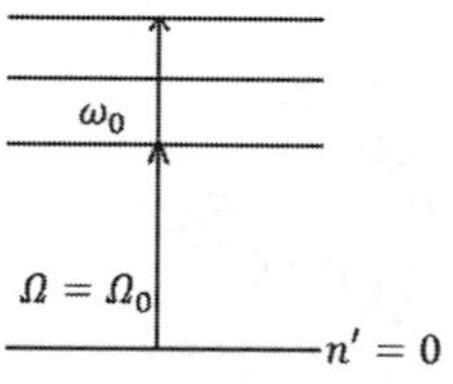

FIGURE 19.7 Describes the zero term of the sum over n in K_1 for $s = 1$, $m = 3$, $n = 3$, $n' = 0$

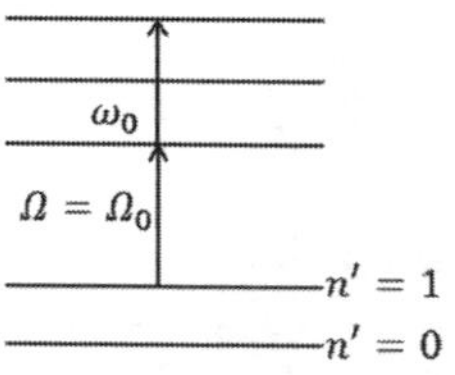

FIGURE 19.8 Describes the zero term of the sum over n in K_1 for $s = 1$, $m = 3$, $n = 3$, $n' = 0$

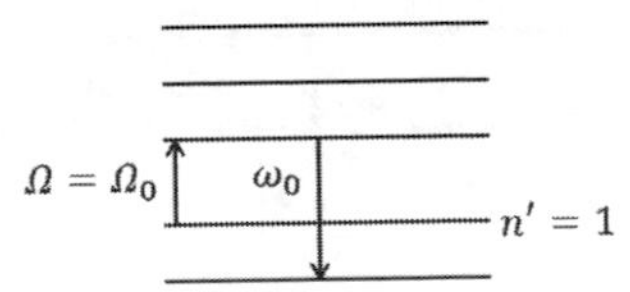

FIGURE 19.9 Describes the zero term of the sum over n in K_1 for $s = 1, m = 1, n = 1, n' = 1$

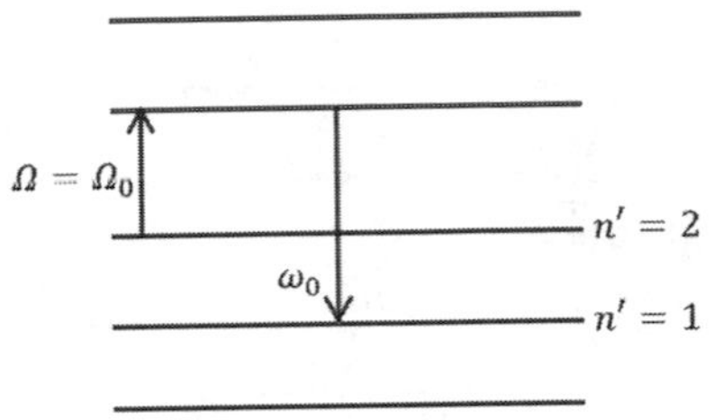

FIGURE 19.10 Describes the zero term of the sum over n in K_1 for $s = 1, m = 1, n = 1, n' = 1$

$$s \mp m = \pm 2 \tag{19.86}$$

This implies, the convenient dimensional resonance has the lowest order ϵ. However, all the contributions are strongly dependent on the temperature (Figures 19.9 and 19.10).

The examples seen above are sufficient to do a general conclusion. For the scaled frequencies,

$$l\Omega_0 = \omega_0 \tag{19.87}$$

the lowest order is proportional to ϵ. However, all the terms are dependent on the temperature. We can conclude that the dimensional resonance occurs at

$$\Omega = \Omega_0 \leq \omega_0 \tag{19.88}$$

This is not observed for the quantum dot of arbitrary radius with

$$\Omega_0 > \omega_0 \tag{19.89}$$

The figures below show the general resonance when

$$\Omega_0 < \omega_0 \tag{19.90}$$

with the transition proportional to ϵ. Here, the absorption of light accompanied by the emission of a phonon can occur only if the energy of the incident photon is larger than that of a phonon,

$$\Omega_0 > \omega_0 \tag{19.91}$$

We observe that the absorption spectrum consists of a **one-phonon line** and at non-zero temperature, the absorption of a photon can be accompanied not only by emission, but also by absorption of one or more phonons (Figures 19.11 and 19.12).

It is found that all dimensional resonances observed are of the order ϵ with respect to the electric field. The temperature dependence is defined by the absorption or emission of the phonon and also the initial condition of the energy level from where there arises a transition. Whether a photon is absorbed or emitted, the initial level from where the transition occurs defines the temperature dependence. The

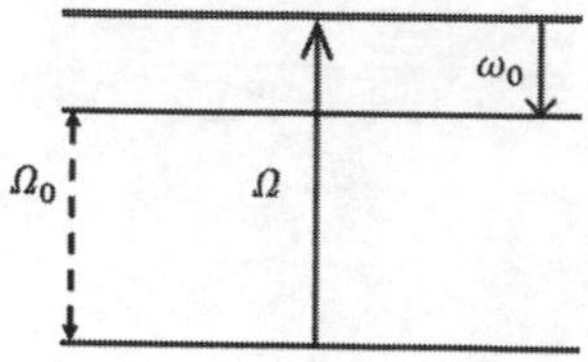

FIGURE 19.11 Describes the general resonance when $\Omega_0 < \omega_0$ and $\Omega = \Omega_0 + \omega_0$

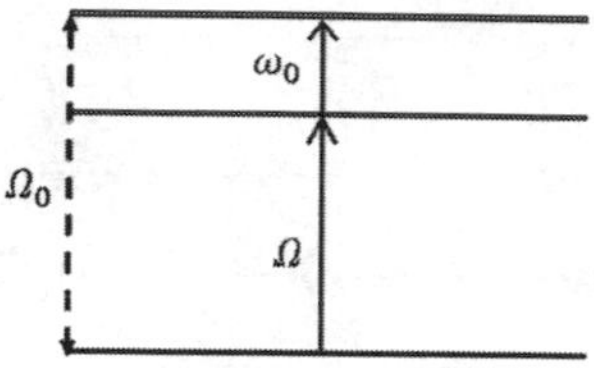

FIGURE 19.12 Describes the general resonance when $\Omega_0 < \omega_0$ and $\Omega = \Omega_0 - \omega_0$

absorption spectrum has the form of stripes whose intensity depends on the resonance character. The most pronounced absorption is at the triple resonance, where values of radiation and oscillatory and optical phonon frequencies are equal.

The practical value of the given results is the fact that with their help we can find also important characteristics of matter and, in particular, the electron–phonon interaction constant, the optical frequency ω_0 and also the parameter that characterizes the electronic confinement potential.

20

Polaronic Kinetics in a Spherical Quantum Dot

We have seen so far that a charge carrier in polar and ionic solids provokes polarization fields and as a consequence results in a large coupling of lattice motion. It is observed that small distortions in the lattice results in the quasi-particle, the so-called phonon with the electron–phonon coupling giving birth to polaronic processes provoking scattering mechanisms by which momentum or kinetic energy of an electron is dissipated. Of course, these scattering processes should limit charge-carrier mobility in the given materials. Therefore, transport properties of polar and ionic solids are influenced by the polaron coupling. It is interesting to predict charge-carrier mobility dependence on the electron–phonon coupling constant although different authors have different relations. The mobility assists in the design of new technological devices, many of which are made of polar and ionic solids and include:

- **battery anodes made of oxides,**
- **transparent conductors serving as electronic displays,**
- **chalcogenides and halides serving as light emission in displays and lighting,**
- **and absorption in photovoltaic solar cells.**

For polar and ionic solids, the dielectric electron-phonon coupling dominates the electronic scattering and may constitute two entirely different effects. This includes the renormalization of the energy spectrum (the mass and the energy of the particle is due to the scattering of the non-stationary coupling with the lattice). It is possible to model without any empirical parameters the scattering process so as to calculate the temperature-dependent absolute-mobility of the polar or ionic solid. Interestingly, since, for different approaches, the scattering process is presented differently, then consequently we are supposed to have different results. For the polaron theory, it is important to have a unique approach via the Feynman functional integration in the density matrix approach.

The kinetic equation in principle may be one of the approaches in the presentation of the problem. For this at first, the Schrödinger integral equation may be approximately solved after which we write entirely independent Boltzmann equations. The more accurate we solve the first part of the problem the better the scattering results. The Boltzmann equations seems not to be valid for a certain class of transport problems. For many phonon-electron scatterings at a short time interval (i.e., when the scattering amplitude from one phonon is dependent on the presence and behavior of phonon systems), then scattering cannot be separated in time so that the Boltzmann equation should be valid except for elastic scatterings with an example being scattering by a random set of delta-function impurities. For the present case, the scattering of optical phonons from electrons is not elastic although at low temperatures it is completely inelastic. Generally, it is assumed that the Boltzmann equation holds only at low temperatures and for weak coupling to phonons.

Our problem will be better solved via the Feynman functional integration in the representation of the density matrix that considers the electron-phonon response for arbitrary electron-phonon coupling strength and temperatures as well as all frequencies of the applied field. The approach with the density matrix appears to give better results.

We again are going to examine the dynamics of an electron in a polar crystal where the electron interacts with lattice vibrations. Here, the Hamiltonian of the system will be dependent on the electronic and phonon coordinates. For brevity, we are going to examine only one electron as well as the Feynman path integral in the representation of density matrix. An approximate expression for the impedance function of a Fröhlich polaron at all frequencies, temperatures, and coupling strengths was obtained in [65] within the path-integral technique.

Assuming the crystal to be isotropic, we evaluate the polaron conductivity σ via Feynman functional integration in the representation of density matrix where we consider the induced current density $\vec{J}(\tau)$ at time moment τ:

$$\vec{J}(\tau) = e\dot{\vec{r}}(\tau) \tag{20.1}$$

Introducing the electrical conductivity $\sigma(\Omega)$ via the relation

$$\vec{J}(\tau) = \sigma(\Omega)\vec{\varepsilon}(\tau) \tag{20.2}$$

with an alternating weak electric field strength:

$$\vec{\varepsilon}(\tau) = \vec{\varepsilon}_0 \exp\{i\Omega\tau\} \tag{20.3}$$

From the current density $\vec{J}(\tau)$ then

$$e\left\langle \dot{\vec{r}}(\tau)\right\rangle = \sigma(\Omega)\vec{\varepsilon}_0 \exp\{i\Omega\tau\} \tag{20.4}$$

and the expectation value of the electron displacement in the direction of the applied field

$$\left\langle \vec{r}(\tau)\right\rangle = \frac{\sigma(\Omega)\vec{\varepsilon}(\tau)}{ie\Omega} \tag{20.5}$$

The electric field is considered sufficiently weak, so that linear-response theory is applicable. We consider the principle of reason by introducing the memory function (kernel of the average path) $\mathcal{G}(\tau - s)$ and we represent electron displacement in terms of time-dependent variables

$$\left\langle \vec{r}(\tau)\right\rangle \equiv -ie\int_{-\infty}^{\tau} \mathcal{G}(\tau - s)\vec{\varepsilon}(s)ds \tag{20.6}$$

This implies that the state of the system at the time moment τ, is defined by all subsequent time moments from $-\infty$ to τ. Such an integral is valid if the system has a memory. If

$$\mathcal{G}(\tau - s) \approx \delta(\tau - s) \tag{20.7}$$

then the system has no memory (it defines the value of the electric field strength at time moment τ).

Performing the following simple transformation

$$\varepsilon_0 \exp\{i\Omega\tau + i\Omega s - i\Omega s\} = -ie\varepsilon_0 \exp\{i\Omega s\}\int_0^{\infty} \mathcal{G}(t)\exp\{-i\Omega t\}dt \tag{20.8}$$

then

$$\mathcal{G}(\tau - s) \to \mathcal{G}(\tau - s)\theta(\tau - s) \tag{20.9}$$

and the **Heaviside step function**:

$$\theta(\tau - s) = \begin{cases} 1, & \tau > s \\ 0, & \tau < s \end{cases} \tag{20.10}$$

entails compelling the system at a starting point to be evolving towards the future. So, for

$$\mathcal{G}(t)\theta(t > 0) \tag{20.11}$$

we consider the following integral from time moment $-\infty$ to time moment $+\infty$:

$$-ie\varepsilon(\tau)\int_{-\infty}^{\infty} \mathcal{G}(t)\exp\{-i\Omega t\}dt = -ie\varepsilon(\tau)\mathcal{G}(\Omega) \tag{20.12}$$

The **admittance** is obtained to be the inverse Fourier transform of $\mathcal{G}(t)$:

$$\mathcal{G}(\Omega) = \int_{-\infty}^{\infty} \mathcal{G}(t)\exp\{-i\Omega t\}dt \tag{20.13}$$

So, the average coordinate can be rewritten

$$\langle \vec{r}(\tau)\rangle = -ie\vec{\varepsilon}(\tau)\mathcal{G}(\Omega) \tag{20.14}$$

and the conductivity

$$\sigma(\Omega) = e^2\Omega\mathcal{G}(\Omega) \tag{20.15}$$

This is one of the main formulae for further evaluation of some physical quantities via Feynman functional integration in the density matrix representation.

We rewrite the expectation value of the coordinate at time τ:

$$\langle \vec{r}(\tau)\rangle = \mathrm{Tr}\{\vec{r}(\tau)\rho(\tau)\} \tag{20.16}$$

where

$$\rho(\tau) = \hat{\mathrm{T}}_{+}\exp\left\{-\frac{i}{\hbar}\int_{t'}^{\tau}\hat{\mathrm{H}}(t)dt\right\}\rho(t')\hat{\mathrm{T}}_{-}\exp\left\{\frac{i}{\hbar}\int_{t'}^{\tau}\hat{\mathrm{H}}'(t)dt\right\} \tag{20.17}$$

Here $\rho(t')$ is the density matrix of the system at some time t', long before the field is turned on and $\hat{\mathrm{T}}_{+}$ and $\hat{\mathrm{T}}_{-}$ are, respectively, chronologically and anti-chronologically time-ordering operators with

$$\hat{\mathrm{T}}\exp\left\{-\frac{i}{\hbar}\int_{t'}^{\tau}\hat{\mathrm{H}}(t)dt\right\} \tag{20.18}$$

being the unitary operator of the development of a state in time with the complete Hamiltonian

$$\hat{\mathrm{H}}(t) = \hat{\mathrm{H}}_0 - e\vec{r}(\tau)\vec{\varepsilon}(\tau) \tag{20.19}$$

where $\hat{\mathrm{H}}_0$ is the Fröhlich polaron Hamiltonian in the absence of a perturbation.

We expand the density matrix $\rho(\tau)$ over the weak field $\vec{\varepsilon}(\tau)$ and limit ourselves only to the linear term. So,

$$\langle\vec{r}(\tau)\rangle = \mathrm{Tr}\{\vec{r}(\tau)\rho(\tau)\} \tag{20.20}$$

or

$$\langle\vec{r}(\tau)\rangle = \mathrm{Tr}\left\{\vec{r}(\tau)\left[1+\frac{i}{\hbar}\int_{t'}^{\tau} er(t)\vec{\varepsilon}(t)dt\right]\exp\left\{-\frac{i}{\hbar}\int_{t'}^{\tau}\hat{\mathrm{H}}_0(t)dt\right\}\right.$$
$$\left.\rho(t)\left[1-\frac{i}{\hbar}\int_{t'}^{\tau} er'(t)\vec{\varepsilon}(t)dt\right]\exp\left\{-\frac{i}{\hbar}\int_{t'}^{\tau}\hat{\mathrm{H}}'_0(t)dt\right\}\right\} \tag{20.21}$$

Considering the equilibrium density matrix

$$\rho_0(\tau) = \exp\left\{-\frac{i}{\hbar}\int_{t'}^{\tau}\hat{\mathrm{H}}_0(t)dt\right\}\rho(t')\exp\left\{\frac{i}{\hbar}\int_{t'}^{\tau}\hat{\mathrm{H}}'_0(t)dt\right\} \tag{20.22}$$

then

$$\langle\vec{r}(\tau)\rangle = \frac{ie}{\hbar}\mathrm{Tr}\left\{\int_{t'}^{\tau}\left[\vec{r}(\tau)\vec{r}(t)-\vec{r}'(\tau)\vec{r}'(t)\right]\vec{\varepsilon}(t)\rho_0(\tau)dt\right\} \tag{20.23}$$

We calculate this via the Feynman functional integration in the representation of the density matrix using the generating functional:

$$g(\tau-t) = \mathrm{Tr}\int_{\vec{r}_1}^{\vec{r}_2}\exp\left\{\frac{i}{\hbar}\left(S_e - S_e' + \Phi\right)\right\}\rho(\vec{r}_1,\vec{r}_1')D\vec{r}D\vec{r}'d\vec{r}_1d\vec{r}_1' \tag{20.24}$$

where the electronic action functional:

$$S_e = \int\left(\frac{m\dot{\vec{r}}^{\,2}(s)}{2} + e\vec{r}(s)\vec{\varepsilon}(s)\right)ds,\quad S_e' = \int\left(\frac{m\dot{\vec{r}}'^{\,2}(s)}{2} + e\vec{r}'(s)\vec{\varepsilon}'(s)\right)ds \tag{20.25}$$

Letting,

$$\vec{\varepsilon}(s) \equiv \vec{\varepsilon}\delta(t-s) + \vec{\xi}\delta(\tau-s),\quad \vec{\varepsilon}'(s) \equiv \vec{\varepsilon}\delta(t-s) - \vec{\xi}\delta(\tau-s) \tag{20.26}$$

$$\mathrm{F} = S_{0e} - S_{0e}' + \Phi \tag{20.27}$$

where, $\vec{\varepsilon}$ and $\vec{\xi}$ are source field coefficients, then

$$g(\tau-t) = \mathrm{Tr}\int_{\vec{r}_1}^{\vec{r}_2}\exp\left\{\frac{i}{\hbar}\left(\mathrm{F} + e\left(\vec{\varepsilon}\vec{r}(t)+\vec{\xi}\vec{r}(\tau)\right) - e\left(\vec{\varepsilon}\vec{r}'(t)-\vec{\xi}\vec{r}'(\tau)\right)\right)\right\}\rho(t')D\vec{r}D\vec{r}'d\vec{r}_1d\vec{r}_1' \tag{20.28}$$

and

$$\left.\frac{\partial^2 g(\tau-t)}{\partial\vec{\varepsilon}\partial\vec{\xi}}\right|_{\vec{\varepsilon}=\vec{\xi}=0}=-\frac{e^2}{\hbar^2}\mathrm{Tr}\int_{\vec{r}_1}^{\vec{r}_2}\left(\vec{r}(t)-\vec{r}'(t)\right)\left(\vec{r}(\tau)+\vec{r}'(\tau)\right)\exp\left\{\frac{i}{\hbar}\mathrm{F}\right\}\rho(t')D\vec{r}D\vec{r}'d\vec{r}_1d\vec{r}_1' \tag{20.29}$$

Comparing

$$\left\langle\vec{r}(\tau)\right\rangle\equiv-ie\int_{-\infty}^{\tau}\mathcal{G}(\tau-s)\vec{\varepsilon}(s)ds \tag{20.30}$$

with

$$\left\langle\vec{r}(\tau)\right\rangle=\frac{ie}{\hbar}\mathrm{Tr}\left\{\int_{t'}^{\tau}\left[\vec{r}(\tau)\vec{r}(t)-\vec{r}'(\tau)\vec{r}'(t)\right]\vec{\varepsilon}(t)\rho_0(\tau)dt\right\} \tag{20.31}$$

then

$$-ie\mathcal{G}(\tau-t)=\frac{ie}{\hbar}\mathrm{Tr}\left[\vec{r}(\tau)\vec{r}(t)-\vec{r}'(\tau)\vec{r}'(t)\right]\rho_0(\tau) \tag{20.32}$$

and

$$\mathcal{G}(\tau-t)=-\frac{1}{\hbar}\left\langle\left[\vec{r}(\tau)\vec{r}(t)-\vec{r}'(\tau)\vec{r}'(t)\right]\right\rangle \tag{20.33}$$

So, from the above formula,

$$\left.\frac{\partial^2}{\partial\vec{\varepsilon}\partial\vec{\xi}}g(\tau-t)\right|_{\vec{\varepsilon}=\vec{\xi}=0}=-\frac{2e^2}{\hbar^2}\left[\vec{r}(\tau)\vec{r}(t)-\vec{r}'(\tau)\vec{r}'(t)\right] \tag{20.34}$$

from where $\mathcal{G}(\tau-t)$ can be represented as the second derivative:

$$\mathcal{G}(\tau-t)=\frac{\hbar}{2e^2}\left.\frac{\partial^2}{\partial\vec{\varepsilon}\partial\vec{\xi}}g(\tau-t)\right|_{\vec{\varepsilon}=\vec{\xi}=0} \tag{20.35}$$

But

$$g(\tau-t)=\mathrm{Tr}\int_{\vec{r}_1}^{\vec{r}_2}\exp\left\{\frac{i}{\hbar}\left(S_e-S_e'+\Phi\left[\vec{r},\vec{r}'\right]\right)\right\}\rho_{t'}(\vec{r}_1,\vec{r}_1')D\vec{r}D\vec{r}'d\vec{r}_1d\vec{r}_1' \tag{20.36}$$

The exponential function under the integrand is not Gaussian and not easily solvable and for it to be easily solvable we introduce a model influence phase Φ_0:

$$\exp\left\{\frac{i}{\hbar}\left(S_e-S_e'+\Phi\right)\right\}=\exp\left\{\frac{i}{\hbar}\left(\mathrm{F}_0+\Phi-\Phi_0\right)\right\}=\exp\left\{\frac{i}{\hbar}\mathrm{F}_0\right\}\exp\left\{\frac{i}{\hbar}\left(\Phi-\Phi_0\right)\right\} \tag{20.37}$$

where,

$$\mathrm{F}_0=S_e-S_e'+\Phi_0 \tag{20.38}$$

It is instructive to note that the model influence phase Φ_0 imitates the full influence phase Φ of the system. This stems from the model Lagrangian L_0 that gives a good fit to the energy as well to the dynamical behaviour of the system of the system if properly chosen.

We do now a series expansion in 20.37:

$$\exp\left\{\frac{i}{\hbar}\left(S_e - S_e' + \Phi\right)\right\} = \exp\left\{\frac{i}{\hbar}\mathrm{F}_0\right\}\left[1 + \frac{i}{\hbar}\left(\Phi - \Phi_0\right) + \left(\frac{i}{\hbar}\right)^2\left(\Phi - \Phi_0\right)^2 + \cdots\right] \tag{20.39}$$

So,

$$g(\tau - t) \cong \mathrm{Tr}\int_{\vec{r}_1}^{\vec{r}_2}\left[1 + \frac{i}{\hbar}\left(\Phi - \Phi_0\right) + \left(\frac{i}{\hbar}\right)^2\left(\Phi - \Phi_0\right)^2 + \cdots\right]\exp\left\{\frac{i}{\hbar}\mathrm{F}_0\right\}\rho(t')D\vec{r}D\vec{r}'d\vec{r}_1 d\vec{r}_1' \tag{20.40}$$

We introduce the integral

$$g_0 = \mathrm{Tr}\int_{\vec{r}_1}^{\vec{r}_2}\exp\left\{\frac{i}{\hbar}\mathrm{F}_0\right\}\rho(t')D\vec{r}D\vec{r}'d\vec{r}_1 d\vec{r}_1' \tag{20.41}$$

that is Gaussian and so can be exactly solvable and also

$$\left\langle \Phi - \Phi_0 \right\rangle_{\mathrm{F}_0} = \mathrm{Tr}\int_{\vec{r}_1}^{\vec{r}_2}\left(\Phi - \Phi_0\right)\exp\left\{\frac{i}{\hbar}\mathrm{F}_0\right\}\rho(t')D\vec{r}D\vec{r}'d\vec{r}_1 d\vec{r}_1' \tag{20.42}$$

then

$$g(\tau - t) \cong g_0\left[1 + \frac{i}{\hbar}\Phi - \Phi_{0\mathrm{F}_0} + \cdots\right] \tag{20.43}$$

We find the model influence phase Φ_0 via the model Lagrangian L_0 in which the electron interacts with the fictitious particle as well as the weak electric field:

$$L_0 = \frac{m\dot{\vec{r}}^2}{2} + e\vec{r}\vec{\varepsilon} + \frac{M\dot{\vec{R}}^2}{2} - \frac{M\omega^2\left(\vec{r} - \vec{R}\right)^2}{2} \tag{20.44}$$

The variable of the fictitious particle $\vec{R}$ can be excluded in the same manner as that of the phonon variable. The influence phase of the fictitious particle can be obtained from

$$g(t - s) = \int D\vec{R}D\vec{R}'d\vec{R}_1 d\vec{R}_1' \exp\left\{\frac{i}{\hbar}\int_s^t\left[L_0\left(\vec{r},\vec{R}\right) - L_e\left(\vec{r}',\vec{R}'\right)\right]dt\right\} \tag{20.45}$$

or

$$\begin{aligned} g(t - s) = \exp\Bigg\{&\frac{M\omega^4}{4}\frac{i}{\hbar}\int_{-\infty}^{\infty}\int_{-\infty}^{\infty}dtds\left[\left(\vec{r}'(t) - \vec{r}(s)\right)^2 - \left(\vec{r}(t) - \vec{r}(s)\right)^2\right]\phi(\omega, t - s) \\ &+ \left[\left(\vec{r}'(t) - \vec{r}'(s)\right)^2 - \left(\vec{r}(t) - \vec{r}'(s)\right)^2\right]\phi^*(\omega, t - s)\Bigg\} \end{aligned} \tag{20.46}$$

From here

$$\Phi_0\left[\vec{r},\vec{r}'\right]=\frac{M\omega^4}{4}\int_{-\infty}^{\infty}\int_{-\infty}^{\infty}dtds\left[\left(\vec{r}'(t)-\vec{r}(s)\right)^2-\left(\vec{r}(t)-\vec{r}(s)\right)^2\right]\phi(\omega,t-s)$$
$$+\left[\left(\vec{r}'(t)-\vec{r}'(s)\right)^2-\left(\vec{r}(t)-\vec{r}'(s)\right)^2\right]\phi^*(\omega,t-s) \tag{20.47}$$

where,

$$\phi^*(\omega,t-s)\equiv \mathrm{I}(\omega,t-s)+i\mathrm{A}(\omega,t-s) \tag{20.48}$$

$$\phi(\omega,t-s)\equiv \mathrm{I}(\omega,t-s)-i\mathrm{A}(\omega,t-s) \tag{20.49}$$

and

$$\mathrm{I}(\omega,t-s)=\frac{1}{\omega}\theta(t-s)\sin\omega(t-s) \tag{20.50}$$

$$\mathrm{A}(\omega,t-s)=\frac{1}{\omega}\mathrm{F}_\omega(0)\cos\omega(t-s) \tag{20.51}$$

We assume the value of the coordinate at some time moment where in such an evaluation, the time may be shifted to infinity. Let us do the following transformation

$$\int_{-\infty}^{\infty}\int_{-\infty}^{\infty}dtds\,\mathrm{F}(t,s)=\int_{-\infty}^{\infty}dt\int_{-\infty}^{t}ds\,\mathrm{F}(t,s)+\int_{-\infty}^{\infty}dt\int_{t}^{\infty}ds\,\mathrm{F}(t,s) \tag{20.52}$$

or

$$\int_{-\infty}^{\infty}\int_{-\infty}^{\infty}dtds\,\mathrm{F}(t,s)=\int_{-\infty}^{\infty}dt\int_{-\infty}^{t}ds\,\mathrm{F}(t,s)+\int_{-\infty}^{\infty}ds\int_{-\infty}^{s}dt\,\mathrm{F}(t,s) \tag{20.53}$$

or

$$\int_{-\infty}^{\infty}\int_{-\infty}^{\infty}dtds\,\mathrm{F}(t,s)=\int_{-\infty}^{\infty}dt\int_{-\infty}^{t}ds\,\mathrm{F}(t,s)+\int_{-\infty}^{\infty}dt\int_{-\infty}^{t}ds\,\mathrm{F}(s,t) \tag{20.54}$$

or

$$\int_{-\infty}^{\infty}\int_{-\infty}^{\infty}dtds\mathrm{F}(t,s)=\int_{-\infty}^{\infty}dt\int_{-\infty}^{t}ds\left[\mathrm{F}(t,s)+\mathrm{F}(s,t)\right] \tag{20.55}$$

So, from here

$$\Phi_0\left[\vec{r},\vec{r}'\right]=\frac{M\omega^4}{8}\int_{-\infty}^{\infty}\int_{-\infty}^{\infty}dtds\left\{\begin{aligned}&\left[2\left(\vec{r}'(t)-\vec{r}(s)\right)^2-\left(\vec{r}(t)-\vec{r}(s)\right)^2\right]\left[\exp\{-i\omega|t-s|\}+2n(\omega)\cos\omega(t-s)\right]\\&-\left(\vec{r}'(t)-\vec{r}'(s)\right)^2\left[\exp\{i\omega|t-s|\}+2n(\omega)\cos\omega(t-s)\right]\end{aligned}\right\} \tag{20.56}$$

where

$$n(\beta,\omega)=\frac{1}{\exp\{\beta\omega\}-1},\quad \beta=\frac{1}{T} \tag{20.57}$$

This is due to the fact that the total distribution of the phonons is assumed to be in equilibrium. We consider now the model Lagrangian L_0 then

$$g_0 = \mathrm{Tr}\int D\vec{r}D\vec{r}'D\vec{R}D\vec{R}'d\vec{r}_1'd\vec{R}'d\vec{r}_1d\vec{R}_1\exp\left\{\frac{i}{\hbar}\int_s^t\left[L_0\left(\vec{r},\vec{R}\right)-L_e\left(\vec{r}',\vec{R}'\right)\right]dt\right\} \tag{20.58}$$

We introduce new variables to diagonalize the model Lagrangian:

$$\vec{\rho}_0 = \frac{m\vec{r}+M\vec{R}}{m+M}, \quad \vec{\rho}_1 = \vec{r}-\vec{R} \tag{20.59}$$

So,

$$L_0\left(\rho_0,\rho_1\right) = \frac{m+M}{2}\dot{\rho}_0^2 + \frac{\mu}{2}\dot{\rho}_1^2 - \frac{\mu\nu^2}{2}\dot{\rho}_1^2 + e\left(b_0\rho_0 + b_1\rho_1\right)\varepsilon\left(t\right) \tag{20.60}$$

where the reduced mass

$$\mu = \frac{mM}{m+M}, \quad b_0 = 1, b_1 = \frac{M}{m} \tag{20.61}$$

The coordinate ρ_0 describes the free motion of the particle with mass $M + m$, while the coordinate ρ_1, the harmonic oscillator with frequency ν. So,

$$g_0 = \mathrm{Tr}\int D\rho_0 D\rho_0' D\rho_1 D\rho_1' d\rho_0 d\rho_0' d\rho_1 d\rho_1' \exp\left\{\frac{i}{\hbar}\int_s^t\left[L_0\left(\rho_0,\rho_1\right)-L_0\left(\rho_0',\rho_1'\right)\right]dt\right\} \tag{20.62}$$

After doing the new change of variables

$$x_0 = \rho_0\sqrt{m+M}, \quad x = \sqrt{\mu}\rho_1 \tag{20.63}$$

then g_0 is conveniently transformed to

$$g_0 = \exp\left\{\frac{i}{2\hbar}\int_{-\infty}^{\infty}\int_{-\infty}^{\infty}\sum_{i=0}^{1}\left(\gamma_i(t)-\gamma_i'(t)\right)\left[\left(\gamma_i(s)+\gamma_i'(s)\right)\mathrm{I}\left(\nu_i,t-s\right)+\left(\gamma_i(s)-\gamma_i'(s)\right)i\mathrm{A}\left(\nu_i,t-s\right)\right]\right\} \tag{20.64}$$

Here, we have two oscillators with scaled frequencies $\nu_0 = 0$ (free motion) and $\nu_1 = \nu$ with respective masses $m + M$ and μ:

$$\gamma_0 = e\varepsilon\sqrt{\frac{\mu}{mM}}, \quad \gamma_1 = e\varepsilon\sqrt{\frac{\mu}{m^2}} \tag{20.65}$$

From here, considering

$$\mathrm{I}_0\left(t-s\right) = \frac{e^2\mu}{m}\left(\frac{1}{M}\mathrm{I}\left(0,t-s\right)+\frac{1}{m}\mathrm{I}\left(\nu,t-s\right)\right) \tag{20.66}$$

$$\mathrm{A}_0\left(t-s\right) = \frac{e^2\mu}{m}\left(\frac{1}{M}\mathrm{A}\left(0,t-s\right)+\frac{1}{m}\mathrm{A}\left(\nu,t-s\right)\right) \tag{20.67}$$

then

$$g_0(t-s)=\exp\left\{\frac{2i}{\hbar}\left[\varepsilon\xi I_0(t-s)+i\xi^2 A_0(t-s)\right]\right\} \tag{20.68}$$

and the zero-order approximation to $\mathcal{G}(\tau - t)$ is then

$$\mathcal{G}_0(t-s)=\frac{\hbar}{2e^2}\frac{\partial^2}{\partial\varepsilon\partial\xi}g_0(t-s)\bigg|_{\varepsilon=\xi=0}=\frac{i}{e^2}I_0(t-s) \tag{20.69}$$

The quantity $\frac{1}{e^2}I_0(t-s)$ imitates a response function and so $\mathcal{G}_0(t-s)$ should be a response function as expected. The temperature dependence in $\mathcal{G}_0(t-s)$ can only be expected via the variation of the parameter ν with temperature. From 20.15 we have

$$\sigma_0(\Omega)=e^2\Omega\mathcal{G}_0(\Omega)=-\frac{ie^2}{m}\frac{\omega^2-\Omega^2}{\Omega(\nu^2-\Omega^2)} \tag{20.70}$$

This describes the mobility of the polaron without scattering on other particles and resonance occurs at the point $\nu = \pm\,\Omega$ when the laser field frequency is synchronized with that of the fictitious particle. In the classical limit

$$\omega=\nu\to 0 \tag{20.71}$$

then

$$\sigma_0(\Omega)=-\frac{ie^2}{m\Omega} \tag{20.72}$$

This is the classical formula.
Obviously,

$$e\dot{x}=\sigma\varepsilon_0\exp\{i\Omega t\} \tag{20.73}$$

So,

$$m\ddot{x}=e\varepsilon=e\varepsilon_0\exp\{i\Omega t\},\quad \dot{x}=\frac{1}{mi\Omega}e\varepsilon_0\exp\{i\Omega t\} \tag{20.74}$$

$$\frac{e^2\varepsilon}{mi\Omega}=\sigma\varepsilon,\quad \sigma=\frac{e^2}{im\Omega} \tag{20.75}$$

We evaluate now g_1. It is convenient to express the influence phase via the frequency since the final expression is in the frequency representation. The transformation of the integral via the frequency enables one to reduce one integral. This can be done via the Fourier transform

$$I(\nu,t-s)=\frac{1}{2\pi}\int d\Omega I(\nu,\Omega)\exp\{i\Omega(t-s)\} \tag{20.76}$$

After evaluating the integral

$$g=\exp\left\{\frac{i}{\hbar}\left(\frac{1}{2}\int_{-\infty}^{\infty}\int_{-\infty}^{\infty}dtds\left(\gamma(t)-\gamma'(t)\right)\left[\left(\gamma(s)+\gamma'(s)\right)\frac{1}{2\pi}\int_{-\infty}^{\infty}d\Omega\mathrm{I}(\nu,\Omega)\exp\{i\Omega(t-s)\}\right.\right.\right.$$
$$\left.\left.\left.+\left(\gamma(s)-\gamma'(s)\right)\frac{1}{2\pi}\int_{-\infty}^{t}d\Omega i\mathrm{A}(\nu,\Omega)\exp\{i\Omega(t-s)\}\right]\right)\right\} \tag{20.77}$$

we have

$$g_1=\exp\left\{\frac{i}{4\pi\hbar}\int d\Omega\left(\gamma(-\Omega)-\gamma'(-\Omega)\right)\left[\left(\gamma(\Omega)+\gamma'(\Omega)\right)\mathrm{I}(\nu,\Omega)+\left(\gamma(\Omega)-\gamma'(\Omega)\right)i\mathrm{A}(\nu,\Omega)\right]\right\} \tag{20.78}$$

or

$$g_1(\vec{r},\vec{r}')=\frac{i}{2\hbar}\sum_j\int\frac{d\vec{\kappa}}{(2\pi)^3}\left|V_{\vec{\kappa}_j}\right|^2\int_{-\infty}^{\infty}\int_{-\infty}^{\infty}dsdt\left[\mathrm{F}_{\vec{\kappa}}(\omega_{\vec{\kappa}},\vec{r},\vec{r}')\phi(\omega_{\vec{\kappa}},t-s)+\left\langle\tilde{\mathrm{F}}_{\vec{\kappa}}(\omega_{\vec{\kappa}},\vec{r},\vec{r}')\right\rangle\phi^*(\omega_{\vec{\kappa}},t-s)\right]-\frac{i}{\hbar}\langle\Phi_0\rangle \tag{20.79}$$

where

$$\mathrm{F}_{\vec{\kappa}}(\omega_{\vec{\kappa}},\vec{r},\vec{r}')=\cos\left(\vec{\kappa},\vec{r}(t)-\vec{r}(s)\right)-\cos\left(\vec{\kappa},\vec{r}'(t)-\vec{r}(s)\right) \tag{20.80}$$

$$\tilde{\mathrm{F}}_{\vec{\kappa}}(\omega_{\vec{\kappa}},\vec{r},\vec{r}')=\cos\left(\vec{\kappa},\vec{r}(t)-\vec{r}'(s)\right)-\cos\left(\vec{\kappa},\vec{r}'(t)-\vec{r}'(s)\right) \tag{20.81}$$

Our goal will be to evaluate the average values of the cosine functions. We introduce an additional productive function:

$$\Psi(\xi,\xi',\eta,\eta')=\left\langle\exp\left\{\vec{\kappa},\xi r(t)-\eta r(s)+\xi' r'(t)-\eta' r'(s)\right\}\right\rangle \tag{20.82}$$

From here

$$\Psi(1,0,1,0)+\Psi(-1,0,-1,0)=2\cos\left(\vec{\kappa},r(t)-r(s)\right) \tag{20.83}$$

$$\xi r(t)-\eta r(s)-\xi' r'(t)-\eta' r'(s)=\int_{-\infty}^{\infty}dt'\left[\left(\xi\delta(t-t')-\eta\delta(s-t')\right)r(t')-\left(\xi'\delta(t-t')+\eta'\delta(s-t')\right)r'(t')\right]dt' \tag{20.84}$$

$$er\varepsilon(t')=\left(\xi\delta(t-t')-\eta\delta(s-t')\right)r(t') \tag{20.85}$$

$$er'\varepsilon'=\left(\xi'\delta(t-t')+\eta'\delta(s-t')\right)r'(t') \tag{20.86}$$

$$e\varepsilon(t')+\hbar\vec{\kappa}\left(\xi\delta(t-t')-\eta\delta(s-t')\right)\equiv e\mathcal{E}(t') \tag{20.87}$$

$$e\varepsilon'(t')+\hbar\vec{\kappa}\left(\xi'\delta(t-t')+\eta'\delta(s-t')\right)\equiv e\mathcal{E}'(t') \tag{20.88}$$

The difference in the productive function is in the new laser strength $e\mathcal{E}(t')$. This implies our result differs only in the new laser field. We put this generating function in the frequency representation:

$$\Psi=\frac{1}{g_0}\exp\left\{\frac{i}{2\hbar4\pi}\int_{-\infty}^{t}d\Omega\left(\mathcal{E}(-\Omega)-\mathcal{E}'(-\Omega)\right)\left[\left(\mathcal{E}(\Omega)+\mathcal{E}'(\Omega)\right)\mathrm{I}_0(\Omega)+\left(\mathcal{E}(\Omega)-\mathcal{E}'(\Omega)\right)i\mathrm{A}_0(\Omega)\right]\right\} \tag{20.89}$$

Here,

$$\vec{\mathcal{E}}(\pm\Omega)=\vec{\varepsilon}\exp\{\mp i\Omega\sigma\}+\vec{\xi}\exp\{\mp i\Omega\tau\}+\frac{\hbar\kappa}{e}\left(\vec{\xi}\exp\{\mp i\Omega t\}-\vec{\eta}\exp\{\mp i\Omega s\}\right) \tag{20.90}$$

$$\vec{\mathcal{E}}'(\pm\Omega)=\vec{\varepsilon}\exp\{\mp i\Omega\sigma\}-\vec{\xi}\exp\{\mp i\Omega\tau\}-\frac{\hbar\kappa}{e}\left(\vec{\xi}\exp\{\mp i\Omega t\}-\vec{\eta}\exp\{\mp i\Omega s\}\right) \tag{20.91}$$

We can now represent the generating function as

$$\Psi=\exp\left\{\int_{-\infty}^{\infty}d\Omega\left(f_1+f_2+f_3\right)\right\} \tag{20.92}$$

Here,

$$f_1=\frac{ie\vec{\kappa}\vec{\xi}}{2\pi}\exp\{i\Omega\tau\}\left[f_{11}^{-}(-t,-s)\mathrm{I}_0(\Omega)+2f_{12}^{+}(-t,-s)\mathrm{A}_0(\Omega)\right] \tag{20.93}$$

$$f_2=\frac{ie\vec{\kappa}\vec{\varepsilon}}{2\pi}\exp\{i\Omega\sigma\}f_{12}^{+}(-\Omega)\mathrm{I}_0(\Omega) \tag{20.94}$$

$$f_3=\frac{i\hbar\vec{\kappa}^2}{4\pi}f_{12}^{+}(+t,+s)\left[f_{11}^{-}(-t,+s)\mathrm{I}_0(\Omega)+f_{12}^{+}(-t,-s)i\mathrm{A}_0(\Omega)\right] \tag{20.95}$$

$$f_{12}^{+}(-t,-s)=(\xi+\xi')\exp\{-i\Omega t\}-(\eta+\eta')\exp\{-i\Omega s\} \tag{20.96}$$

$$f_{11}^{-}(-t,-s)=(\xi-\xi')\exp\{-i\Omega t\}-(\eta-\eta')\exp\{-i\Omega s\} \tag{20.97}$$

From here, the first order change in $\mathcal{G}$ is:

$$\mathcal{G}_1(\tau-\sigma)=\frac{\hbar}{2e^2}\frac{\partial^2}{\partial\varepsilon\partial\xi}g_1(\tau-\sigma)\Big|,\quad g_1=\frac{i}{\hbar}g_0(\tau-\sigma)\langle\Phi-\Phi_0\rangle \tag{20.98}$$

So, the approximate form of $\mathcal{G}$ is:

$$\mathcal{G}(\tau-\sigma)=\mathcal{G}_0(\tau-\sigma)+\mathcal{G}_1(\tau-\sigma) \tag{20.99}$$

The first order change in $\mathcal{G}$ may as well be represented as

$$\mathcal{G}_1(\tau-\sigma)=\mathcal{G}_1^{(\mathrm{I})}(\tau-\sigma)+\mathcal{G}_2^{(\mathrm{II})}(\tau-\sigma) \tag{20.100}$$

where,

$$\mathcal{G}_1^{(\mathrm{I})}(\tau-\sigma)=\frac{i}{2e^2}\frac{\partial^2}{\partial\varepsilon\partial\xi}g_0(\tau-\sigma)\langle\Phi\rangle\Big| \tag{20.101}$$

$$\mathcal{G}_2^{(\mathrm{II})}(\tau-\sigma)=-\frac{i}{2e^2}\frac{\partial^2}{\partial\varepsilon\partial\xi}g_0(\tau-\sigma)\langle\Phi_0\rangle\Big| \tag{20.102}$$

From here and considering the analytic continuation $t = -i\hbar\tau$ (where τ is a real number):

$$0 \le \tau \le \beta \tag{20.103}$$

then the first order change in $\mathcal{G}$:

$$\mathcal{G}_1(\Omega) = -i\left(\frac{m\Omega^2\left(\nu^2-\omega^2\right)}{\left(\omega^2-\Omega^2\right)} + X(\Omega)\right)\mathrm{I}_0^2(\Omega) \tag{20.104}$$

with

$$X(\Omega) = -\frac{\hbar}{2\pi}\sum_j \frac{d\vec{\kappa}}{(2\pi)^2}\left|V_{\vec{\kappa}_j}\right|^2 \kappa_j^2 \int_0^\infty d\tau \mathrm{F}_{\omega_{\vec{\kappa}_j}}^{(\nu)}\left(\tau-\frac{\beta}{2}\right) \tag{20.105}$$

Here,

$$\mathrm{F}_{\omega_{\vec{\kappa}_j}}^{(\nu)}\left(\tau-\frac{\beta}{2}\right) = ¥_{\omega_{\vec{\kappa}_j}}^{(\beta)}(\Omega,\tau)\exp\left\{-\kappa^2\mathrm{M}_\nu^{(\beta)}\left(\tau-\frac{\beta}{2}\right)\right\} \tag{20.106}$$

and

$$¥_{\omega_{\vec{\kappa}_j}}^{(\beta)}(\Omega,\tau) = \frac{1}{\omega_{\vec{\kappa}_j}}\left(1-\exp\{\hbar\Omega\tau\}\right)\left(\exp\left\{\hbar\omega_{\vec{\kappa}_j}\tau\right\} + 2n(\beta,\omega)\cosh\hbar\omega_{\vec{\kappa}_j}\tau\right) \tag{20.107}$$

$$\mathrm{M}_\nu^{(\beta)}\left(\tau-\frac{\beta}{2}\right) = \frac{\hbar\omega^2}{2m\nu^2}\left\{\frac{\nu^2-\omega^2}{\nu\omega^2}\left(\mathrm{F}_\omega(0)-\mathrm{F}_\nu(\tau)\right) + \beta\left[\frac{1}{4} + \frac{\left(\tau-\frac{\beta}{2}\right)^2}{\beta^2}\right]\right\} \tag{20.108}$$

Here,

$$\mathrm{F}_\nu(\tau) = \frac{\cosh\hbar\nu\left(\tau-\frac{\beta}{2}\right)}{\sinh\left(\frac{\hbar\nu\beta}{2}\right)} \tag{20.109}$$

The function in 20.100 depends on $|\tau-\sigma|$ and is the property of retarded functions that depend on the past. The significance of interaction with the past is that the perturbation due to the moving electron (hole) takes "**time**" to propagate in the crystal. So, the potential that the electron feels at any "**time**" depends on its position at previous times and implies the effect the electron has on the crystal propagates at a finite velocity and can make itself felt on the electron at a later time. The contribution of the function $X(\Omega)$ to 20.100 may then be described as a non-local contribution where the system trajectory interacts with itself and this self-interaction is mediated by the phonon cloud as indicative of the factor $\left|V_{\vec{\kappa}_j}\right|^2$ modulated by $\mathrm{F}_{\omega_{\vec{\kappa}_j}}^{(\nu)}\left(\tau-\frac{\beta}{2}\right)$. This is indicative of the fact that the polaron problem mimics a one-particle problem where the interaction, non-local in time or "**retarded**," occurs between the electron and itself.

So, from the above

$$\sigma = e\Omega\mathcal{G}(\Omega) = e^2\Omega\left(\mathcal{G}_0(\Omega) + \mathcal{G}_1(\Omega)\right) \tag{20.110}$$

from where with increase interaction there is enhanced scattering and consequently enhanced resistance. It is appropriate to find the frequency-dependence mobility impedance function $Z(\Omega)$ via the following Feynman approach:

$$e^2\Omega Z(\Omega) \equiv \left(\mathcal{G}_0(\Omega) + \mathcal{G}_1(\Omega)\right)^{-1} \tag{20.111}$$

This expression can be expanded to first order in small $\mathcal{G}_1(\Omega)$ [66] then

$$e^2\Omega Z(\Omega) \equiv \frac{1}{\mathcal{G}_0(\Omega)} - \frac{\mathcal{G}_1(\Omega)}{\mathcal{G}_0^2(\Omega)} \tag{20.112}$$

Considering the Hamiltonian 18.32, then the final expression [65] for the Fröhlich polaron impedance function:

$$Z(\Omega) \equiv \frac{1}{e^2\Omega\mathcal{G}_0(\Omega)} + \frac{i\mathrm{I}_0^2(\Omega)}{e^2\Omega\mathcal{G}_0^2(\Omega)}\left(\frac{m\Omega^2\left(v^2 - \omega^2\right)}{\left(\omega^2 - \Omega^2\right)} + X(\Omega)\right) \tag{20.113}$$

The quantity,

$$\frac{\mathrm{I}_0^2(\Omega)}{e^2\Omega\mathcal{G}_0^2(\Omega)} X(\Omega) \tag{20.114}$$

describes the polaron scattering while the resistance:

$$\frac{m\Omega^2\left(v^2 - \omega^2\right)\mathrm{I}_0^2(\Omega)}{e^2\Omega\left(\omega^2 - \Omega^2\right)\mathcal{G}_0^2(\Omega)} \tag{20.115}$$

From here we can observe that as the interaction is stronger the scattering and the resistance are enhanced.

The resonance is expected at

$$\Omega = \pm\omega \tag{20.116}$$

For this resonance, the electron continuously absorbs energy from the laser field due to synchronization of its oscillation with that of the field.

In 20.113, the quantity $X(\Omega)$ has information on the interaction of the electron with phonons. The total dependence of the model system is described by the function $\mathrm{M}_v^{(\beta)}\left(\tau - \frac{\beta}{2}\right)$. The Feynman frequency-dependence mobility can be obtained from

$$\frac{1}{\mu} = \operatorname{Re} Z(\Omega) \tag{20.117}$$

For low temperatures, the mobility has the dependence,

$$\mu \approx \exp\left\{\frac{\hbar\omega_0}{T}\right\} \tag{20.118}$$

This is a typical behavior of the large-polaron mobility characteristic for weak coupling polaron [38]. In [67] it was shown that for weak coupling:

$$\mu = \frac{e}{2\alpha_F}\left(\exp\{\beta\omega_0\}-1\right)\left(1-\frac{\alpha_F}{6}+O\left(\alpha_F^2\right)\right) \tag{20.119}$$

For sufficiently low temperature, the Feynman et al. polaron mobility [65] takes the form

$$\mu = \left(\frac{\omega}{v}\right)^3 \frac{3e}{4m\omega_0^2\alpha_F\beta}\exp\{\beta\omega_0\}\exp\left\{\frac{v^2-\omega^2}{\omega^2 v}\right\} \tag{20.120}$$

where v and ω are variational functions and α_F obtained from the Feynman polaron model. From the Boltzmann equation for the Feynman polaron model, Kadanoff [68] found the mobility, for low temperatures to be

$$\mu = \left(\frac{\omega}{v}\right)^3 \frac{3e}{2m\omega_0\alpha_F\beta}\exp\{\beta\omega_0\}\exp\left\{\frac{v^2-\omega^2}{\omega^2 v}\right\} \tag{20.121}$$

Pekar considered different screening effects and different wavefunction and found the mobility to be:

$$\mu_{Pekar} \approx \alpha_F^5 \tag{20.122}$$

The weak-coupling perturbation expansion of the low-temperature polaron mobility obtained via the Green's function technique [67] confirms the mobility derived from the Boltzmann equation is asymptotically exact for weak coupling ($\alpha_F \ll 1$) and at low temperatures ($T \ll \omega_0$). In [68], the mobility of 20.121 differs by the factor of $\frac{3}{2\beta\omega_0}$ from that derived using the polaron Boltzmann equation as given by 20.35. From this comparison, the result of [65] is not valid when $T \to \omega_0$. In [65] and in [69] the above discrepancy can be attributed to an interchange of two limits in calculating the impedance. The mobility at low temperatures is mainly limited by the absorption of phonons, while in the FHIP theory, it is the emission of phonons which gives the dominant contribution as $T \to 0$ [70]. For the case of strong interaction, the electron covers the domain of the absolute square of the wavefunction. If the electron strongly interacts with the phonon wave, then the forces compensate and achieve the value zero. This is the screening effect felt during the selection of the wave function.

21

Kinetic Theory of Gases

21.1 Distribution Function

To enhance our understanding of the subject matter of the non-linear kinetics and conductivity, I would like to review some basic notions in the kinetic theory. This chapter will clarify our understanding of the Boltzmann equation and the mobility. So, in this chapter, we examine the kinetic theory of an ordinary gas with electrically neutral atoms or molecules. Here we will study non-equilibrium processes in a perfect gas which will help us to investigate transport processes where a statistical description of a gas is done with the help of a distribution function $f(q, p, t)$ of the gas particles in their phase spaces (q, p) at the moment t for which

$$f(q,p,t)dqdp \tag{21.1}$$

is the mean number of particles whose centre of inertia at time t is located between $q \to q + dq$ and between $p \to p + dp$. The function $f(q, p, t)$ gives a complete description of the macroscopic state of the gas that we do not consider possible non-equilibrium perturbation of the internal degrees of freedom of the gas particles. It enables us to evaluate all physical quantities of the gas particles say, for example, viscosity coefficients or thermal conductivities. It follows that we may evaluate any transport problem if we have the knowledge of the distribution function $f(q, p, t)$ for the given physical situation of interest.

If

$$n(q,t)dq \tag{21.2}$$

is the mean number of gas particles which at time t are located between $q \to q + dq$ irrespective of the momentum p then from 21.1 we have

$$n(q,t) = \int f(q,p,t)dp \tag{21.3}$$

The integration here is over all possible momenta p. If $\chi(q, p, t)$ is a function describing the property of a gas particle then its mean value may be evaluated using the distribution function

$$\chi(q,t) = \frac{1}{n(q,t)} \int f(q,p,t)\chi(q,p,t)dp \tag{21.4}$$

21.2 Principle of Detailed Equilibrium

The translational motion of a gas particle is always a classical motion that is described by the coordinate $\vec{r}$, its centre of inertia and the momentum p (or velocity $\vec{v} = \frac{\vec{p}}{m}$) of its motion. This motion is represented by the classical equation of motion (Newton's equation):

$$m\frac{d^2\vec{r}}{dt^2} = \vec{F}(\vec{r}) \tag{21.5}$$

where m is the mass of the particle and $\vec{F}(\vec{r})$ the force acting on it. The general property of the equation 21.5 is that, it is invariant under the time reversal $t \to -t$. The solution of 21.5 is symmetric under this time reversal $t \to -t$:

$$\vec{r} = \vec{r}(c_1, c_2, \cdots, t), \quad \vec{r} = \vec{r}(t) \tag{21.6}$$

where c_1, c_2, $\cdots$, are constants of integration. Newton's equation generates values of the first kind:

$$\vec{r}(t) = \vec{r}(-t), \quad E = \frac{mv^2}{2} \tag{21.7}$$

and values of the second kind:

$$\vec{v}(-t) = -\vec{v}(t), \quad \vec{p}(-t) = -\vec{p}(t), \quad \vec{M}(-t) = -\vec{M}(t), \quad \vec{M} = [\vec{r}, \vec{p}] \tag{21.8}$$

Here, $\overrightarrow{M}$ is the angular momentum vector. For a polyatomic gas, the distribution function f may also be dependent on the angles that specify the fixed orientation of the axes of the molecules relative to this vector $\vec{M}$. The vibrational motion of the atoms within the molecule is quantized. So, the vibrational state of the molecule is specified by appropriate quantum numbers. At not too high temperatures, under ordinary conditions, the vibrations are not excited at all and the molecule is at its ground state.

It should have been necessary to define these values to be able to introduce the notion of a state under the **time reversal** $t \to -t$. If we know the state of the particle between two collisions i.e., $\Gamma(\vec{p}, \vec{M}, E)$, then $\Gamma^*(-\vec{p}, -\vec{M}, E)$ is the state under the **time reversal** $t \to -t$. Under such a time reversal, it implies a corresponding reversal of all the velocities for which the **reversal** collision of the particles process under the time reversal of the state:

$$\Gamma_1 \to \Gamma_2 \to \Gamma_3 \to \cdots \tag{21.9}$$

This is a **direct process**. The state after a time reversal $t \to -t$ is obtained from 21.9:

$$\cdots \Gamma_5^* \to \Gamma_4^* \to \cdots \to \Gamma_1^* \tag{21.10}$$

This is a **reverse process**.

As we already know, in the kinetic theory, the most important role is played by the distribution function $f(\vec{r}, \Gamma, t)$. It defines the evolution of the number of particles, $n(\vec{r}, \Gamma, t)$. The quantity Γ as we know plays an important role. It represents the integral of motion of each particle which remains constant for the time t of its free motion between two consecutive collisions in the absence of an external field. For each collision, the quantity Γ is subjected generally to variations. For a monatomic gas, Γ represents three components of momentum $\vec{p} = m\overrightarrow{V}$ of an atom: $d\Gamma = d\vec{p}$. For a diatomic gas it represents the momentum $\vec{p}$ and the angular momentum M.

$$d\Gamma = 2\pi d\vec{p} M dM d\Omega_M \tag{21.11}$$

where $d\Omega_M$ is the element of the solid angle for the direction of the vector $\vec{M}$.

It is necessary for us to bring out the relation between different collisions, considering the fact that the equilibrium state is invariant under the time reversal $t \to -t$. Let us consider a container for which its volume is partitioned into two and the separation of the two volumes has a hole as in Figure 21.1. Let a

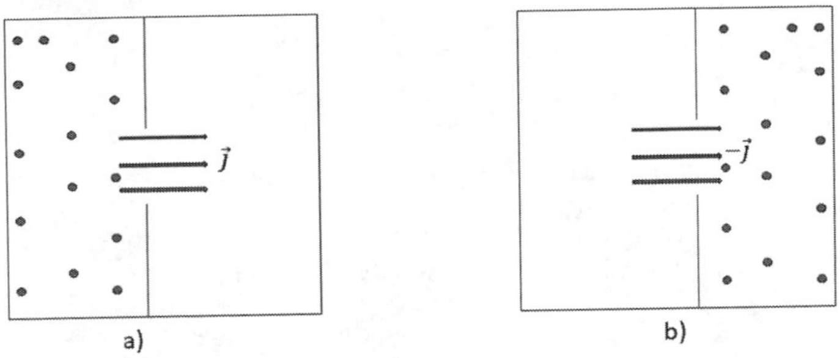

FIGURE 21.1 (a) Showing the direct process and (b) showing the time reversal.

gas be placed in one of the portions. There is a flux of the gas to the empty portion of the volume at any time moment t.

Under the time reversal $t \to -t$ there is a flux $-\vec{j}$ in the reverse direction to that of the initial Figure 21.1. For $\vec{j} \to 0$, a system is transformed to an equilibrium state and the change from the time $t \to -t$ already does not alter anything. Let us for a given equilibrium state consider the collision of two particles initially at Γ and Γ' that emerge after scattering respectively with Γ_1 and Γ'_1.

For the direct process see Figure 21.2.

For the reverse process see Figure 21.3, representing the collision of two particles. They represent simply, say, the collision of particles Γ and Γ' with the transition $\Gamma, \Gamma' \to \Gamma_1, \Gamma'_1$.

The expression $d\nu$ is the number of collisions of the particles and has various useful symmetry properties. This imply the relation between a given collision process and related processes. The equations of motion of the particles should have the following properties:

1. They must be invariant under the time reversal $t \to -t$. Under such a time reversal, the particles retrace their paths in time in a reverse collision. Thus, the number of collisions

$$d\nu * \left(\Gamma_1 *, \Gamma_1' * \to \Gamma *, \Gamma' *\right) = d\nu_1\left(\Gamma_1 *, \Gamma_1' * \to \Gamma *, \Gamma' *\right) \tag{21.12}$$

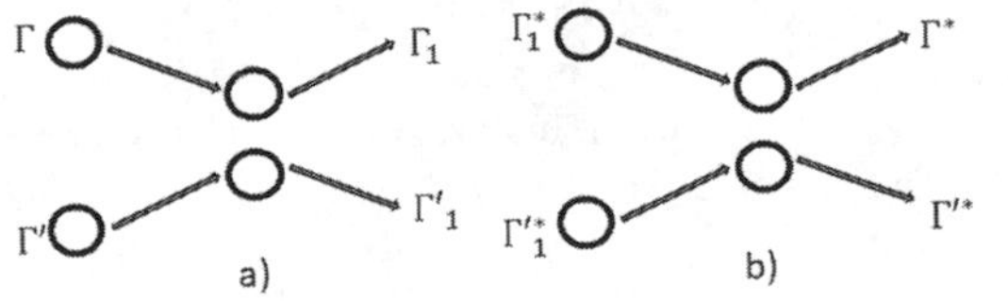

FIGURE 21.2 The direct process where a) correspond to $d\nu(\Gamma, \Gamma' \to \Gamma_1, \Gamma'_1)$ and b) correspond to $d\nu^*\left(\Gamma_1^*, \Gamma_1'^* \to \Gamma^*, \Gamma'^*\right)$.

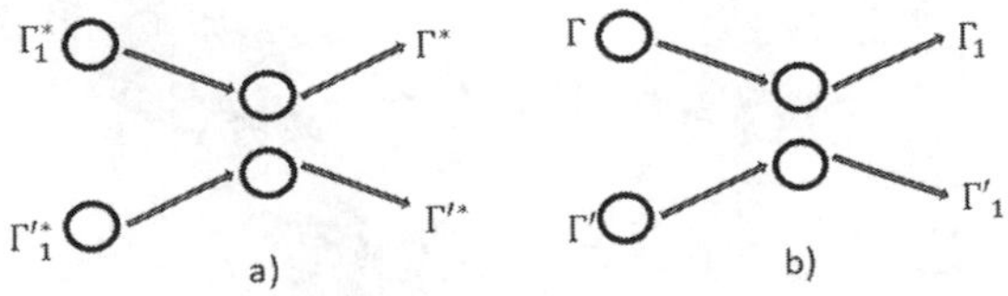

FIGURE 21.3 The reverse process where a) correspond to $\boldsymbol{d\nu_1}(\Gamma, \Gamma' \to \Gamma_1, \Gamma'_1)$ and b) correspond to $d\nu_1\left(\Gamma_1^*, \Gamma_1'^* \to \Gamma^*, \Gamma'^*\right)$.

It follows that

$$dv\left(\Gamma,\Gamma' \to \Gamma_1,\Gamma_1'\right) = dv\left(\Gamma_1*,\Gamma_1'* \to \Gamma*,\Gamma'*\right) \tag{21.13}$$

Thus, the number of direct collisions is equal to the number of reverse collisions. This is the so-called **principle of detailed equilibrium**

2. The equations of motion must be invariant under space inversion

$$\hat{I}\vec{r} = -\vec{r} \tag{21.14}$$

For such a space inversion the signs of all velocities change but the time order remains the same. Let

$$\tilde{\Gamma} = \tilde{\Gamma}\left(t \to -t, \hat{I}\vec{r}\right) \tag{21.15}$$

then it follows that

$$dv\left(\Gamma,\Gamma' \to \Gamma_1,\Gamma_1'\right) = dv\left(\tilde{\Gamma}_1,\tilde{\Gamma}'_1 \to \tilde{\Gamma},\tilde{\Gamma}'\right) \tag{21.16}$$

Since $t \to -t$ and 21.14 then

$$\vec{p} = m\frac{d\vec{r}}{dt} = m\frac{d\hat{I}\vec{r}}{d(-t)} = -m\frac{d\hat{I}\vec{r}}{dt} = m\frac{d\vec{r}}{dt} = \vec{p} \tag{21.17}$$

Thus,

$$\hat{\Theta}\vec{p} = \vec{p} \tag{21.18}$$

and from where

$$dv\left(\vec{p},\vec{p}' \to \vec{p}_1,\vec{p}_1'\right) = dv\left(\vec{p}_1,\vec{p}_1' \to \vec{p},\vec{p}'\right) \tag{21.19}$$

This is for the case if the state is described by an integral of motion $\vec{p}$ and depicted in Figure 21.4.

The transition probability is evaluated from the golden role of quantum mechanics (perturbation theory), i.e.,

$$W = \frac{2\pi}{\hbar}\left|\mathrm{M}_{ik}\right|^2 \tag{21.20}$$

where M_{ik} is the transition amplitude and may be represented via

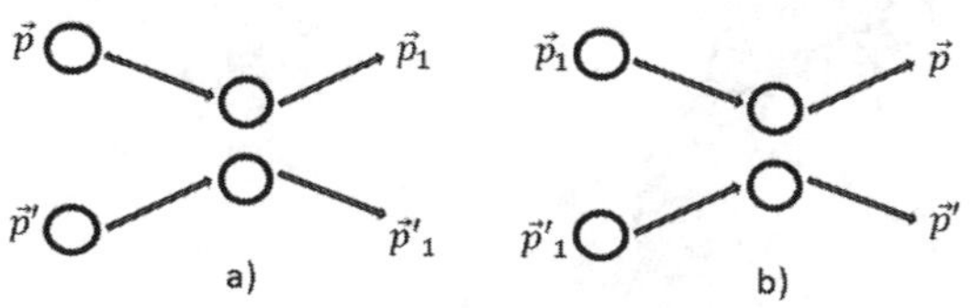

FIGURE 21.4 The collision process if the state is described by an integral of motion that is the momentum $\vec{p}$ where a) correspond to $dv\left(\vec{p},\vec{p}' \to \vec{p}_1,\vec{p}_1'\right)$ and b) corresponds to $dv\left(\vec{p}_1,\vec{p}_1' \to \vec{p},\vec{p}'\right)$.

$$S_{ik} = \sqrt{\frac{2\pi}{\hbar}} \mathrm{M}_{ik} \tag{21.21}$$

That is the scattering matrix or S-matrix (it is a unitary matrix). It may be proven that

$$\sum_k S_{ik} S_{kl} = \sum_k S_{ik}^* S_{kl} = S_{il} \tag{21.22}$$

if $i = l$ then

$$\sum_k S_{ik} S_{ki} = \sum_k S_{ik}^* S_{ki} = \sum_k \left|S_{ik}\right|^2 = 1 \tag{21.23}$$

It follows that the transition probability from the ith state to any other state is equal to 21.1. The scattering probability is

$$\sum_k \left|S_{ik}\right|^2 - \left|S_{ii}\right|^2 \tag{21.24}$$

$$\sum_k \tilde{S}_{ki}^\dagger S_{ki}^\dagger = \sum_k S_{ki}^* S_{ki} = \sum_k \left|S_{ki}\right|^2 - \left|S_{ii}\right|^2 \tag{21.25}$$

and so,

$$\int W\left(\Gamma,\Gamma' \to \Gamma_1,\Gamma'_1\right) d\Gamma_1 d\Gamma'_1 = \int W\left(\Gamma_1,\Gamma'_1 \to \Gamma,\Gamma'\right) d\Gamma_1 d\Gamma'_1 \tag{21.26}$$

21.3 Transport Phenomenon and Boltzmann-Lorentz Kinetic Equation

We find the basic equation in the kinetic theory of gases that will specify the change of the distribution function $f = f\left(\vec{r},\vec{v},t\right)$ with time t. This equation is the counterpart of the Schrödinger equation that specifies the variation of the wave function with time. The resulting transport equation forms the basis of all our transport evaluations. In order that we should know the distribution $f = f\left(\vec{r},\vec{v},t\right)$, it is always necessary to know the kinetic equation for which it is satisfied. Let us examine a particular value of momentum and particular position of a given system. We consider an elementary volume in phase space

$$d\Gamma = dq d\tau_v \tag{21.27}$$

where

$$dq = dxdydz \tag{21.28}$$

and

$$d\tau_v = dV_x dV_y dV_z \tag{21.29}$$

then the number of particles in the given elementary volume

$$f\left(\vec{r},\vec{v},t\right) d\Gamma \tag{21.30}$$

Let us begin with the evaluation of the physical quantities. Consider v_x being the velocity of an electron in the element dq_x, then the mean velocity of the particle with velocity v_x.

$$\langle v_x \rangle = \int f(\vec{r},\vec{v},t) v_x d\tau_v \tag{21.31}$$

and the current density

$$j_x = -e\langle v_x \rangle = -e\int f(\vec{r},\vec{v},t) v_x d\tau_v \tag{21.32}$$

Consider the theory of the kinetic phenomena. We limit ourselves to a weak field:

$$f(\vec{r},\vec{v},t) = f(E) + f_1(\vec{r},\vec{v},t) \tag{21.33}$$

and here $f_1(\vec{r},\vec{v},t)$ is an infinitesimally small term relative to $f(E)$. This implies we concentrate on a system where the deviation from equilibrium is small. Substituting $f(E)$ into 21.32, the resultant gives zero as an equilibrium state cannot generate current. If we consider the fact that the energy

$$E = \frac{mv^2}{2} + V(\vec{r}) \tag{21.34}$$

(where $\frac{mv^2}{2}$ is the kinetic energy and $V(\vec{r})$ the potential energy) is an energy function of the velocity v then it follows that

$$\int f(E) v d\tau_v = 0 \tag{21.35}$$

and consequently

$$j_x = -e\int f_1(\vec{r},\vec{v},t) v_x d\tau_v \tag{21.36}$$

Let us find the equation satisfying the distributive function $f = f(\vec{r},\vec{v},t)$ i.e. the kinetic equation. This equation may be obtained, if is considered that, the kinetic effect is a stationary one and the process is dependent on time moment t only through $\vec{r}$ and $\vec{v}$ i.e.,

$$\frac{df}{dt} = 0 \tag{21.37}$$

This is **Liouville theorem** [71,72] and in plasma physics it is sometimes called **Vlasov equation** (expected kinetic equation) [71,72]. The convective derivative in 21.37 corresponds to differentiation along phase path of the molecule that is determined by its equations of motion. As space plasma is collisionless, except for the ionosphere, we can safely neglect the collision term in the Boltzmann equation and this yields the simplest form of the kinetic equation for plasma, the Vlasov equation in 21.37. The absence of the collision term renders phase space density constant – **Liouville Theorem** – the phase volume can be deformed, but its density is invariant [1] during the dynamic evolution of plasma. This theorem holds for the full distribution function. For the Vlasov equation, the Liouville theorem holds only when the collision and correction terms between particles and microscopic fields are neglected.

Let us examine again 21.37 and suppose we have an occupied state by an electron then $f = 1$. When an external field is applied this state moves through phase space according to a quasi-classical equation. As we follow the trajectory corresponding to the electron, $f = 1$. Suppose we follow an unoccupied state then $f = 0$. The occupation does not change with time. Consider that f is associated with an arbitrary system. Then f does not change with time as we follow a trajectory in phase space. So, relation 21.37 holds and represents a transport equation (Liouville's theorem)when we go through a trajectory in phase space.

For the process to be stationary, it is necessary that the equation should be balanced. It follows that there exists a field and a temperature gradient. The field accelerates the electron and, consequently, the current increases continuously. It is therefore necessary that the equation should be balanced considering the scattering of electrons. The scattering of electrons is by a mechanism that is not included in the applied field. We may also take the following assumptions into consideration:

1. We consider the electronic gas to be sufficiently dilute. In this case we take only two particles into account.
2. The effects of the external force F' on the magnitude of the scattering cross section are ignored.
3. There is no appreciable variation of the distribution function $f(\vec{r},\vec{v},t)$ during the interval of the order of duration of a molecular scattering. $f(\vec{r},\vec{v},t)$ does not vary appreciably over a spatial distance of the order of intermolecular range.
4. We consider scattering between two particles. We neglect possible correlation between particle initial velocities prior to scattering. This is the so-called assumption of **molecular chaos.** This is true when the gas density is sufficiently low. In this case the mean free path L_p is much greater than the range of intermolecular forces. It may be seen that two gas particles originate before their encounter at a relative separation of the order L_p. It follows that L_p is sufficiently large that a correlation between the particles initial velocities is not feasible.

During the scattering process an electron discontinuously changes its momentum. Thus, it makes a discontinuous jump in phase space. If we consider $\left(\frac{\partial f}{\partial t}\right)_{\text{St}}$ being the **term concerned with collision (or scattering) of particles** and $\left(\frac{\partial f}{\partial t}\right)_{\text{fd}}$ the **term concerned with the field-induced motion** then from 21.37 follows

$$\left(\frac{\partial f}{\partial t}\right)_{\text{fd}}+\left(\frac{\partial f}{\partial t}\right)_{\text{St}}=0 \tag{21.38}$$

For the case of electrons, the **field-induced motion term** relates the externally applied electric and magnetic fields while the **collision (or scattering) term** describes the interaction of electrons with lattice imperfections (such as impurities, dislocations, lattice vibrations) and with other electrons.

Let us transform 21.38 where only $\vec{r}$ and $\vec{v}$ have to vary. We consider also the element of phase space $d\Gamma$ at the moment $t + dt$ in which the element is located, i.e., the elementary volume at the neighbourhood of $\vec{r}$ and $\vec{v}$. At the moment t we have the elementary volume at the neighbourhood of $\vec{r}-\vec{v}dt$ and $\vec{v}-\dot{\vec{v}}\,dt$. The charge density does not change (the electron does change its path) and thus

$$f(\vec{r},\vec{v},t+dt)=f\left(\vec{r}-\vec{v}dt,\vec{v}-\dot{\vec{v}}\,dt,t\right) \tag{21.39}$$

and we expand it in a series 21.39. Limiting ourselves only to the first term.

$$\frac{\partial f}{\partial t}=-\left(\frac{\partial f}{\partial \vec{r}},\vec{v}\right)-\left(\frac{\partial f}{\partial \vec{v}},\dot{\vec{v}}\right) \tag{21.40}$$

where

$$\dot{\vec{v}} = \text{acceleration} = \frac{\text{force}}{\text{mass}} = \frac{\vec{F}}{m} \tag{21.41}$$

Relation 21.40 is responsible for the external field:

$$\left(\frac{\partial f}{\partial t}\right)_{\text{fd}} = -\left(\frac{\partial f}{\partial \vec{r}},\vec{v}\right) - \left(\frac{\partial f}{\partial \vec{p}},\vec{F}\right) \tag{21.42}$$

where $\vec{p} = m\vec{V}$.

In 21.40,

$$\frac{d\vec{p}}{dt} = \vec{F} \tag{21.43}$$

i.e. it is the rate of change of momentum with time (at a given point on the trajectory) and is equal to the applied force at the given point. The quantity

$$\vec{v} = \frac{d\vec{r}}{dt} \tag{21.44}$$

in 21.42 is the rate of change of position in the trajectory for a given velocity (momentum). The first term 21.42 is as a result of motion and the second as a result of the external field.

If the electron is in the volume dqv at the moment, t then after scattering it is transported to the volume dqv'. We find the quantum mechanical probability at the time moment, dt. The electron leaves the element of phase space dp into dp'. The change in the number of electrons in element of phase volume relates to the fact that, after scattering of two electrons, the quantity Γ is subjected to a change. It follows that the number of collisions is

$$dv\left(\Gamma,\Gamma' \to \Gamma_1,\Gamma_1'\right) = W\left(\Gamma,\Gamma' \to \Gamma_1,\Gamma_1'\right) f\left(\Gamma\right) f\left(\Gamma'\right) d\Gamma d\Gamma' d\Gamma_1 d\Gamma'_1 \tag{21.45}$$

We are interested in the number of particles entering and leaving the state Γ. Here W is transition probability. It follows that the total number of particles at the exit of Γ is given by:

$$d\Gamma \int W\left(\Gamma,\Gamma' \to \Gamma_1,\Gamma_1'\right) f\left(\Gamma\right) f\left(\Gamma'\right) d\Gamma' d\Gamma_1 d\Gamma'_1 \tag{21.46}$$

which corresponds to the number of particles at the entrance into the element of phase space $d\Gamma$:

$$d\Gamma \int W\left(\Gamma_1,\Gamma'_1 \to \Gamma,\Gamma'\right) f\left(\Gamma_1\right) f\left(\Gamma'_1\right) d\Gamma' d\Gamma_1 d\Gamma'_1 \tag{21.47}$$

If we let for brevity

$$W\left(\Gamma,\Gamma' \to \Gamma_1,\Gamma_1'\right) = W \tag{21.48}$$

$$W\left(\Gamma_1,\Gamma'_1 \to \Gamma,\Gamma'\right) = W'_1 \tag{21.49}$$

$$f\left(\Gamma\right) = f,\quad f\left(\Gamma'\right) = f',\quad f\left(\Gamma_1\right) = f_1,\quad f\left(\Gamma'_1\right) = f'_1 \tag{21.50}$$

then it follows that in the unit of phase volume $d\Gamma$, considering the difference of the number of particles at the exit and entrance then the scattering integral:

$$\left(\frac{\partial f}{\partial t}\right)_{St} = \int\left[W' f_1 f'_1 - W f f'\right] d\Gamma' d\Gamma_1 d\Gamma'_1 \tag{21.51}$$

This is an integro-differential equation that needs to be solved. So, considering 21.26

$$\int W d\Gamma_1 d\Gamma'_1 = \int W' d\Gamma_1 d\Gamma'_1 \tag{21.52}$$

$$\left(\frac{\partial f}{\partial t}\right)_{St} = \int W'\left(f_1 f_1' - f f'\right) d\Gamma' d\Gamma_1 d\Gamma'_1 \tag{21.53}$$

then, the **kinetic (or transport) equation:**

$$\left(\vec{v}, \nabla_{\vec{r}} f\right) + \left(\vec{F}, \nabla_{\vec{p}} f\right) = \int W'\left(f_1 f'_1 - f f'\right) d\Gamma' d\Gamma_1 d\Gamma'_1 \tag{21.54}$$

which is an integral-differential equation and is called the **Boltzmann-Lorenz** equation and derived by Ludwig Boltzmann, the founder of the kinetic theory in 1969. For the scattering of a single particle by an impurity, say, the scattering integral is calculated via the Fermi golden rule given in 11.92.

21.4 Transport Relaxation Time

Let us examine 21.54, where the first term on the LHS is the drift term. The given electrons leave that region of space with velocity $\vec{v}$. If the distribution function varies in space, then the number of electrons that enter the given region will differ from the number that leaves that region. The second term on the LHS of 21.54 describes the change that arises due to the particle's acceleration into different momentum states. Suppose the distribution function varies with momentum. Then the number of particles being accelerated out is different from that accelerated into the given momentum region. The term on the RHS of 21.53 is a change because electrons may be scattered out at a different rate than when they are scattered in.

Let us try to write equation 21.53 in a more convenient form to work with. We introduce a **relaxation-time approximation** which is a time scale set when an external perturbation is switched off for the system to return to its equilibrium state. When the perturbation is maintained over a time scale much longer that the relaxation time then a steady flow of energy, particles, charge and so on is achieved. If we consider small perturbations, then the response can be characterized by the so-called **transport coefficients** between the given quantities.

We now consider the property of 21.53 where if we do the substitution for the statistical distribution

$$f_0(E) = \exp\left\{\frac{\mu - E}{T}\right\} \tag{21.55}$$

given

$$E = \frac{mv^2}{2} + V(\vec{r}) \tag{21.56}$$

then

$$\left(\vec{v}, \nabla_{\vec{r}} f\right) = -\frac{f}{T}\left(\vec{v}, \nabla_{\vec{r}} V\right) = \frac{f}{T}\left(\vec{v}, \vec{F}\right) \tag{21.57}$$

$$\left(\vec{F},\nabla_{\vec{p}} f\right) = -\frac{f}{T}\left(\vec{F},\vec{v}\right) \tag{21.58}$$

It follows that

$$\left(\vec{v},\nabla_{\vec{r}} f\right) = -\left(\vec{F},\nabla_{\vec{p}} f\right) \tag{21.59}$$

and thus, if the distribution function is an equilibrium function then

$$\left(\frac{\partial f}{\partial t}\right)_{\text{fd}} = 0 \tag{21.60}$$

and therefore

$$\left(\frac{\partial f}{\partial t}\right)_{\text{St}} = 0 \tag{21.61}$$

considering

$$\int W'\left(f_{10} f'_{10} - f_0 f'_0\right) d\Gamma_1 d\Gamma' d\Gamma'_1 \tag{21.62}$$

and as $W' \neq 0$ thus

$$f_{10} f'_{10} - f_0 f'_0 = 0 \tag{21.63}$$

from where

$$\ln f_{10} + \ln f'_{10} = \ln f_0 + \ln f'_0 \tag{21.64}$$

which is equivalent to the law of conservation of energy considering 21.55. This follows that the equilibrium function 21.55 satisfies the Boltzmann equation. It follows that for an equilibrium state, the distribution function has the universal form:

$$\rho(q,p) = A \exp\left\{-\frac{E(q,p)}{T}\right\} \tag{21.65}$$

Suppose

$$f = f_0 + f_1 \tag{21.66}$$

Here, f_0 is the equilibrium distribution function (it is not a function of the position) and f_1 is the deviation from the equilibrium position. If we substitute 21.66 into 21.58, then follows the **relaxation-time-approximation** of the scattering term:

$$\left(\frac{\partial f}{\partial t}\right)_{st} = -\frac{f_1}{\tau} \tag{21.67}$$

or

$$\left(\frac{\partial f}{\partial t}\right)_{st} = -\frac{f_1}{\tau} = \frac{f - f_0}{\tau} \tag{21.68}$$

where τ^{-1} is the proportionality coefficient, $[\tau]$ = seconds. If the time moment $t = 0$ is the time from where the external field was put off, then

$$f - f_0 = \left(f - f_0\right)\Big|_{t=0} \exp\left\{-\frac{t}{\tau}\right\} \tag{21.69}$$

It follows that, from the non-equilibrium state, the system attains an equilibrium with an increase in t and thus τ is called the **relaxation time** between molecular collisions. If there exist a field, then there will always exist such a τ that the kinetic equation:

$$\left(\vec{v},\nabla_{\vec{r}} f\right) + \left(\vec{F},\nabla_{\vec{p}} f\right) = -\frac{f_1}{\tau} \tag{21.70}$$

It may be seen from 21.69 that the momentum of an electron is randomized by a characteristic decay time. It should be noted that different phenomena correspond to different relaxation times. The relaxation time approximation is plausible and agrees with a wide range of experiments. However, it is certainly not true in detail.

Suppose, say, that the scattering events are basically elastic; then they will tend to cause decay of any current to the equilibrium value of zero. However, this will not effectively cause the decay of any isotropic deviation from equilibrium of the distribution as a function of energy. Consequently, it is necessary, however, to define different relaxation times for different phenomena under investigation. In addition, the choice of the distribution function for the equilibrium situation is very necessary. It should be noted that if the relaxation approximation is used correctly this yields a good description of the properties of the given system.

21.5 Boltzmann H-Theorem

Consider an isolated gas left to itself. In this case, the gas, as is the case with any isolated macroscopic system, approaches the equilibrium state. It implies that the evolution of the distribution function at the equilibrium state is according to the kinetic equation and must be accompanied by an increase in the entropy of the gas. Here we will be concerned with the statistical rather than the mechanical approach. The first law of thermodynamics will be an absolute postulate. In addition to that, we have the second law of thermodynamics with the principles of increment of entropy for an isolated system and for the case of an arbitrary system is not true the equality

$$dS \geq \frac{dQ}{dT} \tag{21.71}$$

Here, dS is the change in entropy, dQ the heat absorbed and T the absolute temperature. The relation $dS \geq 0$ is the law of statistical physics (second law of thermodynamics). It may be proven using an ideal gas.

Consider the Boltzmann kinetic equation 21.54. For the Boltzmann ideal non-degenerate gas, the entropy S is given by:

$$S = \int f \ln\frac{e}{f}\, d\Gamma\, dV \tag{21.72}$$

from where it follows that

$$\frac{dS}{dt}=\frac{\partial}{\partial t}\int f\ln\frac{e}{f}d\Gamma dV=\int\frac{\partial}{\partial t}\left[f\ln\frac{e}{f}\right]d\Gamma dV \tag{21.73}$$

If we consider 21.73 and 21.53 then follows the operation $\frac{\partial}{\partial t}$:

$$\frac{\partial}{\partial t}=-\left(\vec{v},\nabla_{\vec{r}}\right)-\left(\vec{F},\nabla_{\vec{p}}\right)+\frac{1}{f}\left(\frac{\partial}{\partial t}\right)_{\mathrm{St}} \tag{21.74}$$

and if we apply it to 21.74 then we have

$$\frac{\partial S}{\partial t}=\int\left(-\left(\vec{v},\nabla_{\vec{r}}\right)-\left(\vec{F},\nabla_{\vec{p}}\right)+\frac{\mathrm{St}(f)}{f}\right)f\ln\frac{e}{f}dVd\Gamma \tag{21.75}$$

where

$$St(f)\equiv\left(\frac{\partial f}{\partial t}\right)_{\mathrm{St}} \tag{21.76}$$

We may say beforehand that the statistical equilibrium in a gas is established by the scattering of the gas molecules. It thus follows that the increment in entropy should relate to the scattering part of the variation of the distribution function. This is because the part concerned with the variation of the distribution function related to the free motion of the gas molecules may not be the reason for the variation of the entropy of the gas. We prove this assertion. We examine each integral separately:

$$\int\left(\vec{v},\nabla_{\vec{r}}\right)f\ln\frac{e}{f}dVd\Gamma=\int\left[\mathrm{div}_{\vec{r}}\left(\vec{v}f\ln\frac{e}{f}\right)-f\ln\frac{e}{f}div_{\vec{r}}\vec{v}\right]dVd\Gamma \tag{21.77}$$

It should be noted that from classical mechanics $\vec{v}=\vec{v}(t)\neq\vec{v}(\vec{r})$. It follows from this that the summand under the integral sign in 21.77 is zero.

Consider the case:

$$\int\left(\vec{F},\nabla_{\vec{p}}\right)f\ln\frac{e}{f}dVd\Gamma=\int\left[\mathrm{div}_{\vec{p}}\left(\vec{F}f\ln\frac{e}{f}\right)-f\ln\frac{e}{f}div_{\vec{p}}\vec{F}\right]dVd\Gamma \tag{21.78}$$

Similarly, in the same manner the integral in 21.78 is equal to zero as

$$\mathrm{div}_{\vec{p}}\vec{F}(\vec{r})=0 \tag{21.79}$$

Thus, as $\int div\vec{\chi}dV$ is the flux of the vector $\vec{\chi}$ on the surface enclosing the volume V, then for $V\to\infty$ we have

$$\mathrm{div}\vec{\chi}=0 \tag{21.80}$$

Consequently, if we consider 21.77 and 21.78 in 21.75 then we have

$$\frac{\partial S}{\partial t} = \int W\left(\Gamma,\Gamma' \to \Gamma_1,\Gamma_1'\right)\left(f_1 f_1' - ff'\right)\ln\frac{e}{f}\,d\Gamma d\Gamma' d\Gamma_1 d\Gamma_1' dV \tag{21.81}$$

The expression in 21.81 is not very symmetric as a result of the factor $\ln\frac{e}{f}$. We try to make it symmetric. If $\Gamma \underset{\leftarrow}{\overset{\rightarrow}{}} \Gamma'$ and $\Gamma_1 \underset{\leftarrow}{\overset{\rightarrow}{}} \Gamma_1'$ then we have

$$\int W\left(ff' - f_1 f_1'\right)\ln\frac{e}{f}\,d\Gamma d\Gamma' d\Gamma_1 d\Gamma_1' dV \tag{21.82}$$

If $\Gamma \underset{\leftarrow}{\overset{\rightarrow}{}} \Gamma_1'$ and $\Gamma' \underset{\leftarrow}{\overset{\rightarrow}{}} \Gamma_1'$ then we have

$$\int W\left(ff' - f_1 f_1'\right)\ln\frac{e}{f'}\,d\Gamma d\Gamma' d\Gamma_1 d\Gamma_1' dV \tag{21.83}$$

The integral 21.81–21.83 is the same and thus we may sum them and then divide by 4 in order to obtain a symmetric formula:

$$\frac{dS}{dt} = \frac{1}{4}\int W\left(f_1 f_1' - ff'\right)\left(\ln\frac{e}{f} + \ln\frac{e}{f'} - \ln\frac{e}{f_1} - \ln\frac{e}{f_1'}\right)d\Gamma d\Gamma' d\Gamma_1 d\Gamma_1' dV \tag{21.84}$$

If we let

$$\frac{f_1 f_1'}{ff'} \equiv x \tag{21.85}$$

then

$$\frac{dS}{dt} = \frac{1}{4}\int Wff'\left(x-1\right)\ln x\,d\Gamma d\Gamma' d\Gamma_1 d\Gamma_1' dV \tag{21.86}$$

It may be seen that

$$\left(x-1\right)\ln x > 0 \quad \textit{for} \quad x > 0 \tag{21.87}$$

and

$$\left(x-1\right)\ln x = 0 \quad \textit{for} \quad x = 1 \tag{21.88}$$

and as W, f and f' are positive then it follows that

$$\frac{dS}{dt} \geq 0 \tag{21.89}$$

This is the **Boltzmann H-theorem,** which expresses the law of increment in entropy (the equality sign represents the equilibrium state). This was what it was necessary to prove. The proof of the law of increment in entropy using the kinetic equation was given by Boltzmann and gives the first macroscopic justification of the said law. This law applied to gases is often called the **H**-theorem, where **H** represents the entropy used by Boltzmann.

21.6 Thermal Conductivity

Consider a gas in a containing vessel where it is subjected to a temperature gradient. In order to evaluate the coefficient of the thermal conductivity of the gas, it is necessary to solve the kinetic equation having a temperature gradient. Let us limit ourselves to a stationary state, though we have a non-equilibrium state for which we write the Boltzmann equation. In our case $\frac{df}{dt} = 0$. From this it follows that

$$\left(\vec{F},\nabla_{\vec{p}}\right)f = \int W'\left(f_1 f_1' - ff'\right)\ln\frac{e}{f}d\Gamma' d\Gamma_1 d\Gamma_1' \tag{21.90}$$

Here the solution is not all that simple. To consider dissipative processes (thermal conduction and viscosity) in a slightly inhomogeneous gas we do an approximation. This involves regarding the distribution function in each region of the gas as being not a local-equilibrium function f_0 but one with a slight deviation of $\chi \ll 1$ from f_0:

$$f = f_0\left(1+\chi\right) \tag{21.91}$$

The quantity $\chi \ll 1$is a small correction due to the non-equilibrium state and is an odd function with respect to the velocity $\left(\chi \equiv \chi\left(\vec{v}\right)\right)$. The function χ must satisfy the transport equation as well as certain additional conditions. This is because the equilibrium distribution function f_0 corresponds to given values (in given volume element) of the gas particle number

$$\int f_0 d\Gamma = \int f d\Gamma \tag{21.92}$$

total energy

$$\int E f_0 d\Gamma = \int E f d\Gamma \tag{21.93}$$

and momentum densities

$$\int \vec{p} f_0 d\Gamma = \int \vec{p} f d\Gamma \tag{21.94}$$

So, from 21.92 it follows that

$$\int f_0 \chi d\Gamma = 0 \tag{21.95}$$

and from 21.93 then

$$\int E f_0 \chi d\Gamma = 0 \tag{21.96}$$

Similarly, from 21.94 then

$$\int \vec{p} f_0 \chi d\Gamma = 0 \tag{21.97}$$

The expression in 21.91 corresponds to a small deviation from the equilibrium state. It is a linear approximation as $\chi^2 = O(\chi^2)$. If we substitute 21.91 into 21.90 considering

$$f_{10} f'_{10} - f_0 f'_0 = 0 \tag{21.98}$$

Letting,

$$f_0 = f(\Gamma) \tag{21.99}$$

then

$$\int W'\left[f_{10}f_{01}'\left(1+\chi_1+\chi_1'\right)-f_0f_0'\left(1+\chi+\chi'\right)\right]d\Gamma'd\Gamma_1d\Gamma_1' = f_0\int W'f'\left(1+\chi_1-\chi-\chi'\right)d\Gamma'd\Gamma_1d\Gamma_1' \tag{21.100}$$

If there is no temperature gradient, then the equilibrium part is equal to zero and from this it follows that

$$\chi = \left(\vec{h},\nabla T\right) \tag{21.101}$$

It follows from 21.100 that

$$f_0\int W'f_0'\left(\chi_1-\chi_1'-\chi-\chi'\right)d\Gamma'd\Gamma_1d\Gamma_1' = f_0\nabla T\int W'f_0'\left(\vec{h}_1+\vec{h}_1'-\vec{h}-\vec{h}_1\right)d\Gamma'd\Gamma_1d\Gamma' \tag{21.102}$$

or

$$f_0\int W'f_0'\left(\chi_1-\chi_1'-\chi-\chi'\right)d\Gamma'd\Gamma_1d\Gamma_1' = f_0\nabla TI\left[\vec{h}\right] \tag{21.103}$$

We examine the LHS of 21.90 i.e. $\left(\vec{v},\nabla_{\vec{r}}f\right)$ and neglect $\nabla\chi$ as $\chi << 1$. It should be noted that the chemical potential μ of the gas is a function of the temperature T:

$$\mu = \mu(T) \tag{21.104}$$

If there is no external field, then the energy of the gas is independent of the coordinate but dependent on the temperature. Thus, as

$$f_0 = \exp\left\{\frac{\mu - E(\Gamma)}{T}\right\} \tag{21.105}$$

then

$$\nabla f_0 = f_0\left(\frac{1}{T}\frac{\partial\mu}{\partial T} - \frac{1}{T^2}\mu + \frac{1}{T^2}E(\Gamma)\right)\nabla T \tag{21.106}$$

It should be noted that

$$\left(\frac{\partial\mu}{\partial T}\right)_p = -S, \quad \mu = H - TS \tag{21.107}$$

where S and H are the entropy and enthalpy of the gas, respectively. Thus

$$\vec{v}\nabla f = \vec{v}\frac{\nabla T}{T^2}f_0\left(E(\Gamma)-H\right) = f_0\nabla TI\left[\vec{h}\right] \tag{21.108}$$

from where

$$\frac{\vec{v}}{T^2}\left(E(\Gamma)-H\right)=I\left[\vec{h}\right] \tag{21.109}$$

We examine how with the knowledge of $\vec{h}$ we may evaluate the thermal flux and thermal conductivity. The energy flux in the direction of the ith coordinate:

$$q_i=\int Ev_i f d\Gamma=\int Ev_i f_0\left(1+\chi\right)d\Gamma=q_{i0}+\int Ef_0 v_i\chi d\Gamma \tag{21.110}$$

As

$$q_{i0}=\int Ef_0 v_i d\Gamma=0 \tag{21.111}$$

then

$$q_i=\int Ev_i f_0\chi d\Gamma=\int Ev_i f_0 h_k\frac{\partial T}{\partial x_k}d\Gamma=\frac{\partial T}{\partial x_k}\int Ef_0 v_i h_k d\Gamma \tag{21.112}$$

From here it follows that

$$q_i=-\mathrm{K}_{ik}\frac{\partial T}{\partial x_k} \tag{21.113}$$

where K_{ik} is the tensor of the thermal conductivity:

$$\mathrm{K}_{ik}=\int Ev_i f_0 h_k d\Gamma \tag{21.114}$$

in the general case as the flux moves in the reverse direction to ∇T. Considering the fact that in equilibrium, the gas is isotropic then there are no preferred directions in it and so

$$\mathrm{K}_{ik}=\mathrm{K}\delta_{ik},\quad \mathrm{K}=\frac{\mathrm{K}_{ii}}{3} \tag{21.115}$$

Hence, the energy flux:

$$q=-\mathrm{K}\nabla T \tag{21.116}$$

and the scalar thermal conductivity:

$$\mathrm{K}=-\frac{1}{3T}\int Ef_0\left(\vec{v},\vec{h}\right)d\Gamma \tag{21.117}$$

This quantity is made positive by the transport equation so that the energy flux must be in the opposite direction to the temperature gradient. It is instructive to note that owing to the time-reversal properties of microscopic processes, the transport coefficient, and in particular scalar thermal conductivity, is observed to be not entirely independent.

21.7 Diffusion

We examine the phenomenon of diffusion, which is the process of diluting components of gases because of their initial and non-equilibrium concentrations. In diffusion, we always have the participation of two types of particles which is a difficult problem, but we may simplify it using the following supposition:

1. We consider diffusion in a mixture of gases where the **heavy** gas particles of mass M is much larger than **light** gas particles of mass m so that $M >> m$. The latter is assumed monatomic. The mean thermal translational energy of motion is the same for all particles at a given temperature for all masses:

$$\frac{m\vec{v}^2}{2} = \frac{M\vec{V}^2}{2} = \frac{3}{2}T \tag{21.118}$$

where $\vec{v}$ and $\vec{V}$ are the velocities of particles with the masses m and M respectively. As $M >> m$ then it follows from 21.118 that

$$\sqrt{\frac{\vec{V}^2}{\vec{v}^2}} = \sqrt{\frac{m}{M}} << 1 \tag{21.119}$$

from where

$$\sqrt{\vec{V}^2} << \sqrt{\vec{v}^2} \tag{21.120}$$

We see that the mean speed of the heavy gas particles is much less than that of the light ones and can be approximated as being at rest. So, when a light and a heavy particle collide, the latter is assumed fixed while the velocity of the light particle changes direction though remaining unaltered in magnitude. If we consider 21.120, then it follows that if $\vec{v}$ is the velocity of the particle of mass m before collision and $\vec{v}'$ is the velocity after collision then

$$|\vec{v}| = |\vec{v}'| \quad , \quad f_0(v) = f_0(v') \tag{21.121}$$

2. Consider N_1 and N_2 to be the densities of the number of particles with masses m and M respectively. Thus, N_2 = constant as the heavy particles are uniformly distributed and their concentration is a constant.

$$N_1 << N_2 \tag{21.122}$$

3. We suppose that we have a stationary process, i.e., $N_1(x) \neq N_1(x,t)$, where x is the coordinate and t the time .It follows from here that, it is sufficient to consider only one kinetic equation for light particles:

$$\frac{\partial f}{\partial t} = -\left(\vec{v},\nabla_{\vec{r}} f\right) - \left(\vec{F},\nabla_{\vec{p}} f\right) + \mathrm{St}(f) \tag{21.123}$$

and if we consider property 3), then $\frac{\partial f}{\partial t} = 0$. As the external field is absent, then $\vec{F} = 0$. The concentration changes only in the direction of the ox-axis (one-dimensional problem for $f(x)$). It follows that

$$\left(v_x,\nabla_x f\right) = \mathrm{St}(f) \tag{21.124}$$

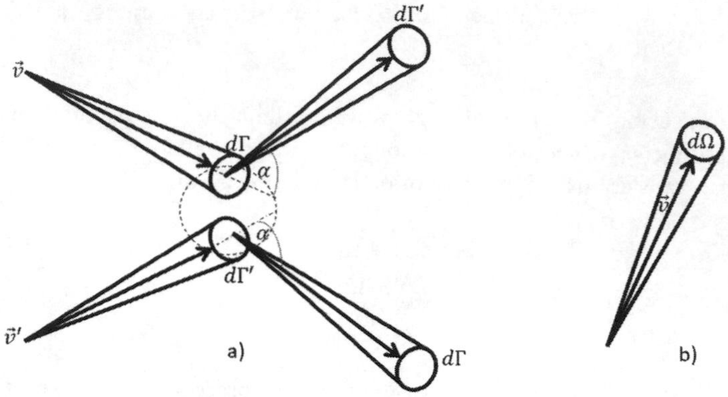

FIGURE 21.5 a) Showing the scattering of light particles (circular ends of the cone) on the heavy particle (broken circle) and α is the scattering angle. The arrows show the directions of the moving light particles. (b) Shows the direction of the velocity $\vec{v}$ of the light particle and the solid angle $d\Omega$.

Consider the number of particles in the state with the velocity $\vec{v}$ i.e.,

$$f(x,\vec{v}) \equiv f(\vec{v}) \tag{21.125}$$

During scattering, f reduces considering the process.

Figure 21.5a) shows the scattering of the light particles on the heavy particle and Figure 21.5b) shows the direction of the velocity of the light particle and the solid angle. The target is that the heavy particle is at rest. Here we may see the exit from the state with velocity $\vec{v}$ and entrance to the state with the velocity $\vec{v}'$. The effective differential scattering cross-section:

$$d\sigma(v,\alpha) = \frac{W(\vec{v}' - \vec{v})}{v} \tag{21.126}$$

If we consider the number of scatterings at the exit of the state with velocity, $\vec{v}'$ then $vf(v)d\Gamma$ is the flux of the particles. In Figure 21.5, $d\Omega$ is the solid angle. The number of scattering particles is equal to the number of particles through the cross-section $d\sigma(\vec{v},\alpha)$ and is equal to

$$vf(v)d\Gamma d\sigma(v,\alpha)N_2 \tag{21.127}$$

which is equal to the number of collisions per second per unit volume of light with heavy particles at the entrance. At the exit we have

$$v'f(v')d\Gamma d\sigma(v',\alpha)N_2 \tag{21.128}$$

Let

$$d\sigma(v,\alpha) = W(v,\alpha)d\Gamma' \tag{21.129}$$

and

$$d\sigma(v',\alpha) = W(v,\alpha)d\Gamma \tag{21.130}$$

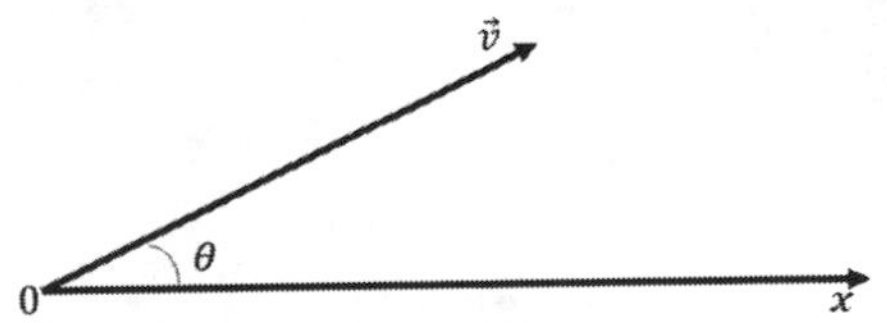

FIGURE 21.6 The angle $\boldsymbol{\theta}$ made by the velocity $\vec{v}$ on the ***ox*** axis.

If we suppose that $d\Gamma = 1$, then the difference of the number of collisions per second per unit volume of light particles with heavy particles at the entrance and exit:

$$vf(v')d\Gamma' W(v,\alpha)N_2 - vf(v)d\Gamma' W(v,\alpha)N_2 \tag{21.131}$$

from where it follows that

$$\mathrm{St}(f) \approx vN_2 \int W(v,\alpha)\left[f(v') - f(v)\right]d\Gamma' \tag{21.132}$$

Suppose the concentration and temperature gradients are not too large. This implies the quantities vary only slightly over distances of the order of the mean free path.

For this case, we consider a state which is at the neighborhood of the equilibrium state, i.e. the Lorentz approximation (1905) (see Figure 21.6 for angle θ):

$$f(v) = f_0(E) + \cos\theta h(E) \tag{21.133}$$

Here, the second summand is a small correction to the local equilibrium distribution function f_0 while E is the energy. If $\theta = 0$, then from 21.133 we have

$$f = f_0 + h(E) \tag{21.134}$$

and if $\theta = \pi$ then

$$f = f_0 - h(E) \tag{21.135}$$

Consider that the particles with velocity $\vec{v}$ in the positive ox-direction are greater than those in the negative ox-direction. It follows that

$$\mathrm{St}(f) = vN_2 \int W(v,\alpha)\left[f_0 - \cos\theta' h(E) - f_0 - \cos\theta h(E)\right]d\Gamma' \tag{21.136}$$

or

$$\mathrm{St}(f) = vN_2 \int W(v,\alpha)\left[\cos\theta' - \cos\theta\right]d\Gamma' \tag{21.137}$$

In Figure 21.7b), ϕ is the azimuthal angle relative to the ox-axis and the momentum relative to the polar axis and v_p is the projection of v on that plane. It follows that

$$\cos\theta' - \cos\theta = \cos(\alpha + \theta) - \cos\theta = -\cos\theta(1 - \cos\alpha) - \sin\theta \sin\alpha \tag{21.138}$$

and if $\theta \to 0$, then

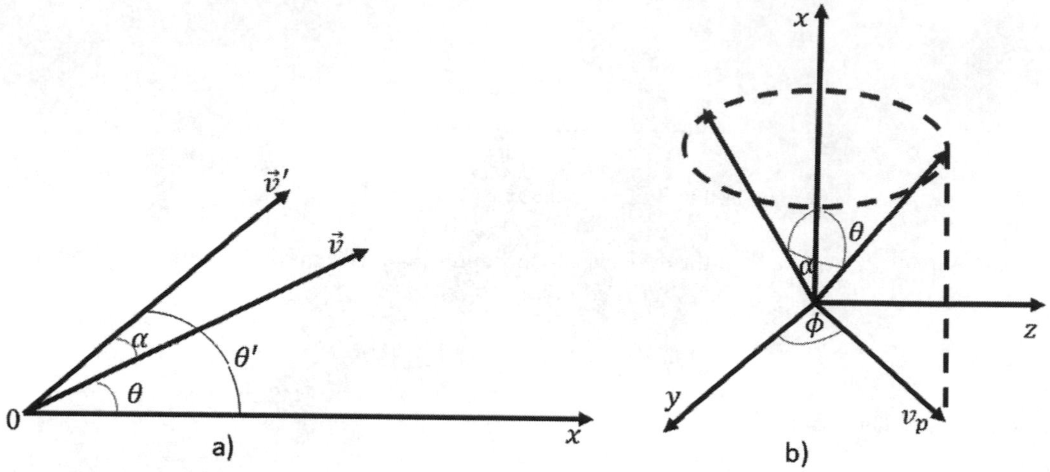

FIGURE 21.7 (a) Showing the angles , θ' made by the velocities $\vec{v}$ and $\vec{v}'$ respectively on the *ox*-axis and the relation with the angle α while in (b), ϕ is the azimuthal angle relative to the *ox*-axis and the momentum relative to the polar axis and v_p is the projection of v on that plane.

$$\mathrm{St}(f) = -vN_2\cos\theta h(E)\int W(v,\alpha)(1-\cos\alpha)d\Gamma = -vN_2\cos\theta h(E)\sigma^*(v,\alpha) \tag{21.139}$$

where $\sigma^*(v,\alpha)$ is the transport effective section. As $v\cos\theta = v_x$ then

$$v_x\nabla_x f = -v_x N_2 h(E)\sigma^*(v,\alpha) \tag{21.140}$$

from where

$$h(E) = -\frac{1}{N_2\sigma^*(v,\alpha)}\frac{\partial f_0}{\partial x} \tag{21.141}$$

Thus

$$f(v) = f_0 + \cos\theta h(E) = f_0 + \frac{\cos\theta}{N_2\sigma^*(v,\alpha)}\frac{\partial f_0}{\partial x} \tag{21.142}$$

The flux of the particles in the *ox*-direction is

$$j_x = \int v_x f(v)d\Gamma = \int v_x\left(f_0(E) - \frac{\cos\theta}{N_2\sigma^*(v,\alpha)}\nabla_x f_0\right)d\Gamma \tag{21.143}$$

As

$$v\cos\theta = v_x,\quad f_0 = f_0(E),\quad E \approx v^2,\quad d\Gamma \approx \sin\theta \tag{21.144}$$

then

$$\int v_x f_0(E)d\Gamma = 0 \tag{21.145}$$

Hence

$$j_x = -\frac{1}{N_2}\int \frac{v\cos^2\theta}{\sigma^*(v)}\nabla_x f_0 d\Gamma \tag{21.146}$$

Considering,

$$f_0 = N_1\left(\frac{m}{2\pi T}\right)^{\frac{3}{2}}\exp\left\{-\frac{mv^2}{T}\right\}\ ,\quad \overline{\cos^2\theta} = \frac{1}{3} \tag{21.147}$$

then

$$j_x = -\frac{1}{3N_2}\int \frac{v}{\sigma^*(v)}N_1\left(\frac{m}{2\pi T}\right)^{\frac{3}{2}}\exp\left\{-\frac{mv^2}{T}\right\}d\Gamma = -\frac{1}{3N_2}\nabla_x N_1\overline{\left(\frac{v}{\sigma^*(v)}\right)} \tag{21.148}$$

If

$$\frac{N_1}{N_2} = C_1(x) \tag{21.149}$$

is the concentration, then

$$j_x = -\frac{1}{3}\nabla_x\left[C_1(x)\overline{\left(\frac{v}{\sigma^*(v)}\right)}\right] = -\frac{1}{3}\overline{\left(\frac{v}{\sigma^*(v)}\right)}\frac{dC_1}{dx} - \frac{1}{3}C_1(x)\frac{\partial}{\partial T}\overline{\left(\frac{v}{\sigma^*(v)}\right)}\frac{dT}{dx} \tag{21.150}$$

Let

$$D = \frac{1}{3N}\overline{\left(\frac{v}{\sigma^*(v)}\right)}\ ,\quad N = N_1 + N_2 \tag{21.151}$$

be the coefficient of diffusion and

$$\lambda = \frac{1}{3ND}C_1(x)\frac{\partial}{\partial T}\overline{\left(\frac{v}{\sigma^*(v)}\right)} = C_1(x)\frac{d}{dT}\ln\overline{\left(\frac{v}{\sigma^*(v)}\right)} \tag{21.152}$$

be the coefficient of thermo-diffusion, then

$$j_x = ND\frac{dC_1}{dx} - ND\lambda\frac{dT}{dx} \tag{21.153}$$

The coefficient λ disturbs the equilibrium state until the diffusion flux is equal to the thermo-diffusion flux. Here, there is equal distribution of the particles, $j_x = 0$ and it follows that

$$\frac{d}{dx}\left[C_1(x)\overline{\left(\frac{v}{\sigma^*(v)}\right)}\right] = 0 \tag{21.154}$$

or

$$C_1(x) = \frac{\text{const}}{\overline{\left(\frac{v}{\sigma^*(v)}\right)}} \tag{21.155}$$

The total density of the number of particles is connected to the pressure p and the temperature T through the relation

$$N = \frac{p}{T} \tag{21.156}$$

Let the pressure of the gas be constant:

$$p = \text{const} \tag{21.157}$$

then from

$$N_1 << N_2 \tag{21.158}$$

we have

$$p = N_2 T \tag{21.159}$$

From 21.148 and 21.153 then we have

$$j_x = -\frac{T}{3N_2T}\frac{d}{dx}\left[N_1(x)\overline{\left(\frac{v}{\sigma^*(v)}\right)}\right] = -\frac{T}{3}\frac{d}{dx}\left[\frac{C_1(x)}{T}\overline{\left(\frac{v}{\sigma^*(v)}\right)}\right] \tag{21.160}$$

or

$$j_x = -\frac{T}{3}\frac{d}{dx}\left[\frac{C_1(x)}{T}\overline{\left(\frac{v}{\sigma^*(v)}\right)}\right] \tag{21.161}$$

We reapply the above theory to the case of an electron interacting with lattice vibrations in the presence of an external perturbation field.

21.8 Electron–Phonon System Equation of Motion

We find the path integral link of the electron–phonon system to the kinetic theory of gases. For this case, we examine the non-equilibrium statistical operator of a system of charge carriers. We take the example of a polar crystal lattice where is located given electrons or impurities submitted to an external field. We consider the Hamiltonian for the subsystem defined in 18.32:

$$\hat{H}_S(\vec{r},q) = \hat{H}_e(\vec{r}) + \hat{H}_L(q) + \hat{H}_{e-L}(\vec{r},q) \tag{21.162}$$

Adding a term due to an external field $\hat{H}_{ext}(\vec{r})$ then the Hamiltonian of the full system:

$$\hat{H}_F(\vec{r},q) = \hat{H}_S(\vec{r},q) + \hat{H}_{ext}(\vec{r}) \tag{21.163}$$

dependent on the electronic $\vec{r}$ and phonon q coordinates, where $\hat{H}_e(\vec{r})$ is the electronic, $\hat{H}_L(q)$ -lattice and $\hat{H}_{e-L}(\vec{r},q)$-electron-lattice interaction Hamiltonians, respectively.

The influence functional will be evaluated with the procedure in equation 18.37 to 18.57. On the basis of this we will introduce the system of generalized quantum kinetic equations of motion describing the evolution of the density matrix or distribution function for the system of charge carriers interacting with their environment and subjected to an external field. From the given quantum kinetic equation of motion, the collision integral will be elegantly deduced.

We consider a concrete example by revisiting the density matrix $\rho(\vec{r},\vec{r}')$ in 14.8. We revisit the expectation value $\bar{F}$ of a physical quantity described by the operator $\hat{F}(\vec{r})$ of the given physical quantity F :

$$\hat{F}(\vec{r}) \equiv \mathrm{Tr}\left[\hat{F}(\vec{r})\int K'^{*}\int K\rho(\vec{r}_1,q_1;\vec{r}'_1,q'_1)\right] \tag{21.164}$$

We eliminate lattice variables by path integration over q:

$$\hat{F}(\vec{r}) = \mathrm{Tr}_e\left[\hat{F}(\vec{r})\int K'^{*}_{e}\int K_e W_{ie}\rho_{ie}(\vec{r}_1,\vec{r}'_1)\right]\mathrm{Tr}_L\left[\int K'^{*}_{L,e-L}\int K_{L,e-L}W_{i'L}\rho_{i'L}(q_1,q'_1)\right] \tag{21.165}$$

Here, Tr_e and Tr_L denote respectively the trace over the electronic and lattice variables:

$$\mathrm{Tr}_L\left[\int K'^{*}_{L,e-L}\int K_{L,e-L}W_{i'L}\rho_{i'L}(q_1,q'_1)\right] = \exp\left\{\frac{i}{\hbar}\Phi\left[\vec{r},\vec{r}'\right]\right\} \tag{21.166}$$

and $\Phi\left[\vec{r},\vec{r}'\right]$ is the functional of the influence phase of the electron–phonon interaction. So, we rewrite 21.165 as:

$$\hat{F}(\vec{r}) = \mathrm{Tr}_e\left[\hat{F}(\vec{r})\int K'^{*}_{e}\int K_e W_{ie}\rho_{ie}(\vec{r}_1,\vec{r}'_1)\exp\left\{\frac{i}{\hbar}\Phi\left[\vec{r},\vec{r}'\right]\right\}\right] \equiv \mathrm{Tr}_e\left[\hat{F}(\vec{r})\rho_e(\vec{r},\vec{r}')\right] \tag{21.167}$$

The pure electronic density matrix is $\rho_e(\vec{r},\vec{r}')$:

$$\rho_e(\vec{r},\vec{r}') = \int K'^{*}_{e}\int K_e W_{ie}\rho_{ie}(\vec{r}_1,\vec{r}'_1)\exp\left\{\frac{i}{\hbar}\Phi\left[\vec{r},\vec{r}'\right]\right\} \tag{21.168}$$

and the reduced density matrix is given by:

$$\rho_e(\vec{r}_2,\vec{r}'_2) = \int d\vec{r}_1 d\vec{r}'_1\rho_{ie}(\vec{r}_1,\vec{r}'_1)\int D\vec{r}D\vec{r}'\exp\left\{\frac{i}{\hbar}\left(S_e\left[\vec{r}\right] - S_e\left[\vec{r}'\right]\right)\right\}F\left[r,r'\right] \tag{21.169}$$

where the influence functional of the phonon subsystem on the electronic density matrix:

$$F\left[r,r'\right] \equiv \exp\left\{\frac{i}{\hbar}\Phi\left[r,r'\right]\right\} \tag{21.170}$$

Here, the initial density matrix is $\rho_{ie}(\vec{r}_1,\vec{r}'_1)$. It is worth noting that 21.168 is expressed only via the variables of the electronic system since the phonon subsystem has been eliminated through path integration. So, equation 21.170 is, in principle, the phonon influence phase in the non-stationary density matrix operator of the subsystem of charge carriers for arbitrary electron–phonon coupling strength.

The equation of motion for the electronic density matrix via the Feynman functional integral for arbitrary electron–phonon coupling constants is treated in reference [61] and generalized to varying strong electric fields. The problem is resolved for a stationary non-linear response of band electrons to constant electric and strong electromagnetic fields.

We examine the dynamics of the electronic density matrix calculated in relation 21.169. We find the equation of motion for the electronic density matrix by first introducing an intermediate time t_c that time slice the time into $[t_1, t_c]$ and $[t_c, t_2]$. We use the semi-group property and compute the density matrix and then evaluate the equation of motion for the electronic density matrix $\rho_e\left[\vec{r}_2,\vec{r}'_2\right]$ that can be written by differentiating the density matrix 21.168 over the time [1,61]:

$$\frac{\partial}{\partial t_2}\rho_e\left(\vec{r}_2,\vec{r}'_2\right)=-\frac{i}{\hbar}\left[\hat{\mathrm{H}}_e\left(\vec{r}_2\right)-\hat{\mathrm{H}}_e^*\left(\vec{r}'_2\right),\rho_e\left(\vec{r}_2,\vec{r}'_2\right)\right]+\mathrm{St}\left(\rho_e\left(\vec{r}_2,\vec{r}'_2\right)\right) \quad (21.171)$$

Here, the scattering integral $\mathrm{St}\left(\rho_e\left(\vec{r}_2,\vec{r}'_2\right)\right)$, due to the electron–phonon interaction is:

$$\mathrm{St}\left(\rho_e\left(\vec{r}_2,\vec{r}'_2\right)\right)\equiv\rho_e^{(1)}\left(\vec{r}_2,\vec{r}'_2\right)=\int d\vec{r}_1 d\vec{r}'_1\,\rho_{ie}\left(\vec{r}_1,\vec{r}'_1\right)\int D\vec{r}D\vec{r}'\exp\left\{\frac{i}{\hbar}\left(\mathrm{S}_e\left[\vec{r}\right]-\mathrm{S}_e\left[\vec{r}'\right]\right)\right\}\mathrm{F}^{(1)}\left[r,r'\right] \quad (21.172)$$

with the quantity $\mathrm{F}^{(1)}[r,r']$ tailoring the change of the electron–phonon interaction electronic density matrix:

$$\mathrm{F}^{(1)}\left[r,r'\right]\equiv\frac{\partial}{\partial t_2}\mathrm{F}\left[r,r'\right] \quad (21.173)$$

In the absence of the electron–phonon interaction, this quantity vanishes and, consequently, the equation of motion in 21.164 yields the Liouville equation [1] for the electronic matrix density. Suppose t_2 achieves the value of t_1 then the scattering integral 21.165 vanishes and is also a consequence of elimination of the phonon variables. This mimic introducing a retarded interaction of the electron with itself [61] and so by iteration, the equation of motion of the scattering integral:

$$\frac{\partial}{\partial t_2}\rho_e^{(k)}\left(\vec{r}_2,\vec{r}'_2\right)=-\frac{i}{\hbar}\left[\hat{\mathrm{H}}_e\left(\vec{r}_2\right)-\hat{\mathrm{H}}_e^*\left(\vec{r}'_2\right),\rho_e^{(k)}\left(\vec{r}_2,\vec{r}'_2\right)\right]+\rho_e^{(k+1)}\left(\vec{r}_2,\vec{r}'_2\right) \quad (21.174)$$

with

$$\rho_e^{(k)}\left(\vec{r}_2,\vec{r}'_2\right)=\int d\vec{r}_1 d\vec{r}'_1\,\rho_{ie}\left(\vec{r}_1,\vec{r}'_1\right)\int D\vec{r}D\vec{r}'\exp\left\{\frac{i}{\hbar}\left(\mathrm{S}_e\left[\vec{r}\right]-\mathrm{S}_e\left[\vec{r}'\right]\right)\right\}\mathrm{F}^{(k)}\left[r,r'\right] \quad (21.175)$$

and

$$\mathrm{F}^{(k)}\left[r,r'\right]=\frac{\partial^k}{\partial t_2^{(k)}}\mathrm{F}\left[r,r'\right] \quad (21.176)$$

There exist three autonomous alternatives to quantization [6,7,9,61,73,74]:

1. Utilizing operators in Hilbert space due to Heisenberg, Schrödinger, Dirac and others;
2. Path integral conceived by Dirac and constructed by Feynman;
3. Phase space formulation due to Wigner quasi-distribution function and Weyl correspondence between quantum mechanical operators in Hilbert space and ordinary c-number functions in phase space.

This last alternative is critical to relate equation 21.171 to that of the transport equation by introducing the **Wigner quasi-distribution function**:

$$W^k(\vec{r},\vec{p},t) = \frac{1}{V}\int d\vec{r}'\exp\left\{-\frac{i}{\hbar}\vec{p}\vec{r}'\right\}\rho\left(\vec{r}+\frac{\vec{r}'}{2},\vec{r}-\frac{\vec{r}'}{2}\middle| t\right) \quad (21.177)$$

that satisfies the following transport equation [61]:

$$\frac{\partial}{\partial t_2}W^k(\vec{r},\vec{p},t_2) = \hat{\mathrm{F}}^k(\vec{r},\vec{p},t_2)W^k(\vec{r},\vec{p},t_2) + W^{(1)k}(\vec{r},\vec{p},t_2) \quad (21.178)$$

Here $\vec{p}$ is the total momentum of the particle that includes the contribution of the external field and $\hat{\mathrm{F}}^k(\vec{r},\vec{p},t_2)$ describes the translational and accelerated motion of the electron:

$$\hat{\mathrm{F}}^k(\vec{r},\vec{p},t_2) = -\hbar m\vec{\kappa}\nabla_{\vec{r}} - \gamma(\vec{r},t_2)\nabla_{\vec{\kappa}} \quad (21.179)$$

The quantity $\vec{\kappa}$ is the momentum of the particle of mass m and $\gamma(\vec{r},t_2)$ relates the force acting on the electron and the scattering integral $W^{(1)k}(\vec{r},\vec{p},t_2)$ that describes the change in $W^k(\vec{r},t_2)$ tailoring the electron–phonon interaction coupled to the external fields:

$$W^{(1)k}(\vec{r},\vec{p},t_2) = \frac{1}{V}\int d\vec{r}'\exp\left\{-\frac{i}{\hbar}\vec{p}\vec{r}'\right\}\rho^{(1)}\left(\vec{r}+\frac{\vec{r}'}{2},\vec{r}-\frac{\vec{r}'}{2}\middle| t_2\right) \quad (21.180)$$

The distribution function $W^k(\vec{r},\vec{p},t)$ is normalized to unity:

$$\int d\vec{r}d\vec{p}W^k(\vec{r},\vec{p},t) = 1 \quad (21.181)$$

Considering the classical limit $\hbar \to 0$, the function $W^k(\vec{r},\vec{p},t)$ reduces to the probability density in the coordinate space $\vec{r}$ that is highly localized and multiplied by delta functions in momentum:

$$W^{(1)k}(\vec{r},\vec{p},t) = \rho^{(1)}(\vec{r},\vec{r},,t)\delta(\vec{p}) \quad (21.182)$$

References

[1] L.C. Fai, M.W. Gary, *Statistical Thermodynamics: Understanding the Properties of Macroscopic Systems*, CRC Press, USA, 2012.

[2] L.C. Fai, *Polaron and Bipolaron States in Quantum Wires and Quantum Dots. Nonlinear Polaron Light Absorption in Quantum Dots*, Faculty of Physics, Department of Theoretical Physics, Moldova State University, Kishinev, 1997, pp. 1–139.

[3] R.P. Feynman, Space-Time Approach to Non-Relativistic Quantum Mechanics, *Reviews of Modern Physics*, 20 (1948) 367–387.

[4] L.S. Schulman, *Techniques and Applications of Path Integration*, Dover Publications Inc., Mineola, New York, 2005.

[5] H. Kleinert, *Path Integrals in Quantum Mechanics, Statistics, Polymer Physics, and Financial Markets*, World Scientific, Singapore, 2009.

[6] P.A.M. Dirac, *Quantum Electrodynamics*, Dover, New York, 1958.

[7] P.A.M. Dirac, The Lagrangian in quantum mechanics, *Phys. Z. Sowjetunion*, 3 (1933) 64–72.

[8] L.C. Fai, *Quantum Field Theory, Feynman Path Integrals and Diagrammatic Techniques in Condensed Matter*, 1st ed., CRC Press, Taylor & Francis Group, Boca Raton, FL, 2019.

[9] R.P. Feynman, A.R. Hibbs, *Quantum Mechanics and Path Integrals*, McGraw-Hill, New York, 1965.

[10] M. Kac, On distributions of certain Wiener functionals, *Transactions of the American Mathematical Society*, 65 (1949) 1–13.

[11] V.P. Maslov, *Perturbation Theory and Asymptotic Methods*, Moscow University Press, Moscow, 1965.

[12] P.A. Horváthy, The Maslov correction in the semiclassical Feynman integral, *Central European Journal of Physics*, 9 (2011) 1–12.

[13] P.A.M. Dirac, *The Principles of Quantum Mechanics*, Clarendon Press, Oxford, England, 1981.

[14] C.M. Bender, S.A. Orszag, *Advanced Mathematical Methods for Scientists and Engineers*, 1st ed., Springer-Verlag, New York, 1999.

[15] P. Schattschneider, T. Schachinger, M. Stöger-Pollach, S. Löffler, A. Steiger-Thirsfeld, K.Y. Bliokh, F. Nori, Imaging the dynamics of free-electron Landau states, *Nature Communications*, 5 (2014) 4586.

[16] L.D. Landau, E.M. Lifshitz, *Quantum Mechanics*, Pergamon, London, 1965.

[17] D. Foata, A combinatorial proof of the Mehler formula, *Journal of Combinatorial Theory, Series A*, 24 (1978) 367–376.

[18] F. Brackx, N. de Schepper, K.I. Kou, F. Sommen, The Mehler Formula for the Generalized Clifford–Hermite Polynomials, *Acta Mathematica Sinica, English Series*, 23 (2007) 697–704.

[19] J.B. Fo, E.S.C. Neto, The Mehler formula and the Green function of the multi-dimensional isotropic harmonic oscillator, *Journal of Physics A: Mathematical and General*, 9 (1976) 683–685.

[20] G. Grynberg, A. Aspect, C. Fabre, C. Cohen-Tannoudji, Complement 2C: The density matrix and the optical Bloch equations, in: A. Aspect, C. Fabre, G. Grynberg (Eds.) *Introduction to Quantum Optics*, Cambridge University Press, Cambridge, 2010, 140–166.

[21] L.C. Fai, V. Teboul, A. Monteil, I. Nsangou, S. Maabou, Polaron in cylindrical and spherical quantum dots, *Condensed Matter Physics*, 7 (2004) 157–166.

[22] E.P. Pokatilov, V.M. Fomin, S.N. Balaban, S.N. Klimin, L.C. Fai, J.T. Devreese, Electron-phonon interaction in cylindrical and planar quantum wires, *Superlattices and Microstructures*, 23 (1998) 331–336.

[23] J. Tempere, W. Casteels, M.K. Oberthaler, S. Knoop, E. Timmermans, J.T. Devreese, Feynman path-integral treatment of the BEC-impurity polaron, *Physical Review B*, 80 (2009) 184504.

[24] T. Ichmoukhamedov, J. Tempere, Feynman path-integral treatment of the Bose polaron beyond the Fröhlich model, *Physical Review A*, 100 (2019) 043605.

[25] A.S. Mishchenko, N. Nagaosa, K.M. Shen, Z.X. Shen, X.J. Zhou, T.P. Devereaux, Polaronic metal in lightly doped high-T c cuprates, *EPL (Europhysics Letters)*, 95 (2011) 57007.

[26] A. Kaminski, S. Das Sarma, Polaron Percolation in Diluted Magnetic Semiconductors, *Physical Review Letters*, 88 (2002) 247202.

[27] A.S. Alexandrov, *Polarons in advanced materials*, Springer, Netherlands, 2007.

[28] G.A. Fiete, G. Zaránd, K. Damle, Effective Hamiltonian for Ga1−xMnxAs in the Dilute Limit, *Physical Review Letters*, 91 (2003) 097202.

[29] J.S. Pan, H.B. Pan, Size-Quantum Effect of the Energy of a Charge Carrier in a Semiconductor Crystallite, *Physica Status Solidi (b)*, 148 (1988) 129–141.

[30] J.C. Marini, B. Stebe, E. Kartheuser, Exciton-phonon interaction in CdSe and CuCl polar semiconductor nanospheres, *Physical Review B*, 50 (1994) 14302–14308.

[31] P.M. Petroff, A.C. Gossard, R.A. Logan, W. Wiegmann, Toward quantum well wires: Fabrication and optical properties, *Applied Physics Letters*, 41 (1982) 635–638.

[32] S.N. Klimin, E.P. Pokatilov, V.M. Fomin, Bulk and Interface Polarons in Quantum Wires and Dots, *Physica Status Solidi (b)*, 184 (1994) 373–383.

[33] L.C. Fai, A. Fomethe, A.J. Fotue, V.B. Mborong, S. Domngang, N. Issofa, M. Tchoffo, Bipolaron in a quasi-0D quantum dot, *Superlattices and Microstructures*, 43 (2008) 44–52.

[34] R.P. Feynman, Slow Electrons in a Polar Crystal, *Physical Review*, 97 (1955) 660–665.

[35] L.D. Landau, Research in the Electron Theory of Crystals, *Phys. Z. Sowjet*, 3 (1933) 664.

[36] S.I. Pekar, Theory Of Polarons In Many-Valley Crystals, *Soviet Physics JETP*, 28 (1969).

[37] S.I. Pekar, Autolocalization of the electron in an inertially polarizable dielectric medium, *Zh. Eksp. Teor. Fiz.*, 16 (1946) 335.

[38] H. Fröhlich, Electrons in lattice fields, *Advances in Physics*, 3 (1954) 325–361.

[39] B.T. Geĭlikman, Adiabatic perturbation theory for metals and the problem of lattice stability, *Soviet Physics Uspekhi*, 18 (1975) 190–202.

[40] H. Fröhlich, H. Pelzer, S. Zienau, Properties of slow electrons in polar materials, *The London, Edinburgh, and Dublin Philosophical Magazine and Journal of Science*, 41 (1950) 221–242.

[41] L. Landau, On the motion of electrons in a crystal lattice, *Phys. Z. Sowjetunion*, 3 (1933) 664.

[42] S.I. Pekar, U.S.A.E. Commission, I. Division of Technical, Research in Electron Theory of Crystals, 1963.

[43] F. Ortmann, F. Bechstedt, K. Hannewald, Characteristics of small- and large-polaron motion in organic crystals, *Journal of Physics: Condensed Matter*, 22 (2010) 465802.

[44] V.D. Lakhno, G.N. Chuev, Structure of a strongly coupled large polaron, *Physics-Uspekhi*, 38 (1995) 273–285.

[45] S. Kokott, S.V. Levchenko, P. Rinke, M. Scheffler, First-principles supercell calculations of small polarons with proper account for long-range polarization effects, *New Journal of Physics*, 20 (2018) 033023.

[46] G.N. Chuev, V.D. Lakhno, *Perspectives of Polarons*, World Scientific Publishing Company, Singapore, 1996.

[47] M. Porsch, A Green's Function Approach to Intermediate- and Strong-Coupling Polaron Theory in the Presence of External Fields, *Physica Status Solidi (b)*, 39 (1970) 477–491.

[48] H. Fröhlich, N.F. Mott, Theory of electrical breakdown in ionic crystals, *Proceedings of the Royal Society of London. Series A - Mathematical and Physical Sciences*, 160 (1937) 230–241.

[49] E.H. Lieb, R. Seiringer, Divergence of the Effective Mass of a Polaron in the Strong Coupling Limit, *Journal of Statistical Physics*, doi: 10.1007/s10955-019-02322-3(2019).

[50] V.M. Fomin, E.P. Pokatilov, On the representation of the non-stationary density matrix of an electron-phonon system by a functional integral, *Physica Status Solidi (b)*, 87 727–732.

[51] S.I. Pekar, *A Research in Electron Theory of Crystals*, Gostekhizdat, Moscow, 1951.

[52] R. Evrard, On the excited states of the polaron, *Physics Letters*, 14 (1965) 295–296.
[53] S. J. Miyake, Strong-Coupling Limit of the Polaron Ground State, *Journal of the Physical Society of Japan*, 38 (1975) 181–182.
[54] T.D. Lee, F.E. Low, D. Pines, The Motion of Slow Electrons in a Polar Crystal, *Physical Review*, 90 (1953) 297–302.
[55] R.L. Frank, E.H. Lieb, R. Seiringer, L.E. Thomas, Bipolaron and N-polaron binding energies, *Physical review letters*, 104 (2010) 210402.
[56] R.L. Frank, E.H. Lieb, R. Seiringer, Binding of Polarons and Atoms at Threshold, *Communications in Mathematical Physics*, 313 (2012) 405–424.
[57] I. Nsangou, L.C. Fai, V.B. Mborong, Magnetopolaron in a cylindrical nanocrystal, *Journal of the Cameroon Academy of Science*, 4 (2004) 255–270.
[58] V.M. Fomin, E.P. Pokatilov, Non-linear absorption of electromagnetic waves by band charge carriers in a laser field, *Physica Status Solidi (b)*, 78 (1976) 831–842.
[59] V.M. Fomin, V.N. Gladilin, J.T. Devreese, E.P. Pokatilov, S.N. Balaban, S.N. Klimin, Phonon-assisted optical transitions in spherical nanocrystals, *Solid State Communications*, 105 (1998) 113–117.
[60] L.C. Fai, V. Teboul, A. Monteil, S. Maabou, Phonon-Assisted Photoluminescence in a Spherical Nanocrystal, *Journal of Applied Spectroscopy*, 72 (2005) 716–722.
[61] V.M. Fomin, E.P. Pokatilov, Non-Linear Transport Properties of Band Electrons at Arbitrary Electron—Phonon Coupling, *Physica Status Solidi (b)*, 97 (1980) 161–174.
[62] R. Fuchs, K.L. Kliewer, Optical Modes of Vibration in an Ionic Crystal Slab, *Physical Review*, 140 (1965) A2076-A2088.
[63] A.K. Sood, J. Menéndez, M. Cardona, K. Ploog, Resonance Raman Scattering by Confined LO and TO Phonons in GaAs-AlAs Superlattices, *Physical Review Letters*, 54 (1985) 2111–2114.
[64] V.M. Fomin, E.P. Pokatilov, Multiphoton absorption by charge carriers in semiconductors in a magnetic field, *Physica Status Solidi (b)*, 79 (1977) 595–604.
[65] R.P. Feynman, R.W. Hellwarth, C.K. Iddings, P.M. Platzman, Mobility of Slow Electrons in a Polar Crystal, *Physical Review*, 127 (1962) 1004–1017.
[66] J.T. Devreese, *Encyclopedia of Applied Physics*, VCH, Weinheim, 1996.
[67] D.C. Langreth, L.P. Kadanoff, Perturbation Theoretic Calculation of Polaron Mobility, *Physical Review*, 133 (1964) A1070-A1075.
[68] L.P. Kadanoff, Boltzmann Equation for Polarons, *Physical Review*, 130 (1963) 1364–1369.
[69] A.S. Alexandrov, J.T. Devreese, *Advances in Polaron Physics*, Springer-Verlag, Berlin Heidelberg, 2010.
[70] F.M. Peeters, J.T. Devreese, Theory of Polaron Mobility, in: H. Ehrenreich, D. Turnbull, F. Seitz (Eds.) *Solid State Physics*, Academic Press, 1984, pp. 81–133.
[71] H. Ralph Lewis, D.C. Barnes, K.J. Melendez, The Liouville theorem and accurate plasma simulation, *Journal of Computational Physics*, 69 (1987) 267–282.
[72] H.R. Lewis, D.C. Barnes, K.J. Melendez, Liouville Theorem and accurate plasma simulation, *Journal of Computational Physics*, 69 (1987) 267–282.
[73] E. Wigner, On the Quantum Correction For Thermodynamic Equilibrium, *Physical Review*, 40 (1932) 749–759.
[74] H. Weyl, Quantenmechanik und gruppentheorie, *Zeitschrift für Physik*, 46 (1927) 1–46.

Index

A

absorption, 169, 171, 238, 337–339, 344–346, 348–351, 364
acceleration, 61, 372, 373
adiabatic, 160, 163, 239, 242, 279
adjoint(s), 8, 10, 14–16
admittance, 353
advanced, 99, 113, 271
airy, 123, 124, 138
alkali, 239, 263, 277
angular, 67, 125–127, 162, 167, 169, 366
anharmonic, 75
annihilation, 265
anomalies, 238
anti-chronological-ordered, 195
anti-resonant, 170
anti-symmetric, 81, 88
asymmetric/asymmetrical, 315, 320
asymptotic/asymptotically, 54, 56, 114, 124, 125, 129, 132, 142, 239, 264, 270, 273, 276, 286, 287, 302, 337, 364
atom(s), 130, 140, 165–168, 241, 365, 366
atomic, 163, 237
azimuthal, 126, 383, 384

B

band(s), 237, 238, 241, 265, 340, 388
band-electron, 239, 264
band-mass, 240–242, 259
band-width, 164
barrier, 60, 114–121, 130
bases, 8, 9, 40, 152, 245
Bessel, 129, 132
bilinear, 13, 241
binary, 5
binding, 166, 239, 243, 279, 330
binomial, 149
bipolaron(s), 218, 238, 239, 276–280, 283, 285–289, 295–297, 299–303, 310, 312–314, 320–323
bits, 5
Bloch, 158, 200, 237
Bohr, 140, 162, 165
Bohr-Sommerfeld, 110, 111, 119, 133
Boltzmann, 191, 197, 220, 246, 351, 364, 365, 370, 373–375, 377, 378
Boltzmann-Lorentz, 369
Boltzmann-Lorenz, 373
Born, 338
Bose-Einstein, 216, 238
Bose-liquid, 238
boson(s), 237–239, 242
bosonic, 238
Broglie, 100, 104, 105, 121, 130, 205, 220, 229, 240, 268, 284, 334, 339
bulk-type, 238

C

calculus, 34, 181, 231
canonical/canonically, 29, 42, 78, 182, 192, 200, 209, 276
carrier(s), 238, 239, 242, 337, 339, 351, 386–388
Cauchy, 200, 212, 215
causal, 38, 97–99
causality, 38, 98, 99
caustic(s), 35, 75, 142
centrifugal, 125, 129, 130
centroid(s), 220–222, 224, 225
chalcogenides, 351
chaos, 371
Chapman, 51
charge-carrier, 351
chloride, 263
chronological/chronologically, 37, 195, 201, 339, 353
chronological-ordered, 195, 201
chronological-ordered-product, 37
closure, 10, 12, 18, 21, 32, 145
c-number(s), 42, 153, 389
coherent, 52
commutation, 187
commutativity, 25
commutator, 25

commute, 24, 25, 119, 142
comparing, 9, 10, 14, 132, 166, 263, 328, 355
condensate, 238
conductivity(ies), 337, 352, 353, 365, 378, 380
cone, 382
confinement(s), 238, 239, 265, 266, 276–279, 288–290, 295, 303, 304, 310, 315, 323, 350
conjugate(s), 6, 8, 13–16, 20, 78, 182
contour, 119
convective, 370
corpuscular, 66
Coulomb, 130, 238, 278, 279, 282, 287, 288, 303, 313, 322, 338, 339
Coulombic, 130–132, 239
Coulomb-mediated, 238
creation, 265, 275
cross-section, 337, 382
crystal(s), 152, 205, 209, 237–243, 246, 259, 262–265, 267, 271, 275–277, 286, 287, 294, 301, 305, 308, 314, 330, 352, 362, 386
crystalline, 140
cyclotron, 87, 239, 323
cylindrical, 265, 304, 310, 315, 320
cylindrical-symmetrical, 277

D

Debye, 339
decoherence, 241
degeneracy(ies), 23, 163
degenerate, 23
delta-function, 65, 213, 351
delta-normalization, 91, 159
diagonalization, 282, 283, 291, 297, 305, 312
diatomic, 366
dielectric, 239, 263, 339, 351
diffraction, 53, 104, 164
diffusion, 51, 380, 381, 385
dipole, 69, 338, 339
Dirac, 10, 12, 29, 33, 44, 62, 65, 112, 122, 135, 168, 170, 180, 213, 324, 328, 389
Dirac-delta-functional, 222
Dirichlet, 34, 61, 63
distinguishable/distinguishability, 3, 166
dot(s), 8, 9, 30, 218, 224, 238, 289, 294–296, 302–304, 310, 315, 319, 320, 337, 349, 351
dualism, 66
duality, 3
Dyson, 154

E

Ehrenfest, 185
eigenbasis, 5, 161
eigenenergy(ies), 95–97, 135, 139, 140, 161, 207
eigenfunction(s), 4, 17, 96–98, 127, 129, 135, 138, 139, 150, 151, 160, 162, 193, 194, 200, 207, 229, 231, 234, 276, 327
eigenmode(s), 63, 73, 205, 280, 282, 290, 291, 305, 311, 312, 316, 321
eigenstate(s), 17–21, 24, 25, 33, 36, 39, 95–98, 150, 160, 161, 193, 196, 197, 207, 231, 232, 248, 251
eigenvalue(s), 2, 7, 8, 15, 17–20, 31, 39, 95, 97, 124, 126, 135, 138, 139, 150, 160, 163, 192, 193, 196, 229, 231, 232, 248, 327
eigenvector(s), 2, 17, 18, 32, 35, 95–97, 207
eigenwave, 139, 140
eikonal, 60, 101, 102, 120
Einstein, 171
electric, 69, 122, 137, 138, 237, 239, 337, 349, 352, 356, 371, 388
electro-magnetic, 67, 104, 169, 205, 337–339, 388
electron, 1–3, 69, 150–152, 166, 209, 237–244, 246, 248, 255, 259, 262, 264–267, 271, 272, 274–276, 278, 287, 290, 294, 304, 305, 310, 315, 324, 329–332, 337–340, 345, 346, 351, 352, 356, 362–364, 370–372, 375, 386, 388, 389
electron–lattice, 249, 255, 265, 341
electron-phonon, 151, 238, 239, 245, 246, 248–251, 253–256, 260, 263, 265, 266, 268, 269, 276, 277, 284, 288, 289, 293, 295, 299, 303, 307, 318, 330, 342, 343, 350, 351, 386–389
electron-phonon-coupling, 238, 239
electron-phonon-coupling-amplitude, 150
electron-phonon-electron, 238
electron-phonon-hole, 239
Euclidean, 191, 200, 201, 203, 206–209, 247
Euler-Lagrange, 34, 60, 67, 81, 88, 182, 185, 201, 203
extrema, 53, 59
extremum, 56, 59, 181, 219

F

Fermi, 168, 373
fermion(s), 238, 239, 242
fermionic, 238
Feynman, 5, 29, 35, 36, 39, 44, 47–50, 66, 77, 79–81, 87, 91, 96, 98, 135, 138, 142, 150–152, 174, 188, 207, 209, 219, 222, 225, 229, 231–233, 235, 237, 239, 242, 245, 246, 257, 263, 267, 273, 275, 276, 280, 283, 287, 289, 291, 292, 302, 305, 324, 329, 331, 338, 339, 351–354, 363, 364, 388, 389
Feynman-Jensen-Peierls, 226, 230
FHIP, 364
FHO, 70, 147, 148, 210, 211, 216, 217
fictitious, 152, 243–245, 257, 258, 264, 267, 270, 278, 280–282, 284, 287, 290, 292, 296, 305, 330–332, 356, 359
flux(es), 1, 339, 367, 376, 380, 382, 384, 385
focal, 74, 75
force-driven, 34, 70
force-free, 34
four-dimensional, 95

Fourier, 32, 53, 54, 63, 72, 82–84, 97, 99, 112, 116, 123, 260, 265, 353, 359
Franck-Condon, 274
Fresnel, 39, 45, 46, 64, 73
Fröhlich, 238, 239, 242, 263, 276, 337, 352, 353, 363
Fröhlich-polarons, 264
functional-evolution-integral, 247
functional-evolution-integral-operator, 248
functionals, 150, 152, 179, 245, 340, 343

G

Gamma, 345
gap, 238
gas(es), 205, 212, 237, 365–367, 369, 371, 375–381, 386
Gaussian, 35, 44, 45, 63, 67, 72, 73, 80–82, 87, 93, 141, 201, 209, 211, 214, 217, 222, 224, 283, 292, 355, 356
generating, 139, 148, 237, 257–259, 269, 270, 285, 293, 294, 300, 307, 308, 313, 318, 322, 343, 354, 360, 361
Gibbs, 196
gratings, 39, 40
Green, 38, 48, 94, 97, 99, 118, 120, 222, 364
ground-state, 150, 151, 265, 276

H

half-plane, 113
half-space, 253
half-waves, 54
halides, 239, 263, 277, 351
Hamilton, 41–43, 49, 78
Hamiltonian(s), 24, 29, 30, 36, 37, 39, 41–44, 49, 50, 77–79, 95–98, 125, 135, 142, 150, 153, 160, 161, 163, 166, 182, 192, 194, 197, 199, 202, 206, 208, 209, 231–235, 238, 239, 246, 247, 265, 266, 278–280, 284, 304, 310, 315, 320, 340, 352, 353, 363, 386, 387
Hamilton-Jacobi, 102, 136, 183, 186
harmonic, 35, 53, 67–72, 74, 82, 86, 109, 127, 128, 135, 137, 138, 140, 147, 150, 169, 183, 201, 204, 209, 210, 214, 215, 217, 223, 224, 239, 241, 245, 251, 258, 267, 270, 282, 291, 305, 324–326, 330, 334, 358
Heaviside, 38, 88, 97, 98, 146, 159, 161, 207, 253, 342, 353
Heisenberg, 29, 30, 33, 35, 36, 38, 43, 48, 92, 98, 140, 145, 146, 165, 192, 195, 196, 389
helium, 166
Helmholtz, 230, 231
Hermite, 139, 141, 148
Hermite-polynomials, 247
Hermitian, 13–15, 18–20, 24, 25, 33, 36, 83, 85, 143, 169, 175, 193, 248
Hermiticity, 143, 177
Hilbert, 29, 389
hole(s), 4, 5, 237, 238, 242, 244, 264, 271, 275, 286, 287, 294, 301, 308, 313, 331, 362, 366
H-theorem, 375, 377
Huygens, 39
hydrogen, 130, 166
hydrogenic, 166

I

idempotent, 10
imaginary, 19, 45, 49, 103, 107, 120, 136, 191, 200, 206–208, 210, 231
impedance, 239, 265, 352, 363, 364
impurity(ies), 241, 337, 338, 351, 371, 373, 386
incoherent, 242
indetermination, 2, 3
indistinguishable, 3
induction, 87
inelastic, 351
interface-like, 238
interface-type, 238
interference, 1–3, 5, 46, 48, 75, 92, 93
intermediate-coupling, 239, 265, 276, 277
intermolecular, 371
ion(s), 140, 166, 237, 264, 339, 344
ionic, 237–239, 242, 262, 277, 351
ionization, 166
ionized, 166, 337
ionosphere, 370
irreducible, 24

J

Jacobian, 74, 84
Jacobi-Hamilton, 101

K

Kadanoff, 364
Kepler, 130
kernel, 16, 19, 39, 48, 332, 352
Kohn, 238
Kolmogoroff, 51
Kronecker, 5

L

Lagrange, 47, 61, 183, 280
Lagrangian, 29, 33, 43, 44, 47–51, 67, 71, 77, 79, 81, 82, 87, 145, 147, 153, 201, 208, 209, 214, 216, 231, 232, 242–244, 255–257, 266–268, 278, 280, 282–284, 290–292, 294, 296–299, 305, 306, 310, 312, 315–317, 320–322, 329, 330, 334, 335, 338, 356, 358
Landau, 87, 237, 239, 242, 264, 276
Landau-Fröhlich-polaron, 264
Landau-Pekar, 276
Laplacian, 126

large-polaron, 264, 364
laser, 337, 339, 340, 344–346, 359, 360, 363
lattice, 140, 150, 152, 205, 209, 237–248, 250, 256, 263–265, 267, 268, 275, 276, 278, 290, 293, 294, 301, 305, 307, 308, 314, 317, 324, 329–331, 340, 341, 345, 351, 352, 371, 386, 387
law, 4, 51, 92, 185, 261, 374, 375, 377
Lee, 276
Legendre, 127
lemma, 54, 55
linear-response, 352
Liouville, 370, 371, 388
LLP, 276
LO, 263
local-equilibrium, 378
longitudinal, 238, 239, 242, 243, 255, 263–266, 279, 315, 316, 320, 337
long-wavelength, 242, 264
LO-phonon, 264
Lorentz, 383
low-temperature, 337, 364
Ludwig, 373

M

macroscopic, 140, 237, 365, 375, 377
magnetic, 43, 87, 89, 94, 126, 137, 202, 231, 239, 323, 324, 329, 330, 332, 333, 371
magneto-polaron, 333
many-body, 39, 150, 151, 166, 239, 279, 330
many-electron, 238
many-particle, 166, 209, 238
Maslov, 35, 74, 75, 142
Maslov-Morse, 75
Maxwell's, 67
meson, 239
mesoscopic, 337, 338
microscopic, 237, 370, 380
mid-point, 43, 44, 94
Minkowski, 191, 208
mobility, 238, 239, 246, 265, 351, 359, 363–365
monatomic, 366, 381
multi-particle, 209
multiphoton, 337, 338
multipolaron, 279
multi-slit, 4
multivalent, 107

N

nanocrystals, 239, 337
nanostructure, 339
n-dimensional, 8, 9
Neumann, 197
Neumann-Liouville, 154
neutron(s), 166, 238
Newton, 30, 149, 185, 365, 366
Newtonian, 185
non-classical, 29, 49
nondegenerate, 77, 110
nonequilibrium, 248
non-Gaussian, 60, 152, 153, 245
nuclear, 67
nucleon, 239
nucleus, 130, 165, 166

O

one-body, 151, 330
one-dimensional, 67, 92, 101, 108, 110, 141, 213, 215, 220, 381
one-electronic, 166
one-particle, 245, 271, 362
one-phonon, 349
operators, 8–10, 13–16, 19, 20, 25, 26, 29, 30, 35, 41, 42, 51, 77, 92, 119, 129, 142, 146, 147, 169, 179, 195, 239, 265, 327, 353, 389
orthogonal, 5–7, 10, 11, 20, 63, 89, 141, 151
orthogonality, 5–7, 21, 33, 40, 143, 327
orthonormal, 7–10, 13, 17, 96, 150, 193, 194, 207, 248
orthonormality, 6, 7, 63, 72
ortho-normalization, 11, 21, 31

P

parabola, 121
parabolic, 238, 243, 264, 265, 277, 278, 290, 295, 304, 310, 315, 320
partition, 106, 142, 191–194, 197, 199, 201–205, 207–210, 212, 214, 215, 217, 220, 221, 224, 229, 230, 232, 239, 240, 243–245, 250, 255, 256, 267–269, 283–285, 292, 293, 299, 306, 307, 312, 317, 326, 329–331, 334, 336
path-integral, 29, 35, 50, 92, 352
Pekar, 239, 242, 264, 276, 364
perturbation, 72, 147, 151, 153–158, 160–171, 173, 174, 177, 209, 231, 242, 252, 259, 271, 286, 301, 308, 313, 325, 337, 338, 353, 362, 364, 365, 368, 373, 386
perturbation-theory, 276
phase-space, 43, 111
phonon(s), 69, 150, 151, 205, 209, 217, 237–239, 241–250, 255, 262–266, 271, 272, 275, 278, 284, 300, 337–339, 344–346, 349–352, 356, 358, 362–364, 387, 388
phonon-electron, 351
phonon-mediated, 152, 238
photoemission, 238
photon(s), 337, 346, 349
photon–phonon, 346
photovoltaic, 351
Pines, 276
Planck, 30, 53
plasma, 337, 339, 370
polar, 125, 126, 150, 237–239, 242, 246, 262, 263, 272, 278, 324, 329, 338, 344, 351, 352, 383, 384, 386

polarizable, 238, 239, 242
polarization, 150, 237, 238, 241, 242, 246, 255, 264, 275, 288, 303, 351
polarization-field, 241
polaron(s), 150, 152, 218, 237–243, 245, 246, 255–257, 259, 261, 263–265, 267, 269–279, 283–286, 288–290, 292–295, 302–309, 315, 316, 318, 319, 323, 324, 330, 333, 337, 351–353, 359, 362–364
polaronic, 237, 239, 277, 351
polyatomic, 366
polymer, 35
polynomial(s), 19, 127, 139, 141, 148
probability, 1–7, 9, 11, 13, 18, 23, 31, 32, 35, 38, 39, 43, 47, 48, 50, 51, 65, 75, 91, 103, 104, 107, 112, 115, 119, 121, 141, 142, 162–164, 167–171, 173, 177, 192–194, 196, 197, 205, 248, 250, 337, 341, 368, 369, 372, 389
Produkt-Ansatz, 274, 275
propagator(s), 36, 38, 39, 48, 52, 60, 62, 63, 65, 71, 72, 74, 81, 82, 87, 95, 97–100, 105, 113, 114, 116, 117, 142, 161, 202, 207
protons, 130, 166

Q

quanta, 337
quasi, 238, 264, 265
quasi-classical/quasi-classically, 35, 46, 53, 54, 60, 65, 67, 72, 81, 82, 100–110, 116, 117, 119, 120, 125, 128–132, 191, 196, 202, 210, 219, 221–224, 326, 370
quasi-classics, 111
quasi-particles, 237, 239, 246, 264, 265, 339, 351
qubits, 5

R

radiation, 169, 205, 337, 339, 344, 345, 350
Raman, 238
reciprocity, 6, 8
relaxation-time, 373
relaxation-time-approximation, 374
renormalization, 351
resonance(s), 164, 169, 170, 239, 345–347, 349, 350, 359, 363
resonant, 164, 165, 169–171
retardation, 99, 225, 257, 269
retarded, 38, 95, 97–100, 113, 114, 152, 245, 259, 271, 286, 294, 301, 308, 313, 318, 323, 330, 344, 362, 388
Richard, 5, 29, 44
Riemann, 37, 42
Riemann-Lebesgue, 54, 55
Ritz, 229, 231, 232, 235
Rodriguez, 127

S

saddle, 58, 59, 116
Schrödinger, 2, 29, 30, 35, 36, 39, 60, 87, 91, 92, 94–99, 101, 103, 107, 108, 112, 114, 118, 122, 126, 129, 131, 132, 135, 136, 145, 147, 153, 159–161, 194, 195, 207, 220, 229, 246–248, 337, 340, 351, 369, 389
Schturm-Liouville, 17, 20
Schwinger, 184, 186
Schwinger-Dyson, 184
self-adjoint, 13, 14, 20, 21
self-consistent, 69, 275
self energy, 239, 242, 263, 276
self-induced, 237, 238
self-interaction, 152, 245, 271, 362
self-localized, 242
self-trapping, 237, 264
semiconductors, 237, 238, 337
semi-group, 51, 52, 388
single-valued, 107, 130
sinusoidal, 170
S-matrix, 29, 369
Smoluchovsky, 51
spectral, 17, 97, 138, 139, 166
spectroscopy, 67, 166, 337
spectrum, 11, 12, 16, 17, 19, 21, 22, 96, 97, 135, 139–141, 166–168, 171, 192, 197, 209, 216, 240, 246, 337, 338, 349–351
sphere, 254, 342
spherical, 127, 289, 295, 302, 319, 337, 351
s-photon, 344
spin, 166, 264
spinors, 3
spintronics, 238
strong-coupling, 239, 242, 275, 276
superconductivity, 238, 239, 277
superposition, 3–5, 7, 11, 13, 52, 92, 103, 106, 113, 291, 297
symmetric, 5, 14, 81, 85, 125, 126, 129, 142, 261, 319, 366, 377
symmetrical, 77, 93, 327
symmetry(ies), 24, 101, 232, 238, 243, 330, 367
system-bath, 242
system-plus-bath, 241
system-plus-environment, 244

T

Taylor, 60, 71, 72, 93, 94, 102, 107, 125, 184, 187, 204, 206, 221, 259, 343
theorem, 185, 271, 314, 370, 371
thermo-diffusion, 385
time-ordered, 146
Trotter, 39
two-electron, 166
twofold, 260, 261, 308, 314, 323
twofold-integration, 271

U

ultracold, 237
univalent, 20, 107, 197
unscreened, 237
upper-bound, 150, 276

V

variational, 30, 43, 150, 152, 166, 225, 229, 231–233, 235, 239, 243, 245, 257, 263, 264, 266, 267, 269, 273, 276, 278, 289, 290, 295, 304, 305, 310, 319, 323, 324, 330, 364
vector(s), 3, 5–11, 13, 14, 17, 36, 38–40, 43, 44, 50, 82, 88, 94–96, 98, 125, 147, 192, 193, 209, 231, 241, 242, 253, 264, 265, 276, 329, 366, 376
viscosity, 365, 378
Vlasov, 370

W

wave-particle-duality, 5
weak-coupling, 242, 276, 337, 364
Wentzel-Kramer-Brillouin, 53, 101, 104, 117
Weyl, 389
Wigner, 389
äWronskian, 100, 115

Z

zero-order, 359
zero-point, 140, 192